Reviews for the fourth edition

'This is the best book on Innovation Management I have found so far. I have been using it for years teaching my engineering students at TU Delft. This book covers important insights from modern evolutionary research; it also provides useful practical knowledge for innovation management.'

Alfred Kleinknecht, Professor of Economics of Innovation
at TU Delft, Netherlands and Visiting Professor,
Université de la Sorbonne, Paris, France.

'I am convinced that it will become a landmark and a classic for Technology and Innovation Management. It is a comprehensive, carefully argued, self-contained presentation of the state-of-the-art of managing innovation. Students will benefit from the lucid exposition of key concepts and excellent teaching support, whilst scholars will find new insights suggestive of further research.'

Peter Augsdorfer, Professor of Corporate Strategy
and Technology Management, University of Ingolstadt,
Germany and Grenoble Ecole de Management, France.

'Tidd and Bessant have an awesome grasp of challenges innovators face in an increasingly knowledge-based and globally distributed world. Follow their search-select-implement-capture stages to understand how to meet these challenges.'

Professor Mari Sako, Said Business School,
University of Oxford, U.K.

Reviews for the third edition

'A limpid and very useful account of what we know about the management of innovation. Must read for executives, scholars and students.'

Yves Doz, Timken Chaired Professor of Global Technology
and Innovation, INSEAD.

"Managing Innovation masterfully synthesizes the extensive literature on this extremely complex, often fragmented topic. The book is enlivened by real-world examples from all over the globe and from a wide variety of industries, it serves as a comprehensive reference for academics as well as a practical guide for managers."

Dorothy A. Leonard, William J. Abernathy Professor of
Business Administration, Harvard Business School

"This book has established itself as a leading textbook in its field. Among its most important strengths is the way innovation is analyzed from many perspectives. Issues and challenges are scrutinized. Resources, organization, and processes are discussed in great detail. The important context for understanding innovation is explored at macro as well as micro levels. In this new edition the authors have integrated important contemporary issues such as the power of networks for innovation and radical disruptive innovation. Since the pedagogical strengths and support in terms of cases and web support have been developed the book qualifies as a leading textbook even more clearly."

Christer Karlsson, Professor of Innovation and
Operations Management, Dean CBS Executive MBA,
Copenhagen Business School, Denmark

"In the knowledge society we live in today, the continuous creation of knowledge is the only way for an organization to survive. Managing Innovation concisely and precisely explains how an organization can keep innovating within and across the organizational boundary, and how such innovation affects the organization and society. With many examples, it is easy to read and understand. Excellent and a must-read for anyone who wants to learn about innovation."

Professor Jiro Nonaka, JAIST (Japan Advanced Institute
of Science and Technology), Japan

Reviews for the second edition

'This is an extraordinary synthesis of the most important things that are understood about innovation, written by some of the world's foremost scholars in this field.'

Clayton M. Christensen, Professor of Business Administration,
Harvard Business School.

'The capacity to innovate is a key source of competitive advantage; but the management of innovation is risky. The authors provide a clear, systematic and integrated framework which will guide students and practising managers alike through a complex field. Updated to address key contemporary themes in knowledge management, networks and new technology, and with an exemplary combination of research and practitioner material, this is probably the most comprehensive guide to innovation management currently available.'

Rob Goffee, Professor of Organizational Behaviour,
London Business School.

'In a highly readable yet challenging text, Tidd, Bessant and Pavitt are true to their subtitle, since they do indeed achieve a rare analytical integration of technological, market and organizational change. Alive to the vital importance of context, they nonetheless reveal generic aspects to the process of innovation. Read this book and you will understand more, and with a little luck, an encounter with a rich example will resonate with experience, hopes and fears and provide a useful guide to action.'

Sandra Dawson, KPMG Professor of Management Studies and Director,
Judge Institute of Management, University of Cambridge.

'This is an excellent book. Not only is it practical and easy to read, it is also full of useful cases and examples, as well as a comprehensive reference to the current literature. I will be recommending it to my entrepreneurship students.'

Professor Sue Birley, Director, The Entrepreneurship Centre,
Imperial College, University of London, UK.

'The first edition of this book was essential reading for anyone trying to get to grips with innovation in theory and practice. This new edition, by embracing the challengees faced in the "new economy", is an ideal companion for the serious innovator. Starting from the view that anyone can develop competencies in innovation this comprehensive text provides managers with essential support as they develop their capability. The second edition contains many case illustrations illuminating both theory and practice in successful innovation and is a "must" for aspiring MBAs.'

David Birchall, Professor and Director of the Centre for Business in the Digital Economy
(CBDE), Henley Management College, UK.

'The authors of this book have managed to capture the essence of leading-edge thinking in the management of techonological innovation and presented the multidimensional nature of the subject in an integrated manner that will be useful for the practitioner and essential reading for students and researchers in the field. This is the book we have been waiting for!'

Professor Carl W. I. Pistorius, Dean, Management of Technology Programme,
University of Pretoria, South Africa.

'Innovation has become widely recognized as a key to competitive success. Leaders of businesses of all sizes and from all industries now put sustained innovation among their top priorities and concerns – but, for many, innovation seems mysterious, unpredictable, apparently unmanageable. Yet it can be managed. This book provides a highly readable account of the best current thinking about building and sustaining innovation. It draws particular attention to important emerging issues, such as the use of networks of suppliers, customers and others outside the firm itself to stimulate innovation, and the role of knowledge and knowledge management to support and sustain it. As the authors say, there is no "one best way" to manage innovation: different situations call for different solutions. But if you want to drive innovation in your own organization, this book will help you to understand the issues that matter and the steps you can take.'

Richard J. Granger, Vice President, Technology & Innovation Management Practice,
Arthur D. Little Inc.

'Innovation has always been a challenge, but never more so nowadays in these turbulent times. This second edition of *Managing Innovation* helps address the practicalities of the challenge and places them firmly in today's new environment, where technology is changing faster and faster. Integrating the multiple aspects of innovation – and not just treating it as a technical issue – is a real benefit this book brings.'

C. John Brady, Director, McKinsey & Company Inc.

'The characteristics of doing business today – rapid change, extreme volatility and high uncertainty – mean that traditional ways of managing technology need to be radically reappraised for any company that sees technical leadership as a critical business differentiator. Through their research work and worldwide network, Joe Tidd, John Bessant and Keith Pavitt have brought together the latest thinking on innovation management, extensively illustrated with real world examples, and with pointers to how successful implementations may emerge in the future. This book is well worth reading for all who want to achieve leadership in technology management.'

David Hughes, Executive Vice President, Technology Management, Marconi plc.

'Innovation is the cornerstone of what makes businesses successful: offering something uniquely better to the consumer. Innovation, while key, is probably the most difficult (maybe even impossible) element of corporate activity to manage or plan. This book does an excellent job of setting out the specification of ways we can think about how to create innovative organizations, without prescribing a "recipe for success".'

Dr Neil MacGilp, Director, Corporate R&D, Procter & Gamble.

MANAGING INNOVATION

Integrating Technological, Market and Organizational Change

Fourth Edition

Joe Tidd and John Bessant

John Wiley & Sons, Ltd

Published in 2009 by John Wiley & Sons Ltd.
The Atrium, Southern Gate, Chichester,
West Sussex PO19 8SQ, England
Telephone (+44) 1243 779777

Email (for orders and customer service enquiries): cs-books@wiley.co.uk
Visit our Home Page on www.wiley.com

Reprinted August 2009, February 2010, October 2010, September 2011

Other Wiley Editorial Offices

John Wiley & Sons Inc., 111 River Street, Hoboken, NJ 07030, USA

Jossey-Bass, 989 Market Street, San Francisco, CA 94103-1741, USA

Wiley-VCH Verlag GmbH, Boschstr. 12, D-69469 Weinheim, Germany

John Wiley & Sons Australia Ltd, 42 McDougall Street, Milton, Queensland 4064, Australia

John Wiley & Sons (Asia) Pte Ltd, 2 Clementi Loop #02-01, Jin Xing Distripark, Singapore 129809

John Wiley & Sons Canada Ltd, 6045 Freemont Blvd, Mississauga, ONT, L5R 4J3, Canada

Wiley also publishes its books in a variety of electronic formats. Some content that appears in print may not be available in electronic books.

Library of Congress Cataloging-in-Publication Data
British Library Cataloguing in Publication Data

Tidd, Joseph, 1960-
Managing innovation : integrating technological, market, and organizational change / Joe Tidd and
  John Bessant. — 4th ed.
      p. cm.
  Includes bibliographical references and index.
  ISBN 978-0-470-99810-6 (pbk. : alk. paper) 1. Technological innovations--Management. 2. Industrial management. 3. Organizational change. I. Bessant, J. R. II. Title.
  HD45.T534 2009
  658.5'14—dc22

                                                                                              2008052045

A catalogue record for this book is available from the British Library

ISBN 978-0-470-99810-6
Typeset in 10/13 Berkeley Book by Thomson Digital, India.
Printed and bound in Great Britain by Scotprint, Haddington, East Lothian.

CONTENTS

ABOUT THE AUTHORS

Joe Tidd, BSc, MBA, DPhil, DIC is a physicist with subsequent degrees in technology policy and business administration. He is Professor of Technology and Innovation Management at SPRU (Science & Technology Policy Research), University of Sussex and visiting Professor at University College London, Copenhagen Business School and Rotterdam School of Management. He was previously Head of the Management of Innovation Specialisation and Director of the Executive MBA Programme at Imperial College and policy adviser to the CBI (Confederation of British Industry), has presented expert evidence to three Select Committee Enquiries held by the House of Commons and House of Lords. He was a researcher for the five-year, $5 million *International Motor Vehicle Program* organized by the Massachusetts Institute of Technology (MIT) in the USA, which resulted in 'lean production'. He has worked on projects for consultants Arthur D. Little, CAP Gemini and McKinsey, and technology-based firms, including American Express Technology, Applied Materials, ASML, BOC Edwards, BT, National Power, NKT and Nortel Networks, Petrobras and Pfizer. He is the winner of the Price Waterhouse Urwick Medal for contribution to management teaching and research, and the Epton Prize from the R&D Society. He has written nine books and more than 70 papers on the management of technology and innovation, and is the Managing Editor of the *International Journal of Innovation Management*.

John Bessant, BSc, PhD is Professor of Innovation & Entrepreneurship at University of Exeter Business School. He previously held the Chair in Innovation and Technology Management at Imperial College Business school, where he also was Research Director. He previously worked at Cranfield, Brighton and Sussex Universities. In 2003 he was awarded a Senior Fellowship with the Advanced Institute for Management Research and was also elected a Fellow of the British Academy of Management. Author of 15 books and many articles, he has acted as advisor to various national governments, international bodies including the United Nations, The World Bank and OECD and companies including LEGO, Novo Nordisk, Mars, UBS and Morgan Stanley.

PREFACE TO THE FOURTH EDITION

We know that those organizations that are consistently successful at managing innovation outperform their peers in terms of growth and financial performance.[1] However, managing innovation is not easy or automatic. It requires skills and knowledge, which are significantly different to the standard management toolkit and experience, because most management training and advice is aimed to maintain stability, hence the most sought after degree is an MBA (Master of Business Administration). As a result, most organizations either simply do not formally manage the innovation process, or manage it in an ad hoc way. Studies confirm that only around 12% of organizations successfully manage innovation, and only half of these organizations do so consistently across time.[2]

Since the first edition of *Managing Innovation* was published in 1997 we have argued that successful innovation management is much more than managing a single aspect, such as creativity, research and development or product development. Our companion texts deal with such issues more fully,[3] but here we continue to promote an integrated approach, which deals with the interactions between changes in markets, technology and organization. In this fourth edition, we have tried to continue our decade-long tradition of differentiating our work from that of others by developing its unique characteristics:

- Strong evidence-based approach to the understanding and practice of managing innovation, drawing upon thousands of research projects, and, new to this edition, 'Research Notes' on the very latest research findings.
- Practical, experience-tested processes, models and tools, including new to this edition 'Views from the Front Line', first-person accounts from practising managers on the challenges they face while managing innovation.

---

[1] Tidd, J. (2006) *From Knowledge Management to Strategic Competence*, second edition, Imperial College Press, London.
[2] Jaruzelski, B. and K. Dehoff (2008) Booz Allen Hamilton Annual Innovation Survey, *Strategy and Business*, **49**.
[3] Bessant, J. and J. Tidd (2007) *Innovation and Entrepreneurship*, John Wiley & Sons, Ltd, Chichester; Isaksen, S. and J. Tidd (2006) *Meeting the Innovation Challenge: Leadership for Transformation and Growth*, John Wiley & Sons, Ltd, Chichester; Bessant, J. (2003) *High-Involvement Innovation*, John Wiley & Sons, Ltd, Chichester.

- Real illustrations and case examples of innovation in action, in manufacturing and services, private and public sectors, including in-text exhibits, and from **www.managing-innovation.com**, full case studies, video and audio podcasts.
- Supported by extensive, growing web resources, at **www.managing-innovation.com**, featuring a comprehensive range of innovation tools, interactive exercises and tests to help apply the learning.

Our understanding of innovation continues to grow, by systematic research, experimentation and the ultimate test of management practice and experience. It is a challenge for all of us interested in innovation to keep abreast of this fast-developing and multi-disciplinary field. In the general field of business research, the 200 or so active research centres worldwide produce some 5000 papers each year, many relevant to managing innovation.[4] In the more specialist field of technology management, the 120-odd research centres worldwide publish around 250 papers per annum.[5] So since we published the first edition, there have been around 3000 pieces of specialist research added, much more general management research relevant to innovation, and a proliferation of books on the subject. One goal of this book is to help make sense of and navigate through this mass of material. Another aim is to encourage action. As we said in the first edition, and still believe strongly, this book is designed to encourage and support practice and organization-specific experimentation and learning, and not to substitute for it.

We welcome feedback and contributions, and have developed an open environment for this purpose at:

www.managing-innovation.com

Joe Tidd and John Bessant
Brighton, East Sussex, U.K.

---

[4] Mangematin, V. and C. Baden-Fuller (2008) Global contests in the production of business knowledge. *Long Range Planning*, **41** (1), 117–39.

[5] Linton, J.D. (2004) Perspective: ranking business schools on the management of technology. *Journal of Product Innovation Management*, **21**, 416–30.

ACKNOWLEDGEMENTS

Any development project is a collective endeavour. We thank all our students, colleagues and users of previous editions for their feedback and suggestions. In particular, we are indebted to those who joined our Web 2.0 development community, including Santiago Acosta, Peter Augsdorfer, David Bennett, Pun-Arj Chairatana, David Charles, Caroline Cantley, Roderick van Domburg, Dries Faems, Saskia Harkema, Pim den Hertog, Yasuhiko Izumimoto, L.J. Lekkerkerk, Breno Nunes, Darius Singh, Loreta Vaicaityte, Lorraine Warren and Simone Zanoni.

We also thank the more formal contributions made by leading academics for our 'Research Notes', especially Dave Francis, Richard Adams, Bob Phelps, Henrik Florén, Steve Flowers, Johan Frishammar, Flis Henwood, Kathrin Moeslein, J. Roland Ortt, Frank Piller and Bettina von Stamm; and the many experienced practising managers of innovation who took time to contribute the insightful 'Views from the Front Line' – Simon Barnes, Stephen Bold, Ian Collins, Richard Dennis, Julian Fallon, Matt Kingdon, John Thesmer, Lester Handley, Rob Perrons, Wouter Zeeman, John Tregaskes, Dorothea Seebode, Gerard Harkin, Roy Sandbach, Anand Lakhani, Patrick McLaughlin, John Gilbert, Lucy Hooberman, David Overton, Suzana Moreira, Bertus de Jager, Pernille Weiss Terkildsen, Michael Vaag, Martin Curley, Helle Vibeke Carstensen, Arne Madsen, Bo Wesley, Francis Bealin-Kelly and Carlos de Pomme.

As with our previous projects with Wiley, we acknowledge the support and enthusiasm of the extended team at our publishers, in particular, Nicole Burnett, Emma Cooper, Steve Hardman, Peter Hudson and Mark Styles.

HOW TO USE THIS BOOK

Features

RESEARCH NOTE

Mohanbir Sawhney, Robert Wolcott and Inigo Arroniz from the Center for Research in Technology and Innovation at the Kellogg School of Management at Northwestern University, USA, interviewed innovation managers at a number of large firms, including Boeing, DuPont, Microsoft, eBay, Motorola and Sony, and from these developed a survey questionnaire which was sent to a further nineteen firms, such as General Electric, Merck and Siemens.

Analysing these data, they derived an "innovation radar" to represent twelve dimensions of business innovation they identified. Their definition of "business innovation" does not focus on new things, but rather anything that creates new value for customers. Therefore creating new things is neither necessary nor sufficient for such value creation. Instead they propose a systematic approach to business innovation, which may take place in twelve different dimensions:

- Offerings – new products or services.
- Platform – derivative offerings based on reconfiguration of components.
- Solutions – integrated offerings which customer value.
- Customers – umet needs o rnew market segments.
- Customer experience – re-design of customer contact and interactions.
- Value capture – redefine the business model and how income is generated.
- Processes – to improve efficiency or effectiveness.
- Organization – change scope or structures.
- Supply chain – changes in sourcing and order fulfillment.

Web Links

Video Podcast – Finnegan's Fish Bar

Audio Podcast – Glasses Direct

Interactive Exercise – 4Ps tool

Case Study – Dimming of the Light Bulb

Research notes summarize the very latest evidence and reviews of contemporary topics

---

1.2 Why innovation matters

What these organizations have in common is that their success derives in large measure from innovation. Whilst competitive advantage can come from size, or possession of assets, etc. the pattern is increasingly coming to favour those organizations which can mobilize knowledge and technological skills and experience to create novelty in their offerings (product/service) and the ways in which they create and deliver those offerings.

Innovation matters, not only at the level of the individual enterprise but increasingly as the wellspring for national economic growth. In a recent book William Baumol pointed out that 'virtually all of the economic growth that has occurred since the eighteenth century is ultimately attributable to innovation'.[15] Innovation is becoming a central plank in national economic policy – for example, the UK Office of Science and Innovation sees it as 'the motor of the modern economy, turning ideas and knowledge into products and services'.[16] An Australian government website puts the case equally strongly – Companies that do not invest in innovation put their future at risk. Their business is unlikely to prosper, and they are unlikely to be able to compete if they do not seek innovative solutions to emerging problems. According to Statistics Canada (2006), the following factors characterize successful small and medium-sized enterprises:

- Innovation is consistently found to be the most important characteristic associated with success.
- Innovative enterprises typically achieve stronger growth or are more successful than those that do not innovate.
- Enterprises that gain market share and increasing profitability are those that are innovative.

Not surprisingly this rationale underpins a growing set of policy measures designed to encourage and nurture innovation at regional and national level.

The survival/growth question poses a problem for established players but a huge opportunity for newcomers to rewrite the rules of the game. Entrepreneurs are risk-takers – but they calculate the costs of taking a bright idea forward against the potential gains if they succeed in doing something different – especially if that involves upstaging the players already in the game.

CASE STUDY 1.1

The changing nature of the music industry

April 1st 2006 . Apart from being a traditional day for playing practical jokes, this was the day on which another landmark in the rapidly changing world of music was reached. 'Crazy' – a track by Gnarls Barkley – made pop history as the UK's first song to top the charts based on download sales alone. Commenting on the fact that the song had been downloaded more than 31,000 times but was only released for sale in the shops on April 3rd, Gennaro Castaldo, spokesman for retailer HMV, said: "This not only represents a watershed in how the charts are compiled, but shows that legal downloads have come of age . . . .if physical copies fly off the shelves at the same rate it could vie for a place as the year's biggest seller".

Real-life **Case Studies** contextualise the topics covered

---

rules will change and leave them vulnerable. Changes along several core environmental dimensions man that the incidence of discontinuities is likely to rise – for example in response to a massive increase in the rate of knowledge production and the consequent increase in the potential for technology-linked instabilities. But there is also a higher level of interactivity amongst these environmental elements - complexity - which leads to unpredictable emergence. (For example, the rapidly growing field of VoIP (voice over internet protocol) communications is not developing along established trajectories towards a well-defined end-point. Instead it is a process of emergence. The broad parameters are visible – the rise of demand for global communication, increasing availability of broadband, multiple peer-to-peer networking models, growing technological literacy amongst users – and the stakes are high, both for established fixed-line players (who have much to lose) and new entrants (such as Skype, recently bought by eBay for $2.6bn). The dominant design isn't visible yet - instead there is a rich fermenting soup of technological possibilities, business models and potential players from which it will gradually emerge.)

1.4: What is innovation?

 One of America's most successful innovators was Thomas Alva Edison who during his life registered over 1000 patents. Products for which his organization was responsible include the light bulb, 35mm cinema film and even the electric chair. Edison appreciated better than most that the real challenge in innovation was not invention – coming up with good ideas – but in making them work technically and commercially. His skill in doing this created a business empire worth, in 1920, around $21.6bn. He put to good use an understanding of the interactive nature of innovation, realizing that both technology push (which he systematized in one of the world's first organized R&D laboratories) and demand pull need to be mobilized.

VIEWS FROM THE FRONT LINE

"There is nothing more difficult to take in hand, more perilous to conduct, or more uncertain in its success, than to take the lead in the introduction of a new order of things."
Niccolo Machiavelli, 'The Prince', 1532

"Anything that won't sell, I don't want to invent. Its sale is proof of utility, and utility is success."

"Everything comes to him who hustles while he waits."

"Genius is one percent inspiration and ninety-nine percent perspiration."

"I never did anything by accident, nor did any of my inventions come by accident; they came by work."

"Make it a practice to keep on the lookout for novel and interesting ideas that others have used successfully. Your idea has to be original only in its adaptation to the problem you are working on."

Thomas A. Edison

Views from the Front Line provide commentary from practising managers of innovation

Web resources

The text is supported by online resources which are indicated throughout with various icons. These resources are then fully detailed in the Weblinks section at the end of each chapter. The website, which will be integral to the use of this textbook, is

www.managing-innovation.com

The icons are used as follows:

| case study | exercises | tools | video | audio |

The **Web Links** section at the end of each chapter gives details of all the resources available online

PART 1

Managing Innovation

CHAPTER 1

Innovation – what it is and why it matters

'A slow sort of country' said the Red Queen. 'Now here, you see, it takes all the running you can do to keep in the same place. If you want to get somewhere else, you must run at least twice as fast as that!'
(Lewis Carroll, Alice through the Looking Glass)

1.1 Introduction

'We always eat elephants . . .' is a surprising claim made by Carlos Broens, founder and head of a successful toolmaking and precision engineering firm in Australia with an enviable growth record. Broens Industries is a small/medium-sized company of 130 employees which survives in a highly competitive world by exporting over 70% of its products and services to technologically demanding firms in aerospace, medical and other advanced markets. The quote doesn't refer to strange dietary habits but to its confidence in *'taking on the challenges normally seen as impossible for firms of our size'* – a capability which is grounded in a culture of innovation in products and the processes which go to produce them.

At the other end of the scale Kumba Resources is a large South African mining company which makes another dramatic claim – *'We move mountains'*. In their case the mountains contain iron ore and their huge operations require large-scale excavation – and restitution of the landscape afterwards. Much of their business involves complex large-scale machinery – and their ability to keep it running and productive depends on a workforce able to contribute their innovative ideas on a continuing basis.[1]

Innovation is driven by the ability to see connections, to spot opportunities and to take advantage of them. When the Tasman Bridge collapsed in Hobart, Tasmania, in 1975 Robert Clifford was running a small ferry company and saw an opportunity to capitalize on the increased demand for ferries – and to differentiate his offering by selling drinks to thirsty cross-city commuters. The same entrepreneurial flair later helped him build a company – Incat – which pioneered the wave-piercing design that helped them capture over half the world market for fast catamaran ferries. Continuing investment in innovation has helped this company from a relatively isolated island build a key niche in highly competitive international military and civilian markets (`www.incat.com.au`).

But innovation is not just about opening up new markets – it can also offer new ways of serving established and mature ones. Despite a global shift in textile and clothing manufacture towards developing countries the Spanish company, Inditex (through its retail outlets under various names including Zara), has pioneered a highly flexible, fast turnaround clothing operation with over 2000 outlets in 52 countries. It was founded by Amancio Ortega Gaona who set up a small operation in the west of Spain in La Coruña – a region not previously noted for textile production – and the first store opened there in 1975. Central to the Inditex philosophy is close linkage between design, manufacture and retailing and its network of stores constantly feeds back information about trends, which are used to generate new designs. Inditex also experiments with new ideas directly on the public, trying samples of cloth or design and quickly getting back indications of what is going to catch on. Despite its global orientation, most manufacturing is still done in Spain, and the company has managed to reduce the turnaround time between a trigger signal for an innovation and responding to it to around 15 days.

Of course, technology often plays a key role in enabling radical new options. Magink is a company set up in 2000 by a group of Israeli engineers and is now part of the giant Mitsubishi concern. Its business is in exploiting the emerging field of digital ink technology – essentially enabling paper-like display technology for indoor and outdoor displays. These have a number of advantages over other displays such as liquid crystal – low cost, high-viewing angles and high visibility even in full sunlight. One of its major new lines of development is in advertising billboards – a market worth $5 billion in the USA alone – where the prospect of 'programmable hoardings' is now opened up. Magink enables high-resolution images that can be changed much more frequently than conventional paper advertising, and permit billboard site owners to offer variable price time slots, much as television does at present.[2]

At the other end of the technological scale there is scope for improvement on an old product, often using old technologies in new ways. People have always needed artificial limbs and the demand has, sadly, significantly increased as a result of high-technology weaponry such as mines. The problem is compounded by the fact that many of those requiring new limbs are also in the poorest regions of the world and unable to afford expensive prosthetics. The chance meeting of a young surgeon, Dr Pramod Karan Sethi, and a sculptor, Ram Chandra, in a hospital in Jaipur, India, has led to the development of a solution to this problem – the Jaipur foot. This artificial limb was developed using Chandra's skill as a sculptor and Sethi's expertise and is so effective that those who wear it can run, climb trees and pedal bicycles. It was designed to make use of low-tech materials and be simple to assemble – for example, in Afghanistan craftsmen hammer the foot together out of spent artillery shells whilst in Cambodia part of the foot's rubber components are scavenged from truck tyres. Perhaps the greatest achievement has been to do all of this for a low cost – the Jaipur foot costs only $28 in India. Since 1975, nearly 1 million people worldwide have been fitted for the Jaipur limb and the design is being developed and refined, for example, using advanced new materials.[3]

Innovation is of course not confined to manufactured products; plenty of examples of growth through innovation can be found in services.[4–6] In banking the UK First Direct organization became the most competitive bank, attracting around 10 000 new customers each month by offering a telephone banking service backed up by sophisticated IT – a model which eventually became the industry standard. A similar approach to the insurance business – Direct Line – radically changed the basis of that market and led to widespread imitation by all the major players in the sector.[7,8] Internet-based retailers such as Amazon have changed the ways in which products as diverse as books, music and travel are sold, whilst firms like eBay have brought the auction house into many living rooms.

Public services such as healthcare, education and social security may not generate profits but they do affect the quality of life for millions of people. Bright ideas well implemented can lead to valued new services and the efficient delivery of existing ones – at a time when pressure on national purse strings is becoming ever tighter.[9] New ideas – whether wind-up radios in Tanzania or micro-credit financing schemes in Bangladesh – have the potential to change the quality of life and the availability of opportunity for people in some of the poorest regions of the world. There's plenty of scope for innovation and entrepreneurship – and at the limit – about real matters of life and death. For example, the Karolinska Hospital in Stockholm has managed to make radical improvements in the speed, quality and effectiveness of its care services – such as cutting waiting lists by 75% and cancellations by 80% – through innovation.[10] Public-sector innovations have included the postage stamp, the National Health Service in the UK, and much of the early development work behind technologies like fibre optics, radar and the Internet.

1.2 Why innovation matters

What these organizations have in common is that their success derives in large measure from innovation. Whilst competitive advantage can come from size, or possession of assets, etc. the pattern is increasingly coming to favour those organizations that can mobilize knowledge and technological skills and experience to create novelty in their offerings (product/service) and the ways in which they create and deliver those offerings.

Innovation matters, not only at the level of the individual enterprise but also increasingly as the well-spring for national economic growth. In a recent book Baumol pointed out that *'virtually all of the economic growth that has occurred since the eighteenth century is ultimately attributable to innovation'*.[11] The magazine *Business Week* regularly features its list of the top innovative firms in the world. It found that the median profit margin of the top 25 firms was 3.4% in the period 1995–2005 whereas the average for other firms in the S&P Global Index was only 0.4%. Similarly the median annual stock return was 14.3% for the innovators and 11.3% for the rest.[12] Another study by the consultancy Innovaro suggested that 'innovation leaders' had strong links between innovative activities and business performance. Its top five firms were Apple, Nokia, Google, Adidas and Reckitt Benckiser – all noted for different but distinctive innovation performance and the increase of their share prices over the year 2006–7 by between 25% and 135%. This was not just short-term success – these firms had sustained share price growth for the preceding seven years.[13]

Importantly innovation and competitive success are not simply about high-technology companies, for example, the German firm Wurth is the largest maker of screws (and other fastenings such as nuts and bolts) in the world with a turnover of $14 billion. Despite low-cost competition from China, the company has managed to stay ahead through an emphasis on product and process innovation across a supplier network similar to the model used by Dell in computers.[14]

Innovation is becoming a central plank in national economic policy – for example, the UK Office of Science and Innovation sees it as *'the motor of the modern economy, turning ideas and knowledge into products and services'*.[15] An Australian government website (**www.dest.gov.au/sectors/science_innovation**) puts the case equally strongly: *'Companies that do not invest in innovation put their future at risk. Their business is unlikely to prosper, and they are unlikely to be able to compete if they do not seek innovative solutions to emerging problems.'*

According to *Statistics Canada*,[16] the following factors characterize successful small- and medium-sized enterprises:

- *Innovation is consistently found to be the most important characteristic associated with success.*
- *Innovative enterprises typically achieve stronger growth or are more successful than those that do not innovate.*
- *Enterprises that gain market share and increasing profitability are those that are innovative.*

Not surprisingly this rationale underpins a growing set of policy measures designed to encourage and nurture innovation at regional and national level.

The survival/growth question poses a problem for established players but provides a huge opportunity for newcomers to rewrite the rules of the game. One person's problem is another's opportunity and the nature of innovation is that it is fundamentally about *entrepreneurship*. The skill to spot opportunities and create new ways to exploit them is at the heart of the innovation process. Entrepreneurs are risk-takers – but they calculate the costs of taking forward a bright idea against the potential gains if they succeed in doing something different – especially if that involves upstaging the players already in the game.

Innovation contributes in several ways. For example, research evidence suggests a strong correlation between market performance and new products.[17, 18] New products help capture and retain market shares, and increase profitability in those markets. In the case of more mature and established products, competitive sales growth comes not simply from being able to offer low prices but also from a variety of non-price factors – design, customization and quality.[7] And in a world of shortening product life cycles – where, for example, the life of a particular model of television set or computer is measured in months, and even complex products like motor cars now take only a couple of years to develop – being able to replace products frequently with better versions is increasingly important.[19] 'Competing in time' reflects a growing pressure on firms not just to introduce new products but also to do so faster than competitors.[20]

At the same time new product development is an important capability because the environment is constantly changing. Shifts in the socioeconomic field (in what people believe, expect, want and earn) create opportunities and constraints. Legislation may open up new pathways, or close down others, for example, increasing the requirements for environmentally friendly products. Competitors may introduce new products, which represent a major threat to existing market positions. In all these ways firms need the capability to respond through product innovation.

Whilst new products are often seen as the cutting edge of innovation in the marketplace, *process* innovation plays just as important a strategic role. Being able to make something no one else can, or to do so in ways that are better than anyone else, is a powerful source of advantage. For example, the Japanese dominance in the late twentieth century across several sectors – cars, motorcycles, shipbuilding, consumer electronics – owed a great deal to superior abilities in manufacturing – something which resulted from a consistent pattern of process innovation. The Toyota production system and its equivalent in Honda and Nissan led to performance advantages of around two to one over average car makers across a range of quality and productivity indicators.[21] One of the main reasons for the ability of relatively small firms like Oxford Instruments or Incat to survive in highly competitive global markets is the sheer complexity of what they make and the huge difficulties a new entrant would encounter in trying to learn and master their technologies.

Similarly, being able to offer better service – faster, cheaper, higher quality – has long been seen as a source of competitive edge. Citibank was the first bank to offer automated telling machinery (ATM) services and developed a strong market position as a technology leader on the back of this process innovation. Benetton is one of the world's most successful retailers, largely due to its sophisticated IT-led production network, which it innovated over a 10-year period,[22] and the same model has been used to great effect by the Spanish firm Zara. Southwest Airlines achieved an enviable position as the most effective airline in the USA despite being much smaller than its rivals; its success was due to process innovation in areas such as reducing airport turnaround times.[23] This model has subsequently become the template for a whole new generation of low-cost airlines whose efforts have revolutionized the once-cosy world of air travel.

Importantly we need to remember that the advantages which flow from these innovative steps gradually get competed away as others imitate. Unless an organization is able to move into further innovation, it risks being left behind as others take the lead in changing their offerings, their operational processes or the underlying models that drive their business. For example, leadership in banking has passed to others, particularly those who were able to capitalize early on the boom in information and communications technologies; in particular many of the lucrative financial services like securities and share dealing have been dominated by players with radical new models such as Charles Schwab.[24] As all retailers adopt advanced IT so the lead shifts to those who are able – like Zara and Benetton – to streamline their production operations to respond rapidly to the signals flagged by the IT systems.

> **BOX 1.1**   **The innovation imperative**
>
> In the mid-1980s a study by Shell suggested that the average corporate survival rate for large companies was only about half as long as that of a human being. Since then the pressures on firms have increased enormously from all directions – with the inevitable result that business life expectancy is reduced still further. Many studies look at the changing composition of key indices and draw attention to the demise of what were often major firms and in their time key innovators. For example, Foster and Kaplan point out that of the 500 companies originally making up the Standard & Poor 500 list in 1957, only 74 remained on the list through to 1997.[24] Of the top 12 companies which made up the Dow Jones Index in 1900 only one – General Electric – survives today. Even apparently robust giants like IBM, GM or Kodak can suddenly display worrying signs of mortality, whilst for small firms the picture is often considerably worse since they lack the protection of a large resource base.
>
> Some firms have had to change dramatically to stay in business. For example, a company founded in the early nineteenth century, which had Wellington boots and toilet paper amongst its product range, is now one of the largest and most successful in the world in the telecommunications business. Nokia began life as a lumber company, making the equipment and supplies needed to cut down forests in Finland. It moved through into paper and from there into the 'paperless office' world of IT – and from there into mobile telephones.
>
> Another mobile phone player – Vodafone Airtouch – grew to its huge size by merging with a firm called Mannesman which, since its birth in the 1870s, has been more commonly associated with the invention and production of steel tubes! TUI owns Thomson (the travel group) in the UK, and is the largest European travel and tourism services company. Its origins, however, lie in the mines of old Prussia where it was established as a public sector state lead mining and smelting company![25]

CASE STUDY 1.1

The changing nature of the music industry

1 April 2006. Apart from being a traditional day for playing practical jokes, this was the day on which another landmark in the rapidly changing world of music was reached. 'Crazy' – a track by Gnarls Barkley – made pop history as the UK's first song to top the charts based on download sales alone. Commenting on the fact that the song had been downloaded more than 31 000 times but was only released for sale in the shops on 3 April, Gennaro Castaldo, spokesman for retailer HMV, said: *'This not only represents a watershed in how the charts are compiled, but shows that legal downloads have come of age . . . . if physical copies fly off the shelves at the same rate it could vie for a place as the year's biggest seller'*.

One of the less visible but highly challenging aspects of the Internet is the impact it has had – and is having – on the entertainment business. This is particularly the case with music. At one level its impacts could be assumed to be confined to providing new 'e-tailing' channels through which you can obtain the latest CD of your preference – for example from Amazon or CD-Now or 100 other websites. These innovations increase the choice and tailoring of the music-purchasing service and demonstrate some of the 'richness/reach' economic shifts of the new Internet game.

But beneath this updating of essentially the same transaction lies a more fundamental shift – in the ways in which music is created and distributed and in the business model on which the whole music industry is currently predicated. In essence the old model involved a complex network where songwriters and artists depended on A&R (artists and repertoire) staff to select a few acts, production staff who would record in complex and expensive studios, other production staff who would oversee the manufacture of physical discs, tapes and CDs, and marketing and distribution staff who would ensure the product was publicized and disseminated to an increasingly global market.

Several key changes have undermined this structure and brought with it significant disruption to the industry. Old competencies may no longer be relevant – whilst acquiring new ones becomes a matter of urgency. Even well-established names like Sony find it difficult to stay ahead when new entrants are able to exploit the economics of the Internet. At the heart of the change is the potential for creating, storing and distributing music in digital format – a problem which many researchers have worked on for some time. One solution, developed by one of the Fraunhofer Institutes in Germany, is a standard based on the Motion Picture Experts Group (MPEG) level 3 protocol – MP3. MP3 offers a powerful algorithm for managing one of the big problems in transmitting music files – that of compression. Normal audio files cover a wide range of frequencies and are thus very large and not suitable for fast transfer across the Internet – especially with a population who may only be using relatively slow modems. With MP3 effective compression is achieved by cutting out those frequencies which the human ear cannot detect – with the result that the files to be transferred are much smaller.

Therefore MP3 files can be moved across the Internet quickly and shared widely. Various programs exist for transferring normal audio files and inputs – such as CDs – into MP3 and back again.

What does this mean for the music business? In the first instance aspiring musicians no longer need to depend on being picked up by A&R staff from major companies who can bear the costs of recording and production of a physical CD. Instead they can use home recording software and either produce a CD themselves or else go straight to MP3 – and then distribute the product globally via newsgroups, chatrooms, etc. In the process they effectively create a parallel and much more direct music industry, which leaves existing players and artists on the sidelines.

Such changes are not necessarily threatening. For many people the lowering of entry barriers has opened up the possibility of participating in the music business, for example, by making and sharing music without the complexities and costs of a formal recording contract and the resources of a major record company. There is also scope for innovation around the periphery, for example in the music publishing sector where sheet music and lyrics are also susceptible to lowering of barriers through the application of digital technology. Journalism and related activities become increasingly open – now music reviews and other forms of commentary are possible via specialist user groups and channels on the web whereas before they were the province of a few magazine titles. Compiling popularity charts – and the related advertising – is also opened up as the medium switches from physical CDs and tapes distributed and sold via established channels to new media such as MP3 distributed via the Internet.

As if this were not enough the industry is also challenged from another source – the sharing of music between different people connected via the Internet. Although technically illegal this practice of sharing between people's record collections has always taken place – but not on the scale which the Internet threatens to facilitate. Much of the established music industry is concerned with legal issues – how to protect copyright and how to ensure that royalties are paid in the right proportions to those who participate in production and distribution. But when people can share music in MP3 format and distribute it globally the potential for policing the system and collecting royalties becomes extremely difficult to sustain.

It has been made much more so by another technological development – that of peer-to-peer or P2P networking. Sean Fanning, an 18-year-old student with the nickname 'the Napster', was intrigued by the challenge of enabling his friends to 'see' and share between their own personal record collections. He argued that if they held these in MP3 format then it should be possible to set up some kind of central exchange program which facilitated their sharing.

The result – the Napster.com site – offered sophisticated software that enabled P2P transactions. The Napster server did not actually hold any music on its files – but every day millions of swaps were made by people around the world exchanging their music collections. Needless to say this posed a huge threat to the established music business since it involved no payment of royalties. A number of high-profile lawsuits followed but whilst Napster's activities have been curbed the problem did not go away. There are now many other sites emulating and extending what Napster started – sites such as Gnutella, Kazaa and Limewire took the P2P idea further and enabled exchange of many different file formats – text, video, etc. In Napster's own case the phenomenally successful site concluded a deal with entertainment giant Bertelsman that paved the way for subscription-based services to provide some revenue stream to deal with the royalty issue.

Expectations that legal protection would limit the impact of this revolution have been dampened by a US Court of Appeal ruling which rejected claims that P2P violated copyright law. Their judgement said, *'History has shown that time and market forces often provide equilibrium in balancing interests, whether the new technology be a player piano, a copier, a tape recorder, a video recorder, a PC, a karaoke machine or an MP3 player'* (*Personal Computer World*, November 2004, p. 32).

Significantly the new opportunities opened up by this were seized not by music industry firms but by computer companies, especially Apple. In parallel with the launch of its successful iPod personal MP3 player Apple opened a site called iTunes which offered users a choice of thousands of tracks for download at 99 cents each. In its first weeks of operation it recorded 1 million hits. In February 2006 the billionth song ('Speed of Sound') was purchased as part of Coldplay's 'X&Y' album by Alex Ostrovsky from West Bloomfield, Michigan. *'I hope that every customer, artist, and music company executive takes a moment today to reflect on what we've achieved together during the past three years,'* said Steve Jobs, Apple's CEO. *'Over one billion songs have now been legally purchased and downloaded around the globe, representing a major force against music piracy and the future of music distribution as we move from CDs to the Internet.'*

This has been a dramatic shift, reaching the point where more singles were bought as downloads in 2005 than as CDs, and where new players are beginning to dominate the game – for example, Tesco and Microsoft. And the changes don't stop there. In February 2006 the Arctic

Monkeys topped the UK album charts and walked off with a fistful of awards from the music business – yet their rise to prominence had been entirely via 'viral marketing' across the Internet rather than by conventional advertising and promotion. Playing gigs around the northern English town of Sheffield, the band simply gave away CDs of their early songs to their fans, who then obligingly spread them around on the Internet. '*They came to the attention of the public via the Internet, and you had chat rooms, everyone talking about them,*' says a slightly worried Gennaro Castaldo of HMV Records. David Sinclair, a rock journalist suggests that '*It's a big wakeup call to all the record companies, the establishment, if you like . . . This lot caught them all napping . . . We are living in a completely different era, which the Arctic Monkeys have done an awful lot to bring about.*'

The writing may be on the wall for the music industry in the same way as the low-cost airline business has transformed the travel business. And behind the music business the next target may be the movie and entertainment industry where there are already worrying similarities. Or the growing computer games sector with shifts towards more small-scale developers emulating the Arctic Monkeys and using viral marketing to build a sales base.

With the rise of the Internet the scope for service innovation has grown enormously – not for nothing is it sometimes called 'a solution looking for problems'. As Evans and Wurster point out, the traditional picture of services being either offered as a standard to a large market (high 'reach' in their terms) or else highly specialized and customized to a particular individual able to pay a high price (high 'richness') is 'blown to bits' by the opportunities of web-based technology. Now it becomes possible to offer both richness and reach at the same time – and thus to create totally new markets and disrupt radically those which exist in any information-related businesses.[26]

The challenge that the Internet poses is not only one for the major banks and retail companies, although those are the stories which hit the headlines. It is also an issue – and quite possibly a survival one – for thousands of small businesses. Think about the local travel agent and the cosy way in which it used to operate. Racks full of glossy brochures through which people could browse, desks at which helpful sales assistants sort out the details of selecting and booking a holiday, procuring the tickets, arranging insurance and so on. And then think about how all of this can be accomplished at the click of a mouse from the comfort of home – and that it can potentially be done with more choice and at lower cost. Not surprisingly, one of the biggest growth areas in dotcom start-ups was the travel sector and whilst many disappeared when the bubble burst, others like lastminute.com and Expedia have established themselves as mainstream players.

Of course, not everyone wants to shop online and there will continue to be scope for the high-street travel agent in some form – specializing in personal service, acting as a gateway to the Internet-based services for those who are uncomfortable with computers, etc. And, as we have seen, the early euphoria around the dotcom bubble has given rise to a much more cautious advance in Internet-based business. The point is that whatever the dominant technological, social or market conditions, the key to creating – and sustaining – competitive advantage is likely to lie with those organizations which continually innovate.

Table 1.1 indicates some of the ways in which enterprises can obtain strategic advantage through innovation.

TABLE 1.1	Strategic advantages through innovation	
Mechanism	**Strategic advantage**	**Examples**
Novelty in product or service offering	Offering something no one else can	Introducing the first . . . Walkman, mobile phone, fountain pen, camera, dishwasher, telephone bank, online retailer . . . to the world
Novelty in process	Offering it in ways others cannot match – faster, lower cost, more customized	Pilkington's float glass process, Bessemer's steel process, Internet banking, online bookselling
Complexity	Offering something which others find it difficult to master	Rolls-Royce and aircraft engines – only a handful of competitors can master the complex machining and metallurgy involved
Legal protection of intellectual property	Offering something others cannot do unless they pay a licence or other fee	Blockbuster drugs like Zantac, Prozac, Viagra
Add/extend range of competitive factors	Move basis of competition, e.g. from price of product to price and quality, or price, quality, choice	Japanese car manufacturing, which systematically moved the competitive agenda from price to quality, to flexibility and choice, to shorter times between launch of new models, and so on – each time not trading these off against each other but offering them all
Timing	First-mover advantage – being first can be worth significant market share in new product fields Fast-follower advantage – sometimes being first means you encounter many unexpected teething problems, and it makes better sense to watch someone else make the early mistakes and move fast into a follow-up product	Amazon, Yahoo – others can follow, but the advantage 'sticks' to the early movers Palm Pilot and other personal digital assistants (PDAs), which have captured a huge and growing share of the market. In fact the concept and design was articulated in Apple's ill-fated Newton product some five years earlier, but problems with software and especially handwriting recognition meant it flopped

(continued)

TABLE 1.1 (Continued)		
Mechanism	**Strategic advantage**	**Examples**
Robust platform design	Offering something which provides the platform on which other variations and generations can be built	Walkman architecture – through minidisk, CD, DVD, MP3 Boeing 737 – over 40 years old, the design is still being adapted and configured to suit different users – one of the most successful aircraft in the world in terms of sales Intel and AMD with different variants of their microprocessor families
Rewriting the rules	Offering something which represents a completely new product or process concept – a different way of doing things – and makes the old ones redundant	Typewriters vs. computer word processing, ice vs. refrigerators, electric vs. gas or oil lamps
Reconfiguring the parts of the process	Rethinking the way in which bits of the system work together, e.g. building more effective networks, outsourcing and coordination of a virtual company	Zara, Benetton in clothing, Dell in computers, Toyota in its supply chain management
Transferring across different application contexts	Recombining established elements for different markets	Polycarbonate wheels transferred from application market like rolling luggage into children's toys – lightweight micro-scooters
Others?	Innovation is all about finding new ways to do things and to obtain strategic advantage, so there will be room for new ways of gaining and retaining advantage	Napster. This firm began by writing software which would enable music fans to swap their favourite pieces via P2P networking across the Internet. Although Napster suffered from legal issues, followers developed a huge industry based on downloading and file sharing. The experiences of one of these firms – Kazaa – provided the platform for successful high-volume Internet telephony and the company established with this knowledge – Skype – was eventually sold to eBay for $2.6 billion

1.3 Old question, new context

Constant revolutionizing of production, uninterrupted disturbance of all social conditions, everlasting uncertainty . . . all old-established national industries have been destroyed or are daily being destroyed. They are dislodged by new industries . . . whose products are consumed not only at home but in every quarter of the globe. In place of old wants satisfied by the production of the country, we find new wants . . . the intellectual creativity of individual nations become common property.

This quote does not come from a contemporary journalist or politician but from the *Communist Manifesto*, published by Karl Marx and Friedrich Engels in 1848! But it serves to remind us that the innovation challenge isn't new – organizations have always had to think about changing what they offer the world and the ways they create and deliver that offering if they are to survive and grow. The trouble is that innovation involves a moving target – not only is there competition amongst players in the game but also the overall context in which the game is played out keeps shifting. And whilst many organizations have some tried and tested recipes for playing the game there is always the risk that the rules will change and leave them vulnerable. Changes along several core environmental dimensions mean that the incidence of discontinuities is likely to rise – for example in response to a massive increase in the rate of knowledge production and the consequent increase in the potential for technology-linked instabilities. But there is also a higher level of interactivity amongst these environmental elements – complexity – which leads to unpredictable emergence. For example, the rapidly growing field of voice over Internet protocol (VoIP) communications is not developing along established trajectories towards a well-defined endpoint. Instead it is a process of *emergence*. The broad parameters are visible – the rise of demand for global communication, increasing availability of broadband, multiple P2P networking models, growing technological literacy amongst users – and the stakes are high, both for established fixed-line players (who have much to lose) and new entrants (such as Skype). The dominant design isn't visible yet – instead there is a rich fermenting soup of technological possibilities, business models and potential players from which it will gradually emerge.

CASE STUDY 1.2

The difficulties of a firm like Kodak illustrate the problem. Founded around 100 years ago the basis of the business was the production and processing of film and the sales and service associated with mass-market photography. Whilst the latter set of competencies are still highly relevant (even though camera technology has shifted), the move away from wet physical chemistry conducted in the dark (coating emulsions on to films and paper) to digital imaging represents a profound change for the firm. It needs – across a global operation and a workforce of thousands – to let go of old competencies which are unlikely to be needed in the future whilst at the same time to rapidly acquire and absorb cutting-edge new technologies in electronics and communication. Although strenuous efforts are being made to shift from being a manufacturer of film to becoming a key player in the digital imaging industry and beyond, the response from stock markets suggests some scepticism as to Kodak's ability to do so.

Table 1.2 summarizes some of the key changes in the context within which the current innovation game is being played out.

TABLE 1.2	**Changing context for innovation**
Context change	**Indicative examples**
Acceleration of knowledge production	OECD estimates that close to $1 trillion is spent each year (public and private sector) in creating new knowledge – and hence extending the frontier along which 'breakthrough' technological developments may happen
Global distribution of knowledge production	Knowledge production is increasingly involving new players especially in emerging market fields like the BRIC (Brazil, Russia, India, China) nations – so the need to search for innovation opportunities across a much wider space. One consequence of this is that 'knowledge workers' are now much more widely distributed and concentrated in new locations, e.g., Microsoft's third-largest R&D Center employing thousands of scientists and engineers is now in Shanghai
Market fragmentation	Globalization has massively increased the range of markets and segments so that these are now widely dispersed and locally varied – putting pressure on innovation search activity to cover much more territory, often far from 'traditional' experiences, such as the 'bottom of the pyramid' conditions in many emerging markets[3]
Market virtualization	Increasing use of the Internet as marketing channel means different approaches need to be developed. At the same time emergence of large-scale social networks in cyberspace pose challenges in market research approaches, e.g., MySpace currently has over 100 million subscribers. Further challenges arise in the emergence of parallel world communities as a research opportunity, e.g., Second Life now has over 6 million 'residents'
Rise of active users	Although users have long been recognized as a source of innovation there has been an acceleration in the ways in which this is now taking place, e.g., the growth of LINUX has been a user-led open community development.[27] In sectors like media the line between consumers and creators is increasingly blurred - for example, You Tube has around 100 million videos viewed each day but also has over 70 000 new videos uploaded every day from its user base.

(continued)

TABLE 1.2	(Continued)
Context change	**Indicative examples**
Development of technological and social infrastructure	Increasing linkages enabled by information and communications technologies around the internet and broadband have enabled and reinforced alternative social networking possibilities. At the same time the increasing availability of simulation and prototyping tools have reduced the separation between users and producers[28, 29]

RESEARCH NOTE   Joseph Schumpeter – the 'Godfather' of innovation studies

One of the most significant figures in this area of economic theory was Joseph Schumpeter who wrote extensively on the subject. He had a distinguished career as an economist and served as Minister for Finance in the Austrian government. His argument was simple: entrepreneurs will seek to use technological innovation – a new product/service or a new process for making it – to get strategic advantage. For a while this may be the only example of the innovation so the entrepreneur can expect to make a lot of money – what Schumpeter calls 'monopoly profits'. But of course other entrepreneurs will see what has been achieved and try to imitate it – with the result that other innovations emerge, and the resulting 'swarm' of new ideas chips away at the monopoly profits until an equilibrium is reached. At this point the cycle repeats itself – our original entrepreneur or someone else looks for the next innovation that will rewrite the rules of the game, and off we go again. Schumpeter talks of a process of 'creative destruction' where there is a constant search to create something new which simultaneously destroys the old rules and establishes new ones – all driven by the search for new sources of profits.[30]

In his view '*[What counts is] competition from the new commodity, the new technology, the new source of supply, the new type of organization . . . competition which . . . strikes not at the margins of the profits and the outputs of the existing firms but at their foundations and their very lives.*'

1.4 What is innovation?

One of America's most successful innovators was Thomas Alva Edison who registered over 1000 patents. Products for which his organization was responsible include the light bulb, 35 mm cinema film and even the electric chair. Edison appreciated better than most that the real challenge in innovation was not invention – coming up with good ideas – but in making those inventions work technically and commercially. His skill in doing this created a business empire worth, in 1920, around $21.6 billion. He put to good use an understanding of the interactive nature of innovation, realizing that both technology push (which he systematized in one of the world's first organized R&D laboratories) and demand pull need to be mobilized.

His work on electricity provides a good example of this. Edison recognized that although the electric light bulb was a good idea it had little practical relevance in a world where there was no power point to plug it into. Consequently, his team set about building up an entire electricity generation and distribution infrastructure, including designing lamp stands, switches and wiring. In 1882 he switched on the power from the first electric power generation plant in Manhattan and was able to light up 800 bulbs in the area. In the years that followed he built over 300 plants all over the world.[31]

As Edison realized, innovation is more than simply coming up with good ideas: it is the *process* of growing them into practical use. Definitions of innovation may vary in their wording, but they all stress the need to complete the development and exploitation aspects of new knowledge, not just its invention. Some examples are given in the Research Note box below.

If we only understand part of the innovation process, then the behaviours we use in managing it are also likely to be only partially helpful – even if well intentioned and executed. For example, innovation is often confused with invention – but the latter is only the first step in a long process of bringing a good

---

RESEARCH NOTE   **What is innovation?**

One of the problems in managing innovation is the variation in what people understand by the term, often confusing it with invention. In its broadest sense the term comes from the Latin *innovare* meaning 'to make something new'. Our view, shared by the following writers, assumes that innovation is a process of turning opportunity into new ideas and of putting these into widely used practice.

- 'Innovation is the successful exploitation of new ideas' – Innovation Unit (2004) UK Department of Trade and Industry.
- 'Industrial innovation includes the technical, design, manufacturing, management and commercial activities involved in the marketing of a new (or improved) product or the first commercial use of a new (or improved) process or equipment' – Chris Freeman (1982) *The Economics of Industrial Innovation*, 2nd edition, Pinter, London.
- '. . . Innovation does not necessarily imply the commercialization of only a major advance in the technological state of the art (a radical innovation) but it includes also the utilization of even small-scale changes in technological know-how (an improvement or incremental innovation)' – Roy Rothwell and Paul Gardiner (1985) Invention, innovation, re-innovation and the role of the user. *Technovation*, **3**, 168.
- 'Innovation is the specific tool of entrepreneurs, the means by which they exploit change as an opportunity for a different business or service. It is capable of being presented as a discipline, capable of being learned, capable of being practised' – Peter Drucker (1985) *Innovation and Entrepreneurship*, Harper & Row, New York.
- 'Companies achieve competitive advantage through acts of innovation. They approach innovation in its broadest sense, including both new technologies and new ways of doing things' – Michael Porter (1990) *The Competitive Advantage of Nations*, Macmillan, London.
- 'An innovative business is one which lives and breathes "outside the box". It is not just good ideas, it is a combination of good ideas, motivated staff and an instinctive understanding of what your customer wants' – Richard Branson (1998) DTI Innovation Lecture.

---

idea to widespread and effective use. Being a good inventor is – to contradict Emerson* – no guarantee of commercial success and no matter how good the better mousetrap idea, the world will only beat a path to the door if attention is also paid to project management, market development, financial management, organizational behaviour, etc. Case study 1.3 gives some examples which highlight the difference between invention and innovation and that completing the journey is far from easy.

CASE STUDY 1.3

Invention and innovation

Some of the most famous inventions of the nineteenth century came from men whose names are forgotten; the actual names we associate with the products are of the entrepreneurs who brought them into commercial use. For example, the vacuum cleaner was invented by one J. Murray Spengler and originally called an 'electric suction sweeper'. He approached a leather goods maker in the town who knew nothing about vacuum cleaners but had a good idea of how to market and sell them – a certain W.H. Hoover. Similarly, a Boston man called Elias Howe produce the world's first sewing machine in 1846. Unable to sell his ideas despite travelling to England and trying there, he returned to the USA to find one Isaac Singer had stolen the patent and built a successful business from it. Although Singer was eventually forced to pay Howe a royalty on all machines made, the name which most people now associate with sewing machines is Singer not Howe. And Samuel Morse, widely credited as the father of modern telegraphy, actually invented only the code which bears his name; all the other inventions came from others. What Morse brought was enormous energy and a vision of what could be accomplished; to realize this he combined marketing and political skills to secure state funding for development work, and to spread the concept of something which for the first time would link people separated by vast distances on the continent of America. Within five years of demonstrating the principle there were over 5000 miles of telegraph wire in the USA, and Morse was regarded as 'the greatest man of his generation'.[31]

Innovation isn't easy . . . .

Although innovation is increasingly seen as a powerful way of securing competitive advantage and a more secure approach to defending strategic positions, success is by no means guaranteed. The history of product and process innovations is littered with examples of apparently good ideas which failed – in some cases with spectacular consequences. For example:

- In 1952 Ford engineers began working on a new car to counter the mid-size models offered by GM and Chrysler – the 'E' car. After an exhaustive search for a name involving some 20 000 suggestions the car was finally named after Edsel Ford, Henry Ford's only son. It was not a success; when the first Edsels came off the production line Ford had to spend an average of $10 000 per car (twice the vehicle's cost) to get them roadworthy. A publicity plan was to have 75 Edsels

---

* 'If a man has good corn, or wood, or boards, or pigs to sell, or can make better chairs or knives, crucibles or church organs than anybody else, you will find a broad-beaten road to his home, though it be in the woods.' (Entry in his journal 1855, Ralph Waldo Emerson).

drive out on the same day to local dealers; in the event the firm only managed to get 68 to go, whilst in another live TV slot the car failed to start. Nor were these teething troubles; by 1958 consumer indifference to the design and concern about its reputation led the company to abandon the car – at a cost of $450 million and 110,847 Edsels.[31]

- During the latter part of the Second World War it became increasingly clear that there would be a big market for long-distance airliners, especially on the transatlantic route. One UK contender was the Bristol Brabazon, based on a design for a giant long-range bomber, which was approved by the Ministry of Aviation for development in 1943. Consultation with BOAC, the major customer for the new airliner, was 'to associate itself closely with the layout of the aircraft and its equipment' but not to comment on issues like size, range and payload! The budget rapidly escalated, with the construction of new facilities to accommodate such a large plane and, at one stage, the demolition of an entire village in order to extend the runway at Filton, near Bristol. Project control was weak and many unnecessary features were included, for example, the mock-up contained 'a most magnificent ladies' powder room' with wooden aluminium-painted mirrors and even receptacles for the various lotions and powders used by the 'modern young lady'. The prototype took six and a half years to build and involved major technical crises with wings and engine design; although it flew well in tests the character of the post-war aircraft market was very different from that envisaged by the technologists. Consequently in 1952, after flying less than 1000 miles, the project was abandoned at considerable cost to the taxpayer. The parallels with the Concorde project, developed by the same company on the same site a decade later, are hard to escape.

- During the late 1990s revolutionary changes were going on in mobile communications involving many successful innovations – but even experienced players can get their fingers burned. Motorola launched an ambitious venture which aimed to offer mobile communications from literally anywhere on the planet – including the middle of the Sahara Desert or the top of Mount Everest! Achieving this involved a $7 billion project to put 88 satellites into orbit, but despite the costs Iridium – as the venture was known – received investment funds from major backers and the network was established. The trouble was that once the novelty had worn off, most people realized that they did not need to make many calls from remote islands or at the North Pole and that their requirements were generally well met with less exotic mobile networks based around large cities and populated regions. Worse, the handsets for Iridium were large and clumsy because of the complex electronics and wireless equipment they had to contain – and the cost of these hi-tech bricks was a staggering $3000! Call charges were similarly highly priced. Despite the incredible technological achievement which this represented the take-up of the system never happened, and in 1999 the company filed for Chapter 11 bankruptcy. Its problems were not over – the cost of maintaining the satellites safely in orbit was around $2 million per month. Motorola who had to assume the responsibility had hoped that other telecommunications firms might take advantage of these satellites, but after no interest was shown they had to look at a further price tag of $50 million to bring them out of orbit and destroy them safely. Even then the plans to allow them to drift out of orbit and burn up in the atmosphere were criticized by NASA for the risk they might pose in starting a nuclear war, because any pieces which fell to earth would be large enough to trigger Russian anti-missile defences since they might appear not as satellite chunks but Moscow-bound missiles!

1.5 A process view of innovation

In this book we will make use of a simple model of innovation as the *process* of turning ideas into reality and capturing value from them. We will explain the model in more detail in the next chapter but it is worth introducing it here. There are four key phases, each of which requires dealing with particular challenges – and only if we can manage the whole process is innovation likely to be successful.

Phase one involves the question of *search*. To take a biological metaphor, we need to generate variety in our gene pool – and we do this by bringing new ideas to the system. These can come from R&D, 'Eureka' moments, copying, market signals, regulations, competitor behaviour – the list is huge but the underlying challenge is the same – how do we organize an effective search process to ensure a steady flow of 'genetic variety' which gives us a better chance of surviving and thriving?

But simply generating variety isn't enough – we need to *select* from that set of options the variants most likely to help us grow and develop. Unlike natural selection where the process is random we are concerned here with some form of *strategic* choice – out of all the things we could do, what are we going to do – and why? This process needs to take into account competitive differentiation – which choices give us the best chance of standing out from the crowd? – and previous capabilities – can we build on what we already have or is this a step into the unknown?

Generating and selecting still leaves us with the huge problem of actually making it happen – committing our scarce resources and energies to doing something different. This is the challenge of *implementation* – converting ideas into reality. The task is essentially one of managing a growing commitment of resources – time, energy, money and above all mobilizing knowledge of different kinds – against a background of uncertainty. Unlike conventional project management the innovation challenge is about developing something which may never have been done before – and the only way we know whether or not we will succeed is by trying it out.

Here the biological metaphor comes back into play – it is a risky business. We are betting – taking calculated risks rather than random throws of the dice but nonetheless gambling – that we can make this new thing happen (manage the complex project through to successful completion) *and* that it will deliver us the calculated value which exceeds or at least equals what we put into it. If it is a new product or service – the market will rush to our stall to buy what we are offering, or if it is a new process, our internal market will buy into the new way of doing things and we will become more effective as a result. If it is a social innovation, can we manage to make the world a better place in ways which justify the investment we put in?

Viewed in this way the innovation task looks deceptively simple. The big question is, of course, how to make it happen? This has been the subject of intensive study for a long period of time – plenty of practitioners have not only left us their innovations but also some of their accumulated wisdom, lessons about managing the process which they have learned the hard way. And a growing academic community has been working on trying to understand in systematic fashion questions about not only the core process but also the conditions under which it is likely to succeed or fail. This includes knowledge about the kinds of things which influence and help/hinder the process – essentially boiling down to having a clear and focused direction (the underpinning 'why' of the selection stage) and creating the organizational conditions to allow focused creativity.

The end effect is that we have a rich – and convergent – set of recipes which go a long way towards helping answer the practising manager's question when confronted with the problem of organizing and managing innovation – 'What do I do on Monday morning?'. Exploring this in greater detail provides the basis for the rest of the book.

VIEWS FROM THE FRONT LINE

There is nothing more difficult to take in hand, more perilous to conduct, or more uncertain in its success, than to take the lead in the introduction of a new order of things.

(Niccolo Machiavelli, *The Prince*, 1532)

Anything that won't sell, I don't want to invent. Its sale is proof of utility, and utility is success.

Everything comes to him who hustles while he waits.

Genius is one percent inspiration and ninety-nine percent perspiration.

I never did anything by accident, nor did any of my inventions come by accident; they came by work.

Make it a practice to keep on the lookout for novel and interesting ideas that others have used successfully. Your idea has to be original only in its adaptation to the problem you are working on.

(Thomas A. Edison)

Managing and innovation did not always fit comfortably together. That's not surprising. Managers are people who like order. They like forecasts to come out as planned. In fact, managers are often judged on how much order they produce. Innovation, on the other hand, is often a disorderly process. Many times, perhaps most times, innovation does not turn out as planned. As a result, there is tension between managers and innovation.

(Lewis Lehro, about the first years at 3M)

In the past, innovation was defined largely by creativity and the development of new ideas. Today the term encompasses coordinated projects directed toward honing these ideas and converting them into developments that boost the bottom line.

(Howard Smith, Computer Sciences Corporation)

To turn really interesting ideas and fledgling technologies into a company that can continue to innovate for years, it requires a lot of disciplines.

(Steve Jobs, Apple Inc.)

Scope for/types of innovation

If innovation is a process we need to consider the output of that process. In what ways can we innovate – what kinds of opportunities exist to create something different and capture value from bringing those ideas into the world?

Sometimes it is about completely new possibilities, for instance, by exploiting radical breakthroughs in technology. For example, new drugs based on genetic manipulation have opened a major new front in the war against disease. Mobile phones, PDAs and other devices have revolutionized where and when we communicate. Even the humble window pane is the result of radical technological innovation – almost all the window glass in the world is made these days by the Pilkington float glass process which moved the industry away from the time-consuming process of grinding and polishing to get a flat surface.

Equally important is the ability to spot where and how new *markets* can be created and grown. Alexander Graham Bell's invention of the telephone didn't lead to an overnight revolution in communications – that depended on developing the market for person-to-person communications. Henry Ford may not have invented the motor car but in making the Model T – 'a car for Everyman' at a price most people could afford – he grew the mass market for personal transportation. And eBay justifies its multi-billion dollar price tag not because of the technology behind its online auction idea but because it created and grew the market.

Innovation isn't just about opening up new markets – it can also offer new ways of serving established and mature ones. Low-cost airlines are still about transportation, but the innovations which firms like Southwest Airlines, easyJet and Ryanair have introduced have revolutionized air travel and grown the market in the process. One challenging new area for innovation lies in the previously underserved markets of the developing world – the 4 billion people who earn less than $2 per day. The potential for developing radically different innovative products and services aimed at meeting the needs of this vast population at what Prahalad calls 'the bottom of the pyramid' is huge – and the lessons learned may impact on established markets in the developed world as well.[3]

And it isn't just about manufactured products; in most economies the service sector accounts for the vast majority of activity so there is likely to be plenty of scope. Lower capital costs often mean that the opportunities for new entrants and radical change are greatest in the service sector. Online banking and insurance have become commonplace but they have radically transformed the efficiencies with which those sectors work and the range of services they can provide. New entrants riding the Internet wave have rewritten the rule book for a wide range of industrial games, for example, Amazon in retailing, eBay in market trading and auctions, Google in advertising, Skype in telephony. Others have used the web to help them transform business models around things like low-cost airlines, online shopping and the music business.[32]

Four dimensions of innovation space

Essentially we are talking about change, and this can take several forms; for the purposes of this book we will focus on four broad categories: (The video of 'Finnegan's Fish Bar' on the website provides an example of how this 4Ps approach can be used to explore opportunities for innovation in a business.)

- 'Product innovation' – changes in the things (products/services) that an organization offers.
- 'Process innovation' – changes in the ways in which they are created and delivered.
- 'Position innovation' – changes in the context in which the products/services are introduced.
- 'Paradigm innovation' – changes in the underlying mental models which frame what the organization does.

Figure 1.1 shows how these '4Ps' provide the framework for a map of the innovation space available to any organization.[33]

For example, a new design of car, a new insurance package for accident-prone babies and a new home entertainment system would all be examples of product innovation. And change in the manufacturing methods and equipment used to produce the car or the home entertainment system, or in the office procedures and sequencing in the insurance case, would be examples of process innovation.

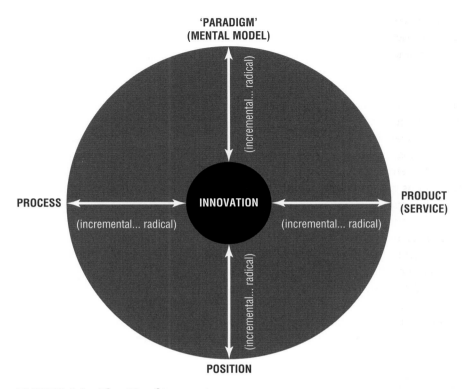

FIGURE 1.1: The 4Ps of innovation space

Sometimes the dividing line is somewhat blurred, for example, a new jet-powered sea ferry is both a product and a process innovation. Services represent a particular case of this where the product and process aspects often merge, for example, is a new holiday package a product or process change?

Innovation can also take place by repositioning the perception of an established product or process in a particular user context. For example, an old-established product in the UK is Lucozade – originally developed in 1927 as a glucose-based drink to help children and invalids in convalescence. These associations with sickness were abandoned by the brand owners, Beechams (now part of GSK), when they relaunched the product as a health drink aimed at the growing fitness market where it is now presented as a performance-enhancing aid to healthy exercise. This shift is a good example of 'position' innovation. In similar fashion Häagen-Dazs were able to give a new and profitable lease of life to an old-established product (ice cream) made with well-known processes. Their strategy was to target a different market segment and to reposition their product as a sensual pleasure to be enjoyed by adults – essentially telling an 'ice cream for grown-ups' story.

Sometimes opportunities for innovation emerge when we reframe the way we look at something. Henry Ford fundamentally changed the face of transportation not because he invented the motor car (he was a comparative latecomer to the new industry) or because he developed the manufacturing process to put one together (as a craft-based specialist industry car making had been established for around 20 years). His contribution was to change the underlying model from one which offered a handmade specialist product to a few wealthy customers to one which offered a car for everyone at a price they could afford. The ensuing shift from craft to mass production was nothing short of a revolution in the way cars (and later countless other products and services) were created and delivered.[21] Of course making the new approach work in practice also required extensive product and process innovation, for example, in component design, in machinery building, in factory layout and particularly in the social system around which work was organized. See Model T case study available on the web.

Recent examples of 'paradigm' innovation – changes in mental models – include the shift to low-cost airlines, the provision of online insurance and other financial services, and the repositioning of drinks like coffee and fruit juice as premium 'designer' products. Although in its later days Enron became infamous for financial malpractice it originally came to prominence as a small gas pipeline contractor which realized the potential in paradigm innovation in the utilities business. In a climate of deregulation and with global interconnection through grid distribution systems, energy and other utilities such as telecommunications bandwidth increasingly became commodities which could be traded much as sugar or cocoa futures.[34]

In their book *Wikinomics*, Tapscott and Williams highlight the wave of innovation which follows the paradigm change to 'mass collaboration' via the Internet which builds on social networks and communities. Companies like LEGO and Adidas (see case studies available on the web) are reinventing themselves by engaging their users as designers and builders rather than as passive consumers, whilst others are exploring the potential of virtual worlds like 'Second Life'.[27, 32] Concerns about global warming and sustainability of key resources such as energy and materials are, arguably, setting the stage for some significant paradigm innovation across many sectors as firms struggle to redefine themselves and their offerings to match these major social issues. Table 1.3 gives examples of innovations mapped on to the 4Ps model.

TABLE 1.3 Some examples of innovations mapped on to the 4Ps model.		
Innovation type	**Incremental – 'do what we do but better'**	**Radical – 'do something different'**
'Product' – what we offer the world	Windows Vista replacing XP – essentially improving on existing software idea VW EOS replacing the Golf – essentially improving on established car design Improved performance incandescent light bulbs	New to the world software, e.g., the first speech recognition program Toyota Prius – bringing a new concept – hybrid engines LED-based lighting, using completely different and more energy efficient principles (see Philips and lightbulb case studies available on the web)
Process – how we create and deliver that offering	Improved fixed-line telephone services Extended range of stock-broking services	Skype and other VoIP systems Online share trading eBay

(continued)

TABLE 1.3 (Continued)		
Innovation type	**Incremental – 'do what we do but better'**	**Radical – 'do something different'**
	Improved auction house operations	Toyota Production System and other 'lean' approaches
	Improved factory operations efficiency through upgraded equipment	Mobile banking in Kenya, Philippines – using phones as an alternative to banking systems
	Improved range of banking services delivered at branch banks	
Position – where we target that offering and the story we tell about it	Häagen Dazs changing the target market for ice cream from children to consenting adults	Addressing underserved markets, e.g., Tata Nano which targets the huge but relatively poor Indian market using the low-cost airline model – target cost is 1 lakh (around $3000)
	Low-cost airlines	
	University of Phoenix and others, building large education businesses via online approaches to reach different markets	'Bottom of the pyramid' approaches using a similar principle – Aravind eye care, Cemex construction products
	Dell and others segmenting and customizing computer configuration for individual users	One laptop per child project – the $100 universal computer
	Banking services targeted at key segments – students, retired people, etc.	Microfinance – Grameen Bank opening up credit for the very poor
Paradigm – how we frame what we do	Bausch and Lomb – moved from 'eye wear' to 'eye care' as its business model, effectively letting go of the old business of spectacles, sunglasses and contact lenses all of which were becoming commodity businesses. Instead it moved into newer high-tech fields like laser surgery equipment, specialist optical devices and research into artificial eyesight	Grameen Bank and other microfinance models – rethinking the assumptions about credit and the poor
		iTunes platform – a complete system of personalized entertainment
		Rolls-Royce – from high-quality aero engines to becoming a service company offering 'power by the hour'

TABLE 1.3 (Continued)		
Innovation type	Incremental – 'do what we do but better'	Radical – 'do something different'
	IBM moving from being a machine maker to a service and solution company – selling off its computer making and building up its consultancy and service side	Cirque du Soleil – redefining the circus experience
	VT moving from being a shipbuilder with roots in Victorian times to a service and facilities management business	

Mapping innovation space

The area indicated by the circle in Figure 1.1 is the potential innovation space within which an organization can operate. (Whether it actually explores and exploits all the space is a question for innovation *strategy* and we will return to this theme later in Chapter 3.) See web for 4Ps interactive exercise.

We can use the model to look at where the organization currently has innovation projects – and where it might move in the future. For example, if the emphasis has been on product and process innovation there may be scope for exploring more around position innovation – which new or underserved markets might we play in? Or around defining a new paradigm, a new business model with which to approach the marketplace.

We can also compare maps for different organizations competing in the same market – and use the tool as a way of identifying where there might be relatively unexplored space which might offer significant innovation opportunities. By looking at where other organizations are clustering their efforts we can pick up valuable clues about how to find relatively uncontested space and focus our efforts on these – as the low-cost airlines did with targeting new and underserved markets for travel.[35]

RESEARCH NOTE	Mapping innovation space

Figure 1.2 shows how the 4Ps approach was applied in a company (R&P Ltd) making garden machinery. The diamond diagram provides an indication of where and how they could construct a broad-ranging 'innovation agenda'. Nine innovation activities were listed on the diamond chart, including:

• Building totally customized products for customer's individual orders (paradigm).
• Using sensors in the next generation of lawn mowers to avoid roots and stones (product).

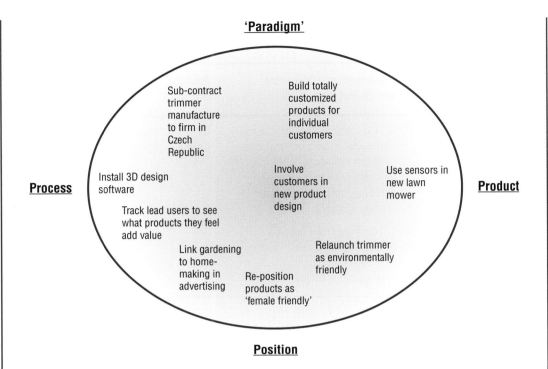

FIGURE 1.2: Suggested innovations mapped on to the 4Ps framework

- Repositioning the company's products as female-friendly as more women are keen gardeners (position).
- Installing 3D design software in the R&D department (process).

The selection of just nine major innovation initiatives gave focus to R&P's innovation management: the firm considered that 'it is important not to try to do too much at once'. Some initiatives, such as relaunching their trimmer as environmentally friendly, require both product and positional innovation. Such interdependencies are clarified by discussion on the placing of an initiative on the diamond diagram. Also, the fact that the senior management group had the 4Ps on one sheet of paper had the effect of enlarging choice – they saw completing the diagram as a tool for helping them think in a systematic way about using the innovation capability of the firm.

Source: based on Francis, D. and J. Bessant (2005) Targeting innovation and implications for capability development. *Technovation*, **25** (3), 171–83.

1.6 Exploring different aspects of innovation

The overall innovation space provides a simple map of the table on which we might place our innovation bets. But before making those bets we should consider some of the other characteristics of innovation which might shape our strategic decisions about where and when to play. These key aspects include:

- Degree of novelty – incremental or radical innovation?
- Platforms and families of innovations.
- Discontinuous innovation – what happens when the rules of the game change?
- Level of innovation – component or architecture?
- Timing – the innovation life cycle.

We will explore these – and the challenges they pose for managing innovation – a little more in the following section.

Incremental innovation – doing what we do but better

A key issue in managing innovation relates to the degree of novelty involved in different places across the innovation space. Clearly, updating the styling on our car is not the same as coming up with a completely new concept car which has an electric engine and is made of new composite materials as opposed to steel and glass. Similarly, increasing the speed and accuracy of a lathe is not the same as replacing it with a computer-controlled laser forming process. There are degrees of novelty in these, running from minor, incremental improvements right through to radical changes which transform the way we think about and use them. Sometimes these changes are common to a particular sector or activity, but sometimes they are so radical and far-reaching that they change the basis of society – for example the role played by steam power in the Industrial Revolution or the ubiquitous changes resulting from today's communications and computing technologies.

As far as managing the innovation process is concerned, these differences are important. The ways in which we approach incremental, day-to-day change will differ from those used occasionally to handle a radical step change in a product or process. But we should also remember that it is the *perceived* degree of novelty which matters; novelty is very much in the eye of the beholder. For example, in a giant, technologically advanced organization like Shell or IBM advanced networked information systems are commonplace, but for a small car dealership or food processor even the use of a simple PC to connect to the Internet may still represent a major challenge.

The reality is that although innovation sometimes involves a discontinuous shift, most of the time it takes place in incremental fashion. Essentially this is product/process improvement along the lines of 'doing what we do, but better' – and there is plenty to commend this approach. For example, the Bic ballpoint pen was originally developed in 1957 but remains a strong product with daily sales of 14 million units worldwide. Although superficially the same shape, closer inspection reveals a host of incremental changes that have taken place in materials, inks, ball technology, safety features, etc. Products are rarely 'new to the world', process innovation is mainly about optimization and getting the bugs out of the system. (Ettlie suggests disruptive or new-to-the-world innovations are only 6% to 10% of all projects labelled innovation.[36]) Studies of incremental process development (such as Hollander's famous study of DuPont rayon plants) suggest that the cumulative gains in efficiency are often much greater over time than those which come from occasional radical changes.[37] Other examples include Tremblay's studies of paper mills,[38] Enos on petroleum refining[39] and Figueredo's of steel plants.[40] For more detailed examples of continuous improvement see Forte, NPI and HBL case studies on web.

Continuous improvement of this kind has received considerable attention in recent years, originally as part of the 'total quality management' movement in the late twentieth century, reflecting the significant gains which Japanese manufacturers were able to make in improving quality and productivity through sustained incremental change.[41] But these ideas are not new – similar principles underpin the famous 'learning curve' effect where productivity improves with increases in the scale of production; the reason for this lies in the learning and continuous incremental problem-solving innovation which accompanies the introduction of a new product or process.[42] More recent experience of deploying 'lean' thinking in manufacturing and services and increasingly between as well as within enterprises underlines further the huge scope for such continuous innovation.[43] See web for example of continuous improvement tools.

Platform innovation

One way in which the continuous incremental innovation approach can be harnessed to good effect is through the concept of 'platforms'. This is a way of creating stretch and space around an innovation and depends on being able to establish a strong basic platform or family which can be extended. Rothwell and Gardiner give several examples of such 'robust designs' which can be stretched and otherwise modified to extend the range and life of the product, including Boeing airliners and Rolls-Royce jet engines.[44] Major investments by large semiconductor manufacturers like Intel and AMD are amortized to some extent by being used to design and produce a family of devices based on common families or platforms such as the Pentium, Celeron, Athlon or Duron chipsets.[45] Car makers are increasingly moving to produce models, which although apparently different in style, make use of common components and floor pans or chassis. Perhaps the most famous product platform is the 'Walkman' originally developed by Sony as a portable radio and cassette system; the platform concept has come to underpin a wide range of offerings from all major manufacturers for this market and deploying technologies such as minidisk, CD, DVD and now MP3 players.

In processes much has been made of the ability to enhance and improve performance over many years from the original design concepts – in fields like steel making and chemicals, for example. Service innovation offers other examples where a basic concept can be adapted and tailored for a wide range of similar applications without undergoing the high initial design costs – as is the case with different mortgage or insurance products. Sometimes platforms can be extended across different sectors – for example, the original ideas behind 'lean' thinking originated in firms such as Toyota in the field of car manufacturing – but have subsequently been applied across many other manufacturing sectors and into both public and private service applications including hospitals, supermarkets and banks.[43]

Platforms and families are powerful ways for companies to recoup their high initial investments in R&D by deploying the technology across a number of market fields. For example, Procter & Gamble invested heavily in its cyclodextrin development for original application in detergents but then were able to use this technology or variants on it in a family of products including odour control ('Febreze'), soaps and fine fragrances ('Olay'), off-flavour food control, disinfectants, bleaches and fabric softening ('Tide', 'Bounce'). They were also able to license out the technology for use in non-competing areas such as industrial-scale carpet care and in the pharmaceutical industry.

If we take the idea of 'position' innovation mentioned earlier then the role of brands can be seen as establishing a strong platform association which can be extended beyond an initial product or service. For example Richard Branson's Virgin brand has successfully provided a platform for entry into a variety of new fields including trains, financial services, telecommunications and food, whilst Stelios Haji-Ioannou has done something similar with his 'easy' brand, moving into cinemas, car rental, cruises and hotels from the original base in low-cost flying.

In their work on what they call 'management innovation' Hamel highlights a number of core organizational innovations (such as 'total quality management') which have diffused widely across sectors.[46]

These are essentially paradigm innovations which represent concepts that can be shaped and stretched to fit a variety of different contexts – for example Henry Ford's original ideas on mass production became applied and adapted to a host of other industries. McDonald's owed much of its inspiration to him in designing its fast-food business and in turn it was a powerful influence on the development of the Aravind eye clinics in India, which bring low-cost eye surgery to the masses.[3]

Discontinuous innovation – what happens when the game changes?

Most of the time innovation takes place within a set of rules of the game which are clearly understood, and involves players trying to innovate by doing what they have been doing (product, process, position, etc.) but better. Some manage this more effectively than others but the 'rules of the game' are accepted and do not change.[47]

However, occasionally something happens which dislocates this framework and changes the rules. By definition these are not everyday events but they have the capacity to redefine the space and the boundary conditions – they open up new opportunities and challenge existing players to reframe what they are doing in the light of new conditions.[48, 49, 50, 51] This is a central theme in Schumpeter's original theory of innovation which he saw as involving a process of 'creative destruction'.[24, 30]

CASE STUDY 1.4

The melting ice industry

Back in the 1880s there was a thriving industry in the north-eastern United States in the lucrative business of selling ice. The business model was deceptively simple – work hard to cut chunks of ice out of the frozen northern wastes, wrap the harvest quickly and ship it as quickly as possible to the warmer southern states – and increasingly overseas – where it could be used to preserve food. In its heyday this was a big industry – in 1886 the record harvest ran to 25 million tons – and it employed thousands of people in cutting, storing and shipping the product. It was an industry with a strong commitment to innovation – developments in ice cutting, snow ploughs, insulation techniques and logistics underpinned the industry's strong growth. The impact of these innovations was significant – they enabled, for example, an expansion of markets to far-flung locations such as Hong Kong, Bombay and Rio de Janeiro where, despite the distance and journey times, sufficient ice remained of cargoes originally loaded in ports like Boston to make the venture highly profitable.[52]

At the same time researchers like the young Carl von Linde were working in their laboratories on the emerging problems of refrigeration. It wasn't long before artificial ice making became a reality – Joseph Perkins had demonstrated that vaporizing and condensing a volatile liquid in a closed system would do the job and in doing so outlined the basic architecture that underpins today's refrigerators. In 1870 Linde published his research and by 1873 a patented commercial refrigeration system was on the market. In the years which followed the industry grew – in 1879 there were 35 plants and 10 years later 222 making artificial ice. Effectively this development sounded the death knell for the ice-harvesting industry – although it took a long time to go under. For a while both industries grew alongside each other, learning and innovating along their different pathways and expanding the overall market for ice – for example, by feeding the growing urban demand to fill domestic 'ice boxes'. But inevitably the new technology took over as the old harvesting model reached the limits of what it could achieve in terms of technological efficiencies. Significantly most of the established ice harvesters were too locked into the old model to make the transition and so went under – to be replaced by the new refrigeration industry dominated by new entrant firms.

Change of this kind can come through the emergence of a new technology – like the ice industry example (see Case study 1.4). Or it can come through the emergence of a completely new market with new characteristics and expectations. In his famous studies of the computer disk drive, steel and hydraulic excavator industries Christensen highlights the problems that arise under these conditions. For example, the disk drive industry was a thriving sector in which the voracious demands of a growing range of customer industries meant there was a booming market for disk drive storage units. Around 120 players populated what had become an industry worth $18 billion by 1995 – and like their predecessors in ice harvesting – it was a richly innovative industry. Firms worked closely with their customers, understanding the particular needs and demands for more storage capacity, faster access times, smaller footprints, etc. But just like our ice industry, the virtuous circle around the original computer industry was broken – in this case not by a radical technological shift but also by the emergence of a new market with very different needs and expectations.[53] See web for patterns of discontinuous innovation exercise.

The key point about this sector was that disruption happened not once but several times, involving different generations of technologies, markets and participating firms. For example, whilst the emphasis in the mini-computer world of the mid-1970s was on high performance and the requirement for storage units correspondingly technologically sophisticated, the emerging market for personal computers had a very different shape. These were much less clever machines, capable of running simpler software and with massively inferior performance – but at a price which a very different set of people could afford. Importantly, although simpler, they were capable of doing most of the basic tasks that a much wider market was interested in – simple arithmetical calculations, word processing and basic graphics. As the market grew so learning effects meant that these capabilities improved – but from a much lower cost base. The result was, in the end, just like that of Linde and his contemporaries on the ice industry – but from a different direction. Of the major manufacturers in the disk drive industry serving the mini-computer market only a handful survived – and leadership in the new industry shifted to new entrant firms working with a very different model.

CASE STUDY 1.5

Technological excellence may not be enough . . .

In the 1970s Xerox was the dominant player in photocopiers, having built the industry from its early days when it was founded on the radical technology pioneered by Chester Carlsen and the Battelle Institute. But despite its prowess in the core technologies and continuing investment in maintaining an edge it found itself seriously threatened by a new generation of small copiers developed by new entrants including several Japanese players. Despite the fact that Xerox had enormous experience in the industry and a deep understanding of the core technology it took them almost eight years of mishaps and false starts to introduce a competitive product. In that time Xerox lost around half its market share and suffered severe financial problems. As Henderson and Clark put it, in describing this case, 'apparently modest changes to the existing technology . . . have quite dramatic consequences'.[54]

In similar fashion in the 1950s the electronics giant RCA developed a prototype portable transistor-based radio using technologies which it had come to understand well. However, it saw little reason to promote such an apparently inferior technology and continued to develop and build its high-range devices. By contrast Sony used it to gain access to the consumer market and to build a whole generation of portable consumer devices – and in the process acquired considerable technological experience which enabled it to enter and compete successfully in higher value, more complex markets.[55]

Discontinuity can also come about by reframing the way we think about an industry – changing the dominant business model and hence the 'rules of the game'. Think about the revolution in flying which the low-cost carriers have brought about. Here the challenge came via a new business model rather than technology – based on the premise that if prices could be kept low a large new market could be opened up. The power of the new way of framing the business was that it opened up a new – and very different – trajectory along which all sorts of innovations began to happen. In order to make low prices pay a number of problems needed solving – keeping load factors high, cutting administration costs, enabling rapid turnaround times at terminals – but once the model began to work it attracted not only new customers but also increasingly established flyers who saw the advantages of lower prices.

What these – and many other examples – have in common is that they represent the challenge of *discontinuous* innovation. None of the industries were lacking in innovation or a commitment to further change. But the ice harvesters, mini-computer disk companies and the established airlines all carried on their innovation on a stage covered with a relatively predictable carpet. The trouble was that shifts in technology, in new market emergence or in new business models pulled this carpet out from under the firms – and created a new set of conditions on which a new game would be played. Under such conditions, it is the new players who tend to do better because they don't have to wrestle with learning new tricks and letting go of their old ones. Established players often do badly – in part because the natural response is to press even harder on the pedal driving the existing ways of organizing and managing innovation. In the ice industry example the problem was not that the major players weren't interested in R&D – on the contrary they worked really hard at keeping a technological edge in insulation, harvesting and other tools. But they were blindsided by technological changes coming from a different field altogether – and when they woke up to the threat posed by mechanical ice making their response was to work even harder at improving their own ice-harvesting and shipping technologies. It is here that the so-called 'sailing ship' effect can often be observed, in which a mature technology accelerates in its rate of improvement as a response to a competing new alternative – as was the case with the development of sailing ships in competition with newly emerging steamship technology.[56]

In similar fashion the problem for the firms in the disk drive industry wasn't that they didn't listen to customers but rather that they listened too well. They built a virtuous circle of demanding customers in their existing market place with whom they developed a stream of improvement innovations – continuously stretching their products and processes to do what they were doing better and better. The trouble was that they were getting close to the wrong customers – the discontinuity which got them into trouble was the emergence of a completely different set of users with very different needs and values.

Table 1.4 gives some examples of such triggers for discontinuity. Common to these from an innovation management point of view is the need to recognize that under discontinuous conditions (which thankfully don't emerge every day) we need different approaches to organizing and managing innovation. If we try and use established models which work under steady state conditions we find – as is the reported experience of many – we are increasingly out of our depth and risk being upstaged by new and more agile players.

TABLE 1.4	**Sources of discontinuity**		
Triggers/ sources of discontinuity	**Explanation**	**Problems posed**	**Examples (of good and bad experiences)**
New market emerges	Most markets evolve through a process of gradual expansion but at certain times completely new markets emerge which cannot be analysed or predicted in advance or explored through using conventional market research/analytical techniques	Established players don't see it because they are focused on their existing markets May discount it as being too small or not representing their preferred target market – fringe/cranks dismissal Originators of new product may not see potential in new markets and may ignore them, e.g. text messaging	Disk drives, excavators, mini-mills.[53] Mobile phone/SMS where market which actually emerged was not the one expected or predicted by originators
New technology emerges	Step change takes place in product or process technology – may result from convergence and maturing of several streams (e.g. industrial automation, mobile phones) or as a result of a single	Don't see it because beyond the periphery of technology search environment Not an extension of current areas but completely new field or approach	Ice harvesting to cold storage[52] Valves to solid-state electronics[57] Photos to digital images

(continued)

TABLE 1.4	**(Continued)**		
Triggers/ sources of discontinuity	**Explanation**	**Problems posed**	**Examples (of good and bad experiences)**
	breakthrough (e.g. LED as white light source)	Tipping point may not be a single breakthrough but convergence and maturing of established technological streams, whose combined effect is underestimated	
		Not-invented-here effect – new technology represents a different basis for delivering value, e.g. telephone vs. telegraphy	
New political rules emerge	Political conditions which shape the economic and social rules may shift dramatically, e.g., the collapse of communism meant an alternative model (capitalist, competition as opposed to central planning) – and many ex-state firms couldn't adapt their ways of thinking	Old mindset about how business is done, rules of the game, etc. are challenged and established firms fail to understand or learn new rules	Centrally planned to market economy, e.g., former Soviet Union Apartheid to post-apartheid South Africa – inward and insular to externally linked[58] Free trade/globalization results in dismantling protective tariff and other barriers and new competition basis emerges[58, 59]
Running out of road	Firms in mature industries may need to escape the constraints of diminishing space for product and process	Current system is built around a particular trajectory and embedded in a steady-state set of innovation routines which militate against widespread search	Medproducts[60] Kodak Encyclopaedia Britannica[26]

(continued)

TABLE 1.4	**(Continued)**		
Triggers/ sources of discontinuity	**Explanation**	**Problems posed**	**Examples (of good and bad experiences)**
	innovation and the increasing competition of industry structures by either exit or by radical reorientation of their business	or risk-taking experiments	Preussag[25] Mannesmann
Sea change in market sentiment or behaviour	Public opinion or behaviour shifts slowly and then tips over into a new model, e.g., the music industry is in the midst of a (technology-enabled) revolution in delivery systems from buying records, tapes and CDs to direct download of tracks in MP3 and related formats	Don't pick up on it or persist in alternative explanations – cognitive dissonance – until it may be too late	Apple, Napster, Dell, Microsoft vs. traditional music industry[61]
Deregulation/ shifts in regulatory regime	Political and market pressures lead to shifts in the regulatory framework and enable the emergence of a new set of rules, e.g., liberalization, privatization or deregulation	New rules of the game but old mindsets persist and existing player unable to move fast enough or see new opportunities opened up	Old monopoly positions in fields like telecommunications and energy were dismantled and new players/combinations of enterprises emerged. In particular, energy and bandwidth become increasingly viewed as commodities. Innovations include skills in trading and distribution – a factor behind the considerable success of Enron in the late 1990s as it

(continued)

TABLE 1.4	(Continued)		
Triggers/ sources of discontinuity	**Explanation**	**Problems posed**	**Examples (of good and bad experiences)**
			emerged from a small gas pipeline business to becoming a major energy trade[34] – unquantifiable chances may need to be taken
Fractures along 'fault lines'	Long-standing issues of concern to a minority accumulate momentum (sometimes through the action of pressure groups) and suddenly the system switches/tips over, e.g., social attitudes to smoking or health concerns about obesity levels and fast foods	Rules of the game suddenly shift and then new pattern gathers rapid momentum wrong-footing existing players working with old assumptions. Other players who have been working in the background developing parallel alternatives may suddenly come into the limelight as new conditions favour them	McDonald's and obesity Tobacco companies and smoking bans Oil/energy companies and global warming Opportunity for new energy sources like wind power, cf. Danish dominance[62]
Unthinkable events	Unimagined and therefore not prepared for events which – sometimes literally – change the world and set up new rules of the game	New rules may disempower existing players or render competencies unnecessary	World Trade Center – 9/11
Business model innovation	Established business models are challenged by a reframing, usually by a new entrant who redefines/reframes the problem and the consequent rules of the game	New entrants see opportunity to deliver product/service via new business model and rewrite rules – existing players have at best to be fast followers	Amazon Charles Schwab[61] Southwest and other low-cost airlines[34, 61, 63]

(continued)

TABLE 1.4	(Continued)		
Triggers/ sources of discontinuity	**Explanation**	**Problems posed**	**Examples (of good and bad experiences)**
Shifts in 'techno-economic paradigm' – systemic changes which impact whole sectors or even whole societies	Change takes place at system level, involving technology and market shifts. This involves the convergence of a number of trends which result in a 'paradigm shift' where the old order is replaced	Hard to see where new paradigm begins until rules become established. Existing players tend to reinforce their commitment to old model, reinforced by 'sailing ship' effects	Industrial Revolution[64–66] Mass production
Architectural innovation	Changes at the level of the system architecture rewrite the rules of the game for those involved at component level	Established players develop particular ways of seeing and frame their interactions, e.g., who they talk to in acquiring and using knowledge to drive innovation – according to this set of views. Architectural shifts may involve reframing but at the component level it is difficult to pick up the need for doing so – and thus new entrants better able to work with new architecture can emerge	Photolithography in chip manufacture[54, 67]

Component/architecture innovation and the importance of knowledge

Another important lens through which to view innovation opportunities is as components within larger systems. Rather like Russian dolls we can think of innovations that change things at the level of components or those that involve change in a whole system. For example, we can put a faster transistor on a microchip on a circuit board for the graphics display in a computer. Or we can change the way several boards are put together in the computer to give it particular capabilities – a games box, an e-book, a media PC. Or we can link the computers in a network to drive a small business or office. Or we can link the networks to others into the Internet. There's scope for innovation at each level – but changes in the higher level systems often have implications for lower down. For example, if cars – as a complex assembly – were suddenly designed to be made out of plastic instead of metal it would still leave scope for car assemblers – but would pose some sleepless nights for producers of metal components! See web for patterns of architecture/component innovation exercise.

Innovation is about knowledge – creating new possibilities through combining different knowledge sets. These can be in the form of knowledge about what is technically possible or what particular configuration would meet an articulated or latent need. Such knowledge may already exist in our experience, based on something we have seen or done before. Or it could result from a process of search – research into technologies, markets, competitor actions, etc. And it could be in explicit form, codified in such a way that others can access it, discuss it, transfer it, etc. – or it can be in tacit form, known about but not actually put into words or formulae.[68]

The process of weaving these different knowledge sets together into a successful innovation is one which takes place under highly uncertain conditions. We don't know what the final innovation configuration will look like (and we don't know how we will get there). Managing innovation is about turning these uncertainties into knowledge – but we can do so only by committing resources to reduce the uncertainty – effectively a balancing act. Figure 1.3 illustrates this process of increasing resource commitment whilst reducing uncertainty.

Viewed in this way we can see that incremental innovation, whilst by no means risk-free, is at least potentially manageable because we are starting from something we know about and developing improvements in it. But as we move to more radical options, uncertainty is higher and we have no prior idea of what we are to develop or how to develop it! Again this helps us understand why discontinuous innovation is so hard to deal with.

A key contribution to our understanding here comes from the work of Henderson and Clark who looked closely at the kinds of knowledge involved in different kinds of innovation.[54] They argue that innovation rarely involves dealing with a single technology or market but rather a bundle of knowledge,

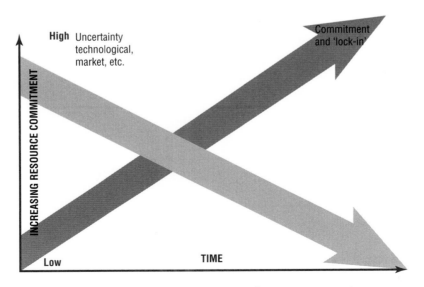

FIGURE 1.3: Innovation, uncertainty and resource commitment

which is brought together into a configuration. Successful innovation management requires that we can get hold of and use knowledge about *components* and also about how those can be put together – what they termed the *architecture* of an innovation.

We can see this more clearly with an example. Change at the component level in building a flying machine might involve switching to newer metallurgy or composite materials for the wing construction or the use of fly-by-wire controls instead of control lines or hydraulics. But the underlying knowledge about how to link aerofoil shapes, control systems, propulsion systems, etc. at the *system* level is unchanged – and being successful at both requires a different and higher order set of competencies.

One of the difficulties with this is that innovation knowledge flows – and the structures which evolve to support them – tend to reflect the nature of the innovation. So if it is at component level then the relevant people with skills and knowledge around these components will talk to each other – and when change takes place they can integrate new knowledge. But when change takes place at the higher system level – 'architectural innovation' in Henderson and Clark's terms – then the existing channels and flows may not be appropriate or sufficient to support the innovation and the firm needs to develop new ones. This is another reason why existing incumbents often fare badly when major system level change takes place – because they have the twin difficulties of learning and configuring a new knowledge system and 'unlearning' an old and established one.

Figure 1.4 illustrates the range of choices, highlighting the point that such change can happen at component or sub-system level or across the whole system.

A variation on this theme comes in the field of 'technology fusion', where different technological streams converge, such that products which used to have a discrete identity begin to merge into new architectures. An example here is the home automation industry, where the fusion of technologies such as computing, telecommunications, industrial control and elementary robotics is enabling a new generation of housing systems with integrated entertainment, environmental control (heating, air conditioning, lighting) and communication possibilities.[69, 70]

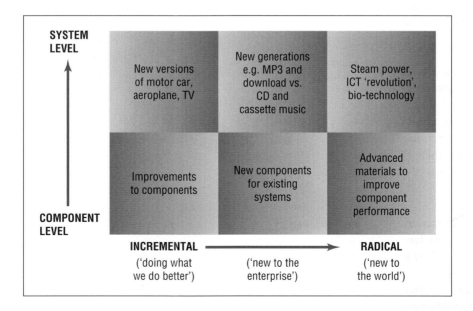

FIGURE 1.4: Dimensions of innovation

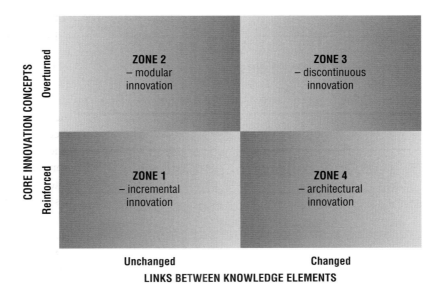

FIGURE 1.5: Component and architectural innovation
Source: Abernathy, W. and J. Utterback (1978) Patterns of industrial innovation. *Technology Review*, **80**, 40–47.

Similarly, a new addition to the range of financial services may represent a component product innovation, but its impacts are likely to be less far-reaching (and the attendant risks of its introduction lower) than a complete shift in the nature of the service package – for example, the shift to direct-line systems instead of offering financial services through intermediaries.

Many businesses are now built on business models that stress integrated solutions – systems of many components which together deliver value to end-users. These are often complex, multi-organization networks – examples might include rail networks, mobile phone systems, major construction projects or design and development of new aircraft like the Boeing Dreamliner or the Airbus A380. Managing innovation on this scale requires development of skills in what Hobday and colleagues call 'the business of systems integration'.[71]

Figure 1.5 highlights the issues for managing innovation. In Zone 1 the rules of the game are clear – this is about steady-state improvement to products or processes and uses knowledge accumulated around core components.

In Zone 2 there is significant change in one element but the overall architecture remains the same. Here there is a need to learn new knowledge but within an established and clear framework of sources and users – for example, moving to electronic ignition or direct injection in a car engine, the use of new materials in airframe components, the use of IT systems instead of paper processing in key financial or insurance transactions. None of these involve major shifts or dislocations.

In Zone 3 we have discontinuous innovation where neither the end state nor the ways in which it can be achieved are known – essentially the whole set of rules of the game changes and there is scope for new entrants.

In Zone 4 we have the condition where new combinations – architectures – emerge, possibly around the needs of different groups of users (as in the disruptive innovation case). Here the challenge is in reconfiguring the knowledge sources and configurations. We may use existing knowledge and recombine it in different ways or we may use a combination of new and old. Examples might be low-cost airlines and direct-line insurance.

The innovation life cycle – different emphasis over time

We also need to recognize that innovation opportunities change over time. In new industries – like today's biotech, Internet-software or nano-materials – there is huge scope for experimentation around new product and service concepts. But more mature industries tend to focus on process innovation or position innovation, looking for ways of delivering products and services more cheaply or flexibly, or for new market segments into which to sell them. In their pioneering work on this theme Abernathy and Utterback developed a model describing the pattern in terms of three distinct phases (see Figure 1.6).[72]

Initially, under the discontinuous conditions, which arise when completely new technology and/or markets emerge, there is what they term a 'fluid phase' where there is high uncertainty along two dimensions:

- The *target* – what will the new configuration be and who will want it?
- The *technical* – how will we harness new technological knowledge to create and deliver this?

No one knows what the 'right' configuration of technological means and market needs will be and so there is extensive experimentation (accompanied by many failures) and fast learning by a range of players including many new entrepreneurial businesses.

Gradually these experiments begin to converge around what they call a 'dominant design' – something which begins to set up the rules of the game. This represents a convergence around the most popular (importantly not necessarily the most technologically sophisticated or elegant) solution to the emerging configuration. At this point a 'bandwagon' begins to roll and innovation options become increasingly channelled around a core set of possibilities – what Dosi calls a 'technological trajectory'.[64] It becomes increasingly difficult to explore outside this space because entrepreneurial interest and the resources which that brings increasingly focus on possibilities within the dominant design corridor.

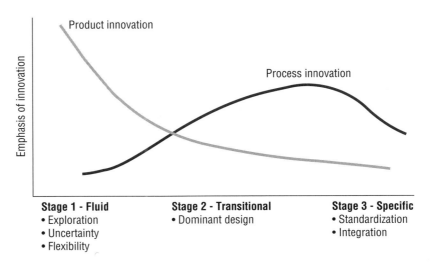

FIGURE 1.6: Abernathy and Utterback's model of innovation life cycle

This can apply to products or processes: in both cases the key characteristics become stabilized and experimentation moves to getting the bugs out and refining the dominant design. For example, the nineteenth-century chemical industry moved from making soda ash (an essential ingredient in making soap, glass and a host of other products) from the earliest days where it was produced by burning vegetable matter, through to a sophisticated chemical reaction which was carried out in a batch process (the Leblanc process), which was one of the drivers of the Industrial Revolution. This process dominated for nearly a century but was in turn replaced by a new generation of continuous processes using electrolytic techniques, which originated in Belgium where they were developed by the Solvay brothers. Moving to the Leblanc process or the Solvay process did not happen overnight; it took decades of work to refine and improve each process, and to fully understand the chemistry and engineering required to get consistent high quality and output.

The same pattern can be seen in products. For example, the original design for a camera is something that goes back to the early nineteenth century and – as a visit to any science museum will show – involved all sorts of ingenious solutions. The dominant design gradually emerged with an architecture which we would recognize – shutter and lens arrangement, focusing principles, back plate for film or plates, etc. But this design was then modified still further, for example, with different lenses, motorized drives, flash technology – and, in the case of George Eastman's work – to creating a simple and relatively 'idiot-proof' model camera (the Box Brownie) which opened up photography to a mass market. More recent development has seen a similar fluid phase around digital imaging devices. See web for product lifecycle analysis.

The period in which the dominant design emerges and emphasis shifts to imitation and development is termed the 'transitional phase' in the Abernathy and Utterback model. Activities move from radical concept development to more focused efforts geared around product differentiation and to delivering it reliably, cheaply, with higher quality and extended functionality.

As the concept matures still further so incremental innovation becomes more significant and emphasis shifts to factors such as cost – which means efforts within the industries that grow up around these product areas tend to focus increasingly on rationalization, on scale economies and on process innovation to drive out cost and improve productivity. Product innovation is increasingly about differentiation through customization to meet the particular needs of specific users. Abernathy and Utterback term this the 'specific phase'.

Finally the stage is set for change – the scope for innovation becomes smaller and smaller whilst outside – for example, in the laboratories and imaginations of research scientists – new possibilities are emerging. Eventually a new technology emerges, which has the potential to challenge all the by now well-established rules – and the game is disrupted. In the camera case, for example, this is happening with the advent of digital photography, which is having an impact on cameras and the overall service package around how we get, keep and share our photographs. In our chemical case this is happening with biotechnology and the emergence of the possibility of no longer needing giant chemical plants but instead moving to small-scale operations using live organisms genetically engineered to produce what we need.

Table 1.5 sets out the main elements of this model.

Although originally developed for manufactured products the model also works for services, for example the early days of Internet banking were characterized by a typically fluid phase with many options and models being offered. This gradually moved to a transitional phase, for example building a dominant design consensus on the package of services offered, the levels and nature of security and privacy support, the interactivity of website. The field has now become mature with much of the competition shifting to marginal issues such as relative interest rates. Similar patterns can be seen in VoIP telephony, online auctions such as eBay and travel and entertainment booking services such as Expedia.

TABLE 1.5	Stages in the innovation life cycle		
Innovation characteristic	**Fluid pattern**	**Transitional phase**	**Specific phase**
Competitive emphasis placed on . . .	Functional product performance	Product variation	Cost reduction
Innovation stimulated by . . .	Information on user needs, technical inputs	Opportunities created by expanding internal technical capability	Pressure to reduce cost, improve quality, etc.
Predominant type of innovation	Frequent major changes in products	Major process innovations required by rising volume	Incremental product and process innovation
Product line	Diverse, often including custom designs	Includes at least one stable or dominant design	Mostly undifferentiated standard products
Production processes	Flexible and inefficient – aim is to experiment and make frequent changes	Becoming more rigid and defined	Efficient, often capital intensive and relatively rigid

We should also remember that there is a long-term cycle involved – mature businesses that have already gone through their fluid and transitional phases do not necessarily stay in the mature phase forever. Rather they become increasingly vulnerable to a new wave of change as the cycle repeats itself – for example, the lighting industry is entering a new fluid phase based on applications of solid-state LED technology but this comes after over 100 years of the incandescent bulb developed by Swan, Edison and others. Their early experiments eventually converged on a dominant product design after which emphasis shifted to process innovation around cost, quality and other parameters – a trajectory that has characterized the industry and led to increasing consolidation amongst a few big players. That may all be about to change driven by a completely new – and much more powerful – technology based on solid-state electronics.

The pattern can be seen in many studies and its implications for innovation management are important. In particular it helps us understand why established organizations often find it hard to deal with the kind of discontinuous change discussed earlier. Organizations build capabilities around a particular trajectory and those who may be strong in the later (specific) phase of an established trajectory often find it hard to move into the new one. (The example of the firms which successfully exploited the transistor in the early 1950s is a good case in point – many were new ventures, sometimes started by enthusiasts in their garage, yet they rose to challenge major players in the electronics industry such as

Raytheon.[57]) This is partly a consequence of sunk costs and commitments to existing technologies and markets and partly because of psychological and institutional barriers. They may respond but in a slow fashion – and they may make the mistake of giving responsibility for the new development to those whose current activities would be threatened by a shift.[73]

Importantly, the 'fluid' or 'ferment' phase is characterized by *coexistence* of old and new technologies and by rapid improvements of both. (It is here that the so-called 'sailing ship' effect mentioned earlier can often be observed, in which a mature technology accelerates in its rate of improvement as a response to a competing new alternative.)

Whilst some research suggests existing incumbents do badly when discontinuous change triggers a new fluid phase, we need to be careful here. Not all existing players do so – many of them are able to build on the new trajectory and deploy/leverage their accumulated knowledge, networks, skills and financial assets to enhance their competence through building on the new opportunity.[53] Equally, whilst it is true that new entrants – often small entrepreneurial firms – play a strong role in this early phase we should not forget that we see only the successful players. We need to remember that there is a strong ecological pressure on new entrants, which means only the fittest or luckiest survive.

It is more helpful to suggest that there is something about the ways in which innovation is *managed* under these conditions that poses problems. Good practice of the 'steady-state' kind described above is helpful in the mature phase but can actively militate against the entry and success in the fluid phase of a new technology.[74] How do enterprises pick up signals about changes if they take place in areas where they don't normally do research? How do they understand the needs of a market which doesn't exist but will shape the eventual package? If they talk to their existing customers the likelihood is that those customers will tend to ask for more of the same, so which new users should they talk to – and how do they find them?

The challenge seems to be to develop ways of managing innovation not only under 'steady-state' but also under the highly uncertain, rapidly evolving and changing conditions resulting from a dislocation or discontinuity. The kinds of organizational behaviour needed here will include things like agility, flexibility, the ability to learn fast and the lack of preconceptions about the ways in which things might evolve – and these are often associated with new small firms. There are ways in which large and established players can also exhibit this kind of behaviour but they often conflict with their normal ways of thinking and working.

Worryingly the source of the discontinuity which destabilizes an industry – new technology, emergence of a new market, rise of a new business model – often comes from outside that industry. So even those large incumbent firms, which take time and resources to carry out research to try and stay abreast of developments in their field, may find that they are wrong-footed by the entry of something that has been developed in a different field. The massive changes in insurance and financial services, which have characterized the shift to online and telephone provision, were largely developed by IT professionals often working outside the original industry.[7] In extreme cases we find what is often termed the 'not-invented-here' (NIH) effect, where a firm finds out about a technology but decides against following it up because it does not fit with its perception of the industry or the likely rate and direction of its technological development. Famous examples of this include Kodak's rejection of the Polaroid process and Western Union's dismissal of Bell's telephone invention. In a famous memo dated 1876 the board commented, 'this "telephone" has too many shortcomings to be seriously considered as a means of communication. The device is inherently of no value to us.'

Managing innovation

This chapter has begun to explore the challenges posed by innovation. It has looked at why innovation matters and opened up some perspectives on what it involves. And it has raised the idea of innovation as a core *process* which needs to be organized and managed in order to enable the renewal of any organization. We talked about this a little earlier in the chapter and Figure 1.7 sets it out as a graphic highlighting the key questions around *managing* innovation.

We've seen that the scope for innovation is wide – in terms of overall innovation space and in the many different ways this can be populated, with both incremental and more radical options. At the limit we have the challenges posed when innovation moves into the territory of discontinuous change and a whole new game begins. We've also looked briefly at concepts such as component and architecture innovation and the critical role that knowledge plays in managing these different forms. Finally we've looked at the issue of timing and of understanding the nature of different innovation types at different stages.

All that gives us a feel for what innovation is and why it matters. But what we now need to do is understand how to organize the innovation process itself. That's the focus of the rest of the book, and we deal with it in the following fashion.

Chapter 2 looks at the process model in more detail and explores the ways in which this generic model can be configured for particular types of organization. It also looks at what we've learned about success and failure in managing innovation – themes which are examined in greater detail in the subsequent chapters.

Part 2 looks at the key contextual issues around successful innovation management. In Chapter 3 we pick up the question: do we have an innovative organization? And examine the role that key concepts such as leadership, structure, communication and motivation play in building and sustaining a culture of focused creativity.

Chapter 4 looks at the question: do we have a clear innovation strategy? And explores this theme in depth. Is there a clear sense of where and how innovation will take the organization forward and is there a roadmap for this? Is the strategy shared and understood – and how can we ensure alignment of the

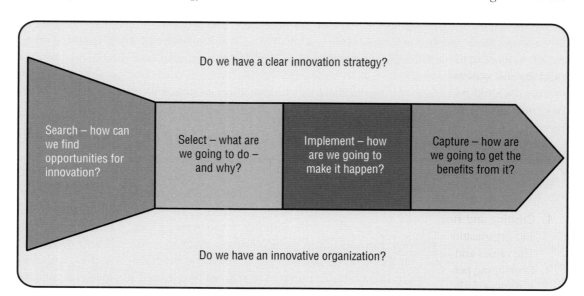

FIGURE 1.7: Simplified model of the innovation process

various different innovation efforts across the organization? What tools and techniques can be used to develop and enable analysis, selection and implementation of innovation?

Part 3 moves on to the first of the core elements in our process model – the 'search' question. Chapter 5 explores the issues around the question of what triggers the innovation process – the multiple sources which we need to be aware of and the challenges involved in searching for and picking up signals from them. Chapter 6 takes up the complementary question: *how* do we carry out this search activity? Which structures, tools and techniques are appropriate under what conditions? How do we balance search around exploration of completely new territory with exploiting what we already know in new forms? In particular it looks at the major challenge of building and sustaining rich networks to enable what has become labelled 'open innovation'.

Part 4 moves into the area of selection in the core process model. Chapter 7 looks at how the innovation decision process works – of all the possible options generated by effective search which ones will we back – and why? Making decisions of this kind is not simple because of the underlying uncertainty involved – so which approaches, tools and techniques can we bring to bear? Chapter 8 picks up another core theme: how to choose and implement innovation options whilst building and capturing value from the intellectual effort involved. Managing intellectual property becomes an increasingly significant issue in a world where knowledge production approaches the $1 trillion/year mark worldwide and where the ability to generate knowledge may be less significant than the ability to trade and use it effectively.

Part 5 looks at the 'implementation' phase, where issues of how we move innovation ideas into reality become central. Chapter 9 examines the ways in which innovation projects of various kinds are organized and managed and explores structures, tools and other support mechanisms to help facilitate this. Chapter 10 picks up the issue of new ventures, both those arising from within the existing organization (corporate entrepreneurship) and those which involve setting up a new entrepreneurial venture.

Part 6 looks at the last phase: how can we ensure that we capture value from our efforts at innovation? Chapter 11 examines questions of adoption and diffusion and the ways we can develop and work with markets for innovation. It picks up on both commercially driven value capture and also the question of 'social entrepreneurship' where concern is less about profits than about creating sustainable social value.

Finally Chapter 12 looks at how we can assess the ways in which we organize and manage innovation and use these to drive a learning process to enable us to do it better next time. The concern here is not just to build a strong innovation management capability but to recognize that – faced with the moving target that innovation represents in terms of technologies, markets, competitors, regulators and so on – the challenge is to create a learning and adaptive approach which constantly upgrades this capability. In other words we are concerned to build 'dynamic capability'.

VIEWS FROM THE FRONT LINE

Where do you see the top three challenges in managing innovation?

1. Creating and sustaining a culture in which innovation can flourish. This includes a physical and organizational space where experimentation, evaluation and examination can take place. The values and behaviours that facilitate innovation have to be developed and sustained.
2. Developing people who can flourish in that environment; people who can question, challenge and suggest ideas as part of a group with a common objective, unconstrained by the day-to-day operational environment.

3. Managing innovation in the midst of a commercial enterprise that is focused on exploitation – maximum benefit from the minimum of resource, that requires repeatability and a right-first-time process approach.

(Patrick McLaughlin, Managing Director, Cerulean – an extended interview with Patrick is on the website)

1. The level at which long-term innovation activities are best conducted, without losing connectedness with the business units at which the innovations should finally be incubated and elaborated
2. Having diverse type of individuals in the company motivated for spending time on innovation-related activities
3. Having the right balance between application-oriented innovation and more fundamental innovation

(Wouter Zeeman, CRH Insulation Europe)

1. Innovation is too often seen as a technically driven issue; in other words the preserve of those strange 'scientific' and 'engineering' people, so it's for them not 'us' the wider community. The challenge is in confronting this issue and hopefully inspiring and changing people's perception so that innovation is OK for all of 'us'.
2. Raising awareness. Coupled with the above people do not fully understand what innovation is or how it applies to their world.
3. Managing in my opinion is either the wrong word or the wrong thing to do; managing implies command and control and whilst important it does not always fit well with the challenge of leading innovation which is far more about inspiring, building confidence and risk taking. Most senior managers are risk averse therefore a solid management background is not always a best fit for the challenge of leading innovation.

(John Tregaskes, Technical Specialist Manager, Serco)

1. Culture – encouraging people to challenge the way we do things and generate creative ideas.
2. Balancing innovation with the levels of risk management and control required in a financial services environment.
3. Ensuring that innovation in one area does not lead to sub-optimization and negative impact in another.

(John Gilbert, Head of Process Excellence, UBS)

1. Alignment of expectations on innovation with senior management. A clear definition of the nature of innovation is required, i.e. radical vs. incremental innovation and the 4Ps. What should be the primary focus?
2. To drive a project portfolio of both incremental (do better) and radical (do different) innovation. How do you get the right balance?
3. To get sufficient, dedicated, human and financial resources up-front.

(John Thesmer, Managing Director, Ictal Care, Denmark)

1. Finding R&D money for far-sighted technology projects at a time when shareholders seem to apply increasing amounts of pressure on companies to deliver short-term results. Every industry needs to keep innovating to stay competitive in the future – and the rate of technological change is accelerating. But companies are being forced to pursue these objectives for less and less money. Managing this difficult balance of 'doing more with less' is a major challenge in our industry, and I am certain that we are not alone.

2. Building a corporate culture that doesn't punish risk-takers. Managers in many organizations seem to be measured almost exclusively according to how well they are performing according to some fairly basic measurements, e.g. sales or number of units. No one would disagree that absorbing new technologies can potentially help to improve these statistics in the long term, but new technologies can be a rather daunting obstacle in the short term. Sometimes technology trials fail. An organization needs to recognize this, and has to lead its teams and managers in a way that encourages a healthy amount of risk without losing control of the big picture.

3. Striking the right balance between in-house R&D and leveraging external innovations. The scope and scale of innovation is growing at a pace that makes it all but unthinkable that any single company can do it all themselves. But which elements should be retained internally vs. which ones can be outsourced? There's never a shortage of people writing papers and books that attempt to address this very topic, but managers in the field are hungrier than ever for useful and practical guidance on this issue.

(Rob Perrons, Shell Exploration, USA)

VIEWS FROM THE FRONT LINE

George Buckley, CEO of 3M, is a PhD chemical engineer by training. 3M has global sales of around $23 billion and historically has aimed to achieve a third of sales from products introduced in the past five years. The famous company culture, the '3M way', includes a policy of allowing employees to spend 15% of their time on their own projects, and has been successfully emulated by other innovative companies such as Google.

He argues that 'Invention is by its very nature a disorderly process, you cannot say I'm going to schedule myself for three good ideas on Wednesday and two on Friday. That's not how creativity works'.[75] After a focus on improving efficiency, quality and financial performance 2001–6, under its new CEO, 3M is now refocusing on its core innovation capability. Buckley believes that the company had become too dominated by formal quality and measurement processes, to the detriment of innovation: '. . . you cannot create in that atmosphere of confinement or sameness, perhaps one of the mistakes we have made as a company . . . is that when you value sameness more than you value creativity, I think you potentially undermine the heart and soul of a company like 3M . . .', and since becoming CEO has significantly increased the spending on R&D from some $1 billion to nearer to $1.5 billion, and is targeting the company's 45 core technologies such as abrasives to nanotechnology, but has sold the non-core pharmaceutical business.

> ## RESEARCH NOTE
>
> Mohanbir Sawhney, Robert Wolcott and Inigo Arroniz from the Center for Research in Technology and Innovation at the Kellogg School of Management at Northwestern University, USA, interviewed innovation managers at a number of large firms, including Boeing, DuPont, Microsoft, eBay, Motorola and Sony, and from these developed a survey questionnaire which was sent to a further 19 firms, such as General Electric, Merck and Siemens.[76]
>
> Analysing these data, they derived an 'innovation radar' to represent the 12 dimensions of business innovation they identified. Their definition of 'business innovation' does not focus on new things, but rather anything that creates new value for customers. Therefore creating new things is neither necessary nor sufficient for such value creation. Instead they propose a systematic approach to business innovation, which may take place in 12 different dimensions:
>
> - Offerings – new products or services.
> - Platform – derivative offerings based on reconfiguration of components.
> - Solutions – integrated offerings which customers value.
> - Customers – unmet needs or new market segments.
> - Customer experience – redesign of customer contact and interactions.
> - Value capture – redefine the business model and how income is generated.
> - Processes – to improve efficiency or effectiveness.
> - Organization – change scope or structures.
> - Supply chain – changes in sourcing and order fulfillment.
> - Presence – new distribution or sales channels.
> - Brand – leverage or reposition.
> - Networking – create integrated offerings using networks.

Summary and further reading

Few other texts cover the technological, market and organizational aspects of innovation in an integrated fashion. Drucker's *Innovation and Entrepreneurship* (Harper & Row, 1985), provides an accessible introduction to the subject, but perhaps relies more on intuition and experience than on empirical research. A number of interesting texts have also been published since the first edition of this book appeared in 1997. Trott's *Innovation Management and New Product Development* (fourth edition, Prentice Hall, 2008), particularly focuses on the management of product development, books by von Stamm (*Managing Innovation, Design and Creativity,* second edition, John Wiley & Sons, Ltd, 2008) and Bruce and Bessant (*Design in Business*, Pearson Education, 2001) have a strong design emphasis and Jones' book targets practitioners in particular (*Innovating at the Edge*, Butterworth Heinemann, 2002). Dogson, Gann and Salter (*The Management of Technological Innovation*, Oxford University Press, 2008) examine innovation strategy and the 'new innovation toolkit', whilst Goffin and Mitchell (*Innovation Management*, Pearson, 2005) also look from a management tools perspective. Brockhoff *et al.* (*The Dynamics of Innovation,* Springer, 1999) and Sundbo and Fugelsang (*Innovation as Strategic Reflexivity*, Routledge,

2002) provide some largely European views, whilst Ettlie's *Managing Technological Innovation* (John Wiley & Sons, Inc., 1999) is based on the experience of US firms, mainly from manufacturing, as are Mascitelli (*The Growth Warriors*, Technology Perspectives, 1999) and Schilling (*Strategic Management of Technological Innovation*, McGraw Hill, 2005). A few books explore the implications for a wider developing country context, notably Forbes and Wield (*From Followers to Leaders*, Routledge, 2002) and Prahalad (*The Fortune at the Bottom of the Pyramid*, Wharton School Publishing, 2006) and a couple look at public policy implications (Branscomb, L. and J. Keller, eds. *Investing in Innovation*, MIT Press, 1999; Dodgson, M. and J. Bessant, *Effective Innovation Policy*, International Thomson Business Press, 1996).

There are several compilations and handbooks covering the field, the best known being *Strategic Management of Technology and Innovation* (Burgelman, R., C. Christensen, and S. Wheelwright, eds., McGraw-Hill, 2004) now in its fourth edition and containing a wide range of key papers and case studies, though with a very strong US emphasis. A more international flavour is present in Dodgson and Rothwell (*The Handbook of Industrial Innovation*, Edward Elgar, 1995) and Shavinina (*International Handbook on Innovation*, Elsevier, 2003). The work arising from the Minnesota Innovation Project also provides a good overview of the field and the key research themes contained within it (Van de Ven, A., *The Innovation Journey*, Oxford University Press, 1999).

Case studies of innovation provide a rich resource for understanding the workings of the process in particular contexts. Good compilations include those of Baden-Fuller and Pitt (*Strategic Innovation*, Routledge, 1996), Nayak and Ketteringham (*Breakthroughs: How leadership and drive create commercial innovations that sweep the world*, Mercury, 1986) and von Stamm (*The Innovation Wave*, John Wiley & Sons, Ltd, 2003), whilst other books link theory to case examples, for example Tidd and Hull (*Service Innovation*, Imperial College Press, 2003). Several books cover the experiences of particular companies including 3M, Corning, DuPont, Toyota and others (Kanter, R., ed., *Innovation: Breakthrough thinking at 3M, DuPont, GE, Pfizer and Rubbermaid*, Harper Business, 1997; Graham, M. and A. Shuldiner, *Corning and the Craft of Innovation*, Oxford University Press, 2001; Kelley, T., J. Littman, and T. Peters, *The Art of Innovation: Lessons in Creativity from Ideo, America's Leading Design Firm*, Currency, 2001). Internet-related innovation is well covered in a number of books mostly oriented towards practitioners, for example, Evans and Wurster (*Blown to Bits: How the New Economics of Information Transforms Strategy*, Harvard Business School Press, 2000), Loudon (*Webs of Innovation*, FT.Com, 2001), Oram (*Peer-to-Peer: Harnessing the Power of Disruptive Technologies*, O'Reilly, 2001) Alderman (*Sonic Boom*, Fourth Estate, 2001) and Pottruck and Pearce (*Clicks and Mortar*, Jossey Bass, 2000). The implications of the Internet for greater user involvement in the innovation process and the emergence of new models is dealt with by von Hippel (*The Democratization of Innovation*, MIT Press, 2005) and others (e.g., Tapscott, D. and A. Williams, *Wikinomics*, Portfolio, 2006).

Most other texts tend to focus on a single dimension of innovation management. In *The Nature of the Innovative Process* (Pinter Publishers, 1988), Dosi adopts an evolutionary economics perspective and identifies the main issues in the management of technological innovation. On the subject of organizational innovation, Galbraith and Lawler (*Organizing for the Future*, Jossey Bass, 1988) summarize recent thinking on organizational structures and processes, although a more critical account is provided by Wolfe (Organizational innovation, *Journal of Management Studies*, 31 (3), 405–432, 1994). For a review of the key issues and leading work in the field of organizational change and learning see Cohen and Sproull (Organizational Learning, Sage, 1996). Bessant (*High Involvement Innovation*, John Wiley & Sons, Ltd, 2003), Boer *et al.* (*CI Changes*, Ashgate, 1999), Imai (*Kaizen*, Random House, 1987) Schroeder and Robinson (*Ideas are Free*, Berrett Koehler, 2004) look at the issue of high involvement incremental innovation.

Most marketing texts fail to cover the specific issues related to innovative products and services, although a few specialist texts examine the more narrow problem of marketing so-called 'high-technology' products, for example, Jolly (*Commercialising New Technologies*, Harvard Business School Press, 1997) and Moore (*Crossing the Chasm*, Harper Business, 1999). Helpful coverage of the core issues are to be found in the chapter, 'Securing the future' in Hamel and Prahalad's *Competing for the Future* (Harvard Business School Press, 1994) and the chapter 'Learning from the market' in Leonard's *Wellsprings of Knowledge* (Harvard Business School Press, 1995). There are also extensive insights into adoption behaviour from a wealth of studies drawn together by Rogers and colleagues (*Diffusion of Innovations*, Free Press, 1995).

Particular themes in innovation are covered by a number of books and journal special issues, for example, services (Best, M., *The New Competitive Advantage*, Oxford University Press, 2001), networks and clusters (Cooke, P. and K. Morgan, *The Intelligent Region: Industrial and Institutional Innovation in Emilia-Romagna*, University of Cardiff, 1991), sustainability (Dodgson, M. and A. Griffiths, Sustainability and innovation – Special issue, *Innovation Management, Policy and Practice*, 2004) and discontinuous innovation (Day, G. and P. Schoemaker, *Wharton on Managing Emerging Technologies*, John Wiley & Sons, Inc., 2000; Foster, R. and S. Kaplan, *Creative Destruction*, Harvard University Press, 2002). Various websites offer news, research, tools, etc., for example AIM (**www.aimresearch.org**) and NESTA (**www.nesta.org.uk**). A full and updated list is available on the website accompanying this book **www.managing-innovation.com**.

Web links

Here are the full details of the resources available on the website flagged throughout the text:

Case studies:
Kumba Resources
Inditex/Zara
Aravind Eye Clinics
Freeplay Radio
Karolinska Hospital
Model T Ford
LEGO
Threadless
Philips Atmosphere provider
The dimming of the light bulb
Continuous improvement cases

Interactive exercises:
Strategic advantage through innovation
Using the 4Ps
Patterns of discontinuous innovation
Architectural and component innovation

Tools:
4Ps for mapping innovation space
Continuous improvement tools and techniques
Product life cycle analysis

 Video podcast:
Finnegans Fish Bar (4Ps)
Patrick Mchaughlin, Cerulean

References

1. **De Jager, B. C. Minnie, C. De Jager, M. Welgemoed, J. Bessant, and D. Francis** (2004) Enabling continuous improvement – an implementation case study. *International Journal of Manufacturing Technology Management*, **15** (4), 315–24.
2. **Port, O.** (2004) The signs they are a changing. *Business Week*, p. 24.
3. **Prahalad, C.K.** (2006) *The Fortune at the Bottom of the Pyramid*, Wharton School Publishing, New Jersey.
4. **Bessant, J. and A. Davies** (2007) Managing service innovation. *DTI Occasional Paper 9: Innovation in Services*, Department of Trade and Industry, London.
5. **Boden, M. and I. Miles, eds** (1998) *Services and the Knowledge-Based Economy*, Continuum, London.
6. **Tidd, J. and F. Hull, eds** (2003) *Service Innovation: Organizational Responses to Technological Opportunities and Market Imperatives*, Imperial College Press, London.
7. **Baden-Fuller, C. and M. Pitt** (1996) *Strategic Innovation*, Routledge, London.
8. **Jones, T.** (2002) *Innovating at the Edge*, Butterworth-Heinemann, London.
9. **Albury, D.** (2004) *Innovation in the Public Sector*, Strategy Unit, Cabinet Office, London.
10. **Kaplinsky, R., F. den Hertog and B. Coriat** (1995) *Europe's Next Step*, Frank Cass, London.
11. **Baumol, W.** (2002) *The Free-Market Innovation Machine: Analyzing the Growth Miracle of Capitalism*, Princeton University Press, Princeton.
12. **Hauptly, D.** (2008) *Something Really New*, AMACOM, New York, p. 4.
13. **Innovaro** (2008) *Innovation Briefing, Innovation Leaders 2008*, www.innovaro.com.
14. *Financial Times* (2008) The nuts and bolts of innovation. *Financial Times*, 5 March.
15. **Department of Trade and Industry** (2003) *Competing in the Global Economy: The Innovation Challenge*, Department of Trade and Industry, London.
16. *Statistics Canada* (2006) Labour Force Survey. *Statistics Canada*, Ottawa.
17. **Souder, W. and J. Sherman** (1994) *Managing New Technology Development*, McGraw-Hill, New York.
18. **Tidd, J., ed.** (2000) *From Knowledge Management to Strategic Competence: Measuring Technological, Market and Organizational Innovation*, Imperial College Press, London.
19. **Rosenau, M., et al., eds** (1996) *The PDMA Handbook of New Product Development*, John Wiley and Sons, Inc., New York.
20. **Stalk, G. and T. Hout** (1990) *Competing against Time: How Time-Based Competition is Reshaping Global Markets*, Free Press, New York.
21. **Womack, J., D. Jones and D. Roos** (1991) *The Machine that Changed the World*, Rawson Associates, New York.

22. **Belussi, F.** (1989) Benetton – a case study of corporate strategy for innovation in traditional sectors. In M. Dodgson, ed., *Technology Strategy and the Firm*, Longman, London.

23. **Pfeffer, J.** (1994) *Competitive Advantage through People*, Harvard Business School Press, Boston, MA.

24. **Foster, R. and S. Kaplan** (2002) *Creative Destruction*, Harvard University Press, Boston, MA.

25. **Francis, D., J. Bessant and M. Hobday** (2003) Managing radical organisational transformation. *Management Decision*, **41** (1), 18–31.

26. **Evans, P. and T. Wurster** (2000) *Blown to Bits: How the New Economics of Information Transforms Strategy*, Harvard Business School Press, Boston, MA.

27. **Von Hippel, E.** (2005) *The Democratization of Innovation*, MIT Press, Cambridge, MA.

28. **Schrage, M.** (2000) *Serious Play: How the World's Best Companies Simulate to Innovate*, Harvard Business School Press, Boston, MA.

29. **Gann, D.** (2004) *Think, Play, Do: The Business of Innovation. Inaugural Lecture*, Imperial College Press, London.

30. **Schumpeter, J.** (1950) *Capitalism, Socialism and Democracy*, Harper and Row, New York.

31. **Bryson, B.** (1994) *Made in America*, Minerva, London.

32. **Tapscott, D. and A. Williams** (2006) *Wikinomics – How Mass Collaboration Changes Everything*, Portfolio, New York.

33. **Francis, D. and J. Bessant** (2005) Targeting innovation and implications for capability development. *Technovation*, **25** (3), 171–83.

34. **Hamel, G.** (2000) *Leading the Revolution*, Harvard Business School Press, Boston, MA.

35. **Ulnwick, A.** (2005) *What Customers Want: Using Outcome-Driven Innovation to Create Breakthrough Products and Services*, McGraw-Hill, New York.

36. **Ettlie, J.** (1999) *Managing Innovation*, John Wiley & Sons, Inc., New York.

37. **Hollander, S.** (1965) *The Sources of Increased Efficiency: A Study of DuPont Rayon Plants*, MIT Press, Cambridge, MA.

38. **Tremblay, P.** (1994) Comparative analysis of technological capability and productivity growth in the pulp and paper industry in industrialised and industrialising countries, PhD thesis, University of Sussex.

39. **Enos, J.** (1962) *Petroleum Progress and Profits; A History of Process Innovation*, MIT Press, Cambridge, MA.

40. **Figuereido, P.** (2001) *Technological Learning and Competitive Performance*, Edward Elgar, Cheltenham.

41. **Imai, K.** (1987) *Kaizen*, Random House, New York.

42. **Arrow, K.** (1962) The economic implications of learning by doing. *Review of Economic Studies*, **29** (2), 155–73.

43. **Womack, J. and D. Jones** (2005) *Lean Solutions*, Free Press, New York.

44. **Rothwell, R. and P. Gardiner** (1983) Tough customers, good design. *Design Studies*, **4** (3), 161–9.

45. **Gawer, A. and M. Cusumano** (2002) *Platform Leadership: How Intel, Microsoft, and Cisco Drive Industry*, Harvard Business School Press, Boston, MA.

46. **Hamel, G.** (2007) *The Future of Management*, Harvard Business School Press, Boston, MA.

47. **Phillips, W. et al.** (2006) Beyond the steady state: managing discontinuous product and process innovation. *International Journal of Innovation Management*, **10** (2), 175–96.

48. **Kim, W. and R. Mauborgne** (2005) *Blue Ocean Strategy: How to Create Uncontested Market Space and Make the Competition Irrelevant*, Harvard Business School Press, Boston, MA.

49. **Hargadon, A.** (2003) *How Breakthroughs Happen*, Harvard Business School Press, Boston, MA.

50. **Leifer, R., C. McDermott, G. O'Conner, L.S. Peters, M. Rice, and R. Veryzer** (2000) *Radical Innovation*, Harvard Business School Press, Boston, MA.

51. **O'Connor, G.C., R. Leifer, A. Paulson, and L.S. Peters** (2008) *Grabbing lightning*, Jossey Bass, San Francisco.

52. **Utterback, J.** (1994) *Mastering the Dynamics of Innovation*, Harvard Business School Press, Boston, MA.

53. **Christensen, C.** (1997) *The Innovator's Dilemma*, Harvard Business School Press, Boston, MA.

54. **Henderson, R. and K. Clark** (1990) Architectural innovation: the reconfiguration of existing product technologies and the failure of established firms. *Administrative Science Quarterly*, **35**, 9–30.

55. **Christensen, C. and M. Raynor** (2003) *The Innovator's Solution: Creating and Sustaining Successful Growth*, Harvard Business School Press, Boston, MA.

56. **Gilfillan, S.** (1935) *Inventing the Ship*, Follett, Chicago.

57. **Braun, E. and S. Macdonald** (1980) *Revolution in Miniature*, Cambridge University Press, Cambridge.

58. **Barnes, J., J. Bessant, N. Dunne, and M. Morris** (2001) Developing manufacturing competitiveness in South Africa. *Technovation*, **21** (5).

59. **Kaplinsky, R., M. Morris, and J. Readman** (2003) The globalisation of product markets and immiserising growth: lessons from the South African furniture industry. *World Development*, **30** (7), 1159–78.

60. **Bessant, J. and D. Francis** (2004) Developing parallel routines for product innovation. *11th PDMA Product Development Conference*, EIASM, Dublin.

61. **Prahalad, C.** (2004) The blinders of dominant logic. *Long Range Planning*, **37** (2), 171–9.

62. **Douthwaite, B.** (2002) *Enabling Innovation*, Zed Books, London.

63. **Day, G. and P. Schoemaker** (2004) Driving through the fog: managing at the edge. *Long Range Planning*, **37** (2), 127–42.

64. **Dosi, G.** (1982) Technological paradigms and technological trajectories. *Research Policy*, **11**, 147–62.

65. **Freeman, C. and C. Perez** (1989) Structural crises of adjustment: business cycles and investment behavior. In G. Dosi (ed.) *Technical Change and Economic Theory*, Frances Pinter, London.

66. **Perez, C.** (2002) *Technological Revolutions and Financial Capital*, Edward Elgar, Cheltenham.

67. **Henderson, R.** (1994) The evaluation of integrative capability: innovation in cardio-vascular drug discovery. *Industrial and Corporate Change*, **3** (3), 607–30.

68. **Polanyi, M.** (1967) *The Tacit Dimension*, Routledge and Kegan Paul, London.

69. **Tidd, J.** (1994) *Home Automation: Market and Technology Networks*, Whurr Publishers, London.

70. **Kodama, F.** (1992) Technology fusion and the new R&D. *Harvard Business Review*, July/August.

71. **Davies, A. and M. Hobday** (2005) *The Business of Projects: Managing Innovation in Complex Products and Systems*, Cambridge University Press, Cambridge.

72. **Abernathy, W. and J. Utterback** (1978) Patterns of industrial innovation. *Technology Review*, **80**, 40–7.

73. **Foster, R.** (1986) *Innovation – The Attacker's Advantage*, Pan Books, London.

74. **Bessant, J. and D. Francis** (2005) Dealing with discontinuity – how to sharpen up your innovation act. In *AIM Executive Briefings*, AIM-ESRC/EPSRC Advanced Institute of Management Research, London.

75. **Hindo, B.** (2007) At 3M: a struggle between efficiency and creativity. *Business Week*, 11 June, Special Section, pp. 8–14.

76. **Sawhney, M., R.C. Wolcott and I. Arroniz** (2006) The 12 different ways for companies to innovate. *MIT Sloan Management Review*, Spring, 75–81.

CHAPTER 2

Innovation as a core business process

2.1 Introduction

Chapter 1 set out a view of innovation as the core renewal process within an organization, refreshing what it offers the world and how it creates and delivers that offering. Innovation is a generic activity associated with survival and growth and at this level of abstraction we can see the underlying process as common to all firms. Figure 2.1 provides a simple map of this and we will use this throughout the book as a framework to help explore the ways in which innovation can be managed. At its heart it involves:

- Searching – scanning the environment (internal and external) for, and processing relevant signals about, threats and opportunities for change.
- Selecting – deciding (on the basis of a strategic view of how the enterprise can best develop) which of these signals to respond to.
- Implementing – translating the potential in the trigger idea into something new and launching it in an internal or external market. Making this happen is not a single event but needs attention to acquiring the knowledge resources to enable the innovation, executing the project under conditions of uncertainty, both of which require extensive problem solving, and launching the innovation into relevant internal or external markets.
- Capturing value from the innovation – both in terms of sustaining adoption and diffusion and also in learning from progressing through this cycle so that the organization can build its knowledge base and improve the ways in which the process is managed.

The challenge facing any organization is to try and find ways of managing this process to provide a good solution to the problem of renewal. Different circumstances lead to many different solutions – for example, large science-based firms like pharmaceutical companies will tend to create solutions which have heavy activities around formal R&D, patent searching, etc., whilst small engineering subcontractors will emphasize rapid implementation capability. Retailers may have relatively small R&D commitments in the formal sense but will stress scanning the environment to pick up new consumer trends, and they are likely to place heavy emphasis on marketing. The case on the website of Tesco's Fresh & Easy store development gives a good example of this. Consumer goods producers may be more concerned with rapid product development and launch, often with variants and repositioning of basic product concepts. Heavy engineering firms involved in products such as power plant are likely to be design intensive, and critically dependent on project management and systems integration aspects of the implementation phase. Public-sector organizations have to configure their innovation process to cope with strong external political and regulatory influences. The cases on the website of the RED and Open Door projects give some insights into these issues.

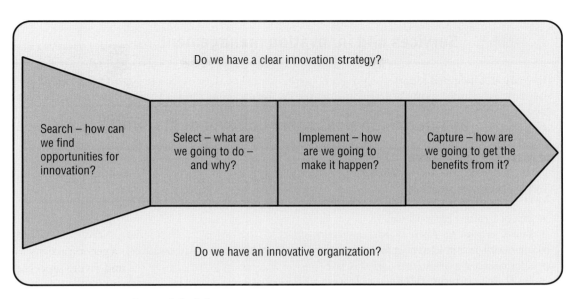

FIGURE 2.1: A simple model of the innovation process

Despite these variations the underlying pattern of phases in innovation remains constant. In this chapter we want to explore the process nature of innovation in more detail, and to look at the kinds of variations on this basic theme. But we also want to suggest that there is some commonality around the things which are managed and the influences that can be brought to bear on them in successful innovation. These 'enablers' represent the levers that can be used to manage innovation in any organization. Once again, how these enablers are actually put together varies between firms, but they represent particular solutions to the general problem of managing innovation. Exploring these enablers in more detail is the basis of the following chapters in the book.

Central to our view is that innovation management is a learned capability. Although there are common issues to be confronted and a convergent set of recipes for dealing with them, each organization must find its own particular solution and develop this in its own context. Simply copying ideas from elsewhere is not enough; these must be adapted and shaped to suit particular circumstances.

2.2 Variations on a theme

Innovations vary widely, in scale, nature, degree of novelty and so on – and so do innovating organizations. But at this level of abstraction it is possible to see the same basic process operating in each case. For example, developing a new consumer product will involve picking up signals about potential needs and new technological possibilities, developing a strategic concept, coming up with options and then working those up into new products which can be launched into the marketplace.

In similar fashion deciding to install a new piece of process technology also follows this pattern. Signals about needs – in this case internal ones, such as problems with the current equipment – and new technological means are processed and provide an input to developing a strategic concept. This then requires identifying an existing option, or inventing a new one which must then be developed to such a

> ### Box 2.1   Services and innovation management
>
> In 2001 an influential report was presented to the annual conference of a key economic sector laying down the innovation challenge in clear terms: *'We are at the brink of change of an unprecedented and exponential kind and magnitude . . .We must be willing and able to discard old paradigms and engender and embrace manifest change . . . These required changes include implementing new customer-centric processes and products, cutting costs and improving service through the application of IT and business process re-engineering and putting in place systems and a culture for sustainable innovation.'* Another study, in 2006, reviewed the capability of firms within this sector to deal with innovation and highlighted problems such as:
>
> - *lack of a culture of innovation*
> - *lack of strategy for where to focus innovation efforts*
> - *innovation is seen to conflict with fee-paying work and is thus not always valued*
> - *a formal innovation process does not exist*
> - *project management skills are very limited*
>
> At first sight these seem typical of statements made regularly about the importance of innovation in a manufacturing economy and the difficulties individual firms – particularly the smaller and less experienced – face in trying to manage the process. But these are in fact *service* sector examples – the first report was to the US Bar Association, the second the result of a survey of 40 professional law firms in the UK trying to prepare for the big changes likely to arise as a result of the Clementi (2004) review.[1]

point that it can be implemented, that is launched, by users within the enterprise – effectively a group of internal customers. The same principles of needing to understand user needs and to prepare the marketplace for effective launch will apply as in the case of product innovation.

Services and innovation

Table 2.1 gives some examples of different types of innovation in services, using the same '4Ps' typology which we introduced in Chapter 1.

Service innovations are often much easier to imitate and the competitive advantages that they offer can quickly be competed away because there are fewer barriers, for example, intellectual property (IP) protection. The pattern of airline innovation on the transatlantic route provides another example – there is a fast pace of innovation but as soon as one airline introduces something like a flat bed, others will quickly emulate it. Arguably the drive to personalization of the service experience will be strong because it is only through such customized experiences that a degree of customer 'lock on' takes place.[2] Certainly the experience of Internet banking and insurance suggests that, despite attempts to customize the experience via sophisticated web technologies there is little customer loyalty and a high rate of churn. However, the lower capital cost of creating and delivering services and their relative simplicity makes co-creation more of an option. Where manufacturing may require sophisticated tools such as computer-aided design and rapid prototyping, services lend themselves to shared experimentation at

TABLE 2.1	Examples of incremental and radical innovations in services	
Type of innovation	**Incremental – 'Do better'**	**Radical – 'Do different'**
Product – service offering to end users	Modified/improved version of an established service offering, e.g. more customized mortgage or savings 'products', add-on features to basic travel experience (e.g. entertainment system), increased range of features in telecomm service	Radical departure, e.g. online retailing
Process – ways of creating and delivering the offering	Lower cost delivery through 'back office' process optimization, waste reduction through lean, six sigma, etc. approaches	Radical shift in process route, e.g. moving from face-to-face contact to online, supermarkets and self-service shopping rather than traditional retailing, hub-and-spoke delivery systems
Position – target market and the 'story' told to those segments	Opening up new market segments, e.g. offering specialist insurance products for students	Radical shift in approach, e.g. opening up new travel markets via low-cost travel innovation, shifting healthcare provision to communities
Paradigm – underlying business model	Rethinking the underlying model, e.g. migrating from insurance agents and brokers to direct and online systems	Radical shift in mindset, e.g. moving from product-based to service-based manufacturing

relatively lower cost. There is growing interest in such models involving active users in design of services, for example in the open source movement around software or in the digital entertainment and communication fields where community and social networking sites such as MySpace, Flickr and YouTube have had a major impact.

Services may appear different because they are often less tangible – but the same underlying innovation model applies. The process whereby an insurance or financial services company launches a new product will follow a path of searching for trigger signals, strategic concept, product and market development and launch. What is developed may be less tangible than a new television set, but the underlying structure to the process is the same. We should also recognize that increasingly what we call manufacturing includes a sizable service component with core products being offered together with supporting services – a website, a customer information or helpline, updates.[3,4] Indeed for many complex

Chris Voss and colleagues from the London Business School and the Advanced Institute for Management Research have been carrying out extensive research on 'experience innovation'. This focuses on how service businesses in particular are using the creation and delivery of novel and rich experiences to attract and retain customers. A study in 2004 examined 50 organizations in the areas of retail, entertainment and sport, theme parks, destinations and hotels, largely from the UK, Europe and the USA. The research identified a repeated cycle of investment and management, vibrant experiences, customer growth, profitability and reinvestment that drives profit, which can be seen as the experience profit cycle. The research also examined how organizations are turning services into destinations, compelling places where people visit for an extended period of time, engage in multiple activities and want to return to.[5]

Subsequent work looking in more detail at examples in the UK and USA addressed the question of how focusing on the customer experience changes the way services and service delivery processes are designed. It examined the process and content of experience design. The study involved eight case studies of design agencies and consultancies that specialize in experience design and nine case studies of experiential service providers. The research showed that companies often use the customer journey and touchpoints approach to design experiences. Innovation took place in five design areas: physical environment; service employees; service delivery process; fellow customers; and back office support. An important part of the design process is collecting customer insights.[6,7]

product systems – such as aircraft engines – the overall package is likely to have a life in excess of 30 or 40 years and the service and support component may represent a significant part of the purchase. At the limit such manufacturers are recognizing that their users actually want to buy some service attribute which is embodied in the product – so aero-engine manufacturers are offering 'power by the hour' rather than simply selling engines. The computer giant IBM transformed its fortunes in this way; it began life as a manufacturer of mainframes, became active in the early days of the personal computer (PC), but increasingly saw its business becoming one of providing solutions and services. Following a traumatic period in the 1990s the company has moved much further into service territory and in 2006 sold off its last remaining PC business to the Chinese firm Lenovo.[8] The Marshalls case study illustrates growing via this services model.

It is important in the context of service innovation to remind ourselves of the definition of innovation – 'the successful exploitation of new ideas'. Whilst this involves invention – the creation of some new or different combination of needs and means, there is much more to getting that invention successfully developed and widely adopted. Central to this is the idea of different kinds of knowledge streams being woven together – about possibilities (for example, opened up by new technology) and needs (whether articulated or latent). Countless studies of innovation highlight its nature as an interactive, coupling process – yet much thinking in policy and management practice defaults to linear views of the process and especially to a knowledge-push model.

In the context of service innovation the search for and use of demand-side knowledge is critical – many services are simultaneously created and consumed and end-user understanding and empathy are essential to success. This is not to say that new knowledge – for example, of technological possibilities – is unimportant but the balance of importance in service innovation may be more in the direction of demand-side knowledge.

One consequence of this different orientation is that much of the language which surrounds discussion of innovation may differ between manufacturing and service contexts. The underlying principles and issues may be the same but the labels may differ. For example, the term 'R&D' used in a manufacturing context conjures images associated with organized research and development. Search involves reviewing established scientific knowledge (in papers, via patent searches, etc.) and identifying interesting lines of enquiry which are followed through via designed experiments in laboratories. Small-scale successes may be further explored in pilot plants or via construction of prototypes and there is a gradual convergence around the final product or process involving an increasing commitment of resources and an increasing involvement of wider skills and knowledge sets. Eventually the new product is launched into the marketplace or the new process adopted and diffused across an internal context.

The Frascati manual (which takes its name from the location in Italy where a 1963 OECD meeting on the topic of innovation took place) is a widely used reference work for developing innovation and technology policy. It defines R&D as 'creative work undertaken on a systematic basis in order to increase the stock of knowledge . . . and the use of this stock of knowledge to devise new applications'.[9] If we look at the challenge of service innovation we can see a similar process taking place – search (albeit with a much stronger demand-side emphasis), experiment and prototyping (which may extend the 'laboratory' concept to pilots and trials with potential end-users) and a gradual scaling up of commitment and activity leading to launch. Service businesses may not have a formal R&D department but they do undertake this kind of activity in order to deliver a stream of innovations. Importantly the knowledge sets with which they work involve a much higher level of user insight and experience.

They are also similar to manufacturing in that much of their innovation-related work is about 'doing what we do but better' – essentially building competitive advantage through a stream of incremental innovations and extensions to original concepts. The distinction made in Frascati between 'routine' – incremental – improvements and R&D also applies in service innovation as the case of incremental innovation at NPI on the website shows.

The extended enterprise

One of the significant developments in business innovation, driven by globalization and enabling technologies, has been the 'outsourcing' of key business processes – IT, call centre management, human resources administration, etc. Although indicative of a structural shift in the economy it has at its heart the same innovation drivers. Even if companies are being 'hollowed out' the challenges facing the outsourcer and its client remain those of process innovation.[10, 11] The underlying business model of outsourcing is based on being able to do something more efficiently than the client and thereby creating a business margin – but achieving this depends critically on the ability to reengineer and then continuously improve on core business processes. And over time the attractiveness of one outsourcer over another increasingly moves from simply being able to execute outsourced standard operations more efficiently and towards being able to offer – or to co-evolve with a client – new products and services. Companies like IBM have been very active in recent years trying to establish a presence – and an underlying discipline – in the field of 'service science'.[12]

The challenge here becomes one of process innovation within outsourcing agencies – how can they develop their capabilities for carrying out processes more effectively (cheaper, faster, higher quality, etc.), and how can they sustain their ability to continue to innovate along this trajectory?

What about not for profit?

The distinction between commercial and not-for-profit organizations may also blur when considering innovation. Whilst private-sector firms may compete for the attentions of their markets through offering new things or new ways of delivering them, public-sector and nonprofit organizations use innovation to help them 'compete' against the challenges of, for example, delivering healthcare, education, law and order.[13] They are similarly preoccupied with process innovation (the challenge of using often scarce resources more effectively or becoming faster and more flexible in their response to a diverse environment) and with product innovation – using combinations of new and existing knowledge to deliver new or improved 'product concepts' – such as decentralized healthcare, community policing or micro-credit banking.

Examples of public-sector innovation remind us that this is fertile and challenging ground for developing innovations.[14] But the underlying model is different – by its nature, public-sector innovation is 'contested' amongst a diverse range of stakeholders.[15, 16] Unlike much private-sector innovation, which is driven by ideas of competition and focused decision making, public-sector innovation has different – and often conflicting – drivers and the rewards and incentives may be absent or different. There is also the problem of 'centre/periphery' relationships – often much innovative experimentation takes place close to where services are delivered, but the 'rules of the game' are set (and the purse strings often controlled) at the centre. A major challenge in public-sector innovation is thus enabling diffusion of successful experiments into the mainstream.[17]

Size matters

Another important influence on the particular ways in which innovation is managed is the size of the organization. Typically smaller organizations possess a range of advantages – such as agility, rapid decision making – but equally limitations such as resource constraints (see Table 2.2). These mean that developing effective innovation management will depend on creating structures and behaviours which play to these – for example, keeping high levels of informality to build on shared vision and rapid decision making but possibly to build network linkages to compensate for resource limitations. The Cerulean case on the website gives a good illustration of this challenge.

We need to be clear that small organizations differ widely. In most economies small firms account for 95% or more of the total business world and within this huge number there is enormous variation, from micro-businesses such as hairdressing and accounting services through to high-technology start-ups. Once again we have to recognize that the generic challenge of innovation can be taken up by businesses as diverse as running a fish and chip shop through to launching a nanotechnology spin-out with millions of pounds in venture capital – but the particular ways in which the process is managed are likely to differ widely. The example of Finnegan's Fish Bar on the website actually looks at the innovation challenges in a fish and chip shop.

For example, small-/medium-sized enterprises (SMEs) often fail to feature in surveys of R&D and other formal indicators of innovative activity. Yet they do engage in innovative activity and carry out research – but this tends to be around process improvement or customer service and often involving tacit rather than formalized knowledge.[18] Much research has been carried out to try and segment the large number of SMEs into particular types of innovator and to explore the contingencies which shape their particular approach to managing innovation. Work by David Birch, for example, looked at those

| TABLE 2.2 | **Advantages and disadvantages for small firm innovators** |

Advantages	Disadvantages
Speed of decision making	Lack of formal systems for management control, e.g. of project times and costs
Informal culture	Lack of access to key resources, especially finance
High quality communications – everyone knows what is going on	Lack of key skills and experience
Shared and clear vision	Lack of long-term strategy and direction
Flexibility, agility	Lack of structure and succession planning
Entrepreneurial spirit and risk taking	Poor risk management
Energy, enthusiasm, passion for innovation	Lack of application to detail, lack of systems
Good at networking internally and externally	Lack of access to resources

SMEs – 'gazelles' – which offered high growth potential (greater than 20%/year) – clearly of interest in terms of job creation and overall economic expansion.[19] But subsequent studies of SMEs and growth suggests the innovation picture is more complex. In particular the idea that high-tech, young and research-intensive SMEs in fast-growing sectors were associated with high economic growth does not appear to hold water. Instead gazelles had relatively little to do with high tech – US figures from the Bureau of Statistics suggest that only 2% of high-growth SMEs are high tech, gazelles were somewhat older than small companies in general and few gazelles were found in fast-growing sectors. Only 5% of gazelles were present in the three fastest-growing US sectors, and the top five sectors in which high-growth SMEs were found were in slow-growth sectors like chemicals, electrical equipment, plastics and paper products.[20] As Birch commented in 2004 '*most people think that companies are like cows – growing a lot when young and then very little thereafter . . . . It turns out we're mistaken. Companies, unlike cows, are regularly "born again" – they take on new management, stumble on a new technology or benefit from a change in the marketplace. Whatever the cause, statistics show older companies are more likely to grow rapidly than even the youngest ones . . .*'.[21]

This perspective is borne out by studies in the OECD and of long-standing SME-led development in areas like Cambridge in the UK.[22] It argues for a more fine-grained view of SMEs and their role as innovators and sources of growth – whilst high-tech research-performing firms of this kind are important , so too are those 'hidden' innovators in more mature sectors or performing process rather than product innovation.

Project-based organizations

For many enterprises the challenge is one of moving towards project-based organization – whether for realizing a specific project (such as construction of a major facility like an airport or a hospital) or for managing the design and build around complex product systems like aero engines, flight simulators or communications networks. Project organization of this kind represents an interesting case, involving a system which brings together many different elements into an integrated whole, often involving different firms, long timescales and high levels of technological risk.[6]

Increasingly they are associated with innovations in project organization and management, for example, in the area of project financing and risk sharing. Although such projects may appear very different from the core innovation process associated with, for example, producing a new soap powder for the mass market, the underlying process is still one of careful understanding of user needs and meeting those. The involvement of users throughout the development process, and the close integration of different perspectives will be of particular importance, but the overall map of the process is the same.

Networks and systems

As we saw in Chapter 1, one of the emerging features of the twenty-first-century innovation landscape is that it is much less of a single enterprise activity. For a variety of reasons it is increasingly a multiplayer game in which organizations of different shapes and sizes work together in networks. These may be regional clusters, or supply chains or product development consortia or strategic alliances which bring competitors and customers into a temporary collaboration to work at the frontier of new technology application. Although the dynamics of such networks are significantly different from those operating in a single organization and the controls and sanctions much less visible, the underlying innovation process challenge remains the same – how to build shared views around trigger ideas and then realize them. Throughout the book we will look at the particular issues raised in trying to manage innovation beyond the boundaries of the organization and Chapter 6 in particular picks up this theme of managing across innovation networks.

One of the key implications of this multiplayer perspective is the need to shift our way of thinking from that of a single enterprise to more of a systems view. Innovation doesn't take place in isolation and if we are to manage it effectively we need to develop skills in thinking about and operating at this system level. Such a system view needs to include other players – customers and suppliers, competing firms, collaborators and beyond that a wider range of actors who influence the ways in which innovation takes place.[23,24]

Variations in national, regional and local context

Thinking about the wider context within which innovation takes place has led to the emergence of the concept of 'innovation systems'. These include the range of actors – government, financial, educational, labour market, science and technology infrastructure, etc. – which represent the context within which organizations operate their innovation process,[25] and the ways in which they are connected. They can be local, regional and national – and the ways in which they evolve and operate vary widely. In some cases there is clear synergy between these elements which create the supportive conditions within which

The power of regional innovation systems

Michael Best's fascinating account of the ways in which the Massachusetts economy managed to reinvent itself several times is one which underlines the importance of innovation systems. In the 1950s the state suffered heavily from the loss of its traditional industries of textiles and shoes but by the early 1980s the 'Massachusetts miracle' led to the establishment of a new high-tech industrial district. It was a resurgence enabled in no small measure by an underpinning network of specialist skills, high-tech research and training centres (the Boston area has the highest concentration of colleges, universities, research labs and hospitals in the world) and by the rapid establishment of entrepreneurial firms keen to exploit the emerging 'knowledge economy'. But in turn this miracle turned to dust in the years between 1986 and 1992 when around one-third of the manufacturing jobs in the region disappeared as the minicomputer and defence-related industries collapsed. Despite gloomy predictions about its future, the region built again on its rich network of skills, technology sources and a diverse local supply base, which allowed rapid new product development to emerge again as a powerhouse in high technology such as special-purpose machinery, optoelectronics, medical laser technology, digital printing equipment and biotech.

Source: Best, M. (2001) *The New Competitive Advantage*, Oxford University Press, Oxford.

innovation can flourish, for example, the regional innovation-led clusters of Baden-Württemberg in Germany, Cambridge in the UK, Silicon Valley and Route 128 in the USA, or the island of Singapore.[22,26,27] Increasingly effective innovation management is being seen as a challenge of connecting to and working with such innovation systems – and this again has implications for how we might organize and manage the generic process – see Case study 2.1. (We discuss national systems of innovation in more depth in Chapter 4.)

Do better/do different

It's not just the sector, size of firm or wider context which moderates the way the innovation process operates. An increasing number of authors draw attention to the need to take the degree of novelty in an innovation into account.[28–30] At a basic level the structures and behaviours needed to help enable incremental improvements will tend to be incorporated into the day-to-day standard operating procedures of the organization. More radical projects may require more specialized attention, for example, arrangements to enable working across functional boundaries. At the limit the organization may need to review the whole bundle of routines which it uses for managing innovation when it confronts discontinuous conditions and the 'rules of the game' change. The website has some examples of organizations confronting this challenge. The video interview with Patrick McLaughlin on Cerulean highlights the difficulties in creating an organization capable of producing radical innovation whilst the Philip's case study explores the issues in creating and executing radically new projects within a large organization.

As we saw in Chapter 1, we can think of innovation in terms of two complementary modes. The first can be termed 'doing what we do but better' – a 'steady state' in which innovation happens but within a defined envelope around which our 'good practice' routines can operate. This contrasts with 'do different' innovation where the rules of the game have shifted (due to major technological, market or political shifts, for example) and where managing innovation is much more a process of exploration and coevolution under conditions of high uncertainty. A number of writers have explored this issue and conclude that under turbulent conditions firms need to develop capabilities for managing both aspects of innovation.[31–33]

Once again the generic model of the innovation process remains the same. Under 'do different' conditions, organizations still need to search for trigger signals – the difference is that they need to explore in much less familiar places and deploy peripheral vision to pick up weak signals early enough to move. They still need to make strategic choices about what they will do – but they will often have vague and incomplete information and the decision making involved will thus be much more risky – arguing for a higher tolerance of failure and fast learning. Implementation will require much higher levels of flexibility around projects – and monitoring and review may need to take place against more flexible criteria than might be applied to 'do better' innovation types.[34]

For established organizations (like Cerulean and Philips) the challenge is that they need to develop the capability to manage both kinds of innovation. Much of the time they will need robust systems for dealing with 'do better' but from time to time they risk being challenged by new entrants better able to capitalize on the new conditions opened up by discontinuity – unless they can develop a 'do different' capability to run in parallel. New entrants don't have this problem when riding the waves of a discontinuous shift, for example, exploiting opportunities opened up by a completely new technology. But they in turn will become established incumbents and face the challenge later if they do not develop the capacity to exploit their initial advantage through 'do better' innovation process and also build capability for dealing with the next wave of change by creating a 'do different' capability.[35]

The challenge is thus – as shown in Figure 2.2 – to develop an ambidextrous capability for managing both kinds of innovation within the same organization. We will return to this theme repeatedly in

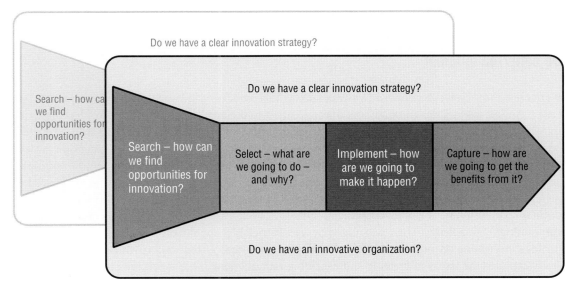

FIGURE 2.2: Managing steady-state and discontinuous innovation

TABLE 2.3	How context affects innovation management		
Context variable	**Modifiers to the basic process**	**Example references discussing these**
Sector	Different sectors have different priorities and characteristics, e.g. scale-intensive, science-intensive	36, 37
Size	Small firms differ in terms of access to resources, etc. and so need to develop more linkages	18, 38–41
National systems of innovation	Different countries have more or less supportive contexts in terms of institutions, policies, etc.	25, 26, 42
Life cycle (of technology, industry, etc.)	Different stages in life cycle emphasize different aspects of innovation, e.g. new-technology industries versus mature established firms	43–46
Degree of novelty – continuous vs. discontinuous innovation	'More of the same' improvement innovation requires different approaches to organization and management to more radical forms. At the limit firms may deploy 'dual structures' or even split or spin off in order to exploit opportunities	28, 47–49
Role played by external agencies such as regulators	Some sectors, e.g. utilities, telecommunications and some public services, are heavily influenced by external regimes which shape the rate and direction of innovative activity. Others – like food or healthcare – may be highly regulated in certain directions	50, 51

the book, exploring the additional or different challenges posed when innovation has to be managed beyond the steady state.

Table 2.3 lists some of the wide range of influences around which organizations need to configure their particular versions of the generic innovation process. The key message in this section is that the same generic process can be observed – the management challenge is configuration. On the website there is an exercise designed to explore different sectoral patterns of innovation.

2.3 Evolving models of the process

The importance of viewing innovation as a process is that this understanding shapes the way in which we try and manage it. Put simply, our mental models shape our actions – we pay attention to, allocate resources to, take decisions about things, according to how we think about them. So if innovation is a process we need to have a clear and shared understanding of what that process involves and how it operates.

This understanding of the core process model has changed a great deal over time. Early models (both explicit and, more importantly, the implicit mental models whereby people managed the process) saw it as a linear sequence of functional activities. Either new opportunities arising out of research gave rise to applications and refinements which eventually found their way to the marketplace ('technology push') or else the market signalled needs for something new, which then drew through new solutions to the problem ('need pull', where necessity becomes the mother of invention).

The limitations of such an approach are clear: in practice innovation is a coupling and matching process where interaction is the critical element.[52,53] Sometimes the 'push' will dominate, sometimes the 'pull', but successful innovation requires interaction between the two. The analogy to a pair of scissors is useful here: without both blades it is difficult to cut. (Chapter 5 explores the issue of sources of innovation and how there is considerable interplay between these two types.)

One of the key problems in managing innovation is that we need to make sense of a complex, uncertain and highly risky set of phenomena. Inevitably we try and simplify these through the use of mental models – often reverting to the simplest linear models to help us explore the management issues which emerge over time. Prescriptions for structuring the process along these lines abound, for example, one of the most-cited models for product innovation is due to Booz, Allen and Hamilton.[54] Many variations exist on this theme – for example, Robert Cooper's work suggests a slightly extended view with 'gates' between stages which permit management of the risks in the process.[55] There is also a British Standard (BS 7000), which sets out a design-centred model of the process.[56]

Much recent work recognizes the limits of linear models and tries to build more complexity and interaction into the frameworks. For example, the Product Development Management Association (PDMA) offers a detailed guide to the process and an accompanying toolkit.[57] Increasingly there is recognition of some of the difficulties around what is often termed the 'fuzzy front end' where uncertainty is highest, but there is still convergence around a basic process structure as a way of focusing our attention.[58] The balance needs to be struck between simplifications and representations which help thinking – but just as the map is not the same as the territory it represents so they need to be seen as frameworks for thinking, not as descriptions of the way the process actually operates.

Most innovation is messy, involving false starts, recycling between stages, dead ends, jumps out of sequence, etc. Various authors have tried different metaphors – for example, seeing the process as a railway journey with the option of stopping at different stations, going into sidings or even, at times, going backwards – but most agree that there is still some sequence to the basic process.[59,60] In an important programme of case-study-based research looking at widely different innovation types, Van de Ven and colleagues explored the limitations of simple models of the process.[61] They drew attention to the complex ways in which innovations actually evolve over time, and derived some important modifiers to the basic model:

• Shocks trigger innovations – change happens when people or organizations reach a threshold of opportunity or dissatisfaction.

- Ideas proliferate – after starting out in a single direction, the process proliferates into multiple, divergent progressions.
- Setbacks frequently arise, plans are overoptimistic, commitments escalate, mistakes accumulate and vicious cycles can develop.
- Restructuring of the innovating unit often occurs through external intervention, personnel changes or other unexpected events.
- Top management plays a key role in sponsoring – but also in criticizing and shaping – innovation.
- Success criteria shift over time, differ between groups and make innovation a political process.
- Innovation involves learning, but many of its outcomes are due to other events that occur as the innovation develops – making learning often 'superstitious' in nature.

They suggest that the underlying structure can be represented by the metaphor of an 'innovation journey', which has key phases of initiation, development and implementation/termination. But the progress of any particular innovation along this journey will depend on a variety of contingent circumstances; depending on which of these apply, different specific models of the process will emerge.

Roy Rothwell was for many years a key researcher in the field of innovation management, working at the Science Policy Research Unit (SPRU) at the University of Sussex. In one of his later papers he provided a useful historical perspective, suggesting that our appreciation of the nature of the innovation process has been evolving from such simple linear models (characteristic of the 1960s) through to increasingly complex interactive models (Table 2.4). His 'fifth-generation innovation' concept sees innovation as a multi-actor process, which requires high levels of integration at both intra- and inter-firm levels and which is increasingly facilitated by IT-based networking.[62] Whilst his work did not explicitly mention the Internet, it is clear that the kinds of innovation management challenge posed by the emergence of this new form fit well with the model. Although such fifth-generation models and the technologies which enable them appear complex, they still involve the same basic process framework.[63]

TABLE 2.4	Rothwell's five generations of innovation models
Generation	**Key features**
First/second	Simple linear models – need pull, technology push
Third	Coupling model, recognizing interaction between different elements and feedback loops between them
Fourth	Parallel model, integration within the company, upstream with key suppliers and downstream with demanding and active customers, emphasis on linkages and alliances
Fifth	Systems integration and extensive networking, flexible and customized response, continuous innovation

Problems of partial models

Mental models are important because they help us frame the issues that need managing – but therein also lies the risk. If our mental models are limited then our approach to managing is also likely to be limited. For example, if we believe that innovation is simply a matter of coming up with a good invention – then we risk managing that part of the process well but fail to consider or deal with other key issues around actually taking that invention through technological and market development to successful adoption.

Examples of such 'partial thinking' here include:

- Seeing innovation as a linear 'technology push' process (in which case all the attention goes into funding R&D with little input from users) or one in which the market can be relied upon to pull through innovation.
- Seeing innovation simply in terms of major 'breakthroughs' – and ignoring the significant potential of incremental innovation. In the case of electric light bulbs, the original Edison design remained almost unchanged in concept, but incremental product and process improvement over the 16 years from 1880 to 1896 led to a fall in price of around 80%.[64]
- Seeing innovation as a single isolated change rather than as part of a wider system (effectively restricting innovation to component level rather than seeing the bigger potential of architectural changes).[65]
- Seeing innovation as product or process only, without recognizing the interrelationship between the two.

Table 2.5 provides an overview of the difficulties which arise if we take a partial view of innovation.

2.4 Can we manage innovation?

It would be hard to find anyone prepared to argue against the view that innovation is important and likely to be more so in the coming years. But that still leaves us with the big question of whether or not we can actually manage what is clearly an enormously complex and uncertain process.

There is certainly no easy recipe for success. Indeed, at first glance it might appear that it is impossible to manage something so complex and uncertain. There are problems in developing and refining new basic knowledge, problems in adapting and applying it to new products and processes, problems in convincing others to support and adopt the innovation, problems in gaining acceptance and long-term use, and so on. Since so many people with different disciplinary backgrounds, varying responsibilities and basic goals are involved, the scope for differences of opinion and conflicts over ends and means is wide. In many ways the innovation process represents the place where Murphy and his associated band of lawmakers hold sway, where if anything can go wrong, there's a very good chance that it will!

But despite the uncertain and apparently random nature of the innovation process, it is possible to find an underlying pattern of success. Not every innovation fails, and some firms (and individuals) appear to have learned ways of responding and managing it such that, while there is never a cast-iron guarantee, at least the odds in favour of successful innovation can be improved. We are

TABLE 2.5	**Problems of partial views of innovation**
If innovation is only seen as . . .	**. . . the result can be**
Strong R&D capability	Technology which fails to meet user needs and may not be accepted
The province of specialists	Lack of involvement of others so that there is a lack of key input from different perspectives
Understanding and meeting customer needs	Lack of technical progression, leading to inability to gain competitive edge
Advances along the technology frontier	Producing products or services which the market does not want or designing processes which do not meet the needs of the user and whose implementation is resisted
The province only of large firms	Weak, small firms with too high a dependence on large customers Disruptive innovation as apparently insignificant small players seize new technical or market opportunities
Only about 'breakthrough' changes	Neglect of the potential of incremental innovation. Also an inability to secure and reinforce the gains from radical change because the incremental performance ratchet is not working well
Only about strategically targeted projects	May miss out on lucky 'accidents' which open up new possibilities
Only associated with key individuals	Failure to utilize the creativity of the remainder of employees, and to secure their inputs and perspectives to improve innovation
Only internally generated	The 'not-invented-here' effect, where good ideas from outside are resisted or rejected
Only externally generated	Innovation becomes simply a matter of filling a shopping list of needs from outside and there is little internal learning or development of technological competence
Only concerning single firms	Excludes the possibility of various forms of inter-organizational networking to create new products, streamline shared processes, etc.

using the term 'manage' here not in the sense of designing and running a complex but predictable mechanism (like an elaborate clock) but rather that we are creating conditions within an organization under which a successful resolution of multiple challenges under high levels of uncertainty is made more likely.

One indicator of the possibility of doing this comes from the experiences of organizations that have survived for an extended period of time. Whilst most organizations have comparatively modest life spans there are some which have survived at least one and sometimes multiple centuries. Looking at the experience of these '100 club' members – firms like 3M, Corning, Procter & Gamble, Reuters, Siemens, Philips and Rolls-Royce – we can see that much of their longevity is down to having developed a capacity to innovate on a continuing basis. They have learned – often the hard way – how to manage the process (both in its 'do better' and 'do different' variants) so that they can sustain innovation.[66–68]

It is important to note the distinction here between 'management' and managers. We are not arguing about who is involved in taking decisions or directing activity, but rather about what has to be done. Innovation is a management question, in the sense that there are choices to be made about resources and their disposition and coordination. Close analysis of many technological innovations over the years reveals that although there are technical difficulties – bugs to fix, teething troubles to be resolved and the occasional major technical barrier to surmount – the majority of failures are due to some weakness in the way the process is managed. Success in innovation appears to depend upon two key ingredients: technical resources (people, equipment, knowledge, money, etc.) and the capabilities in the organization to manage them.

This brings us to the concept of what have been termed 'routines'.[69] Organizations develop particular ways of behaving which become 'the way we do things around here' as a result of repetition and reinforcement. These patterns reflect an underlying set of shared beliefs about the world and how to deal with it, and form part of the organization's culture – 'the way we do things in this organization'. They emerge as a result of repeated experiments and experience around what appears to work well – in other words, they are learned. Over time the pattern becomes more of an automatic response to particular situations, and the behaviour becomes what can be termed a 'routine'.

This does not mean that it is necessarily repetitive, only that its execution does not require detailed conscious thought. The analogy can be made with driving a car: it is possible to drive along a stretch of motorway whilst simultaneously talking to someone else, eating or drinking, listening to, and concentrating on something on the radio or planning what to say at the forthcoming meeting. But driving is not a passive behaviour: it requires continuous assessment and adaptation of responses in the light of other traffic behaviour, road conditions, weather and a host of different and unplanned factors. We can say that driving represents a behavioural routine in that it has been learned to the point of being largely automatic.

In the same way an organizational routine might exist around how projects are managed, or new products researched. For example, project management involves a complex set of activities such as planning, team selection, monitoring and execution of tasks, replanning, coping with unexpected crises, and so on. All of these have to be integrated – and offer plenty of opportunities for making mistakes. Project management is widely recognized as an organizational skill, which experienced firms have developed to a high degree but which beginners can make a mess of. Firms with good project management routines are able to codify and pass them on to others via procedures and systems. Most importantly, the principles are also transmitted into 'the way we run projects around here' by existing members passing on the underlying beliefs about project management behaviour to new recruits.

Over time organizational behaviour routines create and are reinforced by various kinds of artefacts – formal and informal structures, procedures and processes which describe 'the way we do things around here', and symbols which represent and characterize the underlying routines. It could be in the form of a policy, for example, 3M is widely known for its routines for regular and fast product innovation. It has enshrined a set of behaviours around encouraging experimentation into what the company terms 'the 15% policy', in which employees are enabled to work on their own curiosity-driven agenda for up to 15% of their time.[70] These routines are firm-specific, for example, they result from an environment in which the costs of product development experimentation are often quite low.

Levitt and March describe routines as involving established sequences of actions for undertaking tasks enshrined in a mixture of technologies, formal procedures or strategies, and informal conventions or habits.[71] Importantly, routines are seen as evolving in the light of experience that works – they become the mechanisms that 'transmit the lessons of history'. In this sense, routines have an existence independent of particular personnel – new members of the organization learn them on arrival, and most routines survive the departure of individual personnel. Equally, they are constantly being adapted and interpreted such that formal policy may not always reflect the current nature of the routine – as Augsdorfer points out in the case of 3M.[72]

For our purposes the important thing to note is that routines are what makes one organization different from another in how they carry out the same basic activity. We could almost say they represent the particular 'personality' of the firm. Each enterprise learns its own particular 'way we do things around here' in answer to the same generic questions – how it manages quality, how it manages people, etc. 'How we manage innovation around here' is one set of routines which describes and differentiates the responses that organizations make to the question of structuring and managing the generic model described above.

It follows that some routines are better than others in coping with the uncertainties of the outside world, in both the short and the long term. And it is possible to learn from others' experience in this way: the important point is to remember that routines are firm-specific and must be learned. Simply copying what someone else does is unlikely to help, any more than watching someone drive and then attempting to copy them will make a novice into an experienced driver. There may be helpful clues, which can be used to improve the novice's routines, but there is no substitute for the long and experience-based process of learning. Box 2.2 gives some examples where change has been introduced without this learning perspective.

BOX 2.2 **Fashion statements versus behavioural change in organizations**

The problem with routines is that they have to be learned – and learning is difficult. It takes time and money to try new things, it disrupts and disturbs the day-to-day working of the firm, it can upset organizational arrangements and require efforts in acquiring and using new skills. Not surprisingly most firms are reluctant learners – and one strategy which they adopt is to try and short-cut the process by borrowing ideas from other organizations.

Whilst there is enormous potential in learning from others, simply copying what seems to work for another organization will not necessarily bring any benefits and may end up costing a great deal and distracting the organization from finding its own ways of dealing with a particular problem. The temptation to copy gives rise to the phenomenon of particular approaches becoming fashionable – something which every organization thinks it needs in order to deal with its particular problems.

Over the past 40 years we have seen many apparent panaceas for the problems of becoming competitive. Organizations are constantly seeking for new answers to old problems, and the scale of investment in the new fashions of management thinking has often been considerable. The original evidence for the value of these tools and techniques was strong, with case studies and other reports testifying to their proven value within the context of origin. But there is also extensive evidence to suggest that these changes do not always work, and in many cases lead to considerable dissatisfaction and disillusionment.

Examples include:

- advanced manufacturing technology (AMT – robots, flexible machines, integrated computer control, etc.);[73, 74]
- total quality management (TQM)[75,76]
- business process reengineering (BPR)[77–79]
- benchmarking best practice[80, 81]
- quality circles[82, 83]
- networking/clustering[27,84,85]
- knowledge management[86]
- open innovation[87]

What is going on here demonstrates well the principles behind behavioural change in organizations. It is not that the original ideas were flawed or that the initial evidence was wrong. Rather it was that other organizations assumed they could simply be copied, without the need to adapt them, to customize them, to modify and change them to suit their circumstances. In other words, there was no learning, and no progress towards making them become routines, part of the underlying culture within the firm. Chapter 4 picks up this theme in the context of thinking about strategy.

2.5 Learning to manage innovation – building and developing routines across the core process

Successful innovation management routines are not easy to acquire. Because they represent what a particular firm has learned over time, through a process of trial and error, they tend to be very firm-specific. Whilst it may be possible to identify the kinds of thing which Google, Procter & Gamble, Nokia, 3M, Toyota, or others have learned to do, simply copying them will not work. Instead each firm has to find its own way of doing these things – in other words, develop its own particular routines.

In the context of innovation management we can see the same hierarchical relationship in developing capability as there is in learning to drive. Basic skills are behaviours associated with things like planning and managing projects or understanding customer needs. These simple routines need to be integrated into broader abilities which taken together make up an organization's capability in managing innovation. Table 2.6 gives some examples.

TABLE 2.6	**Core abilities in managing innovation**
Basic ability	**Contributing routines**
Recognizing	Searching the environment for technical and economic clues to trigger the process of change
Aligning	Ensuring a good fit between the overall business strategy and the proposed change – not innovating because it is fashionable or as a knee-jerk response to a competitor
Acquiring	Recognizing the limitations of the company's own technology base and being able to connect to external sources of knowledge, information, equipment, etc.
	Transferring technology from various outside sources and connecting it to the relevant internal points in the organization
Generating	Having the ability to create some aspects of technology in-house – through R&D, internal engineering groups, etc.
Choosing	Exploring and selecting the most suitable response to the environmental triggers which fit the strategy and the internal resource base/external technology network
Executing	Managing development projects for new products or processes from initial idea through to final launch
	Monitoring and controlling such projects
Implementing	Managing the introduction of change – technical and otherwise – in the organization to ensure acceptance and effective use of innovation
Learning	Having the ability to evaluate and reflect upon the innovation process and identify lessons for improvement in the management routines
Developing the organization	Embedding effective routines in place – in structures, processes, underlying behaviours, etc.

One last point concerns the negative side of routines. They represent, as we have seen, embedded behaviours, which have become reinforced to the point of being almost second nature – 'the way we do things around here'. Therein lies their strength, but also their weakness. Because they represent in-grained patterns of thinking about the world, they are resilient – but they can also become barriers to thinking in different ways. Thus core capabilities can become core rigidities – when the 'way we do things round here' becomes inappropriate, and when the organization is too committed to the old ways to change.[88] So it becomes important, from the standpoint of innovation management, not only to build routines but also to recognize when and how to destroy them and allow new ones to emerge. This is a particularly important issue in the context of managing discontinuous innovation; we return to it in Chapter 4, in the context of strategy.

Our argument in this book is that successful innovation management is primarily about building and improving effective routines. Learning to do this comes from recognizing and understanding effective routines (whether developed in-house or observed in another enterprise) and facilitating their emergence across the organization. And this learning process implies a building up of capability over time.

It's easy to make the assumption that because there is a rich environment full of potential sources of innovation that every organization will find and make use of these. The reality is, of course, that they differ widely in their ability to innovate – and this capability is clearly not evenly distributed across a population. For example, some organizations may simply be unaware of the need to change, never mind having the capability to manage such change. Such firms (and this is a classic problem of small-firm growth) differ from those which recognize in some strategic way the need to change, to acquire and use new knowledge but lack the capability to target their search or to assimilate and make effective use of new knowledge once identified. Others may be clear what they need but lack capability in finding and acquiring it. And others may have well-developed routines for dealing with all of these issues and represent resources on which less experienced firms might draw – as is the case with some major supply chains focused around a core central player.[89]

Figure 2.3 indicates a simple typology, ranging from firms which are 'unconsciously ignorant' (they don't know that they don't know) through to high-performing knowledge-based enterprises. The distinguishing feature is their capability to organize and manage the innovation process in its entirety, from

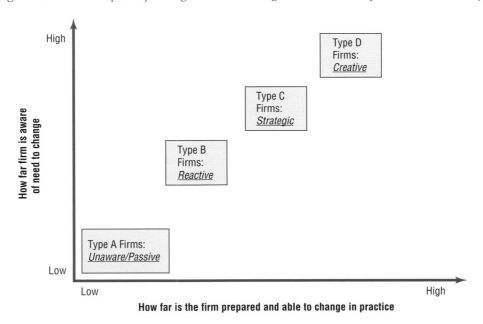

FIGURE 2.3: Groups of firms according to innovation capability[89]
Source: Hobday, M., H. Rush and J. Bessant (2005) Reaching the innovation frontier in Korea: a new corporate strategy dilemma. *Research Policy*, 33, 1433–57

search through selection to effective implementation of new knowledge. Such capability is not a matter of getting lucky once but of having an embedded high-order set of learning routines.

Grouped in this way we can identify simple archetypes that highlight differences in innovation capability. *Type A* firms can be characterized as being 'unconscious' or unaware about the need for innovation. They lack the ability to recognize the need for change in what may be a hostile environment and where technological and market know-how is vital to survival. They do not know where or what they might improve, or how to go about the process of technology upgrading and, as a result, are highly vulnerable. For example, if low-cost competitors enter – or the market demands faster delivery or higher quality – they are often not able to pick up the relevant signals or respond quickly. Even if they do, they may waste scarce resources by targeting the wrong kinds of improvement.

Type B firms recognize the challenge of change but are unclear about how to go about the process in the most effective fashion. Because their internal resources are limited and they often lack key skills and experience, they tend to react to external threats and possibilities, but are unable to shape and exploit events to their advantage. Their external networks are usually poorly developed, for example, most technological know-how comes from their suppliers and from observing the behaviour of other firms in their sector.

Type C firms have a well-developed sense of the need for change and are highly capable in implementing new projects and take a strategic approach to the process of continuous innovation. They have a clear idea of priorities as to what has to be done, when and by whom, and also have strong internal capabilities in both technical and managerial areas and can implement changes with skill and speed. These firms benefit from a consciously developed strategic framework in terms of search, acquisition, implementation and improvement of new knowledge. But they lack the capabilities for radical innovation – to redefine markets through new technology, or to create new market opportunities. They tend to compete within the boundaries of an existing industry and may become 'trapped' in a mature or slow growth sector, despite having exploited technological and market opportunities efficiently within the boundaries of the industry. Sometimes, they are limited in knowing where and how to acquire new knowledge beyond the boundaries of their traditional business.

Type D firms operate at the international knowledge frontier and take a creative and proactive approach to exploiting technological and market knowledge for competitive advantage and do so via extensive and diverse networks. They are at ease with modern strategic frameworks for innovation and take it upon themselves to 'rewrite' the rules of the competitive game with respect to technology, markets and organization. Strong internal resources are coupled with a high degree of absorptive capacity which can enable diversification into other sectors, where their own skills and capabilities bring new advantages and redefine the ways in which firms traditionally compete, or wish to compete.

Some creative firms emerge from traditional and mature sectors to challenge the way business is conducted. For example, Nokia, the Finnish company, moved from pulp and paper into electronics and eventually became a world leader in mobile telecommunications, showing that it was possible to make very high margins in the production of handsets within the developed countries, when most competitors believed it was impossible to achieve this goal (e.g. Ericsson and Motorola originally viewed handsets as low-margin commodity products). It is now in the throes of reinventing itself again, moving from being a mobile phone handset maker to providing an open-source platform on which a wide range of products and services can be built to exploit mobile communications, computing and entertainment. Another example is IBM, which transformed itself from being a 'dinosaur' of the computer industry, to one of the fastest-growing, most highly profitable information technology and consulting services companies in the world.

2.6 Measuring innovation success

Before we move to look at examples of successful routines for innovation management, we should pause for a moment and define what we mean by 'success'. We have already seen that one aspect of this question is the need to measure the overall process rather than its constituent parts. Many successful inventions fail to become successful innovations, even when well planned.[90–93] Equally, innovation alone may not always lead to business success. Although there is strong evidence to connect innovation with performance, success depends on other factors as well. If the fundamentals of the business are weak, then all the innovation in the world may not be sufficient to save it. This argues for strategically focused innovation as part of a 'balanced scorecard' of results measurement.[94, 95]

We also need to consider the time perspective. The real test of innovation success is not a one-off success in the short term but sustained growth through continuous invention and adaptation. It is relatively simple to succeed once with a lucky combination of new ideas and a receptive market at the right time – but it is quite another thing to repeat the performance consistently. Some organizations clearly feel able to do the latter to the point of presenting themselves as innovators, for example, 3M, Sony, IBM, Samsung and Philips, all currently use the term in their advertising campaigns and stake their reputations on their ability to innovate consistently.

In our terms, success relates to the overall innovation process and its ability to contribute consistently to growth. This question of measurement – particularly its use to help shape and improve management of the process – is one to which we will return in Chapter 12.

2.7 What do we know about successful innovation management?

The good news is that there is a knowledge base on which to draw in attempting to answer this question. Quite apart from the wealth of experience (of success and failure) reported by organizations involved with innovation, there is a growing pool of knowledge derived from research. Over the past 80 years or so there have been many studies of the innovation process, looking at many different angles. Different innovations, different sectors, firms of different shapes and sizes, operating in different countries have all come under the microscope and been analysed in a variety of ways (Chapter 9 provides a detailed list of such studies).

From this knowledge base it is clear that there are no easy answers and that innovation varies enormously – by scale, type, sector, etc. Nonetheless, there does appear to be some convergence around our two key points:

- Innovation is a process, not a single event, and needs to be managed as such.
- The influences on the process can be manipulated to affect the outcome – that is, the process can be managed.

Most important, research highlights the concept of success routines, which are learned over time and through experience. For example, successful innovation correlates strongly with how a firm selects and manages projects, how it coordinates the inputs of different functions, how it links up with its customers, etc. Developing an integrated set of routines is strongly associated with successful innovation management, and can give rise to distinctive competitive ability, for example, being able to introduce new products faster than anyone or being able to use new process technology better.[96–98]

The other critical point to emerge from research is that innovation needs managing in an integrated way; it is not enough just to manage or develop abilities in some of these areas. One metaphor (originally developed by researchers at Cranfield University) which helps draw attention to this is to see managing the process in sporting terms; success is more akin to winning a multi-event group of activities (like the pentathlon) than to winning a single high-performance event like the 100 metres.[99]

There are many examples of firms which have highly developed abilities for managing part of the innovation process but which fail because of a lack of ability in others. For example, there are many with an acknowledged strength in R&D and the generation of technological innovation – but which lack the abilities to relate these to the marketplace or to end-users. Others may lack the ability to link innovation to their business strategy. For example, many firms invested in advanced manufacturing technologies – robots, computer aided design, computer controlled machines – during the late twentieth century, but most surveys suggest that only half of these investments really paid off. In the case of the other half the problem was an inability to match the 'gee-whiz' nature of a glamorous technology to their particular needs, and the result was what might be called 'technological jewellery' – visually impressive but with little more than a decorative function.

The concept of capability in innovation management also raises the question of how it is developed over time. This must involve a learning process. It is not sufficient simply to have experiences (good or bad); the key lies in evaluating and reflecting upon them and then developing the organization in such a way that the next time a similar challenge emerges the response is ready. Such a cycle of learning is easy to prescribe but very often missing in organizations – with the result that there often seems to be a great deal of repetition in the pattern of mistakes, and a failure to learn from the misfortunes of others. For example, there is often no identifiable point in the innovation process where a post-mortem is carried out, taking time to try and distil useful learning for next time. In part this is because the people involved are too busy, but it is also because of a fear of blame and criticism. Yet without this pause for thought the odds are that the same mistakes will be repeated.[100, 101] It's important to note that even 'good' innovation management organizations can lose their touch – for example, 3M, for many years a textbook case of how to manage the process found itself in difficulties as a result of overemphasis on incremental innovation (driven by a 'six sigma' culture) at the expense of 'breakthrough' thinking. Its reflection on the problems this posed and commitment to reshaping its innovation management agenda again underlines the importance of learning and of the idea of 'dynamic capability'. (We will return to this theme in Chapter 12.) On the website there is an exercise designed to help explore patterns of success and failure in managing innovation.

VIEW FROM THE FRONT LINE

What factors make for innovation success in your view?

- Encouragement and empowerment from management: for small-scale innovations driven bottom up a clear focus, scope and mechanism is needed to reactively receive and channel ideas or implemented improvements.
- Positive reinforcement of innovative behaviour which encourages others to do the same (e.g. via PR, recognition/reward or just saying thanks).

- Where innovation is driven through large-scale programmes of change, use of a range of tools and a creative environment is crucial to success in generating far-reaching ideas.

(John Gilbert, Head of Process Excellence, UBS)

- Goldilocks resources – not too much, not too little.
- People who are willing to question, to challenge the status quo, who speak out when they are in disagreement, but who are open minded enough to evaluate a new idea.
- Senior management commitment – a visible and constant commitment – to innovation.
- Sufficient slack time to allow idea generation, experimentation and evaluation not directly associated with meeting the given objective.
- Protecting the innovation environment, the space, the resources, the people and the culture from the corrosive effect of a corporate bureaucracy that seeks to exploit existing resource in a repetitive fashion and tries to impose compliance through rule following.
- Recognizing and rewarding innovation, especially 'do-different' innovations.
- Making innovation part of the company culture, not just 'something for product development'.

(Patrick McLaughlin, Managing Director, Cerulean)

- Non-stop motivation for innovation at the managing director level/Not having innovative individuals being accounted for short-term results.
- Build a project-based organization.
- Build a good portfolio management structure.
- Build a funnel or stage-gate system, with gates where projects pass through.
- Ensure a large enough human resource base allocated to innovation-related activities.

(Wouter Zeeman, CRH Insulation Europe)

- No question in my view that innovation success comes from the top of the company, it's all about creating a culture of innovation rather than stagnation. It is essential that the person at the top of the organization is fully behind and demonstrates their support for innovation to succeed.
- A good mix of people and differing skills that they can 'bring to the party' with both the ability and drive to do it and share with others.
- The recognition that we will sometimes get it wrong but that we will learn from this experience and move on to create and develop something that works or improves the current state or/and produce something that is completely new.

(John Tregaskes, Innovation Manager, SERCO)

- Innovation must be an integral part of the company strategy.
- A culture for cooperation and networking with many different external partners, combined with a sincere curiosity towards everything that is new must be found. Be ready to share knowledge because that is the best way to convince others to share with you.
- Make a potential innovation visual to others by early prototyping (physical products) or specific case studies.

(John Thesmer, Managing Director Ictal Care Denmark)

- To make an innovation successful, you have to have a clear understanding of the business drivers and constraints being felt by the people on the 'coal face' – that is, the folks who will make the decision to use your new technology . . . or not. Don't simply launch your technology into the market and wait patiently for it to be adopted. Instead, talk extensively with the end-user and find out firsthand what's working and what is not. Discover for yourself if there are other constraints or issues that might be preventing your technology from taking root. Don't forget that these frontline managers are usually juggling thousands of issues in their minds, and your innovation is just one of them. Your technology might perfectly solve one problem – but it might cause five more that you never thought of. You won't find out what these issues are by staying in the lab or the boardroom. To get answers to these questions, you have to get as close to the end-user as you can.

(Rob Perrons, Shell Exploration, USA)

2.8 Success routines in innovation management

Successful innovators acquire and accumulate technical resources and managerial capabilities over time; there are plenty of opportunities for learning, for example through doing, using, working with other firms, asking the customers, but they all depend upon the readiness of the firm to see innovation less as a lottery than as a process which can be continuously improved.

From the various studies of success and failure in innovation it is possible to construct checklists and even crude blueprints for effective innovation management. A number of models for auditing innovation have been developed in recent years, which provide a framework against which to assess performance in innovation management. Some of these involve simple checklists, others deal with structures, others with the operation of particular subprocesses.[102–104] (We will return to the theme of innovation audits and their role in helping develop capability in Chapter 12.)

For our purposes in exploring innovation management throughout the rest of the book it will be helpful to build on our own simple model and use it to focus attention on key aspects of the innovation management challenge. At its heart we have the generic process described earlier, which sees innovation as a core set of activities distributed over time. (Of course, as we already noted, innovation in real life does not conform neatly to this simple representation – and it is rarely a single event but rather a cycle of activities repeated over time.) The key point is that a number of different actions need to take place as we move through the phases of this model and associated with each are some consistent lessons about effective innovation management routines. The website has an exercise which provides an opportunity to try and map the innovation process within an organization.

Search (see Figure 2.4) – the first phase in innovation involves detecting signals in the environment about potential for change. These could take the form of new technological opportunities, or changing requirements on the part of markets; they could be the result of legislative pressure or competitor action. Most innovations result from the interplay of several forces, some coming from the need for change pulling through innovation and others from the push which comes from new opportunities.

Given the wide range of signals it is important for successful innovation management to have well-developed mechanisms for identifying, processing and selecting information from this turbulent environment. Chapter 5 explores enabling routines associated with successful scanning and processing of relevant signals.

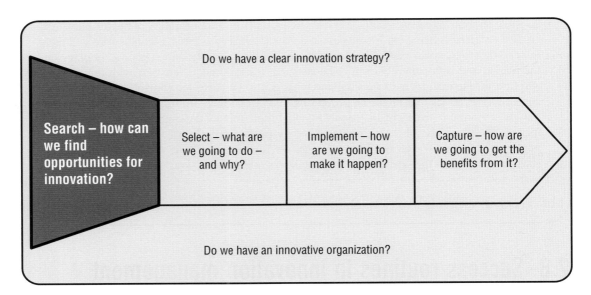

FIGURE 2.4: Search – the first phase in the process model of innovation

Organizations don't, of course, search in infinite space but rather in places where they expect to find something helpful. Over time their search patterns become highly focused and this can – as we have seen – sometimes represent a barrier to more radical forms of innovation. A key challenge in innovation management relates to the clear understanding of what factors shape the 'selection environment' and the development of strategies to ensure their boundaries of this are stretched. Again this theme is picked up in Chapter 5.

Selection (see Figure 2.5) – innovation is inherently risky, and even well-endowed firms cannot take unlimited risks. It is thus essential that some selection is made of the various market and technological opportunities, and that the choices made fit with the overall business strategy of the firm, and build upon established areas of technical and marketing competence. The purpose of this phase is to resolve the inputs into an innovation concept which can be progressed further through the development organization.

Three inputs feed this phase. The first is the flow of signals about possible technological and market opportunities available to the enterprise. The second input concerns the current knowledge base of the firm – its distinctive competence.[105] By this we mean what it knows about its product or service and how that is produced or delivered effectively. This knowledge may be embodied in particular products or equipment, but is also present in the people and systems needed to make the processes work. The important thing here is to ensure that there is a good fit between what the firm currently knows about and the proposed changes it wants to make.

This is not to say that firms should not move into new areas of competence – indeed there has to be an element of change if there is to be any learning. But rather there needs to be a balance and a development strategy. This raises the third input to this phase – the fit with the overall business. At the concept stage it should be possible to relate the proposed innovation to improvements in overall business performance. Thus if a firm is considering investing in flexible manufacturing equipment because the business is moving into markets where increased customer choice is likely to be critical, it will make

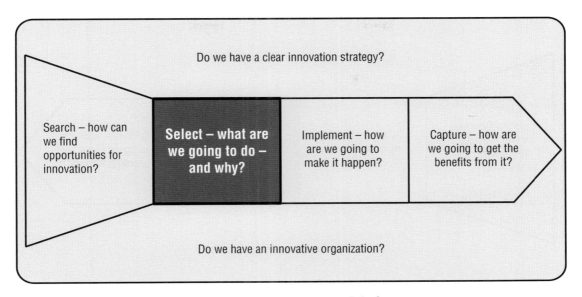

Do we have a clear innovation strategy?

Search – how can we find opportunities for innovation?

Select – what are we going to do – and why?

Implement – how are we going to make it happen?

Capture – how are we going to get the benefits from it?

Do we have an innovative organization?

FIGURE 2.5: Select – the second phase in the process model of innovation

sense. But if it is doing so in a commodity business where everyone wants exactly the same product at the lowest price, then the proposed innovation will not underpin the strategy – and will effectively be a waste of money. Getting close alignment between the overall strategy for the business and the innovation strategy is critical at this stage.

In similar fashion many studies have shown that product innovation failure is often caused by firms trying to launch products which do not match their competence base.[106]

This knowledge base need not be contained within the firm; it is also possible to build upon competencies held elsewhere. The requirement here is to develop the relationships needed to access the necessary complementary knowledge, equipment, resources, etc. Strategic advantage comes when a firm can mobilize a set of internal and external competencies – what Teece calls 'complementary assets' – which make it difficult for others to copy or enter the market.[107] (This theme is picked up in more depth in Chapter 4, and Chapters 7 and 8 explore some of the key routines associated with managing the strategic selection of innovation projects and building a coherent and robust portfolio.)

Implementing (see Figure 2.6) – having picked up relevant trigger signals and made a strategic decision to pursue some of them, the next key phase is actually turning those potential ideas into some kind of reality, for example a new product or service, a change in process, a shift in business model. In some ways this implementation phase can be seen as one which gradually pulls together different pieces of knowledge and weaves them into an innovation. At the early stages there is high uncertainty – details of technological feasibility, of market demand, of competitor behaviour, of regulatory and other influences – all of these are scarce and strategic selection has to be based on a series of 'best guesses'. But gradually over the implementation phase this uncertainty is replaced by knowledge acquired through various routes and at an increasing cost. Technological and market research helps clarify whether or not the innovation is technically possible or if there is a demand for it and if so, its characteristics. As the innovation develops so a continuing thread of problem finding and solving – getting the bugs out of the original concept – takes place, gradually building up relevant knowledge around the innovation.

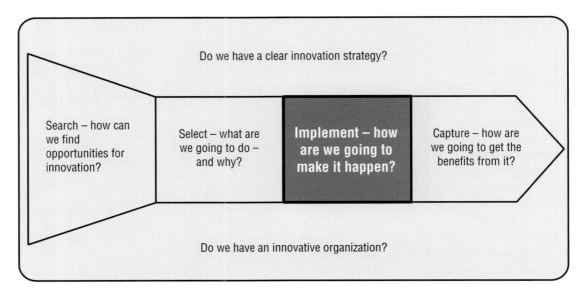

FIGURE 2.6: Implement – the third phase in the process model of innovation

Eventually it is in a form which can be launched into its intended context – internal or external market – and then further knowledge about its adoption (or otherwise) can be used to refine the innovation.

We can explore the implementation phase in a little more detail by considering three core elements: acquiring knowledge; executing the project; and launching and sustaining the innovation. Acquiring knowledge involves combining new and existing knowledge (available within and outside the organization) to offer a solution to the problem. It involves both generation of technological knowledge (via R&D carried out within and outside the organization) and technology transfer (between internal sources or from external sources). As such it represents a first draft of a solution, and is likely to change considerably in its development. The output of this stage in the process is both forward to the next stage of detailed development, and back to the concept stage where it may be abandoned, revised or approved.

Much depends at this stage on the nature of the new concept. If it involves an incremental modification to an existing design, there will be little activity within the invention stage. By contrast, if the concept is totally new, there is considerable scope for creativity. Whilst individuals may differ in terms of their preferred creative style, there is strong evidence to support the view that everyone has the latent capability for creative problem solving.[108,109] Unfortunately, a variety of individual inhibitions and external social and environmental pressures combine and accumulate over time to place restrictions on the exercise of this creative potential. The issue in managing this stage is thus to create the conditions under which this can flourish and contribute to effective innovation.

Another problem with this phase is the need to balance the open-ended environmental conditions, which support creative behaviour, with the somewhat harsher realities involved elsewhere in the innovation process. As with concept testing and development, it is worth spending time exploring ideas and potential solutions rather than jumping on the first apparently workable option.

The challenge in effective R&D is not simply one of putting resources into the system; it is how those resources are used. Effective management of R&D requires a number of organizational routines, including clear strategic direction, effective communication and 'buy-in' to that direction, and integration of effort across different groups.

But not all firms can afford to invest in R&D; for many smaller firms the challenge is to find ways of using technology generated by others or to complement internally generated core technologies with a wider set drawn from outside. This places emphasis on the strategy system discussed above: the need to know which to carry out where and the need for a framework to guide policy in this area. Firms can survive even with no in-house capability to generate technology – but to do so they need to have a well-developed network of external sources to supply it, and the ability to put that externally acquired technology to effective use.

It also requires abilities in finding, selecting and transferring technology from outside the firm. This is rarely a simple shopping transaction although it is often treated as such; it involves abilities in selecting, negotiating and appropriating the benefits from such technology transfer.[110]

Executing the project forms the heart of the innovation process. Its inputs are a clear strategic concept and some initial ideas for realizing the concept. Its outputs are both a developed innovation and a prepared market (internal or external), ready for final launch. This is fundamentally a challenge in project management under uncertain conditions. As we will see in Chapter 9, the issue is not simply one of ensuring certain activities are completed in a particular sequence and delivered against a time and cost budget. The lack of knowledge at the outset and the changing picture as new knowledge is brought in during development means that a high degree of flexibility is required in terms of overall aims and subsidiary activities and sequencing. Much of the process is about weaving together different knowledge sets coming from groups and individuals with widely different functional and disciplinary backgrounds. And the project may involve groups who are widely distributed in organizational and geographical terms – often belonging to completely separate organizations. Consequently the building and managing of a project team, of communicating a clear vision and project plan, of maintaining momentum and motivation, etc. are not trivial tasks.

One way of representing the development stage is as a funnel, moving gradually from broad exploration to narrow focused problem solving and hence to final (and successful) innovation. Unfortunately the apparent rational progress implied in this model is often not borne out in practice; instead various problems emerge, such as the lack of input (or sometimes too much input) from key functions, lack of communication between functions and conflicting goals.

It is during this stage that most of the time, costs and commitment are incurred, and it is characterized by a series of problem-solving loops dealing with expected and unexpected difficulties in the technical and market areas. Although we can represent it as a parallel process, in practice effective management of this stage requires close interaction between marketing-related and technical activities. For example, product development involves a number of functions, ranging from marketing, through design and development to manufacturing, quality assurance and finally back to marketing. Differences in the tasks which each of these functions performs, in the training and experience of those working there and in the timescales and operating pressures under which they work all mean that each of these areas becomes characterized by a different working culture. Functional divisions of this kind are often exaggerated by location, where R&D and design activities are grouped away from the mainstream production and sales operations – in some cases on a completely different site.

Separation of this kind can lead to a number of problems in the overall development process. Distancing the design function from the marketplace can lead to inappropriate designs which do not meet the real customer needs, or which are 'overengineered', embodying a technically sophisticated and elegant solution which exceeds the actual requirement (and may be too expensive as a consequence).

This kind of phenomenon is often found in industries that have a tradition of defence contracting, where work has been carried out on a cost-plus basis involving projects which have emphasized technical design features rather than commercial or manufacturability criteria.

Similarly, the absence of a close link with manufacturing means that much of the information about the basic 'make-ability' of a new design either does not get back to the design area at all or else does so at a stage too late to make a difference or to allow the design to be changed. There are many cases in which manufacturing has wrestled with the problem of making or assembling a product which requires complex manipulation, but where minor design change – for example, relocation of a screw hole – would considerably simplify the process. In many cases such an approach has led to major reductions in the number of operations necessary – simplifying the process and often, as an extension, making it more susceptible to automation and further improvements in control, quality and throughput.

In similar fashion, many process innovations fail because of a lack of involvement on the part of users and others likely to be affected by the innovation. For example, many IT systems, whilst technically capable, fail to contribute to improved performance because of inadequate consideration of current working patterns which they will disrupt, lack of skills development amongst those who will be using them, inadequately specified user needs, and so on.

Although services are often less tangible, the underlying difficulties in implementation are similar. Different knowledge sets need to be brought together at key points in the process of creating and deploying new offerings. For example, developing a new insurance or financial service product requires technical input on the part of actuaries, accountants, IT specialists, etc., but this needs to be combined with information about customers and key elements of the marketing mix – the presentation, the pricing, the positioning of the new service. Knowledge of this kind will lie particularly with marketing and related staff, but their perspective must be brought to bear early enough in the process to avoid creating a new service which no one actually wants to buy.

The 'traditional' approach to this stage was a linear sequence of problem solving, but much recent work in improving development performance (especially in compressing the time required) involves attempts to do much of this concurrently or in overlapping stages. Useful metaphors for these two approaches are the relay race and the rugby team.[111] These should be seen as representing two poles of a continuum: as we shall see in Chapter 9 the important issue is to choose an appropriate level of parallel development.

In conjunction with the technical problem solving associated with developing an innovation, there is also a set of activities associated with preparing the market into which it will be launched. Whether this market is a group of retail consumers or a set of internal users of a new process, the same requirement exists for developing and preparing this market for launch, since it is only when the target market makes the decision to adopt the innovation that the whole innovation process is completed. The process is again one of sequentially collecting information, solving problems and focusing efforts towards a final launch. In particular it involves collecting information on actual or anticipated customer needs and feeding this into the product development process, whilst simultaneously preparing the marketplace and marketing for the new product. It is essential throughout this process that a dialogue is maintained with other functions involved in the development process, and that the process of development is staged via a series of 'gates' which control progress and resource commitment.

A key aspect of the marketing effort involves anticipating likely responses to new product concepts and using this information to design the product and the way in which it is launched and marketed. This process of analysis builds upon knowledge about various sources of what Thomas calls 'market friction'.[112]

Buyer behaviour is a complex subject, but there are several key guidelines which emerge to help shape market development for a new product. The first is the underlying process of adoption of something new; typically this involves a sequence of awareness, interest, trial, evaluation and adoption. Thus simply making people aware, via advertising, of the existence of a new product, will not be sufficient; they need to be drawn into the process through the other stages. Converting awareness to interest, for example, means forging a link between the new product concept and a personal need (whether real or induced via advertising). Chapter 9 deals with this issue in greater depth.

Successful implementation of internal (process) innovations also requires skilled change management. This is effectively a variation on the marketing principles outlined above, and stresses communication, involvement and intervention (via training, etc.) to minimize resistance to change – again essentially analogous to Thomas's concept of 'market friction'. Chapter 9 discusses this theme in greater detail and presents some key enabling routines for the implementation phase. The website includes a case study of AMP, a large UK organization, and how it managed the introduction of major internal change.

Understanding user needs has always been a critical determinant of innovation success and one way of achieving this is by bringing users into the loop at a much earlier stage. The work of Eric von Hippel and others has shown repeatedly that early involvement and allowing users to play an active role in the innovation process leads to better adoption and higher quality innovation. It is, effectively, the analogue of the early involvement/parallel-working model mentioned above – and with an increasingly powerful set of tools for simulation and exploration of alternative options there is growing scope for such an approach.[113,114]

Where there is a high degree of uncertainty – as is the case with discontinuous innovation conditions – there is a particular need for adaptive strategies which stress the coevolution of innovation with users, based on a series of 'probe and learn' experimental approaches. The role here for early and active user involvement is critical.

Capture (see Figure 2.7) – The purpose of innovating is rarely to create innovations for their own sake but rather to capture some kind of value from them – be it commercial success, market share, cost reduction or – as in social innovation – changing the world. History abounds with examples of innovations

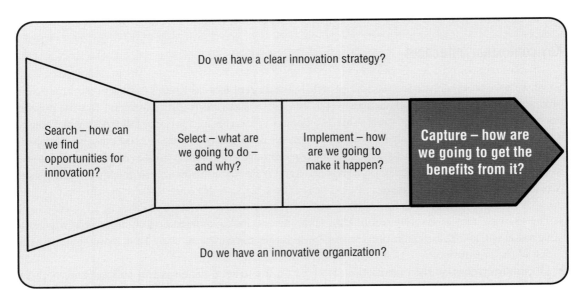

FIGURE 2.7: Capture – the fourth phase in the process model of innovation

which succeeded at a technical level but which failed to deliver value – or achieved it briefly, only to have the advantage competed away by imitators. Capturing value from the process is a critical theme and one to which we will return in Chapter 11. There are many ways in which this can be done, from formal methods such as patenting through to much less formal, like the use of tacit knowledge. And central to the discussion is the concept of 'complementary assets' – what other elements around the system in which the innovation is created and delivered are hard for others to access or duplicate? This gives rise to the idea of what Teece[107] termed 'appropriability regimes' – how easy or hard is it to extract value from investments in innovation?

An inevitable outcome of the launch of an innovation is the creation of new stimuli for restarting the cycle. If the product/service offering or process change fails, this offers valuable information about what to change for next time. A more common scenario is what Rothwell and Gardiner call 'reinnovation': essentially building upon early success but improving the next generation with revised and refined features. In some cases, where the underlying design is sufficiently 'robust' it becomes possible to stretch and reinnovate over many years and models.[115]

But although opportunities emerge for learning and development of innovations and the capability to manage the process which created them, they are not always taken up by organizations. Amongst the main requirements in this stage is the willingness to learn from completed projects. Projects are often reviewed and audited, but these reviews may take the form of an exercise in 'blame accounting' and in trying to cover up mistakes and problems. The real need is to capture all the hard-won lessons, from both success and failure, and feed these through to the next generation. Nonaka and Kenney provide a powerful argument for this perspective in their comparison of product innovation at Apple and Canon.[116] Much of the current discussion around the theme of knowledge management represents growing concern about the lack of such 'carry-over' learning – with the result that organizations are often 'reinventing the wheel' or repeating previous mistakes.

Learning can be in terms of technological lessons learned, for example, the acquisition of new processing or product features , which add to the organization's technological competence. But learning can also be around the capabilities and routines needed for effective product innovation management. In this connection some kind of structured audit framework or checklist is useful.

Key contextual influences

So far we have been considering the core generic innovation process as a series of stages distributed over time and have identified key challenges which emerge in their effective management. But the process doesn't take place in a vacuum – it is subject to a range of internal and external influences which shape what is possible and what actually emerges. Rothwell distinguishes between what he terms 'project-related factors' – essentially those which we have been considering so far – and 'corporate conditions' – which set the context in which the process is managed.[62] For the purposes of the book we will consider two sets of such contextual factors:

- The strategic context for innovation. How far is there a clear understanding of the ways in which innovation will take the organization forward? And is this made explicit, shared and 'bought into' by the rest of the organization?
- The innovativeness of the organization. How far do the structure and systems support and motivate innovative behaviour? Is there a sense of support for creativity and risk taking, can people communicate across boundaries, is there a 'climate' conducive to innovation?

2.9 Beyond the steady state

The model we have been developing in this chapter is very much about the world of repeated, continuous innovation where there is the underlying assumption that we are 'doing what we do but better'. This is not necessarily only about incremental innovation, it is possible to have significant step changes in product/service offering, process, etc., but these still take place within an established envelope. The 'rules of the game', in terms of for example technological possibilities, market demands, competitor behaviour, political context, are fairly clear and although there is scope for pushing at the edges, the space within which innovation happens is well defined.

Central to this model is the idea of learning through trial and error to build effective routines which can help improve the chances of successful innovation. Because we get a lot of practice at such innovation it becomes possible to talk about a 'good' (if not 'best') practice model for innovation management which can be used to audit and guide organizational development.

However, we also need to take into account that innovation is sometimes discontinuous in nature. Things happen that lie outside the 'normal' frame and result in changes to the 'rules of the game'. Under these conditions doing more of the same 'good practice' routines may not be enough and may even be inappropriate to dealing with the new challenges. Instead we need a different set of routines – not to use instead of but as well as those we have developed for 'steady-state' conditions. It is likely to be harder to identify and learn these, in part because we don't get so much practice – it is difficult to make a routine out of something which happens only occasionally. But we can observe some of the basic elements of the complementary routines associated with successful innovation management under discontinuous conditions. These tend to be linked with highly flexible behaviour involving agility, tolerance for ambiguity and uncertainty, emphasis on fast learning through quick failure – very much characteristics that are often found in small entrepreneurial firms. Table 2.7 lists some elements of the two complementary models.

TABLE 2.7	**Different innovation management archetypes**	
	Type 1 – Steady-state archetype	**Type 2 – Discontinuous-innovation archetype**
Interpretive schema – how the organization sees and makes sense of the world	There is an established set of 'rules of the game' by which other competitors also play	No clear 'rules of the game' – these emerge over time but cannot be predicted in advance
	Particular pathways in terms of search and selection environments and technological trajectories exist and define the 'innovation space' available to all players in the game	Need high tolerance for ambiguity – seeing multiple parallel possible trajectories
		'Innovation space' defined by open and fuzzy selection environment
	Strategic direction is highly path dependent	Probe and learn experiments needed to build information about emerging patterns and allow dominant design to emerge

(continued)

	Type 1 – Steady-state archetype	Type 2 – Discontinuous-innovation archetype
TABLE 2.7 **(Continued)**		
		Strategic direction is highly path independent
Strategic decision making	Makes use of decision-making processes which allocate resources on the basis of risk management linked to the above 'rules of the game'	High levels of risk taking since no clear trajectories – emphasis on fast and lightweight decisions rather than heavy commitment in initial stages
	(Does the proposal fit the business strategic directions? Does it build on existing competence base?)	Multiple parallel bets, fast failure and learning as dominant themes. High tolerance of failure but risk is managed by limited commitment. Influence flows to those prepared to 'stick their neck out' – entrepreneurial behaviour
	Controlled risks are taken within the bounds of the 'innovation space'	
	Political coalitions are significant influences maintaining the current trajectory	
Operating routines	Operates with a set of routines and structures/procedures that embed them, which are linked to these 'risk rules', e.g. stage-gate monitoring and review for project management	Operating routines are open-ended, based around managing emergence
		Project implementation is about 'fuzzy front end', light touch strategic review and parallel experimentation
	Search behaviour is along defined trajectories and uses tools and techniques for R&D, market research, etc. which assume a known space to be explored – search and selection environment	Probe and learn, fast failure and learn rather than managed risk.
		Search behaviour is about peripheral vision, picking up early warning through weak signals of emerging trends
	Network building to support innovation – e.g. user involvement, supplier partnership, etc. – is on basis of developing close and strong ties	Linkages are with heterogeneous population and emphasis less on established relationships than on weak ties

As we will see throughout the book, a key challenge in managing innovation is the ability to create ways of dealing with both sets of challenges – and if possible to do so in 'ambidextrous' fashion, maintaining close links between the two rather than spinning off completely separate ventures.

Summary and further reading

In this chapter we've looked at the challenge of managing innovation as a core business process concerned with renewing what the organization offers and the ways in which it creates and delivers that offering. The process has a number of elements and we will explore these in more detail in the rest of the book. We have also looked at the question of routines – repeated and learned patterns of behaviour which become 'the way we do things around here' since it is these which constitute the core of innovation management capability. Finally we looked at some of the lessons learned around success routines – what does experience teach us about how to organize and manage innovation?

A number of writers have looked at innovation from a process perspective and some good examples can be found in K. Brockhoff *et al.* (*The Dynamics of Innovation*, Springer, 1999); J. Ettlie (*Managing Innovation*, John Wiley & Sons, Inc., 1999); A. Hargadon (*How Breakthroughs Happen*, Harvard Business School Press, 2003); T. Jones (*Innovating at the Edge*, Butterworth Heinemann, 2002); R. Loveridge and M. Pitt (*The Strategic Management of Technological Innovation*, John Wiley & Sons, Ltd, 1990); P. Trott (*Innovation Management and New Product Development*, Prentice-Hall, 2004) and M. Jelinek and J. Litterer (*Organizing for Technology and Innovation*, McGraw-Hill, 1994). Case studies provide a good lens through which this process can be seen and there are several useful collections including C. Baden-Fuller and M. Pitt's *Strategic Innovation* (Routledge, 1996); R. Burgelman and R. Rosenbloom's *Research on Technological Innovation* (JAI Press, 1993); M. Gallagher and S. Austin's *Continuous Improvement Casebook* (Kogan Page, 1997); R. Kanter's *Innovation: Breakthrough Thinking at 3M, DuPont, GE, Pfizer and Rubbermaid* (Harper Business, 1997); R. Leifer *et al.*, *Radical Innovation* (Harvard Business School Press, 2000); B. Von Stamm, *Managing Innovation, Design and Creativity* (John Wiley & Sons, Ltd, 2008) and R. Weisberg, Case studies of innovation: ordinary thinking, extraordinary outcomes, in *International Handbook of Innovation* (Elsevier, 2003).

Some books cover company histories in detail and give an insight into the particular ways in which firms develop their own bundles of routines, for example, M. Graham and A. Shuldiner, *Corning and the Craft of Innovation* (Oxford University Press, 2001), E. Gundling, *The 3M way to Innovation* (Kodansha International, 2000) and T. Kelley *et al.*, *The Art of Innovation: Lessons in Creativity from IDEO* (Currency, 2001).

Autobiographies of key innovation leaders provide a similar – if sometimes personally biased – insight into this. For example, J. Dyson, *Against the Odds* (Orion, 1997), A. Groves, *Only the Paranoid Survive* (Bantam, 1999) and J. Welch, *Jack! What I've learned from leading a great company and great people* (Headline, 2001). In addition several websites, such as the Product Development Management Association (`www.pdma.org`), carry case studies on a regular basis.

Many books and articles focus on particular aspects of the process, for example, on technology strategy: P. Adler, Technology strategy: a guide to the literature. (*Research in Technological Innovation, Management and Policy*, **4**, 25–151, 1989); R. Burgelman *et al.*, eds, *Strategic Management of Technology and Innovation* (McGraw-Hill Irwin, 2004); M. Dodgson, ed., *Technology and the Firm: Strategies, Management and Public Policy* (Longman, 1990); and D. Ford and M. Saren, *Technology Strategy for Business* (International Thomson Business Press, 1996).

On product or service development: R. Cooper, *Winning at New Products* (Kogan Page, 2001); M. Rosenau *et al.*, *The PDMA Handbook of New Product Development* (John Wiley & Sons, Inc., 1996); J.Tidd and F. Hull, eds, *Service Innovation: Organizational Responses to Technological Opportunities and Market Imperatives* (Imperial College Press, 2003); S. Wheelwright and K. Clark, *Revolutionizing Product Development* (Free Press, 1992).

On process innovation: T. Davenport, *Process Innovation: Re-Engineering Work through Information Technology* (Harvard University Press, 1992); G. Pisano, *The Development Factory: Unlocking the Potential of Process Innovation* (Harvard Business School Press, 1996); and M. Zairi, *Process Innovation Management* (Butterworth Heinemann, 1999).

On technology transfer: M. Saad, *Development through Technology Transfer* (Intellect, 2000).

On implementation: D. Bennett and M. Kerr, A systems approach to the implementation of total quality management (*Total Quality Management*, **7**, 631–65, 1996).

On learning: K. Ayas, *Design for Learning for Innovation* (Eburon, 1997); M. Boisot, M. (1995). Is your firm a creative destroyer? (*Research Policy*, **24**, 489–506, 1995); W. Cohen. and D. Levinthal, Absorptive capacity (*Administrative Science Quarterly*, **35** (1), 128–52, 1990); D. Leonard-Barton, *Wellsprings of Knowledge* (Harvard Business School Press, 1995); and S. A. Zahra and G. George, Absorptive capacity: a review (*Academy of Management Review*, **27**, 185–94, 2002).

For a good review and critique from the academic standpoint, which raises a number of issues concerned with managing innovation, see P. Clark, *Organizational Innovations* (Sage, 2002), N. King, Modelling the innovation process (*Journal of Occupational and Social Psychology*, **65**, 89–100, 1992), R. Loveridge and M. Pitt, *The Strategic Management of Technological Innovation* (John Wiley & Sons, Ltd, 1990) and A. Van de Ven *et al.*, *Research on the Management of Innovation* (Harper and Row, 1989).

Websites such as AIM (**www.aimresearch.org**) and NESTA (**www.nesta.org**) regularly report academic research around innovation and NESTA also run an excellent briefing on innovation-related themes (NESTA-ID).

Web links

Here are the full details of the resources available on the website flagged throughout the text:

 Case studies:
Tesco Fresh & Easy
Open Door
NHS RED
Zara
Marshalls
NPI
Cerulean
Philips
Coloplast
Corning
3M
AMP

Interactive exercises:
Sector innovation patterns
The way we do things round here
Success and failure review
Innovation success and failure
Mapping the innovation process

 Video podcast:
Finnegan's Fish Bar
Bill's
Cerulean

 Audio podcast:
Philips
Cerulean

References

1. **Bessant, J. and A. Davies** (2007) Managing service innovation. DTI Occasional Paper 9: Innovation in Services, Department of Trade and Industry, London.
2. **Vandermerwe, S.** (2004). *Breaking Through: Implementing Customer Focus in Enterprises*, Palgrave Macmillan, London.
3. **Davies, A. and M. Hobday** (2005) *The Business of Projects: Managing Innovation in Complex Products and Systems*, Cambridge: Cambridge University Press.
4. **Davies, A., T. Brady, and M. Hobday** (2007) Charting a path toward integrated solutions. *Sloan Management Review*, **47** (3), 39–48.
5. **Voss, C.** (2004) *Trends in the Experience and Service Economy*, Advanced Institute of Management/ London Business School, London.
6. **Voss, C., A. Roth and D. Chase** (2007) Experience, service operations strategy, and services as destinations: foundations and exploratory investigation. *Production and Operations Management*, **17**(3), 244–66.
7. **Voss, C. and L. Zomerdijk** (2007) *Innovation in Experiential Services – An Empirical View in 'Innovation in Services'*, Department of Trade and Industry, London.
8. **Garr, D.** (2000) *IBM Redux: Lou Gerstner and the Business Turnaround of the Decade*, Harper Collins, New York.
9. **OECD** (1987) *Science and Technology Indicators*, Organization for Economic Cooperation and Development, Paris.
10. **Sako, M. and A. Tierney** (2005) *Sustainability of Business Service Outsourcing: The Case of Human Resource Outsourcing (HRO)*, Advanced Institute for Management Research, London.
11. **Sako, M., R. Griffiths, and L. Abramovsky** (2004) *Offshoring of Business Services and its Impact on the UK Economy*, Advanced Institute for Management Research, London.
12. **Maglio, P., J. Spohrer, D. Seidman, and J. Ritsko** (2008) Service science, management and engineering (Special Issue). *IBM Systems Journal*, **47** (1), Special Issue.
13. **Albury, D.** (2004) Innovation in the Public Sector, Strategy Unit, Cabinet Office, London.

14. **Bessant, J.** (2005) Enabling continuous and discontinuous innovations: some reflections from a private sector perspective. *Public Money and Management*, **25** (1).

15. **Hartley, J.** (2005) Innovation in governance and public services: past and present. *Public Money and Management*, **25** (1), 27–34.

16. **Hartley, J. and J. Downe** (2007) The shining lights? Public service awards as an approach to service improvement. *Public Administration*, **85** (2), 329–53.

17. **Mulgan, G.** (2007) *Ready or Not? Taking Innovation in the Public Sector Seriously*, NESTA, London.

18. **Hoffman, K., M. Parejo, J. Bessant, and L. Perren** (1997) Small firms, R&D, technology and innovation in the UK. *Technovation*, **18**, 39–55.

19. **Birch, D.** (1987) *Job Creation in America*, Free Press, New York.

20. **OECD** (2002) *High Growth SMEs and Employment*, Organization for Economic Cooperation and Development, Paris.

21. *Fortune* (2004) Interview, Top ten minds. *Fortune*.

22. **Garnsey, E. and E. Stam** (2008) Entrepreneurship in the knowledge economy. In J. Bessant and T. Venables, eds, *Creating Wealth from Knowledge*, Cheltenham, Edward Elgar.

23. **AIM** (2004) *i-works: How High Value Innovation Networks Can Boost UK Productivity*, ESRC/EPSRC Advanced Institute of Management Research, London.

24. **Conway, S. and F. Steward** (1998) Mapping innovation networks. *International Journal of Innovation Management*, **2** (2), 165–96.

25. **Lundvall, B.** (1990) *National Systems of Innovation: Towards a Theory of Innovation and Interactive Learning*, Frances Pinter, London.

26. **Nelson, R.** (1993) *National Innovation Systems: A Comparative Analysis*, Oxford University Press, New York.

27. **Best, M.** (2001) *The New Competitive Advantage*, Oxford University Press, Oxford.

28. **Leifer, R., C. McDermott, G. O'Conner, L. Peters, M. Rice and R. Veryzer** (2000) *Radical Innovation*, Harvard Business School Press, Boston, MA.

29. **Tushman, M. and C. O'Reilly** (1996) Ambidextrous organizations: managing evolutionary and revolutionary change. *California Management Review*, **38** (4), 8–30.

30. **Phillips, W., H. Noke, J. Bessant, and R. Lanning** (2006) Beyond the steady state: managing discontinuous product and process innovation. *International Journal of Innovation Management*, **10** (2), 175–96.

31. **Hamel, G.** (2000) *Leading the Revolution*, Harvard Business School Press, Boston. MA.

32. **Kaplan, S., F. Murray, and R. Henderson** (2003) Discontinuities and senior management: assessing the role of recognition in pharmaceutical firm response to biotechnology. *Industrial and Corporate Change*, **12** (2), 203.

33. **Christensen, C. and M. Raynor** (2003) *The Innovator's Solution: Creating and Sustaining Successful Growth*, Harvard Business School Press, Boston, MA.

34. **Bessant, J. and D. Francis** (2005) *Dealing with Discontinuity – How to Sharpen Up Your Innovation Act*, AIM Executive Briefings, AIM-ESRC/EPSRC Advanced Institute of Management Research, London.

35. **Birkinshaw, J. and C. Gibson** (2004) Building ambidexterity into an organization. *Sloan Management Review*, **45** (4), 47–55.

36. **Tidd, J. and F. Hull, eds** (2003) *Service Innovation: Organizational Responses to Technological Opportunities and Market Imperatives*, Imperial College Press, London.

37. **Pavitt, K.** (1984) Sectoral patterns of technical change: towards a taxonomy and a theory. *Research Policy*, **13**, 343–73.

38. **Oakey, R.** (1991) High technology small firms; their potential for rapid industrial growth. *International Small Business Journal*, **9**, 30–42.

39. **Rothwell, R.** (1978) Small and medium sized manufacturing firms and technological innovation. *Management Decision*, **16** (6), 362–370.

40. **Rothwell, R.** (1983) Innovation and firm size: a case of dynamic complementarity. *Journal of General Management*, **8** (3), 5–25.

41. **Rothwell, R. and M. Dodgson** (1993) SMEs: their role in industrial and economic change. *International Journal of Technology Management* (Special issue on small firms), 8–22.

42. **Pavitt, K.** (2000) *Technology, Management and Systems of Innovation*, Edward Elgar, London.

43. **Abernathy, W. and J. Utterback** (1978) Patterns of industrial innovation. *Technology Review*, **80**, 40–7.

44. **Utterback, J.** (1994) *Mastering the Dynamics of Innovation*, Harvard Business School Press, Boston, MA.

45. **Tushman, M. and P. Anderson** (1987) Technological discontinuities and organizational environments. *Administrative Science Quarterly*, **31** (3), 439–65.

46. **Perez, C.** (2002) *Technological Revolutions and Financial Capital*, Edward Elgar, Cheltenham.

47. **Christensen, C.** (1997) *The Innovator's Dilemma*, Harvard Business School Press, Cambridge, MA.

48. **Christensen, C., S. Anthony, and E. Roth** (2007) *Seeing What's Next*, Harvard Business School Press, Boston, MA.

49. **Bessant, J., R. Lamming, H. Noke, and W. Phillips** (2005) Managing innovation beyond the steady state. *Technovation*, **25** (12), 1366–76.

50. **IPTS** (1998) *The Impact of EU Regulation on Innovation of European Industry*, IPTS/ European Union, Seville.

51. **Whitley, R.** (2000) The institutional structuring of innovation strategies: business systems, firm types and patterns of technical change in different market economies. *Organization Studies*, **21** (5), 855–86.

52. **Freeman, C. and L. Soete** (1997) *The Economics of Industrial Innovation*, 3rd edition, MIT Press, Cambridge, MA.

53. **Coombs, R., P. Saviotti, and V. Walsh** (1985) *Economics and Technological Change*, Macmillan, London.

54. **Booz, Allen and Hamilton** (1982) *New Product Management for the 1980s*, Booz, Allen and Hamilton Consultants.

55. **Cooper, R.** (2001) *Winning at New Products*, 3rd edition, Kogan Page, London.

56. **BSI** (2008) *Design Management Systems. Guide to Managing Innovation*, British Standards Institute, London.

57. **Griffin, A., M. Rosenau, G. Castellion, and N. Anschuetz** (1996) *The PDMA Handbook of New Product Development*, John Wiley and Sons, Inc., New York.

58. **Koen, P.A., G. Ajamian, R. Burkart, A. Clamen, J. Davidson, R. D'Amoe, C. Elkins, K. Herald, M. Incorvia, A. Johnson, R. Karol, R. Seibert, A. Slavejkov, and K. Wagner** (2001) New concept development model: providing clarity and a common language to the 'fuzzy front end' of innovation. *Research Technology Management*, **44** (2), 46–55.

59. **Souder, W. and J. Sherman** (1994) *Managing New Technology Development*, McGraw-Hill, New York.

60. **Van de Ven, A.** (1999) *The Innovation Journey*, Oxford University Press, Oxford.

61. **Van de Ven, A., H. Angle, and M. Poole** (1989) *Research on the Management of Innovation*, Harper and Row, New York.

62. **Rothwell, R.** (1992) Successful industrial innovation: critical success factors for the 1990s. *R&D Management*, **22** (3), 221–39.

63. **Dodgson, M., D. Gann, and A. Salter** (2002) The intensification of innovation. *International Journal of Innovation Management*, **6** (1), 53–83.

64. **Bright, A.** (1949) *The Electric Lamp Industry: Technological Change and Economic Development from 1800 to 1947*, Macmillan, New York.

65. **Henderson, R. and K. Clark** (1990) Architectural innovation: the reconfiguration of existing product technologies and the failure of established firms. *Administrative Science Quarterly*, **35**, 9–30.

66. **Graham, M. and A. Shuldiner** (2001) *Corning and the Craft of Innovation*, Oxford University Press, Oxford.

67. **Gundling, E.** (2000) *The 3M Way to Innovation: Balancing People and Profit*, Kodansha International, New York.

68. **de Geus, A.** (1996) *The Living Company*, Harvard Business School Press, Boston, MA.

69. **Arrow, K.** (1962) Economic welfare and the allocation of resources for invention. In R. Nelson, ed., *The Rate and Direction of Inventive Activity*, Princeton University Press, Princeton.

70. **Kanter, R., ed.** (1997) *Innovation: Breakthrough Thinking at 3M, DuPont, GE, Pfizer and Rubbermaid*, New York, Harper Business.

71. **Levitt, B. and J. March** (1988) Organisational learning. *Annual Review of Sociology*, **14**, 319–40.

72. **Augsdorfer, P.** (1996) *Forbidden Fruit*, Avebury, Aldershot.

73. **Ettlie, J.** (1988) *Taking Charge of Manufacturing*, Jossey-Bass, San Francisco.

74. **Bessant, J.** (1991) *Managing Advanced Manufacturing Technology: The Challenge of the Fifth Wave*, NCC-Blackwell, Oxford/Manchester.

75. **Knights, D. and D. McCabe** (1996) Do quality initiatives need management? *The TQM Magazine*, **8**, 24–6.

76. **Knights, D. and D. McCabe** (1999) Are there no limits to authority? TQM and organizational power. *Organizational Studies*, Spring.

77. **Grover, V. and S. Jeong** (1995) The implementation of business process re-engineering. *Journal of Management Information Systems*, **12** (1), 109–44.

78. **Davenport, T.** (1992) *Process Innovation: Re-engineering Work Through Information Technology*, Harvard University Press, Boston, MA.

79. **Davenport, T.** (1995) Will participative makeovers of business processes succeed where reengineering failed? *Planning Review*, 24.

80. **Camp, R.** (1989) *Benchmarking – The Search for Industry Best Practices that Lead to Superior Performance*, Quality Press, Milwaukee, WI.

81. **Zairi, M.** (1996) *Effective Benchmarking: Learning from the Best*, Chapman and Hall, London.

82. **Hill, F.** (1986) Quality circles in the UK: a longitudinal study. *Personnel Review*, **15** (3), 25–34.

83. **Lillrank, P. and N. Kano** (1990) *Continuous Improvement: Quality Control Circles in Japanese Industry*, University of Michigan Press, Ann Arbor.

84. **Swann, P., M. Prevezer, and D. Stout, eds** (1998) *The Dynamics of Industrial Clustering*, Oxford University Press, Oxford.

85. **Newell, S. and J. Swan** (1987) Trust and inter-organizational networking. *Human Relations*, **53** (10).

86. **Swan, J.** (2003) Knowledge, networking and innovation: developing an understanding of process. In L. Shavinina, ed., *International Handbook of Innovation*, Elsevier, New York.

87. **Dahlander, L. and D. Gann** (2008) How open is innovation? In J. Bessant and T. Venables, eds., *Creating Wealth from Knowledge*, Cheltenham, Edward Elgar.

88. **Leonard-Barton, D.** (1995) *Wellsprings of Knowledge: Building and Sustaining the Sources of Innovation*, Harvard Business School Press, Boston, MA.

89. **Hobday, M., H. Rush, and J. Bessant** (2005) Reaching the innovation frontier in Korea: a new corporate strategy dilemma. *Research Policy*, **33**, 1433–57.

90. **Robertson, A.** (1974) *The Lessons of Failure*, Macdonald, London.

91. **Lilien, G. and E. Yoon** (1989) Success and failure in innovation – a review of the literature. *IEEE Transactions on Engineering Management*, **36** (1), 3–10.

92. **Ernst, H.** (2002) Success factors of new product development: a review of the empirical literature. *International Journal of Management Reviews*, **4** (1), 1–40.

93. **Voss, C.** (1988) Success and failure in AMT. *International Journal of Technology Management*, **3** (3), 285–97.

94. **Kaplan, R.** (1988) *Relevance Lost*, Harvard Business School Press, Cambridge, MA.

95. **Adams, R., R. Phelps, and J. Bessant** (2006) Innovation management measurement: a review. *International Journal of Management Reviews*, **8** (1), 21–47.

96. **Smith, P. and D. Reinertsen** (1991) *Developing Products in Half the Time*, Van Nostrand Reinhold, New York.

97. **Tidd, J.** (1989) *Flexible Automation*, Frances Pinter, London.

98. **Jaikumar, R.** (1986) Post-industrial manufacturing. *Harvard Business Review*, November/December.

99. **Goffin, K. and R. Mitchell** (2005) *Innovation Management*, Pearson, London.

100. **Leonard-Barton, D.** (1992) The organization as learning laboratory. *Sloan Management Review*, **34** (1), 23–38.

101. **Rush, H., T. Brady, and M. Hobday** (1997) *Learning between Projects in Complex Systems*, Centre for the study of Complex Systems.

102. **Chiesa, V., P. Coughlan, and C. Voss** (1996) Development of a technical innovation audit. *Journal of Product Innovation Management*, **13** (2), 105–36.

103. **Francis, D.** (2001) *Developing Innovative Capability*, University of Brighton, Brighton.

104. **Johne, A. and P. Snelson** (1988) Auditing product innovation activities in manufacturing firms. *R&D Management*, **18** (3), 227–33.

105. **Prahalad, C. and G. Hamel** (1990) The core competence of the corporation. *Harvard Business Review*, **68** (3), 79–91.

106. **Cooper, R.** (2000) *Product Leadership*, Perseus Press, New York

107. **Teece, D.** (1998) Capturing value from knowledge assets: the new economy, markets for know-how, and intangible assets. *California Management Review*, **40** (3), 55–79.

108. **Rickards, T.** (1997) *Creativity and Problem Solving at Work*, Gower, Aldershot.

109. **Leonard, D. and W. Swap** (1999) *When Sparks Fly: Igniting Creativity in Groups*, Harvard Business School Press, Boston, MA.

110. **Dodgson, M. and J. Bessant** (1996) *Effective Innovation Policy*, International Thomson Business Press, London.

111. **Clark, K. and T. Fujimoto** (1992) *Product Development Performance*, Harvard Business School Press, Boston, MA.

112. **Thomas, R.** (1993) *New Product Development: Managing and Forecasting for Strategic Success*, John Wiley & Sons, Inc., New York.
113. **Von Hippel, E.** (2005) *The Democratization of Innovation*, MIT Press, Cambridge, MA.
114. **Dodgson, M., D. Gann, and A. Salter** (2005) *Think, Play, Do: Technology and Organization in the Emerging Innovation Process*, Oxford University Press, Oxford.
115. **Rothwell, R. and P. Gardiner** (1985) Invention, innovation, re-innovation and the role of the user. *Technovation*, **3**, 167–86.
116. **Nonaka, I. and M. Kenney** (1991) Towards a new theory of innovation management. *Journal of Engineering and Technology Management*, **8**, 67–83.

PART 2

Context

In this section we look at the key contextual issues around successful innovation management. In Chapter 3 we pick up the question: do we have an innovative organization? We examine the role which key concepts such as leadership, structure, communication and motivation play in building and sustaining a culture of focused creativity.

Chapter 4 looks at the question: do we have a clear innovation strategy? Is there a clear sense of where and how innovation will take the organization forward and is there a roadmap for this? Is the strategy shared and understood – and how can we ensure alignment of the various different innovation efforts across the organization? We examine the central role that capabilities and positions have in developing an innovation strategy.

CHAPTER 3

Building the innovative organization

'Innovation has nothing to do with how many R&D dollars you have . . . . it's not about money. It's about the people you have, how you're led, and how much you get it.'

(Steve Jobs, interview with *Fortune Magazine,* 1981)[1]

'People are our greatest asset.' This phrase – or variations on it – has become one of the clichés of management presentations, mission statements and annual reports throughout the world. Along with concepts like 'empowerment' and 'team working', it expresses a view of people being at the creative heart of the enterprise. But very often the reader of such words – and particularly those 'people' about whom they are written – may have a more cynical view, seeing organizations still operating as if people were part of the problem rather than the key to its solution.

In the field of innovation this theme is of central importance. It is clear from a wealth of psychological research that every human being comes with the capability to find and solve complex problems, and where such creative behaviour can be harnessed amongst a group of people with differing skills and perspectives extraordinary things can be achieved. We can easily think of examples. At the individual level, innovation has always been about exceptional characters who combine energy, enthusiasm and creative insight to invent and carry forward new concepts. James Dyson with his alternative approaches to domestic appliance design; Spence Silver, the 3M chemist who discovered the non-sticky adhesive behind 'Post-it' notes; and Shawn Fanning, the young programmer who wrote the Napster software and almost single-handedly shook the foundations of the music industry, are good illustrations of this.

Innovation is increasingly about teamwork and the creative combination of different disciplines and perspectives. Whether it is in designing a new car in half the time usually taken, bringing a new computer concept to market, establishing new ways of delivering old services such as banking, insurance or travel services, or putting men and women routinely into space, success comes from people working together in high-performance teams.

This effect, when multiplied across the organization, can yield surprising results. In his work on US companies, Pfeffer notes the strong correlation between proactive people management practices and the performance of firms in a variety of sectors.[2] A comprehensive review for the UK Chartered Institute of Personnel and Development suggested that '. . . more than 30 studies carried out in the UK and US since the early 1990s leave no room to doubt that there is a correlation between people management and business performance, that the relationship is positive, and that it is cumulative: the more and the more effective the practices, the better the result'.[3] Similar studies confirm the pattern in German firms.[4] In a knowledge economy where creativity is at a premium, people really are the most important assets which a firm possesses. The management challenge is how to go about building the kind of organizations in which such innovative behaviour can flourish.

This chapter deals with the creation and maintenance of an innovative organizational context, one whose structure and underlying culture – pattern of values and beliefs – support innovation. It is easy to find prescriptions for innovative organizations which highlight the need to eliminate stifling bureaucracy, unhelpful structures, brick walls blocking communication and other factors stopping good ideas getting through. But we must be careful not to fall into the chaos trap – not all innovation works in organic, loose, informal environments or 'skunk works' – and these types of organization can sometimes act against the interests of successful innovation. We need to determine appropriate organization – that is, the most suitable organization given the operating contingencies. Too little order and structure may be as bad as too much.

TABLE 3.1	Components of the innovative organization	
Component	**Key features**	**Example references**
Shared vision, leadership and the will to innovate	Clearly articulated and shared sense of purpose Stretching strategic intent 'Top management commitment'	31, 32, 69, 74
Appropriate structure	Organization design which enables creativity, learning and interaction. Not always a loose 'skunk works' model; key issue is finding appropriate balance between 'organic and mechanistic' options for particular contingencies	9–15, 106
Key individuals	Promoters, champions, gatekeepers and other roles which energize or facilitate innovation	9, 43, 44
Effective team working	Appropriate use of teams (at local, cross-functional and inter-organizational level) to solve problems, requires investment in team selection and building	59, 60, 67
High-involvement innovation	Participation in organization-wide continuous improvement activity	53, 56
Creative climate	Positive approach to creative ideas, supported by relevant motivation systems	69, 73, 74, 75
External focus	Internal and external customer orientation Extensive networking	81, 95, 98

Equally, 'innovative organization' implies more than a structure; it is an integrated set of components that work together to create and reinforce the kind of environment which enables innovation to flourish. Studies of innovative organizations have been extensive, although many can be criticized for taking a narrow view, or for placing too much emphasis on a single prescription like 'team working' or 'loose structures'. Nevertheless it is possible to draw out from these a set of components which appear linked with success; these are outlined in Table 3.1, and explored in the subsequent discussion.

3.1 Shared vision, leadership and the will to innovate

Innovation is essentially about learning and change and is often disruptive, risky and costly. So, as Case study 3.1 shows, it is not surprising that individuals and organizations develop many different cognitive, behavioural and structural ways of reinforcing the status quo. Innovation requires energy to overcome this inertia, and the determination to change the order of things. We see this in the case of individual inventors who champion their ideas against the odds, in entrepreneurs who build businesses through risk-taking behaviour and in organizations which manage to challenge the accepted rules of the game.

The converse is also true – the 'not-invented-here' problem, in which an organization fails to see the potential in a new idea, or decides that it does not fit with its current pattern of business. In other cases the need for change is perceived, but the strength or saliency of the threat is underestimated, such as the difficulties which IBM experienced in the early 1990s in responding to the emerging 'client–server' and network shift in computing away from mainframes, and this is a good example of a firm which believed it had seen and assessed the threat which nearly drove the company out of business. Similarly, General Motors found it difficult to appreciate and interpret the information about

CASE STUDY 3.1

Missing the boat . . .

On 10 March 1875 Alexander Graham Bell called to his assistant, 'Mr Watson, come here, I want you' – the surprising thing about the exchange being that it was the world's first telephone conversation. Excited by their discovery, they demonstrated their idea to senior executives at Western Union. Their written reply, a few days later, suggested that 'after careful consideration of your invention, which is a very interesting novelty, we have come to the conclusion that it has no commercial possibilities . . . we see no future for an electrical toy. . .' Within four years of the invention there were 50 000 telephones in the USA and within 20 years there were five million. In the same time the company which Bell formed, American Telephone and Telegraph (ATT), grew to become the largest corporation in the USA, with stock worth $1000 per share. The original patent (number 174455) became the single most valuable patent in history.

Source: Bryson, B. (1994) *Made in America*, Minerva, London.

Japanese competition, preferring to believe that their access in US markets was due to unfair trade policies rather than recognizing the fundamental need for process innovation, which the 'lean manufacturing' approach pioneered in Japan was bringing to the car industry.[5] Christensen, in his studies of disk drives,[6] and Tripsas and Gravetti, in their analysis of the problems Polaroid faced in making the transition to digital imaging, provide powerful evidence to show the difficulties established firms have in interpreting the signals associated with a new and potentially disruptive technology.[7]

This is also where the concept of 'core rigidities' becomes important.[8] We have become used to seeing core competencies as a source of strength within the organization, but the downside is that the mindset which is being highly competent in doing certain things can also block the organization from changing its mind. Thus ideas which challenge the status quo face an uphill struggle to gain acceptance; innovation requires considerable energy and enthusiasm to overcome barriers of this kind. One of the concerns in successful innovative organizations is finding ways to ensure that individuals with good ideas are able to progress them without having to leave the organization to do so.[9] Chapter 10 discusses the theme of 'intrapreneurship' in more detail.

Changing mindset and refocusing organizational energies requires the articulation of a new vision, and there are many cases where this kind of leadership is credited with starting or turning round organizations. Examples include Jack Welch of GE, Bill Gates (Microsoft), Steve Jobs (Pixar/Apple),[10] Andy Groves (Intel) and Richard Branson (Virgin).[11] Whilst we must be careful of vacuous expressions of 'mission' and 'vision', it is also clear that in cases like these there has been a clear sense of, and commitment to, shared organizational purpose arising from such leadership.

'Top management commitment' is a common prescription associated with successful innovation; the challenge is to translate the concept into reality by finding mechanisms which demonstrate and reinforce the sense of management involvement, commitment, enthusiasm and support. In particular, there needs to be long-term commitment to major projects, as opposed to seeking short-term returns. Since much of innovation is about uncertainty, it follows that returns may not emerge quickly and that there will be a need for 'patient money'. This may not always be easy to provide, especially when demands for shorter term gains by shareholders have to be reconciled with long-term technology development plans. One way of dealing with this problem is to focus not only on returns on investment but also on other considerations like future market penetration and growth or the strategic benefits which might accrue to having a more flexible or responsive production system (Chapter 4 discusses this theme in more detail). Box 3.1 and Case study 3.2 provide examples of such leadership.

Part of this pattern is also top management acceptance of risk. Innovation is inherently uncertain and will inevitably involve failures as well as successes. Successful technology management thus requires that the organization be prepared to take risks and to accept failure as an opportunity for learning and development. This is not to say that unnecessary risks should be taken – rather, as Robert Cooper suggests, the inherent uncertainty in innovation should be reduced where possible through the use of information collection and research.[12]

We should not confuse leadership and commitment with always being the active change agent. In many cases innovation happens in spite of the senior management within an organization, and success emerges as a result of guerrilla tactics rather than a frontal assault on the problem. Much has been made of the dramatic turnaround in IBM's fortunes under the leadership of Lou Gerstner who took the ailing giant firm from a crisis position to one of leadership in the IT services field and an acknowledged pioneer of e-business. But closer analysis reveals that the entry into e-business was the result of a bottom-up team initiative led by a programmer called Dave Grossman. It was his frustration with the lack of response from

---

BOX 3.1   Innovation leadership and climate

Organizations have traditionally conceived of leadership as an heroic attribute, appointing a few 'real' leaders to high-level senior positions in order to get them through difficult times. However, many observers and researchers are becoming cynical about this approach and are beginning to think about the need to recognize and utilize a wider range of leadership practices. Leadership needs to be conceived of as something that happens across functions and levels. New concepts and frameworks are needed in order to embrace this more inclusive approach to leadership.

For example, there is a great deal of writing about the fundamental difference between leadership and management. This literature abounds and has generally promoted the argument that leaders have vision and think creatively ('doing different'), while managers are merely drones and just focus on doing things better. This distinction has led to a general devaluation of management. Emerging work on styles of creativity and management suggests that it is useful to keep preference distinct from capacity. Creativity is present both when doing things differently and doing things better. This means that leadership and management may be two constructs on a continuum, rather than two opposing characteristics.

Our particular emphasis is on resolving the unnecessary and unproductive distinction that is made between leadership and management. When it comes to innovation and transformation, organizations need both sets of skills. We develop a model of innovation leadership that builds on past work, but adds some recent perspectives from the fields of change and innovation management, and personality and social psychology. This multidimensional view of leadership raises the issue of context as an important factor, beyond concern for task and people. This approach suggests the need for a third factor in assessing leadership behaviour, in addition to the traditional concerns for task and people. Therefore we integrate three dimensions of leadership: concern for task; concern for people; and concern for change.

One of the most important roles that leaders play within organizational settings is to create the climate for innovation. We identify the critical dimensions of the climate for innovation, and suggest how leaders might nurture these contexts for innovation.

Source: Isaksen, S. and J. Tidd (2006) *Meeting the Innovation Challenge: Leadership for Transformation and Growth,* John Wiley & Sons, Ltd, Chichester.

---

his line managers that eventually led to the establishment of a broad coalition of people within the company who were able to bring the idea into practice and establish IBM as a major e-business leader. The message for senior management is as much about leading through creating space and support within the organization as it is about direct involvement.

The contribution that leaders make to the performance of their organizations can be significant. Upper echelons theory argues that decisions and choices by top management have an influence on the performance of an organization (positive or negative!), through their assessment of the environment, strategic decision making and support for innovation. The results of different studies vary, but the reviews of research on leadership and performance suggest leadership directly influences around 15% of the differences found in performance of businesses, and contributes around an additional 35% through the choice of business strategy.[13] So directly and indirectly leadership can account for half of the variance in performance observed across organizations. The mechanisms through which leaders

CASE STUDY 3.2

The vision thing – How leadership contributes to transformational change

Moving from a diverse and clumsy conglomerate with origins in the wood and paper industry to the market leader position in mobile telephones is not easy. Yet the story of Nokia is one of managed transformation from a nineteenth-century timber firm to the fifth largest company in Europe, with 44 000 people employed in 14 countries, over a third of whom work on R&D or product design. Much of this transition – which, like many transformations, contained an element of luck – is attributed to the energy and vision of the CEO, Jorma Ollila, who took up this role in 1992 from the mobile phone division.

The transition was not easy – a series of problems, including logistics and availability of chips meant that the phone division made serious losses in 1995 and the stock value was cut in half. In order to meet this challenge Ollila effectively 'bet the company' disposing of almost all of its non-telecoms businesses (which ranged from television sets to toilet paper!) so that by 1995 90% of Nokia was concerned with telecommunications.

A similar pattern can be seen with the case of Siemens. Again with roots in the nineteenth century, Siemens grew to be one of the great names in electrical engineering and a major force in the German economy. But recent years have seen concerns about the company, criticizing it for a lack of focus and for being slow and unresponsive. Faced with this developing picture the company appointed a new board member in 1998 – Edward Krubasik – who came from outside the firm. Restructuring under his leadership has led to the divestment of nearly £10 billion of old businesses and to the repositioning of Siemens as a major IT and software player. In 1999 profits surged and the sales price tripled and 60% of the business is concerned with software. Perhaps most significant as an indicator of this new vision is the fact that Siemens employed 27 000 software engineers in 2000 – more even than Microsoft!

Source: Francis, D., J. Bessant and M. Hobday (2003) Managing radical organisational transformation. *Management Decision,* **41** (1), 18–31.

can influence performance include strategic decision making, often made under conditions of ambiguity and complexity, with limited information. At higher levels of management the problems to be solved are more likely to be ill-defined, demanding leaders to conceptualize more.

Researchers have identified a long list of characteristics that might have something to do with being effective in certain situations, which typically include the following traits:[14]

- bright, alert and intelligent
- seek responsibility and take charge
- skillful in their task domain
- administratively and socially competent
- energetic, active and resilient
- good communicators

Although these lists may describe some characteristics of some leaders in certain situations, measures of these traits yield highly inconsistent relationships with being a good leader.[15] In short, there is no brief and universal list of enduring traits that all good leaders must possess.

Studies in different contexts identify not only the technical expertise of leadership influencing group performance, but also broader cognitive ability, such as creative problem-solving and information-processing skills. For example, studies of groups facing novel, ill-defined problems, confirm that both expertise and cognitive-processing skills are key components of creative leadership, and are both associated with effective performance of creative groups.[16] Moreover, this combination of expertise and cognitive capacity is critical for the evaluation of others' ideas. A study of scientists found that they most valued their leader's inputs at the early stages of a new project, when they were formulating their ideas, and defining the problems, and later at the stage where they needed feedback and insights to the implications of their work. Therefore a key role of creative leadership in such environments is to provide feedback and evaluation, rather than to simply generate ideas.[17] This evaluative role is critical, but is typically seen as not being conducive to creativity and innovation, where the conventional advice is to suspend judgement to foster idea generation. Also, it suggests that the conventional linear view that evaluation follows idea generation may be wrong. Evaluation by creative leadership may precede idea generation and conceptual combination. This approach can support more conventional methods such as brainstorming and structured problem solving.

The quality and nature of the leader–member exchange (LMX) has also been found to influence the creativity of subordinates.[18] A study of 238 knowledge workers from 26 project teams in high-technology firms identified a number of positive aspects of LMX, including monitoring, clarifying and consulting, but also found that the frequency of negative LMX was as high as the positive, around a third of respondents reporting these.[19] Therefore LMX can either enhance or undermine subordinates' sense of competence and self-determination. However, analysis of exchanges perceived to be negative and positive revealed that it was typically how something was done rather than what was done, which suggests that task and relationship behaviours in leadership support and LMX are intimately intertwined, and that negative behaviours can have a disproportionate negative influence.

Intellectual stimulation by leaders has a stronger effect on organizational performance under conditions of perceived uncertainty. Intellectual stimulation includes behaviours that increase others' awareness of and interest in problems, and develops their propensity and ability to tackle problems in new ways. It is also associated with commitment to an organization.[20] Stratified system theory (SST) focuses on the cognitive aspects of leadership, and argues that conceptual capacity is associated with superior performance in strategic decision making where there is a need to integrate complex information and think abstractly in order to assess the environment. It also is likely to demand a combination of these problem-solving capabilities and social skills, as leaders will depend upon others to identify and implement solutions.[21] This suggests that under conditions of environmental uncertainty the contribution of leadership is not simply, or even primarily, to inspire or build confidence, but rather to solve problems and make appropriate strategic decisions.

Rafferty and Griffin propose other subdimensions to the concept of transformational leadership that may have a greater influence on creativity and innovation, including articulating a vision and inspirational communication.[20] They define a vision as 'the expression of an idealized picture of the future based around organizational values', and inspirational communication as 'the expression of positive and encouraging messages about the organization, and statements that build motivation and confidence'. They found that the expression of a vision has a negative effect on followers' confidence, unless accompanied with inspirational communication. Independent of vision, inspirational communication was associated with commitment and interpersonal helping behaviours. Mission awareness, associated with transformational leadership, has been found to predict the success of R&D projects, the degree to which depends on the stage of the project. For example, in the planning and conceptual stage mission awareness explained 67% of the subsequent project success.[22] Leadership clarity is associated with clear team objectives, high levels of participation, commitment to excellence, and support for innovation.[23] Leadership clarity, partly mediated by good team processes, is a good predictor of team innovation. The effects are significant, with good team processes predicting up to 37% of the variance in team innovation, and 17% of clarity of leadership. Conversely, a lack of clarity about or over leadership is negatively associated with team innovation. This suggests that the tendency to focus on the style of leadership may be premature, and that the initial focus should be on maximizing leadership clarity and minimizing leadership conflict, which appears to be critical in all cases. The specific style of leadership, however, is more contingent upon the team context and nature of task.

The creative leader needs to do much more than simply provide a passive, supportive role, to encourage creative followers. Perceptual measures of leaders' performance suggest that in a research environment the perception of leader's technical skill is the single best predictor of research group performance, with correlation in the 0.40–0.53 range.[24] Keller found that the type of project moderates the relationships between leadership style and project success, and found that transformational leadership was a stronger predictor in research projects than in development projects.[25] This strongly suggests that certain qualities of transformational leadership may be most appropriate under conditions of high complexity, uncertainty or novelty. Indeed, studies comparing the effects of leadership styles in research and administrative environments have found that transformational leadership has a greater impact on performance in a research environment than in an administrative, although the effect is positive in both cases, whereas a transactional style has a positive effect in an administrative context, but a negative effect in a research context.[26]

3.2 Appropriate organization structure

No matter how well developed the systems are for defining and developing innovative products and processes they are unlikely to succeed unless the surrounding organizational context is favourable. Achieving this is not easy, it involves creating the organizational structures and processes which enable technological change to thrive. For example, rigid hierarchical organizations in which there is little integration between functions and where communication tends to be top-down and one-way in character are unlikely to be very supportive of the smooth information flows and cross-functional cooperation recognized as being important success factors.

Much of the literature recognizes that organizational structures are influenced by the nature of tasks to be performed within the organization. In essence the less programmed and more uncertain the tasks, the greater the need for flexibility around the structuring of relationships.[27] For example, activities such as production, order processing, purchasing are characterized by decision making which is subject to little variation. (Indeed in some cases these decisions can be automated through employing particular decision rules embodied in computer systems.) But others require judgement and insight and vary considerably from day to day – and these include those decisions associated with innovation. Activities of this kind are unlikely to lend themselves to routine, structured and formalized relationships, but instead require flexibility and extensive interaction. Several writers have noted this difference between what have been termed 'programmed' and 'non-programmed' decisions and argued that the greater the level of non-programmed decision making, the more the organization needs a loose and flexible structure.[28]

Considerable work was done on this problem in the late 1950s by researchers Tom Burns and George Stalker, who outlined the characteristics of what they termed 'organic' and 'mechanistic' organizations.[29] The former are essentially environments suited to conditions of rapid change whilst the latter are more suited to stable conditions – although these represent poles on an ideal spectrum they do provide useful design guidelines about organizations for effective innovation. Other studies include those of Rosabeth Moss-Kanter,[30] and Hesselbein et al.[31]

We should be careful, however, in assuming that innovation is simply confined to R&D laboratories (where versions of this form of organization have often been present). Increasingly, innovation is becoming a corporate-wide task, involving production, marketing, administration, purchasing and many other functions. This provides strong pressure for widespread organizational change towards more organic models.

The relevance of Burns and Stalker's model can be seen in an increasing number of cases where organizations have restructured to become less mechanistic. For example, General Electric in the USA underwent a painful but ultimately successful transformation, moving away from a rigid and mechanistic structure to a looser and decentralized form.[11] ABB, the Swiss–Swedish engineering group, developed a particular approach to their global business based on operating as a federation of small businesses, each of which retained much of the organic character of small firms.[32] Other examples of radical changes in structure include the Brazilian white goods firm Semco and the Danish hearing aid company Oticon.[33] But again we need to be careful – what works under one set of circumstances may diminish in value under others. Whilst models such as that deployed by ABB helped at the time, later developments meant that these proved less appropriate and were insufficient to deal with new challenges emerging elsewhere in the business.

Related to this work has been another strand which looks at the relationship between different environments and organizational form. Once again the evidence suggests that the higher the uncertainty and complexity in the environment, the greater the need for flexible structures and processes to deal with it.[34] This partly explains why some fast-growing sectors, for example, electronics or biotechnology, are often associated with more organic organizational forms, whereas mature industries often involve more mechanistic arrangements.

One important study in this connection was that originally carried out by Lawrence and Lorsch looking at product innovation. Their work showed that innovation success in mature industries like food packaging and growing sectors like plastics depended on having structures which were sufficiently differentiated (in terms of internal specialist groups) to meet the needs of a diverse marketplace. But success also depended on having the ability to link these specialist groups together effectively so as to respond quickly to market signals; they reviewed several variants on coordination mechanisms, some of which were more or less effective than others. Better coordination was associated with more flexible structures capable of rapid response.[35]

We can see clear application of this principle in the current efforts to reduce 'time to market' in a range of businesses.[36] Rapid product innovation and improved customer responsiveness are being achieved through extensive organizational change programmes involving parallel working, early involvement of different functional specialists, closer market links and user involvement, and through the development of team working and other organizational aids to coordination.

Another strand of work, which has had a strong influence on the way we think about organizational design, was that originated by Joan Woodward associated with the nature of the industrial processes being carried out.[37] Her studies suggested that structures varied between industries with a relatively high degree of discretion (such as small batch manufacturing) through to those involving mass production where more hierarchical and heavily structured forms prevailed. Significantly, the process industries, although also capital intensive, allowed a higher level of discretion.

Other variables and combinations, which have been studied for their influence on structure, include size, age and company strategy.[38] The extensive debate on organization structure began to resolve itself into a 'contingency' model in the 1970s. In essence this view argues that there is no single 'best' structure, but that successful organizations tend to be those which develop the most suitable 'fit' between structure and operating contingencies. For example, it makes sense to structure an operation like McDonald's in a mechanistic and highly controlled form, in order to be able to replicate this model across the world and to deliver similar standards of product and service. But trying to develop a new computer operating system or genetically engineer a new drug would not be possible in such a structure.

Similarly, structures which enable a large international firm to carry out R&D simultaneously in several countries and to integrate their efforts through a series of procedures would be largely irrelevant and excessively bureaucratic for a small, high-tech start-up firm.

The Canadian writer Henry Mintzberg drew much of the work on structure together and proposed a series of archetypes which provide templates for the basic structural configurations into which firms are likely to fall.[39] These categories – and their implications for innovation management – are summarized in Table 3.2. Case study 3.3 gives an example of the importance of organizational structure and the need to find appropriate models.

TABLE 3.2	Mintzberg's structural archetypes	
Organization archetype	**Key features**	**Innovation implications**
Simple structure	Centralized organic type – centrally controlled but can respond quickly to changes in the environment. Usually small and often directly controlled by one person. Designed and controlled in the mind of the individual with whom decision-making authority rests. Strengths are speed of response and clarity of purpose. Weaknesses are the vulnerability to individual misjudgement or prejudice and resource limits on growth	Small start-ups in high technology – 'garage businesses' are often simple structures. Strengths are in energy, enthusiasm and entrepreneurial flair – simple structure innovating firms are often highly creative. Weaknesses are in long-term stability and growth, and overdependence on key people who may not always be moving in the right business direction
Machine bureaucracy	Centralized mechanistic organization, controlled centrally by systems. A structure designed like a complex machine with	Machine bureaucracies depend on specialists for innovation, and this is channelled into the overall design of the system. Examples include fast

(continued)

TABLE 3.2	(Continued)	
Organization archetype	**Key features**	**Innovation implications**
	people seen as cogs in the machine. Design stresses the function of the whole and specialization of the parts to the point where they are easily and quickly interchangeable. Their success comes from developing effective systems which simplify tasks and routinize behaviour. Strengths of such systems are the ability to handle complex integrated processes like vehicle assembly. Weaknesses are the potential for alienation of individuals and the build up of rigidities in inflexible systems	food (McDonald's), mass production (Ford) and large-scale retailing (Tesco), in each of which there is considerable innovation, but concentrated on specialists and impacting at the system level. Strengths of machine bureaucracies are their stability and their focus of technical skills on designing the systems for complex tasks. Weaknesses are their rigidities and inflexibility in the face of rapid change, and the limits on innovation arising from non-specialists
Divisionalized form	Decentralized organic form designed to adapt to local environmental challenges. Typically associated with larger organizations, this model involves specialization into semi-independent units. Examples would be strategic business units or operating divisions. Strengths of such a form are the ability to attack particular niches (regional, market, product, etc.) whilst drawing on central support. Weaknesses are the internal frictions between divisions and the centre	Innovation here often follows a 'core and periphery' model in which R&D of interest to the generic nature is carried out in central facilities whilst more applied and specific work is carried out within the divisions. Strengths of this model include the ability to concentrate on developing competency in specific niches and to mobilize and share knowledge gained across the rest of the organization. Weaknesses include the 'centrifugal pull' away from central R&D towards applied local efforts and the friction and competition between divisions which inhibits sharing of knowledge
Professional bureaucracy	Decentralized mechanistic form, with power located with individuals but coordination via standards. This kind of organization is characterized by relatively high levels of professional skills, and	This kind of structure typifies design and innovation consulting activity within and outside organizations. The formal R&D, IT or engineering groups would be good examples of this, where technical and specialist excellence is valued. Strengths of

(continued)

TABLE 3.2	(Continued)	
Organization archetype	**Key features**	**Innovation implications**
	is typified by specialist teams in consultancies, hospitals or legal firms. Control is largely achieved through consensus on standards ('professionalism') and individuals possess a high degree of autonomy. Strengths of such an organization include high levels of professional skill and the ability to bring teams together	this model are in technical ability and professional standards. Weaknesses include difficulty of managing individuals with high autonomy and knowledge power
Adhocracy	Project type of organization designed to deal with instability and complexity. Adhocracies are not always long-lived, but offer a high degree of flexibility. Team-based, with high levels of individual skill but also ability to work together. Internal rules and structure are minimal and subordinate to getting the job done. Strengths of the model are its ability to cope with high levels of uncertainty and its creativity. Weaknesses include the inability to work together effectively due to unresolved conflicts, and a lack of control due to lack of formal structures or standards	This is the form most commonly associated with innovative project teams – for example, in new product development or major process change. The NASA project organization was one of the most effective adhocracies in the programme to land a man on the moon; significantly the organization changed its structure almost once a year during the 10-year programme, to ensure it was able to respond to the changing and uncertain nature of the project. Strengths of adhocracies are the high levels of creativity and flexibility – the 'skunk works' model advocated in the literature. Weaknesses include lack of control and overcommitment to the project at the expense of the wider organization
Mission-oriented	Emergent model associated with shared common values. This kind of organization is held together by members sharing a common and often altruistic purpose – for example, in voluntary and charity organizations. Strengths are high commitment and the ability of individuals to take	Mission-driven innovation can be highly successful, but requires energy and a clearly articulated sense of purpose. Aspects of total quality management and other value-driven organizational principles are associated with such organizations, with a quest for continuous improvement driven from within rather than in response to external stimulus. Strengths lie in the

(continued)

TABLE 3.2	(Continued)	
Organization archetype	Key features	Innovation implications
	initiatives without reference to others because of shared views about the overall goal. Weaknesses include lack of control and formal sanctions	clear sense of common purpose and the empowerment of individuals to take initiatives in that direction. Weaknesses lie in overdependence on key visionaries to provide clear purpose, and lack of 'buy-in' to the corporate mission

CASE STUDY 3.3

The emergence of mass production

Perhaps the most significant area in which there is a change of perspective is in the role of human resources. Early models of organization were strongly influenced by the work of Frederick Taylor and his principles of 'scientific management'. These ideas – used extensively in the development of mass production industries like automobile manufacture – essentially saw the organization problem as one which required the use of analytical methods to arrive at the 'best' way of carrying out the organization's tasks. This led to an essentially mechanistic model in which people were often seen as cogs in a bigger machine, with clearly defined limits to what they should and shouldn't do. The image presented by Charlie Chaplin in *Modern Times* was only slightly exaggerated; in the car industry the average task cycle for most workers was less than two minutes.

The advantages of this system for the mass production of a small range of goods were clear: productivity increases often ran into three figures with the adoption of this approach. For example, Ford's first assembly line, installed in 1913 for flywheel assembly, saw the assembly time fall from 20 man-minutes to five, and by 1914 three lines were being used in the chassis department to reduce assembly time from around 12 hours to less than two. But its limitations lay in the ability of the system to change, and in the capacity for innovation. By effectively restricting innovation to a few specialists, an important source of creative problem solving, in terms of product and process development, was effectively cut off.

The experience of Ford and others highlights the point that there is no single 'best' kind of organization; the key is to ensure congruence between underlying values and beliefs and the organization which enables innovative routines to flourish. For example, whilst the 'skunk works' model may be appropriate to US product development organizations, it may be inappropriate in Japan where a more disciplined and structured form is needed. Equally some successful innovative organizations are based on team working whereas others are built around key individuals – in both cases reflecting underlying beliefs about how innovation works in those particular organizations. Similarly successful innovation can take place within strongly bureaucratic organizations just as well as in those in which there is a much looser structure – providing that there is underlying congruence between these structures and the innovative behavioural routines.

The increasing importance of innovation and the consequent experience of high levels of change across the organization have begun to pose a challenge for organizational structures normally configured for stability. Thus traditional machine bureaucracies – typified by the car assembly factory – are becoming more hybrid in nature, tending towards what might be termed a 'machine adhocracy' with creativity and flexibility (within limits) being actively encouraged. The case of 'lean production' with its emphasis on team working, participation in problem solving, flexible cells and flattening of hierarchies is a good example, where there is significant loosening of the original model to enhance innovativeness.[40]

The key challenge here for managing innovation remains one of *fit* – of getting the most appropriate structural form for the particular circumstances. Another view of structure is that it is an artefact of what people believe and how they behave; if there is good fit, structure will enable and reinforce innovative behaviour. If it is contradictory to these beliefs – for example, restricting communication, stressing hierarchy – then it is likely to act as a brake on creativity and innovation.

3.3 Key individuals

Another important element is the presence of key enabling figures. The uncertainty and complexity involved in innovation mean that many promising inventions die before they make it to the outside world. One way round this problem is if there is a key individual (or sometimes a group of people) who is prepared to champion its cause and to provide some energy and enthusiasm to help it through the organizational system. Such key figures or project champions have been associated with many famous innovations – for example, the development of Pilkington's float glass process or Edwin Land and the Polaroid photographic system.[41] Case study 3.4 gives an example.

There are, in fact, several roles which key figures can play, which have a bearing on the outcome of a project. First, there is the source of critical technical knowledge – often the inventor or team leader responsible for an invention. They will have the breadth of understanding of the technology behind the innovation and the ability to solve the many development problems likely to emerge in the long haul from laboratory or drawing board to full scale. The contribution here is not only of technical knowledge; it also involves inspiration when particular technological problems appear insoluble, and motivation and commitment.

CASE STUDY 3.4

Bags of ideas – the case of James Dyson

In October 2000 the air inside Court 58 of the Royal Courts of Justice in London rang with terms like 'bagless dust collection', 'cyclone technology', 'triple vortex' and 'dual cyclone' as one of the most bitter of patent battles in recent years was brought to a conclusion. On one side was Hoover, a multinational firm with the eponymous vacuum suction sweeper at the heart of a consumer appliance empire. On the other a lone inventor – James Dyson – who had pioneered a new approach to the humble task of house cleaning and then seen his efforts threatened by an apparent imitation by Hoover. Eventually the court ruled in Dyson's favour.

This represented the culmination of a long and difficult journey which Dyson travelled in bringing his ideas to a wary marketplace. It began in 1979 when Dyson was using, ironically, a Hoover Junior vacuum cleaner to dust the house. He was struck by the inefficiency of a system, which effectively reduced its capability to suck the more it was used since the bag became clogged with dust. He tried various improvements such as a finer mesh filter bag but the results were not promising. The breakthrough came with the idea of using industrial cyclone technology applied in a new way – to the problem of domestic cleaners.

Dyson was already an inventor with some track record and one of his products was a wheelbarrow which used a ball instead of a front wheel. In order to spray the black dust paint in a powder coating plant a cyclone was installed – a well-established engineering solution to the problem of dust extraction. Essentially a mini-tornado is created within a shell and the air in the vortex moves so fast that particles of dust are forced to the edge where they can be collected whilst clean air moves to the centre. Dyson began to ask why the principle could not be applied in vacuum cleaners – and soon found out. His early experiments – with the Hoover – were not entirely successful but eventually he applied for a patent in 1980 for a vacuum cleaning appliance using cyclone technology.

It took another four years and 5127 prototypes and even then he could not patent the application of a single cyclone since that would only represent an improvement on an existing and proven technology. He had to develop a dual cyclone system which used the first to separate out large items of domestic refuse – cigarette ends, dog hairs, cornflakes, etc. – and the second to pick up the finer dust particles. But having proved the technology he found a distinct cold shoulder on the part of the existing vacuum cleaner industry represented by firms like Hoover, Philips and Electrolux. In typical examples of the 'not-invented-here' effect they remained committed to the idea of vacuum cleaners using bags and were unhappy with bagless technology. (This is not entirely surprising since suppliers such as Electrolux make a significant income on selling the replacement bags for its vacuum cleaners.)

Eventually Dyson began the hard work of raising the funds to start his own business – and it gradually paid off. Launched in 1993 – 14 years after the initial idea – Dyson now runs a design-driven business worth around £530 million and has a number of product variants in its vacuum cleaner range; other products under development aim to re-examine domestic appliances like washing machines and dishwashers to try and bring similar new ideas into play. The basic dual cyclone cleaner was one of the products identified by the UK Design Council as one of its 'millennium products'.

Perhaps the greatest accolade though is the fact that the vacuum cleaner giants like Hoover eventually saw the potential and began developing their own versions. Dyson has once again shown the role of the individual champion in innovation – and that success depends on more than just a good idea. Edison's famous comment that it is '1% inspiration and 99% perspiration' seems an apt motto here!

Source: Dyson, J. (1997) *Against the Odds*, Orion, London.

Influential though such technical champions might be, they may not be able to help an innovation progress unaided through the organization. Not all problems are technical in nature; other issues such as procuring resources or convincing sceptical or hostile critics elsewhere in the organization may need to be dealt with. Here our second key role emerges – that of organizational sponsor.

Typically this person has power and influence and is able to pull the various strings of the organization (often from a seat on the board); in this way many of the obstacles to an innovation's progress can be removed or the path at least smoothed. Such sponsors do not necessarily need to have a detailed technical knowledge of the innovation (although this is clearly an asset), but they do need to believe in its potential.

Recent exploration of the product development process has highlighted the important role played by the team members and in particular the project team leader. There are close parallels to the champion model: influential roles range from what Clark and Fujimoto call 'heavyweight' project managers who are deeply involved and have the organizational power to make sure things come together, through to the 'lightweight' project manager whose involvement is more distant. Research on Japanese product development highlights the importance of the *shusha* or team leader; in some companies (such as Honda) the *shusha* is empowered to override even the decisions and views of the chief executive![42] The important message here is to match the choice of project manager type to the requirements of the situation – and not to use the 'sledgehammer' of a heavyweight manager for a simple task.

Key roles are not just on the technical and project management side: studies of innovation (going right back to Project SAPPHO and its replications) also highlighted the importance of the 'business innovator', someone who could represent and bring to bear the broader market or user perspective.[43]

Although innovation history is full of examples where such key individuals – acting alone or in tandem – have had a marked influence on success, we should not forget that there is a downside as well. Negative champions – project assassins – can also be identified, whose influence on the outcome of an innovation project is also significant but in the direction of killing it off. For example, there may be internal political reasons why some parts of an organization do not wish for a particular innovation to progress – and through placing someone on the project team or through lobbying at board level or in other ways a number of obstacles can be placed in its way. Equally, our technical champion may not always be prepared to let go of their pet idea, even if the rest of the organization has decided that it is not a sensible direction in which to progress. Their ability to mobilize support and enthusiasm and to surmount obstacles within the organization can sometimes lead to wrong directions being pursued, or the continued chasing up what many in the organization see as a blind alley.

One other type of key individual is worth mentioning, that of the 'technological gatekeeper'. Innovation is about information and, as we saw above, success is strongly associated with good information flow and communication. Research has shown that such networking is often enabled by key individuals within the organization's informal structure who act as 'gatekeepers' – collecting information from various sources and passing it on to the relevant people who will be best able or most interested to use it. Thomas Allen, working at MIT, made a detailed study of the behaviour of engineers during the large-scale technological developments surrounding the Apollo rocket programme. His studies highlighted the importance of informal communications in successful innovation, and drew particular attention to gatekeepers – who were not always in formal information management positions but who were well connected in the informal social structure of the organization – as key players in the process.[44]

This role is becoming of increasing importance in the field of knowledge management where there is growing recognition that enabling effective sharing and communication of valuable knowledge resources is not simply something which can be accomplished by advanced IT and clever software – there is a strong interpersonal element.[45] Such approaches become particularly important in distributed or virtual teams where 'managing knowledge spaces' and the flows across them are of significance.[46]

3.4 High involvement in innovation

Whereas innovation is often seen as the province of specialists in R&D, marketing, design or IT, the underlying creative skills and problem-solving abilities are possessed by everyone. If mechanisms can be found to focus such abilities on a regular basis across the entire company, the resulting innovative potential is enormous. Although each individual may only be able to develop limited, incremental innovations, the sum of these efforts can have far-reaching impacts.

A good illustration of this is the 'quality miracle' which was worked by the Japanese manufacturing industry in the post-war years, and which owed much to what they term *kaizen* – continuous improvement. Firms like Toyota and Matsushita receive millions of suggestions for improvements every year from their employees – and the vast majority of these are implemented.[47] Western firms have done much to close this gap in recent years. For example, a study of firms in the UK, which have acquired the 'Investors in People' (IiP) award (an externally assessed review of employee involvement practices), shows a correlation between this and higher business performance. Such businesses have a higher rate of return on capital (RRC), higher turnover/sales per employee and have higher profits per employee.

Individual case studies confirm this pattern in a number of countries. As one UK manager[3] put it, 'Our operating costs are reducing year on year due to improved efficiencies. We have seen a 35% reduction in costs within two and a half years by improving quality. There are an average of 21 ideas per employee today compared to nil in 1990. Our people have accomplished this.' Case study 3.5 gives another example.

Although high-involvement schemes of this kind received considerable publicity in the late twentieth century, associated with total quality management and lean production, they are not a new concept. For example, Denny's Shipyard in Dumbarton, Scotland had a system which asked workers (and rewarded them for) 'any change by which work is rendered either superior in quality or more economical in cost' – back in 1871. John Patterson, founder of the National Cash Register Company in the USA, started a suggestion and reward scheme aimed at harnessing what he called 'the hundred-headed brain' around 1894.

CASE STUDY 3.5

High involvement in innovation

At first sight XYZ systems does not appear to be anyone's idea of a 'world-class' manufacturing outfit. Set in a small town in the Midlands with a predominantly agricultural industry, XYZ employs around 30 people producing gauges and other measuring devices for the forecourts of filling stations. Its products are used to monitor and measure levels and other parameters in the big fuel tanks underneath the stations, and on the tankers which deliver to them. Despite its small size (although it is part of a larger but decentralized group) XYZ has managed to command around 80% of the European market. Its processes are competitive against even large manufacturers; its delivery and service level the envy of the industry. It has a fistful of awards for its quality and yet manages to do this across a wide range of products some dating back 30 years, which still need service and repair. XYZ uses technologies from complex electronics and remote sensing right down to basics – they still make a wooden measuring stick, for example.

Its success can be gauged from profitability figures but also from the many awards it has received, and continues to receive, as one of the best factories in the UK.

Yet if you go through the doors of XYZ you would have to look hard for the physical evidence of how the company achieved this enviable position. This is not a highly automated business – it would not be appropriate. Nor is it laid out in modern facilities; instead they have clearly made much of their existing environment and organized it and themselves to best effect.

Where does the difference lie? Fundamentally in the approach taken with the workforce. This is an organization where training matters – investment is well above the average and everyone receives a significant training input, not only in their own particular skills area but also across a wide range of tasks and skills. One consequence of this is that the workforce is very flexible; having been trained to carry out most of the operations, they can quickly move to where they are most needed. The payment system encourages such cooperation, with its simple structure and emphasis on payment for skill, quality and team working. The strategic targets are clear and simple, and are discussed with everyone before being broken down into a series of small manageable improvement projects in a process of policy deployment. All around the works there are copies of the 'bowling chart' which sets out simply – like a tenpin bowling score sheet – the tasks to be worked on as improvement projects and how they could contribute to the overall strategic aims of the business. And if they achieve or exceed those strategic targets – then everyone gains through a profit-sharing and employee ownership scheme.

Being a small firm there is little in the way of hierarchy but the sense of team working is heightened by active leadership and encouragement to discuss and explore issues together – and it doesn't hurt that the operations director practises a form of MBWA – management by walking about!

Perhaps the real secret lies in the way in which people feel enabled to find and solve problems, often experimenting with different solutions and frequently failing – but at least learning and sharing that information for others to build on. Walking round the factory it is clear that this place isn't standing still – whilst major investment in new machines is not an everyday thing, little improvement projects – *kaizens* as they call them – are everywhere. More significant is the fact that the operations director is often surprised by what he finds people doing – it is clear that he has not got a detailed idea of which projects people are working on and what they are doing. But if you ask him if this worries him the answer is clear – and challenging. 'No, it doesn't bother me that I don't know in detail what's going on. They all know the strategy, they all have a clear idea of what we have to do (via the "bowling charts"). They've all been trained, they know how to run improvement projects and they work as a team. And I trust them . . .'

Since much of such employee involvement in innovation focuses on incremental change it is tempting to see its effects as marginal. Studies show, however, that when taken over an extended period it is a significant factor in the strategic development of the organization.[48]

Underpinning such continuous incremental innovation is the organizational culture to support and encourage over the long term. This simple point has been recognized in a number of different fields, all of which converge around the view that higher levels of participation in innovation represent a competitive advantage. For example:

- In the field of quality management it became clear that major advantages could accrue from better and more consistent quality in products and services. Crosby's work on quality costs suggested the scale of the potential savings (typically 20–40% of total sales revenue), and the experience of many Japanese manufacturers during the post-war period provide convincing arguments in favour of this approach.[49]
- The concept of 'lean thinking' has diffused widely during the past 20 years and is now applied in manufacturing and services as diverse as chemicals production, hospital management and supermarket retailing.[50] It originally emerged from detailed studies of assembly plants in the car industry which highlighted significant differences between the best and the average plants along a range of dimensions, including productivity, quality and time. Efforts to identify the source of these significant advantages revealed that the major differences lay not in higher levels of capital investment or more modern equipment, but in the ways in which production was organized and managed. The authors of the study concluded:

> . . . our findings were eye-opening. The Japanese plants require one-half the effort of the American luxury-car plants, half the effort of the best European plant, a quarter of the effort of the average European plant, and one-sixth the effort of the worst European luxury car producer. At the same time, the Japanese plant greatly exceeds the quality level of all plants except one in Europe – and this European plant required four times the effort of the Japanese plant to assemble a comparable product . . .

- Central to this alternative model was an emphasis on team working and participation in innovation.
- The principles underlying 'lean thinking' had originated in experiences with what were loosely called 'Japanese manufacturing techniques'.[51] This bundle of approaches (which included umbrella ideas like 'just-in-time' and specific techniques like poke yoke) were credited with having helped Japanese manufacturers gain significant competitive edge in sectors as diverse as electronics, motor vehicles and steel making.[52] Underpinning these techniques was a philosophy that stressed high levels of employee involvement in the innovation process, particularly through sustained incremental problem solving – *kaizen*.[53]

The transferability of such ideas between locations and into different application areas has also been extensively researched. It is clear from these studies that the principles of 'lean' manufacturing can be extended into supply and distribution chains, into product development and R&D and into service activities and operations.[54] Nor is there any particular barrier in terms of national culture: high-involvement approaches to innovation have been successfully transplanted to a number of different locations (see Case study 3.6).[55]

So there is a considerable weight of experience now available to support the view that enhanced performance can and does result from increasing involvement in innovation through 'high-involvement innovation' (HII). Table 3.3 provides evidence of the effects on performance. But there is also a secondary effect, which should not be underestimated: the more people are involved in change, the more receptive they become to change itself. Since the turbulent nature of most organizational environments is such that increasing levels of change are becoming the norm, involvement of employees in HII programmes may provide a powerful aid to effective management of change.

Growing recognition of the potential has moved the management question away from whether or not to try out employee involvement to one of 'how to make it happen?' The difficulty is less about

TABLE 3.3	Performance of IiP companies against others		
	Average company	Investors company	Gain
Return on capital	9.21%	16.27%	77%
Turnover/sales per employee	£64912	£86625	33%
Profit per employee	£1815	£3198	76%

Source: Pfeffer, J. (1998) The Human Equation: Building Profits by Putting People First, Harvard Business School Press, Boston, MA.

CASE STUDY 3.6

Diffusion of high-involvement innovation

How far has this approach diffused? Why do organizations choose to develop it? What benefits do they receive? And what barriers prevent them moving further along the road towards high involvement?

Questions like these provided the motivation for a large survey carried out in a number of European countries and replicated in Australia during the late 1990s. It was one of the fruits of a cooperative research network which was established to share experiences and diffuse good practice in the area of high-involvement innovation. The survey involved over 1000 organizations in a total of seven countries and provides a useful map of the take-up and experience with high-involvement innovation. (The survey only covered manufacturing although follow-up work is looking at services as well.) Some of the key findings were:

- Overall around 80% of organizations were aware of the concept and its relevance, but its actual implementation, particularly in more developed forms, involved around half of the firms.
- The average number of years which firms had been working with high-involvement innovation on a systematic basis was 3.8, supporting the view that this is not a 'quick fix' but something to be undertaken as a major strategic commitment. Indeed, those firms which were classified as 'CI innovators' – operating well-developed high-involvement systems – had been working on this development for an average of nearly seven years.
- High involvement is still something of a misnomer for many firms, with the bulk of efforts concentrated on shop-floor activities as opposed to other parts of the organization. There is a clear link between the level of maturity and development of high involvement here – the 'CI innovators' group was much more likely to have spread the practices across the organization as a whole.
- Motives for making the journey down this road vary widely but cluster particularly around the themes of quality improvement, cost reduction and productivity improvement.

- In terms of the outcome of high-involvement innovation there is clear evidence of significant activity, with an average per capita rate of suggestions of 43 per year of which around half were actually implemented. This is a difficult figure since it reflects differences in measurement and definition but it does support the view that there is significant potential in workforces across a wide geographical range – it is not simply a Japanese phenomenon. Firms in the sample also reported indirect benefits arising from this including improved morale and motivation, and a more positive attitude towards change.

  What these suggestions can do to improve performance is, of course, the critical question and the evidence from the survey suggests that key strategic targets were being impacted upon.

- On average improvements of around 15% were reported in process areas like quality, delivery, manufacturing lead time, and overall productivity, and there was also an average of 8% improvement in the area of product cost. Of significance is the correlation between performance improvements reported and the maturity of the firm in terms of high-involvement behaviour. The 'CI innovators' – those which had made most progress towards establishing high involvement as 'the way we do things around here' were also the group with the largest reported gains – averaging between 19 and 21% in the above process areas.

Performance areas (% change)	UK	SE	N	NL	FI	DK	Australia	Average across sample (n = 754 responses)
Productivity improvement	19	15	20	14	15	12	16	15
Quality improvement	17	14	17	9	15	15	19	16
Delivery performance improvement	22	12	18	16	18	13	15	16
Lead time reduction	25	16	24	19	14	5	12	15
Product cost reduction	9	9	15	10	8	5	7	8

- Almost all high-involvement innovation activities take place on an 'in-line' basis – that is, as part of the normal working pattern rather than as a voluntary 'off-line' activity. Most of this activity takes place in some form of group work although around a third of the activity is on an individual basis.

- To support this there is widespread use of tools and techniques particularly those linked to problem finding and solving which around 80% of the sample reported using. Beyond this there is extensive use of tools for quality management, process mapping and idea generation, although more specialized techniques like statistical process control or quality function deployment are less widespread. Perhaps more significant is the fact that even with the case of general problem-finding and -solving tools only around a third of staff had been formally trained in their use.

Source: Adapted from Huselid, M. (1995) The impact of human resource management practices on turnover, productivity and corporate financial performance. *Academy of Management Journal*, **38**, 647–56.

getting started than about keeping it going long enough to make a real difference. Many organizations have experience of starting the process – getting an initial surge of ideas and enthusiasm during a 'honeymoon' period – and then seeing it gradually ebb away until there is little or no HII activity. We shouldn't really be surprised at this – clearly if we want to change the ways in which people think and behave on a long-term basis then it's going to need a strategic development programme to make it happen. A quick 'sheep dip' of training plus a bit of enthusiastic arm waving from the managing director isn't likely to do much in the way of fundamentally changing 'the way we do things around here' – the underlying culture – of the organization.

A roadmap for the journey

Research on implementing HII suggests that there are a number of stages in this journey, progressing in terms of the development of systems and capability to involve people and also in terms of the bottom-line benefits.[56] Each of these takes time to move through, and there is no guarantee that organizations will progress to the next level. Moving on means having to find ways of overcoming the particular obstacles associated with different stages (see Figure 3.1). The case of Hosiden provides an example of challenges organizations face at these different stages.

The first stage – level 1 – is what we might call 'unconscious HII'. There is little, if any, HII activity going on, and when it does happen it is essentially random in nature and occasional in frequency. People do help to solve problems from time to time – for example, they will pull together to iron out problems with a new system or working procedure, or getting the bugs out of a new product. But there is no formal attempt to mobilize or build on this activity, and many organizations may actively restrict the opportunities for it to take place. The normal state is one in which HII is not looked for, not recognized, not supported – and often, not even noticed. Not surprisingly, there is little impact associated with this kind of change.

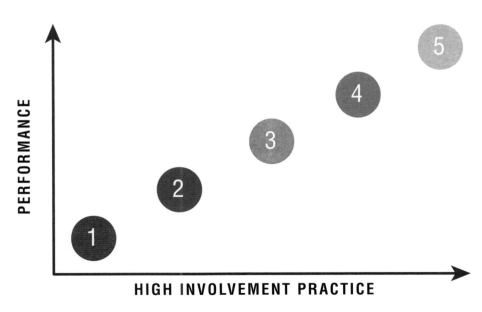

FIGURE 3.1: The five-stage high-involvement innovation model

Level 2, on the other hand, represents an organization's first serious attempts to mobilize HII. It involves setting up a formal process for finding and solving problems in a structured and systematic way – and training and encouraging people to use it. Supporting this will be some form of reward/recognition arrangement to motivate and encourage continued participation. Ideas will be managed through some form of system for processing and progressing as many as possible and handling those that cannot be implemented. Underpinning the whole set-up will be an infrastructure of appropriate mechanisms (teams, task forces or whatever), facilitators and some form of steering group to enable HII to take place and to monitor and adjust its operation over time. None of this can happen without top management support and commitment of resources to back that up.

Level 2 is all about establishing the habit of HII within at least part of the organization. It certainly contributes improvements but these may lack focus and are often concentrated at a local level, having minimal impact on more strategic concerns of the organization. The danger in such HII is that, once having established the habit of HII, it may lack any clear target and begin to fall away. In order to maintain progress there is a need to move to the next level of HII – concerned with strategic focus and systematic improvement.

Level 3 involves coupling the HII habit to the strategic goals of the organization such that all the various local-level improvement activities of teams and individuals can be aligned. In order to do this two key behaviours need to be added to the basic suite – those of strategy deployment and of monitoring and measuring. Strategy (or policy) deployment involves communicating the overall strategy of the organization and breaking it down into manageable objectives towards which HII activities in different areas can be targeted. Linked to this is the need to learn to monitor and measure the performance of a process and use this to drive the continuous improvement cycle.

Level 3 activity represents the point at which HII makes a significant impact on the bottom line – for example, in reducing throughput times, scrap rates, excess inventory, etc. It is particularly effective in conjunction with efforts to achieve external measurable standards (such as ISO 9000) where the disciplines of monitoring and measurement provide drivers for eliminating variation and tracking down root cause problems. The majority of 'success stories' in HII can be found at this level – but it is not the end of the journey.

One of the limits of level 3 HII is that the direction of activity is still largely set by management and within prescribed limits. Activities may take place at different levels, from individuals through small groups to cross-functional teams, but they are still largely responsive and steered externally. The move to level 4 introduces a new element – that of 'empowerment' of individuals and groups to experiment and innovate on their own initiative.

Clearly this is not a step to be taken lightly, and there are many situations where it would be inappropriate – for example, where established procedures are safety critical. But the principle of 'internally directed' HII as opposed to externally steered activity is important, since it allows for the open-ended learning behavior, which we normally associate with professional research scientists and engineers. It requires a high degree of understanding of, and commitment to, the overall strategic objectives, together with training to a high level to enable effective experimentation. It is at this point that the kinds of 'fast-learning' organizations described in some 'state-of-the-art' innovative company case studies can be found – places where everyone is a researcher and where knowledge is widely shared and used.

Level 5 is a notional end-point for the journey – a condition where everyone is fully involved in experimenting and improving things, in sharing knowledge and in creating an active learning organization. Table 3.4 illustrates the key elements in each stage. In the end the task is one of building a shared set of values which bind people in the organization together and enable them to participate in its development. As one manager put it in a UK study, '. . . we never use the word empowerment! You can't empower people – you can only create the climate and structure in which they will take responsibility. . .'[30] Case study 3.7 provides an illustration of Dutton Engineering, but for another excellent example of high-involvement innovation in action, see the video of Veedor Root.

TABLE 3.4	Stages in the evolution of HII capability
Stage of development	**Typical characteristics**
1. 'Natural'/background HII	Problem-solving random
	No formal efforts or structure
	Occasional bursts punctuated by inactivity and non-participation
	Dominant mode of problem solving is by specialists
	Short-term benefits
	No strategic impact
2. Structured HII	Formal attempts to create and sustain HII
	Use of a formal problem-solving process
	Use of participation
	Training in basic HII tools
	Structured idea management system
	Recognition system
	Often parallel system to operations
3. Goal-oriented HII	All of the above, plus formal deployment of strategic goals
	Monitoring and measurement of HII against these goals
	In-line system
4. Proactive/empowered HII	All of the above, plus responsibility for mechanisms, timing, etc. devolved to problem-solving unit
	Internally directed rather than externally directed HII
	High levels of experimentation
5. Full HII capability – the learning organization	HII as the dominant way of life
	Automatic capture and sharing of learning
	Everyone actively involved in innovation process
	Incremental and radical innovation.

CASE STUDY 3.7

Creating high involvement innovation conditions

Dutton Engineering does not, at first sight, seem a likely candidate for world class. A small firm with 28 employees, specializing in steel cases for electronic equipment, it ought to be amongst the ranks of hand-to-mouth metal-bashers of the kind you can find all round the world. Yet Dutton has been doubling its turnover, sales per employee have doubled in an eight-year period, rejects are down from 10% to 0.7%, and over 99% of deliveries are made within 24 hours – compared to only 60% being achieved within one week a few years ago. This transformation has not come overnight – the process started in 1989 – but it has clearly been successful and Dutton are now held up as an example to others of how typical small engineering firms can change.

At the heart of the transformation which Ken Lewis, the original founder and architect of the change, has set in train is a commitment to improvements through people. The workforce is organized into four teams who manage themselves, setting work schedules, dealing with their own customers, costing their own orders and even setting their pay! The company has moved from traditional weekly pay to a system of 'annualized hours' where they contract to work for 1770 hours in year – and tailor this flexibly to the needs of the business with its peaks and troughs of activity. There is a high level of contribution to problem solving, encouraged by a simple reward system which pays £5–15 for bright ideas, and by a bonus scheme whereby 20% of profits are shared.

Source: Lewis, K. and S. Lytton (2000) *How to Transform Your Company*, Management Books 2000, London.

3.5 Effective team working

'It takes five years to develop a new car in this country. Heck, we won World War 2 in four years . . .' Ross Perot's critical comment on the state of the US car industry in the late 1980s captured some of the frustration with existing ways of designing and building cars. In the years that followed significant strides were made in reducing the development cycle, with Ford and Chrysler succeeding in dramatically reducing time and improving quality. Much of the advantage was gained through extensive team working; as Lew Varaldi, project manager of Ford's Team Taurus project put it, '*. . . it's amazing the dedication and commitment you get from people . . . we will never go back to the old ways because we know so much about what they can bring to the party . . .*'[57]

These days cars – and an increasing range of complex products and projects – are designed and built with lead times measured in months not years, and with increasing 'stretch' in terms of features and functionality. Complex service packages are designed and delivered in highly customized fashion and configured and reconfigured to suit the changing needs of a wide range of users. Public-sector services like utilities, transport, healthcare and policing are all adapting to deal with radical new demands and emerging challenges. All of this puts a premium on the kind of team working suggested above, which builds on the principle that innovation is primarily about combining different perspectives in solving problems.

Experiments indicate that groups have more to offer than individuals in terms of both fluency of idea generation and in flexibility of solutions developed. Focusing this potential on innovation tasks is the prime driver for the trend towards high levels of team working – in project teams, in cross-functional and inter-organizational problem-solving groups and in cells and work groups where the focus is on incremental, adaptive innovation.

Considerable work has been done on the characteristics of high-performance project teams for innovative tasks, and the main findings are that such teams rarely happen by accident.[58] They result from a combination of selection and investment in team building, allied to clear guidance on their roles and tasks, and a concentration on managing group process as well as task aspects. A variety of studies has been carried out aimed at identifying key drivers and barriers to effective performance, and they share the conclusion that effective team building is a critical determinant of project success.[59] For example, research within the Ashridge Management College team-working programme developed a model for 'superteams' which includes components of building and managing the internal team and also its interfaces with the rest of the organization.[60]

Holti, Neumann and Standing provide a useful summary of the key factors involved in developing team working.[61] Although there is considerable current emphasis on team working, we should remember that teams are not always the answer. In particular, there are dangers in putting nominal teams together where unresolved conflicts, personality clashes, lack of effective group processes and other factors can diminish their effectiveness. Tranfield *et al.* look at the issue of team working in a number of different contexts and highlight the importance of selecting and building the appropriate team for the task and the context.[62]

Teams are increasingly being seen as a mechanism for bridging boundaries within the organization – and indeed, in dealing with inter-organizational issues. Cross-functional teams can bring together the different knowledge sets needed for tasks like product development or process improvement – but they also represent a forum where often deep-rooted differences in perspectives can be resolved.[63] Lawrence and Lorsch in their pioneering study of differentiation and integration within organizations found that interdepartmental clashes were a major source of friction and contributed much to delays and difficulties in operations. Successful organizations were those which invested in multiple methods for integrating across groups – and the cross-functional team was one of the most valuable resources.[35] But, as we indicated above, building such teams is a major strategic task – they will not happen by accident, and they will require additional efforts to ensure that the implicit conflicts of values and beliefs are resolved effectively.

Teams also provide a powerful enabling mechanism for achieving the kind of decentralized and agile operating structure which many organizations aspire to. As a substitute for hierarchical control, self-managed teams working within a defined area of autonomy can be very effective. For example, Honeywell's defence avionics factory reports a dramatic improvement in on-time delivery – from below 40% in the 1980s to 99% in 1996 – to the implementation of self-managing teams.[64] In the Netherlands one of the most successful bus companies is Vancom Zuid-Limburg, which has improved both price and non-price performance and has high customer satisfaction ratings. Again they attribute this to the use of self-managing teams and to the reduction in overhead costs. In their system one manager supervises over 40 drivers where the average for the sector is a ratio of 1:8. Drivers are also encouraged to participate in problem finding and solving in areas like maintenance, customer service and planning.[65]

Key elements in effective high-performance team working include:

- clearly defined tasks and objectives
- effective team leadership
- good balance of team roles and match to individual behavioural style
- effective conflict resolution mechanisms within the group
- continuing liaison with external organization.

Teams typically go through four stages of development, popularly known as 'forming, storming, norming and performing'.[66] That is, they are put together and then go through a phase of resolving internal differences and conflicts around leadership, objectives, etc. Emerging from this process is a commitment to shared values and norms governing the way the team will work, and it is only after this stage that teams can move on to effective performance of their task. Common approaches to team building can support innovation, but are not sufficient.

Central to team performance is the make-up of the team itself, with good matching between the role requirements of the group and the behavioural preferences of the individuals involved. Belbin's work has been influential here in providing an approach to team role matching. He classifies people into a number of preferred role types – for example, 'the plant' (someone who is a source of new ideas), 'the resource investigator', 'the shaper' and the 'completer/finisher'. Research has shown that the most effective teams are those with diversity in background, ability and behavioural style. In one noted experiment highly talented but similar people in 'Apollo' teams consistently performed less well than mixed, average groups.[67]

With increased emphasis on cross-boundary and dispersed team activity, a series of new challenges are emerging. In the extreme case a product development team might begin work in London, pass on to their US counterparts later in the day who in turn pass on to their far Eastern colleagues – effectively allowing a 24-hour non-stop development activity. This makes for higher productivity potential – but only if the issues around managing dispersed and virtual teams can be resolved. Similarly the concept of sharing knowledge across boundaries depends on enabling structures and mechanisms.[68]

Many people who have attempted to use groups for problem solving find out that using groups is not always easy, pleasurable or effective. Table 3.5 summarizes some of the positive and negative aspects of using groups for innovation.

When considering the use of small groups the leader or facilitator needs to evaluate the liabilities and assets of using groups. The goal is to maximize the positive aspects of group involvement while minimizing the liabilities. For example, as the facilitator or group leader can increase the productive use of diversity the likelihood of individual dominance should decrease. In general, if there is a need to provide for participation to increase acceptance, if the information is widely held, if there is a need to build on and synthesize the diverse range of experiences and perspectives or if it is important to develop and strengthen the group's ability to learn, you may choose to involve a group.

Teams are one of the basic building blocks of every organization. After individuals, they may be considered the most important resource in any organization. Teams conduct so much real, day-to-day work within organizations. This explains the interest in high-performance work systems, electronic groupware, small-group facilitation skills, and a host of other strategies for improving the way groups work. One of the reasons that teams are so essential within organizations is the growing complexity of tasks. Increasingly complex tasks frequently surpass the cognitive capabilities of individuals and necessitate a team approach.

Before we continue, it would be helpful to explore what we mean by a team. Many people use the words group and team interchangeably. In general, the word group refers to an assemblage of people who may just be near to each other. Groups can be a number of people that are regarded as some sort of unity or are classed together on account of any sort of similarity. For us, a team means a combination of individuals who come together or who have been brought together for a common purpose or goal in their organization. A team is a group that must collaborate in their professional work in some enterprise or on some assignment and share accountability or responsibility for obtaining results. There are a variety of ways to differentiate working groups from teams. One senior executive with whom we have worked described groups as individuals with nothing in common, except a zip/postal code. Teams, however, were characterized by a common vision.

| TABLE 3.5 | **Potential assets and liabilities of using a group** |

Potential assets of using a group	Potential liabilities of using a group
1. Greater availability of knowledge and information	1. Social pressure toward uniform thought limits contributions and increases conformity
2. More opportunities for cross-fertilization; increasing the likelihood of building and improving upon ideas of others	2. Group think: groups converge on options, which seem to have greatest agreement, regardless of quality
3. Wider range of experiences and perspectives upon which to draw	3. Dominant individuals influence and exhibit an unequal amount of impact upon outcomes
4. Participation and involvement in problem solving increases understanding, acceptance, commitment, and ownership of outcomes	4. Individuals are less accountable in groups allowing groups to make riskier decisions
5. More opportunities for group development; increasing cohesion, communication and companionship	5. Conflicting individual biases may cause unproductive levels of competition; leading to 'winners' and 'losers'

Source: S. Isaksen and J. Tidd (2006) *Meeting the Innovation Challenge,* John Wiley & Sons, Ltd, Chichester.

Our own work on high-performance teams, consistent with previous research, suggests a number of characteristics that promote effective teamwork:[69]

- *A clear, common and elevating goal.* Having a clear and elevating goal means having understanding, mutual agreement and identification with respect to the primary task a group faces. Active teamwork towards common goals happens when members of a group share a common vision of the desired future state. Creative teams have clear and common goals. 'The most important factor accounting for my team's creative success was, undoubtedly, each member's drive to attain the end goal, knowing the benefits that would be derived from the results.' The goals were clear and compelling, but also open and challenging. Less creative teams have conflicting agendas, different missions, and no agreement on the end result. 'Everyone did their own thing without keeping in mind the overall objective that the group was charged to achieve.' The tasks for the least creative teams were tightly constrained, considered routine, and were overly structured.
- *Results-driven structure.* Individuals within high-performing teams feel productive when their efforts take place with a minimum of grief. Open communication, clear coordination of tasks, clear roles and accountabilities, monitoring performance, providing feedback, fact-based judgement, efficiency and strong impartial management combine to create a results-driven structure.

- *Competent team members*. Competent teams are composed of capable and conscientious members. Members must possess essential skills and abilities, a strong desire to contribute, be capable of collaborating effectively, and have a sense of responsible idealism. They must have knowledge in the domain surrounding the task (or some other domain which may be relevant) as well as with the process of working together. Creative teams recognize the diverse strengths and talents and use them accordingly. 'Each individual brought a cornucopia of experience and insight. All of this, together with the desire to meet the end goal was the key to success.' Less creative teams have inadequate skill sets and are unable to effectively utilize their diversity.

- *Unified commitment*. Having a shared commitment relates to the way the individual members of the group respond. Effective teams have an organizational unity: members display mutual support, dedication and faithfulness to the shared purpose and vision, and a productive degree of self-sacrifice to reach organizational goals. Creative teams 'play hard and work even harder.' Team members enjoy contributing and celebrating their accomplishments. 'All team members were motivated to do the best job possible in reaching the end goal, so everyone was willing to pitch in to get the job done.' There is a high degree of enthusiasm and commitment to get the job done. Less creative teams lack that kind of motivation. There is a lack of initiative, ideas and follow through on suggestions. Less creative teams had a 'lack of motivation and the inability to recognize the value provided by the end result.'

- *Collaborative climate*. Productive teamwork does not just happen. It requires a climate that supports cooperation and collaboration. This kind of situation is characterized by mutual trust – trust in the goodness of others. Organizations desiring to promote teamwork must provide a climate within the larger context that supports cooperation. Creative teams have an environment that encourages new ideas and allows the development of new ways of working. 'No matter what the disagreements, we all knew that we had to bring our ideas together to get the job done.' Everyone feels comfortable discussing ideas, offering suggestions because '. . . ideas are received in a professional and attentive manner . . . people feel free to brainstorm to improve others' ideas without the authors' feelings getting hurt.' In less creative teams new ideas are not attended to or encouraged because '. . . individuals place their own priorities before the teams.' They are characterized by not being able to discuss multiple solutions to a problem because team members cannot listen to any opinion other than their own. In these teams, members are '. . . expected to follow what had always been done and finish as quickly as possible.'

- *Standards of excellence*. Effective teams establish clear standards of excellence. They embrace individual commitment, motivation, self-esteem, individual performance and constant improvement. Members of teams develop a clear and explicit understanding of the norms upon which they will rely.

- *External support and recognition*. Team members need resources, rewards, recognition, popularity and social success. Being liked and admired as individuals and respected for belonging and contributing to a team is often helpful in maintaining the high level of personal energy required for sustained performance. With the increasing use of cross-functional and inter-departmental teams within larger complex organizations, teams must be able to obtain approval and encouragement.

- *Principled leadership*. Leadership is important for teamwork. Whether it is a formally appointed leader or leadership of the emergent kind, the people who exert influence and encourage the accomplishment of important things usually follow some basic principles. Principled leadership includes the management of human differences, protecting less able members and providing a level playing field to encourage contributions from everyone. This is the kind of leadership that promotes legitimate compliance to competent authority. In creative teams the '. . . leader leads by example, encouraging new ideas and sharing best practices.' Leaders provide clear guidance, support and encouragement, and keep everyone working together and moving forward. Leaders also work to obtain support and resources from within and outside the group. In less creative teams, the leader '. . . creates a situation where everyone is confused and afraid to ask questions.' Leaders 'tear down people's ideas,' 'set a tone of distrust,' and 'stifle others who have ideas and energy to succeed.' They '. . . keep all control, but take no action.'

- *Appropriate use of the team.* Teamwork is encouraged when the tasks and situations really call for that kind of activity. Sometimes the team itself must set clear boundaries on when and why it should be deployed. One of the easiest ways to destroy a productive team is to overuse it or use it when it is not appropriate to do so.

- *Participation in decision making.* One of the best ways to encourage teamwork is to engage the members of the team in the process of identifying the challenges and opportunities for improvement, generating ideas and transforming ideas into action. Participation in the process of problem solving and decision making actually builds teamwork and improves the likelihood of acceptance and implementation.

- *Team spirit.* Effective teams know how to have a good time, release tension and relax their need for control. The focus at times is on developing friendship, engaging in tasks for mutual pleasure and recreation. This internal team climate extends beyond the need for a collaborative climate. Creative teams have the ability to work together without major conflicts in personalities. There is a high degree of respect for the contributions of others. Communication is characterized by 'The willingness of team members to listen to one another and honour the opinions of all team members.' Members of these teams report that they know their roles and responsibilities and that this provides freedom to develop new ideas. Less creative teams are characterized by an 'unwillingness to communicate with one another because people do not make the effort to understand each other.' There are instances of animosity, jealousy and political posturing.

- *Embracing appropriate change.* Teams often face the challenges of organizing and defining tasks. In order for teams to remain productive, they must learn how to make necessary changes to procedures. When there is a fundamental change in how the team must operate, different values and preferences may need to be accommodated. Productive teams learn how to use the full spectrum of their members' creativity.

There are also many challenges to the effective management of teams. We have all seen teams that have 'gone wrong'. As a team develops, there are certain aspects or guidelines that might be helpful to keep them on track. Hackman has identified a number of themes relevant to those who design, lead and facilitate teams. In examining a variety of organizational work groups, he found some seemingly small factors that if overlooked in the management of teams will have large implications that tend to destroy the capability of a team to function. These small and often hidden 'tripwires' to major problems include:[70]

- *Group versus team.* One of the mistakes that is often made when managing teams is to call the group a team, but to actually treat it as nothing more than a loose collection of individuals. This is similar to making it a team 'because I said so'. It is important to be very clear about the underlying goal structure. Organizations are often surprised that teams do not function too well in their environment. Of course, they often fail to examine the impact of competition in their rating or review process. People are often asked to perform tasks as a team, but then have all evaluation of performance based on an individual level. This situation sends conflicting messages, and may negatively affect team performance. Teams include mutual accountability for agreed goals and working approach, something that may not necessarily be present in all groups.

- *Ends versus means.* Managing the source of authority for groups is a delicate balance. Just how much authority can you assign to the team to work out its own issues and challenges? Those who convene teams often 'over manage' them by specifying the results as well as how the team should obtain them. The end, direction or outer limit constraints ought to be specified, but the means to get there ought to be within the authority and responsibility of the group. Teamwork is often underutilized because the desired ends are unclear and unspecified. As a result, teams are often given too much guidance on the means (the how) rather than sufficient emphasis on the ends (the what and why). Effective teams are given clear indications of what is the acceptable outcome and end goal and responsibility for working out how to get there.

- *Structured freedom.* It is a major mistake to assemble a group of people and merely tell them in general and unclear terms what needs to be accomplished and then let them work out their own details. At times, the belief is that if teams are to be creative, they ought not be given any structure. It turns out that most groups would find a little structure quite enabling, if it were the right kind. Teams generally need a well-defined task. They need to be composed of an appropriately small number to be manageable but large enough to be diverse. They need clear limits as to the team's authority and responsibility, and they need sufficient freedom to take initiative and make good use of their diversity. It's about striking the right kind of balance between structure, authority and boundaries – and freedom, autonomy and initiative.

- *Support structures and systems.* Often challenging team objectives are set, but the organization fails to provide adequate support in order to make the objectives a reality. In general, high-performing teams need a reward system that recognizes and reinforces excellent team performance. They also need access to good quality and adequate information, as well as training in team-relevant tools and skills. Good team performance is also dependent on having an adequate level of material and financial resources to get the job done. Calling a group a team does not mean that they will automatically obtain all the support needed to accomplish the task.

- *Assumed competence.* Many organizations have a great deal of faith in their selection systems. Facilitators, and others who manage or lead groups, cannot assume that the group members have all the competence they need to work effectively as a team, simply because they have been selected to join any particular organization. Technical skills, domain-relevant expertise and experience and abilities often explain why someone has been included within a group. These are often not the only competencies individuals need for effective team performance. Members will undoubtedly require explicit coaching on skills needed to work well in a team. Coaching and other supportive interventions are best done during the launch, at a natural break in the task, or at the end of a performance or review period. The start-up phase is probably the most important time frame to provide the necessary coaching or training.

CASE STUDY 3.8

Organizational climate for innovation at Google

Google appear to have learned a few lessons from other innovative organizations, such as 3M. Technical employees are expected to spend 20% of their time on projects other than their core job, and similarly managers are required to spend 20% of their time on projects outside the core business, and 10% to completely new products and businesses. This effort devoted to new, non-core business is not evenly allocated weekly or monthly, but when possible or necessary. These are contractual obligations, reinforced by performance reviews and peer pressure, and integral to the 25 different measures of and targets for employees. Ideas progress through a formal qualification process which includes prototyping, pilots and tests with actual users. The assessment of new ideas and projects is highly data-driven and aggressively empirical, reflecting the IT basis of the firm, and is based on rigorous experimentation within 300 employee user panels, segments of Google's 132 million users and trusted third parties. The approach is essentially evolutionary in the sense that many ideas are encouraged, most fail but some are successful, depending on the market response. The generation and market testing of many alternatives, and tolerance of (rapid) failure, are central to the process. In this way the company claims to generate around 100 new products each year, including hits such as Gmail, AdSense and Google News.

However, we need to be careful to untangle cause and effect, and determine how much of this is transferable to other companies and contexts. Google's success to date is predicated on dominating the global demand for search engine services through an unprecedented investment in technology infrastructure – estimated at over a million computers. Its business model is based upon 'ubiquity first, revenues later', and is still reliant on search-based advertising. The revenues generated in this way have allowed it to hire the best, and to provide the space and motivation to innovate. Despite this, it is estimated to have only 120 or so product offerings, and the most recent blockbusters have all been acquisitions: YouTube for video content; DoubleClick for web advertising; and Keyhole for mapping (now Google Earth). In this respect it looks more like Microsoft than 3M.

Source: Bala Iyer and Thomas H. Davenport (2008) Reverse engineering Google's innovation machine. *Harvard Business Review*, April, 58–68.

3.6 Creative climate

Microsoft's only factory asset is the human imagination.

(Bill Gates)

Many great inventions came about as the result of lucky accidental discoveries – for example, Velcro fasteners, the adhesive behind 'Post-it' notes or the principle of float glass manufacturing. But as Louis Pasteur observed, 'chance favours the prepared mind' and we can usefully deploy our understanding of the creative process to help set up the conditions within which such 'accidents' can take place.

Two important features of creativity are relevant in doing this. The first is to recognize that creativity is an attribute which everyone possesses – but their preferred style of expressing it varies widely.[71] Some people are comfortable with ideas which challenge the whole way in which the universe works, whilst others prefer smaller increments of change – ideas about how to improve the jobs they do or their working environment in small incremental steps. (This explains in part why so many 'creative' people – artists, composers, scientists – are also seen as 'difficult' or living outside the conventions of acceptable behaviour.) This has major implications for how we manage creativity within the organization: innovation, as we have seen, involves bringing something new into widespread use, not just inventing it. Whilst the initial flash may require a significant creative leap, much of the rest of the process will involve hundreds of small problem-finding and -solving exercises – each of which needs creative input. And though the former may need the skills or inspiration of a particular individual the latter require the input of many different people over a sustained period of time. Developing the light bulb or the Post-it note or any successful innovation is actually the story of the combined creative endeavours of many individuals. Ken Robinson discusses some of the common attributes and approaches to creativity.

Organizational structures are the visible artefacts of what can be termed an innovative culture – one in which innovation can thrive. Culture is a complex concept, but it basically equates to the pattern of shared values, beliefs and agreed norms which shape behaviour – in other words, it is 'the way we do things round here' in any organization. Schein suggests that culture can be understood in terms of three linked levels, with the deepest and most inaccessible being what each individual believes about the world – the 'taken for granted' assumptions. These shape individual actions and the collective and socially negotiated version of these behaviours defines the dominant set of norms and values for the group. Finally, behaviour in line with these norms creates a set of artefacts – structures, processes, symbols, etc. – which reinforce the pattern.[72]

Given this model it is clear that management cannot directly change culture, but it can intervene at the level of artefacts – by changing structures or processes – and by providing models and reinforcing preferred styles of behaviour. Such 'culture change' actions are now widely tried in the context of change programmes towards total quality management and other models of organization which require more participative culture.

A number of writers have looked at the conditions under which creativity thrives or is suppressed.[73] Kanter[74] provides a list of environmental factors which contribute to stifling innovation; these include:

- dominance of restrictive vertical relationships
- poor lateral communications
- limited tools and resources
- top-down dictates
- formal, restricted vehicles for change
- reinforcing a culture of inferiority (i.e. innovation always has to come from outside to be any good)
- unfocused innovative activity
- unsupporting accounting practices.

The effect of these is to create and reinforce the behavioural norms which inhibit creativity and lead to a culture lacking in innovation. It follows from this that developing an innovative climate is not a simple matter since it consists of a complex web of behaviours and artefacts. And changing this culture is not likely to happen quickly or as a result of single initiatives (such as restructuring or mass training in a new technique).

Instead, building a creative climate involves systematic development of organizational structures, communication policies and procedures, reward and recognition systems, training policy, accounting and measurement systems and deployment of strategy. Mechanisms for doing so in various different kinds of organizations and in different national cultures are described by a number of authors including Cook and Rickards.[75]

Of particular relevance in this area is the design of effective reward systems. Many organizations have reward systems which reflect the performance of repeated tasks rather than encourage the development of new ideas. Progress is associated with 'doing things by the book' rather than challenging and changing things. By contrast, innovative organizations look for ways to reward creative behaviour, and to encourage its emergence. Examples of reward systems include the establishment of a 'dual ladder' which enables technologically innovative staff to progress within the organization without needing to move across to management posts.[76]

One aspect of this worth highlighting concerns the emerging idea of 'intrapreneurship' – internal entrepreneurship.[9] In an organization with a supportive and innovative culture, individuals with bright ideas can progress them with support and encouragement from the system. For example, in 3M the culture encourages individuals to follow up interesting ideas and allows them up to 15% of their time for such activity. If things look promising, there are internal venture funds to enable a more thorough exploration – and if individuals think they can build a business out of the idea, 3M will back it and give them the responsibility to run it. In this way the company has grown through organic, 'intrapreneurial' means. Chapter 10 explores this theme in more detail.

VIEWS FROM THE FRONT LINE

Creating innovation energy

Innovation – it's the corporate world's latest plaything. But it's more than a buzzword. It's commercially critical, it helps organizations to grow during boom times and can help companies to stay alive in tough times. In the 21st century it's not an overstatement to say that in most commercial sectors, to stand still is to die. That's why almost every organization accepts the business imperative to innovate.

So why do some succeed while others fail? What organizational characteristics set the winners apart from the losers? Is innovation a matter of luck or size?

At ?What *If*! we've spent 16 years working on thousands of innovation projects with some of the largest and most successful organizations across the globe. We've rolled our sleeves up and worked late into the night on incremental innovation projects and market changing initiatives. We've met companies that are brilliant at innovation and others that, no matter how hard they try, just can't make it work. We've had a unique and privileged perspective on innovation having worked across so many sectors and in so many countries.

The good news is that there is a clear pattern that determines if your organization has the DNA to spawn innovation, the bad news is that there is no business concept that describes this pattern, this 'magic key'. In fact it's worse than that – traditional business concepts, as basic as strategy, thinking things through carefully – can often do more harm than good. Innovation is as much about trying things out, deliberating, not being too careful. Our collective brains don't have the computing power to use conventional strategic approaches to get to the answer.

So what is this 'pattern' behind successful innovation? We call it *Innovation Energy*. In a nutshell it's the confluence of three forces: an individual's attitude; a group's behavioural dynamic; and the support an organization provides. There is a sweet spot that some organizations either stumble upon or deliberately seek out, this sweet spot is best understood as more of a social or human science than a business concept. At its heart innovation is all about people.

'It's all about people'. That's a great sound bite and we've all heard it a million times before. We all know that it's people, not processes that make things happen. But while most companies are pretty good at constructing processes they are often shockingly bad at getting the most out of the human energy. How often have you heard leaders say, 'Our greatest asset is our people'? Yet those same leaders coop their 'greatest assets' in grey office blocks, suppress them with corporate stuffiness and bury them with hundreds of emails a day. But work doesn't have to sap energy. It can create it. Innovation Energy is the force generated when a group of people work together with the right attitude and behaviours, in an organization structured to help make things happen.

Energy doesn't just happen. Think about what gets you fired up – your favourite football team, playing with your children, having a cause to fight for. Life without the right stimulus leaves you sluggish and lethargic. It's the same in business except multiplied by the amount of people. Put 50 colleagues together and the difference between collective inertia and collective energy is immense. You either charge each other up or bring each other down. So that energy needs managing – more than any other resource. It makes the difference between innovation success or failure.

The three elements of the equation

So let's break down the *attitude, behaviours* and *structures* needed to manage Innovation Energy.

Attitude

The plain fact is that innovation requires us to think very nimbly about our jobs, about what we do with our time. Innovation is by its very nature both threatening and exhilarating. Not everyone in an organization skips into work with a nimble mindset – we all know that cynics lurk in every department and in every team. Innovation teams need a majority of people with the right attitude and others need to be at least 'neutral'.

Our experience within large corporations is that money rarely motivates or affects 'attitude'. Most people we have met who can make a difference to their company's innovation profile are at heart motivated by wanting to do something good, to leave a mark, to be recognized as a key part of a team. It's simple, obvious stuff but look more deeply and the job of management is to answer the question: Why should my people care so much that they'll work through the night, argue against the grain, stick their heads above the parapet? The only reason is that they like what the body corporate is 'going for'. It feels good and they feel good being part of it. This is why issues of vision and purpose are so central to innovation. They provide the lifeblood of innovation energy.

But just how do you get people fired up about a company's bold vision? Well, a crisis will do it. If everyone truly understands what will happen if nothing changes, if the burning platform is made real, that can be the catalyst that galvanizes people behind the need to innovate.

In the early 1990s the Norwegian media company Schibsted recognized that being a traditional newspaper company would not be sustainable over time, so they decided to adopt quite a Darwinian approach to innovation, declaring 'It is not the strongest of the species that survive, nor the most intelligent but the one most responsive to change'. The company invested heavily in new media, making a conscious effort to see themselves as a media company rather than a newspaper company. In the process they effectively cannibalized their old business model to make way for a new one. In 2007 the company was one of the most successful media companies in Scandinavia making over £1 billion in revenue. And, more critically, by 2009 nearly 60% of their earnings are projected to come from their online businesses.

But ambition isn't enough. Companies need to engage their people on a personal level. This means making sure that each individual in the organization has their own 'Ah ha!' moment.

At ?What *If!* we see this all the time, and the power of converting someone from a 'So what' mentality to a 'So that's why we're doing this!' realization is amazing. This often happens when senior management are connected with real people, their consumers. Put a managing director whose company has been making the same inhalers, the same way, for 20 years face to face with a frustrated asthmatic, too embarrassed to use his 'puffers' in front of his children and the revelations are electrifying.

Companies that are really successful at innovating are the ones that manage to tap into people's innate desire to be part of something bigger, a common purpose.

This purpose is always explicit and often disarmingly simple. The people at IKEA aren't in business to sell flat-pack furniture; they are working towards providing 'A Better Everyday Life for Many People'. While over at Apple, Steve Jobs' challenge to his team is to create and sell products 'so good you'll want to lick them'. These companies have managed to engage and unify everyone from the boardroom to the shop floor behind their common purpose: they make coming to work worthwhile.

Behaviours

Behaviour beats process every day of the week. Every single interaction we have sets up a powerful and lasting expectation of just what a conversation or meeting is going to be like in the future. Without realizing it, we're all hard wired before we go into a meeting room – with some folks we'll take risks, with others we'll hold back. So breaking established behaviour patterns is an incredibly powerful force. For this reason companies need to be very prescriptive, sometimes more than feels comfortable, about how they want their people to behave around innovation.

Many of the learnt behaviours that have helped us succeed at work are actually opposite to innovation behaviours. We need to suspend judgement and replace it with what we call *greenhousing* – building ideas collaboratively. We need to suspend the number of heavy PowerPoint charts and replace with real consumer experiences as they grapple with our crudely made prototypes.

The most useful innovation behaviours are *freshness* (trying new stuff out), *greenhousing* (building an idea through collaboration), *realness* (quickly making an idea into the form a customer will buy it as), *bravery* (guts to disagree) and *signalling* (helping a group navigate between creative and analytical behaviour). Let's dwell on this last behaviour. We have found that it's essential to have at least one person with sufficient emotional intelligence to be able to comment on the dynamics of the group. We call this 'signalling' and it's a real art. This is what it sounds like – 'guys, let's step back a bit, we're drilling so deep into the economics of the idea that we're killing it'. Without this behaviour the line between analysis and creativity becomes blurred and innovation collapses.

The problem is that many organizations fall into the trap of prescribing behaviour using a series of bland and ultimately meaningless value statements. 'Integrity', 'Passion', 'Customer First' shout the posters in reception, but they don't translate into action. We have come across many CEOs who are prisoners of a zealous values campaign – trapped with a random set of words that they cannot in their heart support but dare not in public deny. Their silence is deafening.

Innovation needs what's okay and what's not okay to be very clearly articulated, and the most effective way to do this is by telling stories.

Curt Carlson of the Stanford Research Institute (SRI) in California has a hard-hitting story: he asks whether you'd dive into a pool with a single poo in it. The answer is clearly no, it doesn't matter how big the pool is, if someone has left just one small nasty thing in it no one is going to jump in! The story is a crude but an effective way of reminding his people that cynicism is innovation's biggest enemy. All it takes is one raised eyebrow or dismissive sneer to kill a budding idea. This story gets repeated time and again and it sends a clear message about a specific behaviour that will not be tolerated within the organization. Everyone at SRI knows that it is not OK to behave, however subtly, in an undermining way.

Other companies use stories to celebrate good behaviour. The best stories are ones that specifically identify a person, relate their actions, detail the pay-off and then explain the 'so what' – what exactly it was that made the person's action special and noteworthy.

At Xilinx, one of the leading players in the global semiconductor industry, the chairman Wim Roelandts shares a story about a team within the organization who worked for months on a project that in the end did not deliver the desired results. Upon the failure of the project Roelandts very publicly assigned the team involved to work on another high-profile project. As he explains, 'As a technology company the projects that are most likely to fail are the most difficult projects, so if you only reward successful projects no one will ever want to take on the difficult ones. You have to reward failure and genuinely believe that if people learn from their mistakes, then failure is a good thing'.

These types of stories are motivational and are easily understood by everyone in the organization. Storytelling is much more powerful than any mission statement or set of values listed on a credo card or posters with value statements that attempt to brighten our corridors. If used effectively, stories help turn behaviours into habits. Once this happens the organization begins to create its own sustainable source of energy that is almost impossible for any competitor to steal or replicate.

Organizational support for innovation

Innovation Energy is not just a matter of harnessing the right attitude and the right behaviour, it's vital that the organization supports and directs innovation. The most innovative companies are organized like a river, with a clear path that flows much faster than one full of obstacles and tributaries. They have simple and focused structures and processes (that can be broken) that are there to free people, not to get in the way.

There are many ways to block and unblock the river: rewards, resources, communication, flexible process, environment and leadership. Let's look at the last two.

The physical environment of a business has a major influence on energy. Working space provides a great opportunity to create the right energy for your organization, but it's also a potential bear trap just waiting to kill energy dead in its tracks. Too often it is the buildings policy of a business rather than any strategic goal that dictates their structure! Many organizations are housed in grey, generic office blocks with rows of uniform desks and dividers; but what we've found is that people who work in grey, generic and uniform offices tend to come up with grey, generic and uniform ideas. The companies that have created energizing spaces that bring their brands to life and their people together reap the biggest rewards.

When designing their new headquarters in Emeryville, California, the film studio Pixar started from the inside out to ensure a cross-pollination of ideas among the diverse specialities that work within the company. The key to ensuring cross-pollination in the large aircraft hangar-like space is the 'heart' of the building – the large, open centre space where the left brain (techies) and the right brain (creatives) of the company can bump into one another even though they are housed in separate areas. To force people into the shared space the 'heart' houses the mailroom, cafeteria, games room and screening room. This very clever use of space breaks down barriers and prevents people from only fraternizing with the people in their immediate teams.

However, creative structures and clever buildings will count for very little if the organization does not have the right type of leadership. The leadership of a company is absolutely essential to that organization's ability to innovate. The leaders need to have the ambition, share in the purpose and role model the desired behaviours: it is up to them to keep the innovation energy flowing.

The best leaders have focus, and crucially, enable their people to focus. Too many times we have seen companies trying to focus on too many things and, as a result, getting very little success with any of them. It's rather like having too many planes in the air but not enough runways to land them all. The planes are the ideas and the runways are the commercial abilities of a company to make those ideas happen. By its very nature innovation needs a lot of white space around it, it needs a lot of unscheduled time because you just never know where an idea is going or how much time you need to put behind it; so if your diary is absolutely jam-packed with things to do you'll never be able to innovate and never be able to be truly creative.

Behind most stories of great new innovations you will find a story about focus, and innovative leaders are those leaders who cut the number of planes in the air and simply focus on landing very few, but critical things.

Innovative leaders are also very honest about their strengths and limitations and they are unafraid to make any gaps in their strengths public. Some people are born enthusiasts – they are brilliant at emphasizing the positive and cheering people on. Others make great taskmasters – they do not shirk from giving people bad news or telling people something isn't good enough. A team or company run solely by enthusiasts might be an inspiring place to work but chances are it won't be commercially successful. And companies or teams run solely by taskmasters might deliver results but will ultimately be an exhausting place to work. It is important to find the balance between the two types of leadership and the only way to do this is to be honest about your skills and limitations. If you're not prepared to be open about what you're not very good at you don't allow anyone with complementary skills to step in and fill the gaps.

Great leadership is as much about honesty and humility as it is about focus and inspiration.

The Innovation Energy sweet spot

Innovation Energy is the power behind productive change. It can mean the difference between innovating successfully or running out of steam. Innovation Energy can be generated, harnessed and managed by engendering the right attitude, behaviours and structures within your organization. It can turn fading companies into powerhouses of industry. Get it right and you create a stimulating, productive, fun place to work. You'll attract and recruit talented people – bright sparks that will add to the energy and make success all the more likely.

Innovation Energy. It's powerful stuff!

Matt Kingdon, www.whatifinnovation.com. Matt is chairman and chief enthusiast of ?What *If!* an innovation consultancy he co-founded in 1992.

Climate versus culture

Climate is defined as the recurring patterns of behaviour, attitudes and feelings that characterize life in the organization. At the individual level of analysis the concept is called psychological climate. At this level, the concept of climate refers to the intrapersonal perception of the patterns of behaviour, attitudes and feelings as experienced by the individual. When aggregated the concept is called a work unit or organizational climate. These are the objectively shared perceptions that characterize life within a defined work unit or in the larger organization. Climate is distinct from culture in that it is more observable at a surface level within the organization and more amenable to change and improvement efforts. Culture refers to the deeper and more enduring values, norms and beliefs within the organization.

The two terms, culture and climate, have been used interchangeably by many writers, researchers and practitioners. We have found that the following distinctions may help those who are concerned with effecting change and transformation in organizations:

- *Different levels of analysis.* Culture is a rather broad and inclusive concept. Climate can be seen as falling under the more general concept of culture. If your aim is to understand culture, then you need to look at the entire organization as a unit of analysis. If your focus is on climate, then you can use individuals and their shared perceptions of groups, divisions or other levels of analysis. Climate is recursive or scalable.
- *Different disciplines involved.* Culture is within the domain of anthropology and climate falls within the domain of social psychology. The fact that the concepts come from different disciplines means that different methods and tools are used to study them.
- *Normative versus descriptive.* Cultural dimensions have remained relatively descriptive, meaning that one set of values or hidden assumptions were neither better nor worse than another. This is because there is no universally held notion or definition of the best society. Climate is often more normative in that we are more often looking for environments that are not just different, but better for certain things. For example, we can examine different kinds of climates and compare the results against other measures or outcomes like innovation, motivation, growth, etc.
- *More easily observable and influenced.* Climate is distinct from culture in that it is more observable at a surface level within the organization and more amenable to change and improvement efforts.

What is needed is a common sense set of levers for change that leaders can exert direct and deliberate influence over.

Climate and culture are different: traditionally studies of organizational culture are more qualitative, whereas research on organizational climate is more quantitative, but a multidimensional approach helps to integrate the benefits of each perspective.

Research indicates that organizations exhibit larger differences in practices than values, for example, the levels of uncertainty avoidance.

Table 3.6 summarizes some research of how climate influences innovation. Many dimensions of climate have been shown to influence innovation and entrepreneurship, but here we discuss six of the most critical factors.

TABLE 3.6	Climate factors influencing innovation		
Climate factor	**Most innovative (score)**	**Least innovative (score)**	**Difference**
Trust and openness	253	88	165
Challenge and involvement	260	100	160
Support and space for ideas	218	70	148
Conflict and debate	231	83	148
Risk taking	210	65	145
Freedom	202	110	92

Source: Derived from Scott Isaksen and Joe Tidd (2006) *Meeting the Innovation Challenge*, John Wiley & Sons, Ltd, Chichester.

Trust and openness

The trust and openness dimension refers to the emotional safety in relationships. These relationships are considered safe when people are seen as both competent and sharing a common set of values. When there is a strong level of trust, everyone in the organization dares to put forward ideas and opinions. Initiatives can be taken without fear of reprisals and ridicule in case of failure. The communication is open and straightforward. Where trust is missing, count on high expenses for mistakes that may result. People also are afraid of being exploited and robbed of their good ideas.

Trust can make decision making more efficient as it allows positive assumptions and expectations to be made about competence, motives and intentions, and thereby economizes on cognitive resources and information processing. Trust can also influence the effectiveness of an organization through structuring and mobilizing.

Trust helps to structure and shape the patterns of interaction and coordination within and between organizations. Trust can also motivate employees to contribute, commit and cooperate, by facilitating knowledge and resource sharing and joint problem solving. When trust and openness are too low you may see people hoarding resources (i.e., information, software, materials, etc.). There may also be a lack of feedback on new ideas for fear of having concepts stolen. Management may not distribute the resources fairly among individuals or departments. One cause for this condition can be that management does not trust the capabilities and/or integrity of employees. It may help to establish norms and values that management can follow regarding the disbursement of resources, and a means to assure that resources are wisely used.

However, trust can bind and blind. If trust and openness are too high, relationships may be so strong that time and resources at work are often spent on personal issues. It may also lead to a lack of questioning each other that, in turn, may lead to mistakes or less productive outcomes. Cliques may form where there are isolated 'pockets' of high trust. One cause of this condition may be that people have gone through a traumatic organizational experience together and survived (e.g. downsizing, a significant product launch). In this case it may help to develop forums for interdepartmental and intergroup exchange of information and ideas.

Trust is partly the result of individuals' own personality and experience, but can also be influenced by the organizational climate. For example, we know that the nature of rewards can affect some components of trust. Individual competitive rewards tend to reduce information sharing and raise suspicions of others' motives, whereas group or cooperative rewards are more likely to promote information sharing and reduce suspicions of motives. Similarly, the frequency of communication within an organization influences trust, and in general the higher the frequency of communication, the higher the levels of trust. In a climate of low communication, the level of trust is much more dependent on the general attitudes of individuals towards their peers.

Trust is also associated with employees having some degree of role autonomy. Role autonomy is the amount of discretion that employees have in interpreting and executing their jobs. Defining roles too narrowly constrains the decision-making latitude. Role autonomy can also be influenced by the degree to which organizational socialization encourages employees to internalize collective goals and values, for example, a so-called 'clan' culture focuses on developing shared values, beliefs and goals among members of an organization so that appropriate behaviours are reinforced and rewarded, rather than specifying task-related behaviours or outcomes. This approach is most appropriate when tasks are difficult to anticipate or codify, and it is difficult to assess performance. Individual characteristics will also influence role autonomy, including the level of experience, competence and power accumulated over time working for the organization.

Trust may exist at the personal and organizational levels, and researchers have attempted to distinguish different levels, qualities and sources of trust. For example, the following bases of organizational trust have been identified:

- Contractual – honouring the accepted or legal rules of exchange, but can also indicate the absence of other forms of trust
- Goodwill – mutual expectations of commitment beyond contractual requirements
- Institutional – trust based on formal structures
- Network –personal, family or ethnic/religious ties
- Competence – based on reputation for skills and know-how
- Commitment – mutual self-interest, committed to the same goals.

These types of trust are not necessarily mutually exclusive, although over-reliance on contractual and institutional forms may indicate the absence of the other bases of trust. In the case of innovation, problems may occur where trust is based primarily on the network, rather than competence or commitment.

Challenge and involvement

Challenge and involvement is the degree to which people are involved in daily operations, long-term goals and visions. High levels of challenge and involvement mean that people are intrinsically motivated and committed to making contributions to the success of the organization. The climate has a dynamic, electric and inspiring quality. People find joy and meaningfulness in their work, and therefore they invest much energy. In the opposite situation, people are not engaged and feelings of alienation and indifference are present. The common sentiment and attitude is apathy and lack of interest in work and interaction is both dull and listless.

If challenge and involvement are too low, you may see that people are apathetic about their work, are not generally interested in professional development, or are frustrated about the future of the organization. One of the probable causes for this might be that people are not emotionally charged about the vision, mission, purpose and goals of the organization. One of the ways to improve the situation might be to get people involved in interpreting the vision, mission, purpose and goals of the organization for themselves and their work teams.

On the other hand, if the challenge and involvement are too high you may observe that people are showing signs of 'burn out', they are unable to meet project goals and objectives, or they spend 'too many' long hours at work. One of the reasons for this is that the work goals are too much of a stretch. A way to improve the situation is to examine and clarify strategic priorities.

Building and maintaining a challenging climate involves systematic development of organizational structures, communication policies and procedures, reward and recognition systems, training policy, accounting and measurement systems and deployment of strategy. Leaders who focus on work challenge and expertise rather than formal authority result in climates that are more likely to be assessed by members as being innovative and high performance. Studies suggest that output controls such as specific goals, recognition and rewards have a positive association with innovation. A balance must be maintained between creating a climate in which subordinates feel supported and empowered, with the need to provide goals and influence the direction and agenda. Leaders who provide feedback that is high on developmental potential, for example, provide useful information for subordinates to improve, learn and develop, which results in higher levels of creativity.

Intellectual stimulation is one of the most underdeveloped components of leadership, and includes behaviours that increase others' awareness of and interest in problems, and develops their propensity and ability to tackle problems in new ways. Intellectual stimulation by leaders can have a profound effect on organizational performance under conditions of perceived uncertainty, and is also associated with commitment to an organization.

However, innovation is too often seen as the province of specialists in R&D, marketing, design or IT, but the underlying creative skills and problem-solving abilities are possessed by everyone. If mechanisms can be found to focus such abilities on a regular basis across the entire organization, the resulting innovative potential is enormous. Although each individual may only be able to develop limited, incremental innovations, the sum of these efforts can have far-reaching impacts.

Since much of such employee involvement in innovation focuses on incremental change it is tempting to see its effects as marginal. Studies show, however, that when taken over an extended period it is a significant factor in the strategic development of the organization. For example, a study of firms in the UK that have acquired the 'Investors in People' award (an externally assessed review of employee involvement practices) showed a correlation between this and higher business performance. On average, these businesses increased their sales and profits per employee by three-quarters. Another study involved over 1000 organizations in a total of seven countries, and found that those that had formal employee involvement programmes, for example, featuring support and training in idea generation and problem finding and solving, reported performance gains of 15–20%. But there is also an important secondary effect of high involvement: the more people are involved in change, the more receptive they become to it. Since the turbulent nature of most organizational environments is such that increasing levels of change are becoming the norm, greater formal involvement of employees may provide a powerful aid to effective management of change.

CASE STUDY 3.9

Increasing challenge and involvement in an electrical engineering division

The organization was a division of a large, global electrical power and product supply company headquartered in France. The division was located in the South East of the USA and had 92 employees. Its focus was to help clients automate their processes particularly within the automotive, pharmaceutical, microelectronics and food and beverage industries. For example, this division would make the robots that put cars together in the automotive industry or provide public filtration systems.

When this division was merged with the parent company, it was losing about $8 million a year. A new general manager was brought in to turn the division around and make it profitable quickly.

An assessment of the organization's climate identified that it was strongest on the debate dimension but was very close to the stagnated norms when it came to challenge and involvement, playfulness and humour, and conflict. The quantitative and qualitative assessment results were consistent with their own impressions that the division could be characterized as conflict driven, uncommitted to producing results, and people were generally despondent. The leadership decided, after some debate, that they should target challenge and involvement, which was consistent with their strategic emphasis on a global initiative on employee commitment. It was clear to them that they also needed to soften the climate and drive a warmer, more embracing, communicative and exuberant climate.

The management team re-established training and development and encouraged employees to engage in both personal and business-related skills development. They also provided mandatory safety training for all employees. They committed to increase communication by holding monthly all-employee meetings, sharing quarterly reviews on performance, and using cross-functional strategy review sessions. They implemented mandatory 'skip level' meetings to allow more direct interaction between senior managers and all levels of employees. The general manager held 15-minute meetings with all employees at least once a year. All employee suggestions and recommendations were invited and feedback and recognition was required to be immediate. A new monthly recognition and rewards programme was launched across the division for both managers and employees that was based on peer nomination. The management team formed employee review teams to challenge and craft the statements in the hopes of encouraging more ownership and involvement in the overall strategic direction of the business.

In 18 months the division showed a $7 million turnaround, and in 2003 won a worldwide innovation award. The general manager was promoted to a national position.

Source: Scott Isaksen and Joe Tidd (2006) *Meeting the Innovation Challenge*, John Wiley & Sons, Ltd, Chichester.

Support and space for ideas

Idea time is the amount of time people can (and do) use for elaborating new ideas. In the high idea-time situation, possibilities exist to discuss and test impulses and fresh suggestions that are not planned or included in the task assignment and people tend to use these possibilities. When idea time is low, every minute is booked and specified. The time pressure makes thinking outside the instructions and planned routines impossible. Research confirms that individuals under time pressure are significantly less likely to be creative.

If there is insufficient time and space for generating new ideas you may observe that people are only concerned with their current projects and tasks. They may exhibit an unhealthy level of stress. People see professional development and training as hindrances to their ability to complete daily tasks and projects. You may also see that management avoids new ideas because they will take time away from the completion of day-to-day projects and schedules. One of the possible reasons for this could be that project schedules are so intense that they do not allow time to refine the process to take advantage of new ideas. Individuals are generally not physically or mentally capable of performing at 100%. A corrective action could be to develop project schedules that allow time for modification and development.

Conversely, if there is too much time and space for new ideas you may observe that people are showing signs of boredom, that decisions are made through a slow, almost bureaucratic, process because there are too many ideas to evaluate, or the management of new ideas becomes such a task that short-term tasks and projects are not adequately completed. Individuals, teams and managers may lack the skills to handle large numbers of ideas and then converge on the most practical idea(s) for implementation. You may be able to provide training in creativity and facilitation, especially those tools and skills of convergence or focusing.

This suggests that there is an optimum amount of time and space to promote creativity and innovation. The concept of organizational slack was developed to identify the difference between resources currently needed and the total resources available to an organization. When there is little environmental uncertainty or need for change, and the focus is simply on productivity; too much organizational slack represents a static inefficiency. However, when innovation and change is needed, slack can act as a dynamic shock absorber, and allows scope for experimentation. This process tends to be self-reinforcing due to positive feedback between the environment and organization.

When successful, an organization generates more slack, which provides greater resource (people, time, money) for longer term, significant innovation; however, when an organization is less successful, or suffers a fall in performance, it tends to search for immediate and specific problems and their solution, which tends to reduce the slack necessary for longer term innovation and growth.

The research confirms that an appropriate level of organizational slack is associated with superior performance over the longer term. For high-performance organizations the relationship between organizational slack and performance is an inverted 'U' shape, or curvi-linear: too little slack, for example being too lean or too focused, does not allow sufficient time or resource for innovation, but too much slack provides little incentive or direction to innovation. However, for low-performance organizations any slack is simply absorbed and therefore simply represents an inefficiency rather than an opportunity for innovation and growth. Managers too often view time as a constraint or measure of outcomes, rather than as a variable to influence, which can both trigger and facilitate innovation and change. By providing some, but limited, time and resources, individuals and groups can minimize the rigidity that comes from work overload, and the laxness that stems from too much slack.

Idea time helps to generate new ideas, but support is needed to assess and develop these ideas. In a supportive climate, ideas and suggestions are received in an attentive and kind way by bosses and workmates. People listen to each other and encourage initiatives. Possibilities for trying out new ideas are created. The atmosphere is constructive and positive.

When idea support is low, the reflexive 'no' prevails. Every suggestion is immediately refuted by a counter-argument. Fault finding and obstacle raising are the usual styles of responding to ideas. Where there is little idea support, people shoot each other's ideas down, keep ideas to themselves, and idea-suggestion systems are not well utilized. It could be that, based on past experience, people don't think anything will be done. You may need to carefully plan a relaunch of your suggestion system with a series of case studies of what has been acted upon and why.

A supportive climate is vital for gaining information, material resources, organizational slack and political support. This can reduce the energy wasted by individuals through non-legitimate acquisition and support strategies, such as bootlegging (within the organization), or moonlighting (outside the organization). Without appropriate support for new ideas, potential innovators grow frustrated: 'if they speak out too loudly, resentment builds towards them; if they play by the rules and remain silent, resentment builds inside them'. It is not sufficient simply to have a policy or process of support, it is necessary for managers to provide the time and resources to generate and test new ideas.

However, some situations may have too much idea support. In these cases you may observe that people are only deferring judgement. Nothing is getting done and there are too many options because appropriate judgement is not being applied. Too many people are working in different directions. One of the reasons for this condition may be that people are avoiding conflict and staying 'too open'. You may need to help people apply affirmative judgement so that a more balanced approach to evaluation prevails.

In many cases innovation happens in spite of the senior management within an organization, and success emerges as a result of guerrilla tactics rather than a frontal assault on the problem. The message for senior management is as much about leading through creating space and support within the organization as it is about direct involvement.

Conflict and debate

Conflict in an organization refers to the presence of personal, interpersonal or emotional tensions. Although conflict is a negative dimension, all organizations have some level of personal tension.

Conflicts can occur over tasks, processes or relationships. Task conflicts focus on disagreements about the goals and content of work, the 'what?' needs to be done and 'why?' Process conflicts are around 'how?' to achieve a task, means and methods. Relationship or affective conflicts are more emotional, and are characterized by hostility and anger. In general, some task and process conflict is constructive, helping to avoid groupthink, and to consider more diverse opinions and alternative strategies. However, task and process conflict only have a positive effect on performance in a climate of openness and collaborative communication, otherwise it can degenerate into relationship conflict or avoidance. Relationship conflict is generally energy sapping and destructive, as emotional disagreements create anxiety and hostility.

If the level of conflict is too high, groups and individuals dislike or hate each other and the climate can be characterized by 'warfare'. Plots and traps are common in the life of the organization. There is gossip and back-biting going on. You may observe gossiping at water coolers (including character assassination), information hoarding, open aggression, or people lying or exaggerating about their real needs. In these cases, you may need to take initiative to engender cooperation among key individuals or departments.

If conflict is too low you may see that individuals lack any outward signs of motivation or are not interested in their tasks. Meetings are more about 'tell' and not consensus. Deadlines may not be met. It could be that too many ineffective people are entrenched in an overly hierarchical structure. It may be necessary to restructure and identify leaders who possess the kinds of skills that are desired by the organization.

So the goal is not necessarily to minimize conflict and maximize consensus, but to maintain a level of constructive conflict consistent with the need for diversity and a range of different preferences and styles of creative problem solving. Group members with similar creative preferences and problem-solving styles are likely to be more harmonious but much less effective than those with mixed preferences and styles. So if the level of conflict is constructive, people behave in a more mature manner. They have psychological insight and exercise more control over their impulses and emotions.

Debate focuses on issues and ideas (as opposed to conflict which focuses on people and their relationships). Debate involves the productive use and respect for diversity of perspectives and points of view. Debate involves encounters, exchanges or clashes among viewpoints, ideas and differing experiences and knowledge. Many voices are heard and people are keen on putting forward their ideas. Where debates are missing, people follow authoritarian patterns without questioning. When the score on the debate dimension is too low you may see constant moaning and complaining about the way things are, rather than how the individual can improve the situation. Rather than open debate, you may see more infrequent and quiet one-on-one conversation in hallways. In these conditions, there will be a lack of willingness by individuals to engage others in conversation regarding new ideas, thoughts or concepts. One of the reasons for this situation is that people may have had bad experiences when they have interacted in the past. It may help to clarify the rationale of debate in the organization and begin to model the behaviour.

However, if there is too much debate you are likely to see more talk than implementation. Individuals will speak with little or no regard for the impact of their statements. The focus on conversation and debate becomes more on individualistic goals than on cooperative and consensus-based action. One reason for this may be too much diversity or people holding very different value systems. In these situations it may be helpful to hold structured or facilitated discussions and affirm commonly held values.

The mandate for legitimating challenge to the dominant vision may come from the top – such as Jack Welch's challenge to 'destroy your business' memo. Perhaps building on their earlier experiences Intel now has a process called 'constructive confrontation', which essentially encourages a degree of dissent. The company has learned to value the critical insights which come from those closest to the action rather than assume senior managers have the 'right' answers every time.

CASE STUDY 3.10

Developing a creative climate in a medical technology company

A Finnish-based global healthcare organization had 55 000 employees and $50 billion in revenue. Its mission was to develop, manufacture and market products for anaesthesia and critical care.

The senior management team of one division conducted an assessment, and found that they had been doing well on quality and operational excellence initiatives in manufacturing and had improved their sales and marketing results, but were still concerned that there were many other areas on which they could improve, in particular creativity and innovation.

'We held a workshop with the senior team to present the results and engage them to determine what they needed to do to improve their business. We met with the CEO prior to the workshop to highlight the overall results and share the department comparisons. She was not surprised by the results but was very interested to see that some of the departments had different results.'

During the workshop, the team targeted challenge and involvement, freedom, idea time and idea support as critical dimensions to improve to enable them to meet their strategic objectives. The organization was facing increasing competition in its markets and significant advances in technology. Although major progress had been made in the manufacturing area, they needed to improve their product development and marketing efforts by broadening involvement internally and cross-functionally and externally by obtaining deep consumer insight. The main strategy they settled upon was to 'jump start' their innovation in new product development for life support.

Key personnel in new product development and marketing were provided training in creative problem solving, and follow-up projects were launched to apply the learning to existing and new projects.

One project was a major investment in reengineering their main product line. Clinicians were challenged with the current design of the equipment. The initial decision was to redesign the placement of critical control valves used during surgery. The project leader decided to use a number of the tools to go out and clarify the problem with the end-users, involving project team members from research and development as well as marketing. The result was a redefinition of the challenge and the decision to save the millions of dollars involved in the reengineering effort and instead develop a new tactile tool to help the clinicians' problem of having their hands full. Since the professionals in the research and development lab were also directly involved in obtaining and interpreting the consumer insight data, they understood the needs of the end-users and displayed an unusually high degree of energy and commitment to the project.

'We also observed a much greater amount of cross-functional and informal working across departments. Some human resource personnel were replaced and new forms of reward and recognition were developed. Not only was there more consumer insight research going on, but there were more and closer partnerships created with clinicians and end-users of the products. During this period of time the CEO tracked revenue growth and profitability of the division and reported double-digit growth.'

Source: Scott Isaksen and Joe Tidd (2006) *Meeting the Innovation Challenge*, John Wiley & Sons, Ltd, Chichester.

Risk taking

Tolerance of uncertainty and ambiguity constitutes risk taking. In a high risk-taking climate, bold new initiatives can be taken even when the outcomes are unknown. People feel that they can 'take a gamble' on some of their ideas. People will often 'go out on a limb' and be first to put an idea forward.

In a risk-avoiding climate there is a cautious, hesitant mentality. People try to be on the 'safe side'. They decide 'to sleep on the matter'. They set up committees and they cover themselves in many ways before making a decision. If risk taking is too low, employees offer few new ideas or few ideas that are well outside of what is considered safe or ordinary. In risk-avoiding organizations people complain about boring, low-energy jobs and are frustrated by a long, tedious process used to get ideas to action.

These conditions can be caused by the organization not valuing new ideas, or having an evaluation system that is bureaucratic, or people being punished for 'drawing outside the lines'. It can be remedied by developing a company plan that would speed 'ideas to action'.

Conversely, if there is too much risk taking, you will see that people are confused. There are too many ideas floating around, but few are sanctioned. People are frustrated because nothing is getting done. There are many loners doing their own thing in the organization and no evidence of teamwork. These conditions can be caused by individuals not feeling they need a consensus or buy-in from others on their team in their department or organization. A remedy might include some team building and improving the reward system to encourage cooperation rather than individualism or competition.

A recent study of organizational innovation and performance confirms the need for this delicate balance between risk and stability. Risk taking is associated with a higher relative novelty of innovation (how different it was to what the organization had done before), and absolute novelty (how different it was to what any organization had done before), and that both types of novelty are correlated with financial and customer benefits. However, the same study concludes that 'incremental, safe, widespread innovations may be better for internal considerations, but novel, disruptive innovations may be better for market considerations . . . absolute novelty benefits customers and quality of life, relative innovation benefits employee relations (but) risk is detrimental to employee relations'.

This is consistent with the real options approach to investing in risky projects, because investments are sequential and managers have some influence on the timing, resourcing and continuation or abandonment of projects at different stages. By investing relatively small amounts in a wide range of projects, a greater range of opportunities can be explored. Once uncertainty has been reduced, only the most promising projects should be allowed to continue. The goal is not to calculate or optimize, but rather to help to identify risks and payoffs, key uncertainties, decision points and future opportunities that might be created. Combined with other methods, such as decision trees, a real options approach can be particularly effective where high volatility demands flexibility, placing a premium on the certainty of information and timing of decisions.

Research on new product and service development has identified a broad range of strategies for dealing with risk. Both individual characteristics and organizational climate influence perceptions of risk and propensities to avoid, accept or seek risks. Formal techniques such as failure mode and effects analysis (FMEA), potential problem analysis (PPA) and fault tree analysis (FTA) have a role, but the broader signals and support from the organizational climate are more important than the specific tools or methods used. To help assess your own appetite for risk and creativity, try the self-assessment questionnaire online.

Freedom

Freedom is described as the independence in behaviour exerted by the people in the organization. In a climate with much freedom, people are given autonomy to define much of their own work. They are able to exercise discretion in their day-to-day activities. They take the initiative to acquire and share information, make plans and decisions about their work. In a climate with little freedom, people work within strict guidelines and roles. They carry out their work in prescribed ways with little room to redefine their tasks.

If there is not enough freedom people demonstrate very little initiative for suggesting new and better ways of doing things. They may spend a great deal of time and energy obtaining permission and gaining support (internally and externally) or perform all their work 'by the book' and focus too much on the exact requirements of what they are told to do. One of the many reasons could be that the leadership practices are very authoritarian or overly bureaucratic. It might be helpful to initiate a leadership improvement initiative including training, 360° feedback with coaching, skills of managing up, etc.

RESEARCH NOTE   Routines for organizing innovation

Nelson and Winter's (1982) concept of routines, as regular and predictable behavioural patterns, is central to evolutionary economics and studies of innovation. By definition, such routines:

- are regular and predictable
- are collective, social and tacit
- guide cognition, behaviour and performance
- promise to bridge (economic and cognition) theory and (management and organizational) practices
- 'the way we do things around here'.

In his review of the research, Becker (2005) suggested that the term 'recurrent interaction patterns' might provide a more precise term for organizational routines, understood as behavioural regularities. He argues that in practice routines can:

- enable coordination
- provide a degree of stability in behaviour
- enable tasks to be executed subconsciously, economizing on limited cognitive resources
- bind knowledge, including tacit knowledge.

However, in practice (and in management research) routines are very difficult to observe, measure or manage. For these reasons we focus less on the routines themselves, or individual cognition, and more on their influence in collective practice and on performance. Based upon the real-time observation of product and project development in two contrasting organizations it was found that routines play three limited but important roles: as prior and authoritative representations of action, such as standard templates, handbooks and processes; as part of a system of authority, specifications and conformance, such as formal decision points and criteria; and as a template for mandatory post-hoc representations of performed actions and their outcomes, such as audits and benchmarks (Hales and Tidd, 2009). Routines did not directly influence or prescribe actions or behaviours, but rather local instances of work practice and the ubiquitous management and production of knowledges were found in mundane interactions. Hales and Tidd believe that this is more relevant and realistic than the broader interpretations of routines found in much of the innovation and economics literature.

References

Becker, M.C. (2005) Organizational routines – a review of the literature. *Industrial and Corporate Change*, **13**, 643–77; Hales & Tidd, 2009, 18(4). The practice of routines and representations in design and development. *Industrial and Corporate Change*; Nelson, R.R. and S. Winter (1982) *An Evolutionary Theory of Economic Change*, Harvard University Press, Boston, MA.

If there is too much freedom you may observe people going off in their own independent directions. They have an unbalanced concern weighted toward themselves rather than the work group or organization. People may do things that demonstrate little or no concern for important policies/procedures, performing tasks differently and independently redefining how they are done each time. In this case people may not know the procedures, they could be too difficult to follow or the need to conform may be too low. You may start to reward improvement of manuals, process improvements and ways to communicate and share best practices to help correct the situation.

3.7 Boundary spanning

A recurring theme in this book is the extent to which innovation has become an open process involving richer networks across and between organizations. This highlights a long-established characteristic of successful innovating organizations – an orientation which is essentially open to new stimuli from outside.[77] Whether the signals are of threats or opportunities, these organizations have approaches that pick them up and communicate them through the organization. Developing a sense of external orientation – for example, towards key customers or sources of major technological developments – and ensuring that this pervades organizational thinking at all levels are of considerable importance in building an innovative organization.

One of the consistent themes in the literature on innovation success and failure concerns the need to understand user needs. Developing this sense of customer requirements is essential in dealing with the external marketplace as we will discuss in detail in Chapter 9, but it is also a principle which can be usefully extended within the organization. By developing a widespread awareness of customers – both internal and external – quality and innovation can be significantly improved. This approach, which forms one of the cornerstones of 'total quality management' thinking, contrasts sharply with the traditional model in which problems were passed on between sequential elements in the innovation process, and where there was no provision for feedback or mutual adjustment.[78]

Of course, not all industries have the same degree of customer involvement – and in many the dominant focus is more on technology. This does not mean that customer focus is an irrelevant concept: the issue here is one of building relationships which enable clear and regular communication, providing inputs for problem solving and shared innovation.[79]

But the idea of extending involvement goes far beyond customers and end-users. Open innovation requires building such relationships with an extended cast of characters, including suppliers, collaborators, competitors, regulators and multiple other players.[80]

All of the above discussion presumes that the organization in question is a single entity, a group of people organized in a particular fashion towards some form of collective purpose. But increasingly we are seeing the individual enterprise becoming linked with others in some form of collective – a supply chain, an industrial cluster, a cooperative learning club or a product development consortium. Studies exploring this aspect of inter-firm behaviour include learning in shared product development projects,[81] in complex product system configuration,[82] in technology fusion,[83] in strategic alliances,[84] in regional small-firm clusters,[85] in sector consortia,[86] in 'topic networks',[87] and in industry associations.[88]

Consider some examples:

- Studies of 'collective efficiency' have explored the phenomenon of clustering in a number of different contexts.[89] From this work it is clear that the model is not just confined to parts of Italy, Spain and Germany, but diffused around the world – and under certain conditions, extremely effective. For example, one town (Sialkot) in Pakistan plays a dominant role in the world market for specialist surgical instruments made of stainless steel. From a core group of 300 small firms, supported by 1500 even smaller suppliers, 90% of production (1996) was exported and took a 20% share of the world market, second only to Germany. In another case the Sinos valley in Brazil contains around 500 small-firm manufacturers of specialist, high-quality leather shoes. Between 1970 and 1990 their share of the world market rose from 0.3 to 12.5% and in 2006 they exported some 70% of total production. In each case the gains are seen as resulting from close interdependence in a cooperative network.

- Similarly, there has been much discussion about the merits of technological collaboration, especially in the context of complex product systems development.[90] Innovation networks of this kind offer significant advantages in terms of assembling different knowledge sets and reducing the time and costs of development – but are again often difficult to implement.[91]

- Much has been written on the importance of developing cooperative rather than adversarial supply chain relationships.[92] But it is becoming increasingly clear that the kind of 'collective efficiency' described above can operate in this context and contribute not only to improved process efficiency (higher quality, faster speed of response, etc.) but also to shared product development. The case of Toyota is a good illustration of this – the firm has continued to stay ahead despite increasing catch-up efforts on the part of Western firms and the consolidation of the industry. Much of this competitive edge can be attributed to its ability to create and maintain a high-performance knowledge-sharing network.[93]

- Networking represents a powerful solution to the resource problem – no longer is it necessary to have all the resources for innovation (particularly those involving specialized knowledge) under one roof provided you know where to obtain them and how to link up with them. The emergence of powerful information and communications technologies have further facilitated the move towards 'open innovation' and 'virtual organizations' are increasingly a feature of the business landscape.[94]

- Networking is increasingly offered as a way forward for industrial development. National and regional governments are trying to emulate the cluster effect which has proved so successful amongst small firms in Italy and Spain. Large firms are trying to develop their supply chains towards greater efficiency through supply chain learning and development programmes. Complex product systems are increasingly the subject of shared development projects amongst consortia of technology providers. But experience and research suggest that without careful management of these – and the availability of a shared commitment to deal with them – the chances are that such networks will fail to perform effectively.[95]

- Studies of learning behaviour in supply chains suggest considerable potential – one of the most notable examples being the case of the *kyoryokukai* (supplier associations) of Japanese manufacturers in the second half of the twentieth century.[96] Hines reports on other examples of supplier associations (including those in the UK), which have contributed to sustainable growth and development in a number of sectors particularly engineering and automotive.[92] Imai, in describing product development in Japanese manufacturers observes: '[Japanese firms exhibit] an almost fanatical devotion towards learning – both within organizational membership and with outside members of the inter-organizational network.'[97] Lamming[98] (p. 206) identifies such learning as a key feature of lean supply, linking it with innovation in supply relationships. Marsh and Shaw describe collaborative learning experiences in the wine industry including elements of supply chain learning (SCL), whilst the AFFA study reports on other experiences in the agricultural and food sector in Australia.[99] Case studies of SCL in the Dutch and UK food industries, the construction sector and aerospace provide further examples of different modes of SCL organization.[100] Humphrey *et al.* describe SCL emergence in a developing country context (India).[101]

However, obtaining the benefits of networking is not an automatic process – it requires considerable efforts in the area of coordination. Effective networks have what systems theorists call 'emergent properties' – that is, the whole is greater than the sum of the parts. But the risk is high that simply throwing together a group of enterprises will lead to suboptimal performance with the whole being considerably less than the sum of the parts due to friction, poor communications, persistent conflicts over resources or objectives, etc.

Research on inter-organizational networking suggests that a number of core processes need managing in a network, effectively treating it as if it were a particular form of organization.[102] For example, a network with no clear routes for resolving conflicts is likely to be less effective than one which has a clear and accepted set of norms – a 'network culture' – which can handle the inevitable conflicts that emerge.

Building and operating networks can be facilitated by a variety of enabling inputs, for example, the use of advanced information and communications technologies may have a marked impact on the effectiveness with which information processing takes place. In particular research highlights a number of enabling elements which help build and sustain effective networks, including:

- *Key individuals* – creating and sustaining networks depends on putting energy into their formation and operation. Studies of successful networks identify the role of key figures as champions and sponsors, providing leadership and direction, particularly in the tasks of bringing people together and giving a system-level sense of purpose.[103] Increasingly the role of 'network broker' is being played by individuals and agencies concerned with helping create networks on a regional or sectoral basis.
- *Facilitation* – another important element is providing support to the process of networking but not necessarily acting as members of the network. Several studies indicate that such a neutral and catalytic role can help, particularly in the set-up stages and in dealing with core operating processes like conflict resolution.
- *Key organizational roles* – mirroring these individual roles are those played by key organizations – for example, a regional development agency organizing a cluster or a business association bringing together a sectoral network. Gereffi and others talk about the concept of network governance and identify the important roles played by key institutions such as major customers in buyer-driven supply chains.[104] Equally their absence can often limit the effectiveness of a network, for example, in research on supply-chain learning the absence of a key governor limited the extent to which inter-organizational innovation could take place.[105]

We return to the theme of innovation and knowledge networks in Chapter 6.

3.8 Beyond the steady state

As we saw in Chapter 1, increasingly organizations have to deal with the challenge not only of managing innovation in the 'steady state' doing what they do, but better, but also under discontinuous 'do different' conditions. Research suggests that those organizations that are able to thrive and exploit innovative opportunities under these conditions are agile, fast moving and tolerant of high levels of risk and failure. For this reason it is often new entrant firms which do well and existing incumbents tend to suffer when discontinuities emerge. This does not mean that new entrants always win but rather that we need two different organizational models to deal with the two different sets of conditions. Table 3.7 outlines these.

This poses two related challenges for innovation management. For new-entrant enterprises trying to take advantage of new opportunities opened up by discontinuity the issues are particularly about

TABLE 3.7	**Two different innovation management archetypes**	
	Type 1 – Steady state archetype	**Type 2 – Discontinuous innovation archetype**
Interpretive schema – how the organization sees and makes sense of the world	There is an established set of 'rules of the game' by which other competitors also play	No clear 'rules of the game' – these emerge over time but cannot be predicted in advance
	Particular pathways in terms of search and selection environments and technological trajectories exist and define the 'innovation space' available to all players in the game	Need high tolerance for ambiguity – seeing multiple parallel possible trajectories
		'Innovation space' defined by open and fuzzy selection environment
		Probe and learn experiments needed to build information about emerging patterns and allow dominant design to emerge
	Strategic direction is highly path-dependent	Highly path-independent
Strategic decision making	Makes use of decision-making processes which allocate resources on the basis of risk management linked to the above 'rules of the game'. (Does the proposal fit the business strategic directions? Does it build on existing competence base?)	High levels of risk-taking since no clear trajectories – emphasis on fast and lightweight decisions rather than heavy commitment in initial stages
		Multiple parallel bets, fast failure and learning as dominant themes
	Controlled risks are taken within the bounds of the 'innovation space'	High tolerance of failure but risk is managed by limited commitment
	Political coalitions are significant influences maintaining the current trajectory	Influence flows to those prepared to 'stick their neck out' – entrepreneurial behaviour

(continued)

TABLE 3.7	**(Continued)**	
	Type 1 – Steady state archetype	**Type 2 – Discontinuous innovation archetype**
Operating routines	Operates with a set of routines and embedded structures/procedures which are linked to these 'risk rules', e.g. stage-gate monitoring and review for project management Search behaviour is along defined trajectories and uses tools and techniques for R&D, market research, etc. which assume a known space to be explored – search and selection environment Network building to support innovation, e.g. user involvement, supplier partnership, etc., is on basis of developing close and strong ties	Operating routines are open ended, based around managing emergence Project implementation is about 'fuzzy front end', light touch strategic review and parallel experimentation Probe and learn, fast failure and learn rather than managed risk Search behaviour is about peripheral vision, picking up early warning through weak signals of emerging trends Linkages are with heterogeneous population and emphasis less on established relationships than on weak ties

using their agility to probe, learn and reconfigure in search of the dominant design which will eventually emerge. But doing so is resource intensive and hard to sustain; for this reason many new-entrant firms fail because they back the wrong horse or run out of resources. If they do manage to ride the emerging wave until a dominant design emerges they then face the challenge of building a business – essentially moving into the 'mature phase' by building routines and structures to support incremental development.

At the other extreme established organizations run the risk of being too slow to respond or too set in their organizational ways to manage the transition effectively. Their challenge is to find ways of retaining both sets of characteristics within an organization – ambidextrous capability – and there are many different approaches to dealing with the challenge. At one end of the spectrum are models which spin or split off the 'type 2' activities in some form of separate venture, whilst at the other end there are models which attempt to build ambidextrous capability within the organization, for example through the intrapreneurship model of 3M and others (see Case study 3.11). We discuss this theme in more detail in Chapter 10.

If we revisit our model of the innovative organization we can see some specific themes relevant to building such capability (see Table 3.8).

TABLE 3.8	**Components of the innovative organization under discontinuous conditions**	

Component	Key features	Example references
Shared vision, leadership and the will to innovate	Top level support for difficult decisions or radical new directions	107–109
	Different perspectives – often coming from outside the organization or the sector	
	Willingness to let go of the past	
Appropriate structure	Type 1 and type 2 models and finding balance – the ambidextrous challenge	110, 111
	Corporate venturing models, skunk works and other modes	
Key individuals	Key roles of gatekeepers to extend peripheral vision and champions to promote risk taking	112
	New roles to facilitate internal venturing – for example, Shell's 'Gamechangers'	
	Emphasis on intrapreneurship	
Effective team working	Increasing emphasis on bringing together different perspectives and fast-forming temporary teams	68, 113
	Increasing boundary crossing within and between organizations – virtual and dispersed team working	
Continuing and stretching individual development	Training to think and work 'out of the box'	114
	Development of alternative perspectives via formal training, secondment, etc.	

(continued)

TABLE 3.8 (Continued)		
Component	**Key features**	**Example references**
Extensive communication	Need to develop channels for unorthodox ideas to flow	115, 116
	Need capacity to deal with 'off-message' signals	
High involvement in innovation	Internal programmes which seek out and capture new ideas from across the organization and harness entrepreneurial energy to take them forward	111
External focus	Extensive networking required to extend peripheral vision	80, 117, 118
	Move beyond existing and effective value networks to open up new options	
	Open innovation	
Creative climate	Fostering open environment receptive to new and often challenging ideas	69, 73
	Development of intrapreneurship rather than forcing people to leave to exploit new opportunities which they see and believe in	
Learning organization	Increasing emphasis on 'probe and learn' and high failure/fast learning	119–121
	Extending learning across boundaries and into networks	

CASE STUDY 3.11

Building an innovative organization – the case of 3M

3M is a well-known organization employing around 70 000 people in around 200 countries across the world. Its $15 billion of annual sales come from a diverse product range involving around 50 000 items serving multiple markets but building on core technical strengths, some of which like coatings can be traced back to the company's foundation. The company has been around for just over 100 years and during that period has established a clear reputation as a major innovator. Significantly the company paints a consistent picture in interviews and in publications – innovation success is a consequence of creating the culture in which it can take place – it becomes 'the way we do things around here' in a very real sense. This philosophy is borne out in many anecdotes and case histories – the key to their success has been to create the conditions in which innovation can arise from any one of a number of directions, including lucky accidents, and there is a deliberate attempt to avoid putting too much structure in place since this would constrain innovation.

Elements in this complex web include:

- Recognition and reward – throughout the company there are various schemes which acknowledge innovative activity, for example, the Innovator's Award which recognizes effort rather than achievement.
- Reinforcement of core values – innovation is respected, for example, there is a 'hall of fame' whose members are elected on the basis of their innovative achievements.
- Sustaining 'circulation' – movement and combination of people from different perspectives to allow for creative combinations – a key issue in such a large and dispersed organization.
- Allocating 'slack' and permission to play – allowing employees to spend a proportion of their time in curiosity-driven activities which may lead nowhere but which have sometimes given them breakthrough products.
- Patience – acceptance of the need for 'stumbling in motion' as innovative ideas evolve and take shape. Breakthroughs like Post-its and 'Scotchgard' were not overnight successes but took two to three years to 'cook' before they emerged as viable prospects to put into the formal system.
- Acceptance of mistakes and encouragement of risk taking – a famous quote from a former CEO is often cited in this connection: 'Mistakes will be made, but if a person is essentially right, the mistakes he or she makes are not as serious, in the long run, as the mistakes management will make if it's dictatorial and undertakes to tell those under its authority exactly how they must do their job . . . Management that is destructively critical when mistakes are made kills initiative, and it is essential that we have many people with initiative if we are to continue to grow.'
- Encouraging 'bootlegging' – giving employees a sense of empowerment and turning a blind eye to creative ways which staff come up with to get around the system – acts as a counter to rigid bureaucratic procedures.
- Policy of hiring innovators – recruitment approach is looking for people with innovator tendencies and characteristics.
- Recognition of the power of association – deliberate attempts not to separate out different functions but to bring them together in teams and other groupings.
- Encouraging broad perspectives – for example, in developing their overhead projector business it was close links with users developed by getting technical development staff to make sales calls that made the product so user friendly and therefore successful.
- Strong culture – dating back to 1951 of encouraging informal meetings and workshops in a series of groups, committees, etc., under the structural heading of the Technology Forum – established

'to encourage free and active interchange of information and cross-fertilization of ideas'. This is a voluntary activity although the company commits support resources – it enables a company-wide 'college' with fluid interchange of perspectives and ideas.

- Recruiting volunteers – particularly in trying to open up new fields; involvement of customers and other outsiders as part of a development team is encouraged since it mixes perspectives.

A more detailed case study of 3M is available from our web resources.

Summary and further reading

The field of organizational behaviour is widely discussed and there are some good basic texts, such as D. Buchanan and A. Huczynski, *Organizational Behaviour* (sixth edition, FT Prentice Hall, 2007), which provides an excellent synthesis of the main issues, with a good balance of managerial and more critical social science approaches. Specific issues surrounding the development of innovative organizations are well treated by R. Leifer *et al.*, *Radical Innovation* (Harvard Business School Press, 2000), and R. Kanter, *World Class* (Simon & Schuster, 1996). We address the relationships between leadership, innovation and organizational renewal more fully in our book *Meeting the Innovation Challenge: Leadership for Transformation and Growth*, by Scott Isaksen and Joe Tidd (John Wiley & Sons, Ltd, 2006).

Many books and articles look at specific aspects, for example: the development of creative climates, Lynda Gratton, *Hot Spots: Why some companies buzz with energy and innovation, and others don't* (Prentice Hall, 2007); team working by T. DeMarco and T. Lister, *Peopleware: Productive projects and teams* (Dorset House, 1999); or continuous improvement, John Bessant's *High-Involvement Innovation* (John Wiley & Sons, Ltd, 2003). R. Katz, *The Human Side of Managing Technological Innovation* (Oxford University Press, 2003) is an excellent collection of readings, and A.H. Van de Ven, D. Polley, H.L. Angle and M.S. Poole, *The Innovation Journey* (Oxford University Press, 2008) provides a comprehensive review of a seminal study in the field, and includes a discussion of individual, group and organizational issues. Case studies of innovative organizations focus on many of the issues highlighted in this chapter, and good examples include E. Gundling, *The 3M Way to Innovation: Balancing people and profit* (Kodansha International, 2000) and *Corning and the Craft of Innovation* by M. Graham and A. T. Shuldiner (Oxford University Press, 2001).

The 'beyond boundaries' issue of networking is covered by several writers, most following the popular notion of 'open innovation'. The notion was most recently popularized by Henry Chesbrough in *Open Innovation* (Harvard Business School Press, 2003), and has since spawned many similar discussions, but a more serious review of the evidence and research can be found in *Open Innovation: Researching a New Paradigm* (edited by H. Chesbrough, W. Vanhaverbeke and J. West, Oxford University Press, 2008). Finally the theme of 'beyond the steady state' and its implications for organizing for innovation is picked up in several pieces, including G. Day and P. Schoemaker, *Wharton on Managing Emerging Technologies* (John Wiley & Sons Inc., 2000) and T. Jones, *Innovating at the Edge* (Butterworth Heinemann, 2002).

Web links

Here are the full details of the resources available on the website flagged throughout the text:

 Case studies:
  Hosiden
  3M

 Interactive exercises:
  Creativity questionnaire

 Tools:
  Brainstorming
  Problem solving
  Team building

 Video podcast:
  Veedor Root

 Audio podcast:
  Ken Robinson on creativity

References

1. **Kirkpatrick, D**. (1998) The second coming of Apple. *Fortune*, **138**, 90.
2. **Pfeffer, J.** (1998) *The Human Equation: Building Profits by Putting People First*, Harvard Business School Press, Boston, MA.
3. **Caulkin, S.** (2001) *Performance through People*, Chartered Institute of Personnel and Development, London.
4. **Huselid, M.** (1995) The impact of human resource management practices on turnover, productivity and corporate financial performance. *Academy of Management Journal*, **38**, 647–56.
5. **Womack, J., D. Jones and D. Roos** (1991) *The Machine that Changed the World*, Rawson Associates, New York.
6. **Christenson, C.** (1997) *The Innovator's Dilemma*, Harvard Business School Press, Boston, MA.
7. **Tripsas, M. and G. Gavetti** (2000) Capabilities, cognition and inertia: evidence from digital imaging. *Strategic Management Journal*, **21**, 1147–61.
8. **Leonard-Barton, D.** (1995) *Wellsprings of Knowledge: Building and Sustaining the Sources of Innovation*, Harvard Business School Press, Boston, MA.
9. **Pinchot, G.** (1999) *Intrapreneuring in Action – Why You Don't Have to Leave a Corporation to Become an Entrepreneur*, Berrett-Koehler, New York.
10. **Catmull, E.** (2008) How Pixar Fosters Collective Creativity. *Harvard Business Review*, **86** (9), 64–72.
11. **Moody, F.** (1995) *I Sing the Body Electronic*, Hodder & Stoughton, London.
12. **Cooper, R.** (2001) *Winning at New Products*, 3rd edition, Kogan Page, London.
13. **Bowman, E.H. and Helfat, C.E.** (2001) Does corporate strategy matter? *Strategic Management Journal*, **22**, 1–23.

14. **Clark, K.E. and M.B. Clark** (1990) *Measures of Leadership*, The Center for Creative Leadership, Greensboro, NC; **Clark, K.E., M.B. Clark and D.P. Campbell** (1992). *Impact of Leadership*, The Center for Creative Leadership, Greensboro, NC.

15. **Mann, R.D.** (1959) A review of the relationships between personality and performance in small groups. *Psychological Bulletin*, **56**, 241–70.

16. **Connelly, M.S., J.A. Gilbert, S.J. Zaccaro, K.V. Threlfall, M.A. Marks, and M.D. Mumford** (2000) Exploring the relationship of leader skills and knowledge to leader performance. *The Leadership Quarterly*, **11**, 65–86; **Zaccaro, S.J., J.A. Gilbert, K.K. Thor and M.D. Mumford** (2000) Assessment of leadership problem-solving capabilities. *The Leadership Quarterly*, **11**, 37–64.

17. **Farris, G.F.** (1972) The effect of individual role on performance in creative groups. *R&D Management*, **3**, 23–8; **Ehrhart, M.G. and K.J. Klein** (2001) Predicting followers' preferences for charismatic leadership: the influence of follower values and personality. *The Leadership Quarterly*, **12**, 153–80.

18. **Scott, S.G. and R.A. Bruce** (1994) Determinants of innovative behavior: a path model of individual innovation in the workplace. *Academy of Management Journal*, **37** (3), 580–607.

19. **Amabile, T.M., E.A. Schatzel, G.B. Moneta and S.J. Kramer** (2004) Leader behaviors and the work environment for creativity: perceived leader support. *The Leadership Quarterly*, **15** (1), 5–32.

20. **Rafferty, A.E. and M.A. Griffin** (2004) Dimensions of transformational leadership: conceptual and empirical extensions. *The Leadership Quarterly*, **15** (3), 329–54.

21. **Mumford, M.D., S.J. Zaccaro, F.D. Harding, T.O. Jacobs, and E.A. Fleishman** (2000) Leadership skills for a changing world: solving complex social problems. *The Leadership Quarterly*, **11**, 11–35.

22. **Pinto, J. and D. Slevin** (1989) Critical success factors in R&D projects. *Research-Technology Management*, **32**, 12–18; **Podsakoff, P.M., S.B. Mackenzie, J.B. Paine and D.G. Bachrach** (2000) Organizational citizenship behaviors: a critical review of the theoretical and empirical literature and suggestions for future research. *Journal of Management*, **26** (3), 513–63.

23. **West, M.A., C.S. Borrill, J.F. Dawson, F. Brodbeck, D.A. Shapiro, and B. Haward** (2003) Leadership clarity and team innovation in health care. *The Leadership Quarterly*, **14** (4–5), 393–410.

24. **Andrews, F.M. and G.F. Farris** (1967) Supervisory practices and innovation in scientific teams. *Personnel Psychology*, **20**, 497–515; **Barnowe, J.T.** (1975) Leadership performance outcomes in research organizations. Organizational *Behavior and Human Performance*, **14**, 264–80; **Elkins, T. and R.T. Keller** (2003) Leadership in research and development organizations: a literature review and conceptual framework. *The Leadership Quarterly*, **14**, 587–606.

25. **Keller, R.T.** (1992) Transformational leadership and performance of research and development project groups. *Journal of Management*, **18**, 489–501.

26. **Berson, Y. and J.D. Linton** (2005) An examination of the relationships between leadership style, quality, and employee satisfaction in R&D versus administrative environments. *R&D Management*, **35** (1), 51–60.

27. **Thompson, J.** (1967) *Organizations in Action*, McGraw-Hill, New York.

28. **Perrow, C.** (1967) A framework for the comparative analysis of organizations. *American Sociological Review*, **32**, 194–208.

29. **Burns, T. and G. Stalker** (1961) *The Management of Innovation*, Tavistock, London.

30. **Kanter, R.** (1984) *The Change Masters*, Unwin, London.

31. **Hesselbein, F., M. Goldsmith and R. Beckhard, eds** (1997) *Organization of the Future*, Jossey Bass/The Drucker Foundation, San Francisco.

32. **Champy, J. and N. Nohria, eds** (1996) *Fast Forward*, Harvard Business School Press, Boston, MA.

33. **Semler, R.** (1993) *Maverick*, Century Books, London; **Kaplinsky, R., F. den Hertog and B. Coriat** (1995) *Europe's Next Step*, Frank Cass, London.

34. **Miles, R. and C. Snow** (1978) *Organizational Strategy, Structure and Process*, McGraw-Hill, New York; **Lawrence, P. and P. Dyer** (1983) *Renewing American Industry*, Free Press, New York.

35. **Lawrence, P. and J. Lorsch** (1967) *Organization and Environment*, Harvard University Press, Boston, MA.

36. **Stalk, G. and T. Hout** (1990) *Competing against Time: How Time-Based Competition is Reshaping Global Markets*, Free Press, New York.

37. **Woodward, J.** (1965) *Industrial Organization: Theory and Practice*, Oxford University Press, Oxford.

38. **Child, J.** (1980) *Organisations*, Harper & Row, London.

39. **Mintzberg, H.** (1979) *The Structuring of Organizations*, Prentice-Hall, Englewood Cliffs, NJ.

40. **Adler, P.** (1992) The learning bureaucracy: NUMMI. In B. Staw and L. Cummings, eds, *Research in Organizational Behavior*, JAI Press, Greenwich, CT.

41. **Nayak, P. and J. Ketteringham** (1986) *Breakthroughs: How Leadership and Drive Create Commercial Innovations that Sweep the World*, Mercury, London; **Kidder, T.** (1981) *The Soul of a New Machine*, Penguin, Harmondsworth.

42. **Clark, K. and T. Fujimoto** (1992) *Product Development Performance*, Harvard Business School Press, Boston, MA.

43. **Rothwell, R.** (1992) Successful industrial innovation: critical success factors for the 1990s. *R&D Management*, **22** (3), 221–39.

44. **Allen, T.** (1977) *Managing the Flow of Technology*, MIT Press, Boston, MA.

45. **Blackler, F.** (1995) Knowledge, knowledge work and organizations. *Organization Studies*, **16** (6), 1021–46; **Sapsed, J., J. Bessant, D. Partington, D. Tranfield, and M. Young** (2002) Teamworking and knowledge management: a review of converging themes. *International Journal of Management Reviews*, **4** (1).

46. **Duarte, D. and N. Tennant Snyder** (1999) *Mastering Virtual Teams*, Jossey Bass, San Francisco.

47. **Kaplinsky, R.** (1994) *Easternization: The Spread of Japanese Management Techniques to Developing Countries*, Frank Cass, London; **Schroeder, D. and A. Robinson** (1991) America's most successful export to Japan – continuous improvement programs. *Sloan Management Review*, **32** (3), 67–81.

48. **Figuereido, P.** (2001) *Technological Learning and Competitive Performance*, Edward Elgar, Cheltenham.

49. **Deming, W.** (1986) *Out of the Crisis*, MIT Press, Boston, MA; **Crosby, P.** (1977) *Quality is Free*, McGraw-Hill, New York; **Dertouzos M, R. Lester and L. Thurow** (1989) *Made in America: Regaining the Productive Edge*, MIT Press, Boston, MA; **Garvin, D.** (1988) *Managing Quality*, Free Press, New York.

50. **Womack, J. and D. Jones** (1997) *Lean Thinking*, Simon & Schuster, New York.

51. **Schonberger, R.** (1982) *Japanese Manufacturing Techniques: Nine Hidden Lessons in Simplicity*, Free Press, New York.

52. **Shingo, S.** (1983) *A Revolution in Manufacturing: The SMED System*, Productivity Press, Boston, MA; **Suzaki, K.** (1988) *The New Manufacturing Challenge*, Free Press, New York.

53. **Lillrank, P. and N. Kano** (1990) *Continuous Improvement; Quality Control Circles in Japanese Industry*, University of Michigan Press, Ann Arbor.

54. **Caffyn, S.** (1998) *Continuous Improvement in the New Product Development Process*, Centre for Research in Innovation Management, University of Brighton, Brighton; **Lamming, R.** (1993) *Beyond Partnership*, Prentice-Hall, London; **Owen, M. and J. Morgan** (2000) *Statistical Process Control in the Office*, Greenfield Publishing, Kenilworth.

55. **Ishikure, K.** (1988) Achieving Japanese productivity and quality levels at a US plant. *Long Range Planning*, **21** (5), 10–17; **Wickens, P.** (1987) *The Road to Nissan: Flexibility, Quality, Teamwork*, Macmillan, London.

56. **Bessant, J.** (2003) *High Involvement Innovation,* John Wiley & Sons, Ltd, Chichester.

57. **Peters, T.** (1988) *Thriving on Chaos*, Free Press, New York.

58. **Forrester, R. and A. Drexler** (1999) A model for team-based organization performance. *Academy of Management Executive*, **13** (3), 36–49; **Conway, S. and R. Forrester** (1999) *Innovation and Teamworking: Combining Perspectives Through a Focus on Team Boundaries*, University of Aston Business School, Birmingham.

59. **Thamhain, H. and D. Wilemon** (1987) Building high performing engineering project teams. IEEE *Transactions on Engineering Management*, EM-34 (3), 130–7.

60. **Bixby, K.** (1987) *Superteams*, Fontana, London.

61. **Holti, R., J. Neumann and H. Standing** (1995) *Change Everything at Once: The Tavistock Institute's Guide to Developing Teamwork in Manufacturing*, Management Books 2000, London.

62. **Tranfield, D., I. Parry, S. Wilson, S. Smith, and M. Foster** (1998) Teamworked organizational engineering: getting the most out of teamworking. *Management Decision*, **36** (6), 378–84.

63. **Jassawalla, A. and H. Sashittal** (1999) Building collaborative cross-functional new product teams. *Academy of Management Executive*, **13** (3), 50–3.

64. **DTI** (1996) *UK Software Purchasing Survey*, Department of Trade and Industry, London.

65. **Van Beusekom, M.** (1996) *Participation Pays! Cases of Successful Companies with Employee Participation*, Netherlands Participation Institute, The Hague.

66. **Tuckman, B. and N. Jensen** (1977) Stages of small group development revisited. *Group and Organizational Studies*, **2**, 419–27.

67. **Belbin, M.** (2004) *Management Teams – Why they Succeed or Fail*, Butterworth-Heinemann, London.

68. **Smith, P. and E. Blanck** (2002) From experience: leading dispersed teams. *Journal of Product Innovation Management*, **19**, 294–304.

69. **Isaksen, S. and J. Tidd** (2006) *Meeting the Innovation Challenge: Leadership for Transformation and Growth*, John Wiley & Sons, Ltd, Chichester.

70. **Hackman J. R. ed.** (1990). *Groups That Work (And Those That Don't): Creating Conditions for Effective Teamwork*, Jossey Bass, San Francisco.

71. **Kirton, M.** (1989) *Adaptors and Innovators*, Routledge, London.

72. **Schein, E.** (1984) 'Coming to a new awareness of organizational culture. *Sloan Management Review*, Winter, 3–16.

73. **Leonard, D. and W. Swap** (1999) *When Sparks Fly: Igniting Creativity in Groups*, Harvard Business School Press, Boston, MA; **Amabile, T.** (1998) How to kill creativity. *Harvard Business Review*, September/October, 77–87.

74. **Kanter, R. ed.** (1997) *Innovation: Breakthrough thinking at 3M, DuPont, GE, Pfizer and Rubbermaid*, Harper Business, New York.

75. **Cook, P.** (1999) *Best Practice Creativity*, Gower, Aldershot; **Rickards, T.** (1997) *Creativity and Problem Solving at Work*, Gower, Aldershot.

76. **Badawy, M.** (1997) *Developing Managerial Skills in Engineers and Scientists*, John Wiley & Sons, Inc., New York.

77. **Carter, C. and B. Williams** (1957) *Industry and Technical Progress*, Oxford University Press, Oxford.

78. **Oakland, J.** (1989) *Total Quality Management*, Pitman, London.

79. **Schonberger, R.** (1990) *Building a Chain of Customers*, Free Press, New York.

80. **Chesbrough, H.** (2003) *Open Innovation: The New Imperative for Creating and Profiting from Technology*, Harvard Business School Press, Boston, MA.

81. **Bozdogan, K.** (1998) Architectural innovation in product development through early supplier integration. *R&D Management*, **28** (3), 163–73; **Oliver, N. and M. Blakeborough** (1998) Innovation networks: the view from the inside. In J. Grieve Smith and J. Michie, eds, *Innovation, Cooperation and Growth*, Oxford University Press, Oxford.

82. **Miller, R.** (1995) Innovation in complex systems industries: the case of flight simulation. *Industrial and Corporate Change*, **4** (2), 363–400.

83. **Tidd, J.** (1997) Complexity, networks and learning: integrative themes for research on innovation management. *International Journal of Innovation Management*, **1** (1), 1–22.

84. **Simonin, B.** (1999) Ambiguity and the process of knowledge transfer in strategic alliances. *Strategic Management Journal*, **20**, 595–623; **Szulanski, G.** (1996) Exploring internal stickiness: impediments to the transfer of best practice within the firm. *Strategic Management Journal*, **17**, 5–9; **Hamel, G., Y. Doz and C. Prahalad** (1989) Collaborate with your competitors – and win. *Harvard Business Review*, **67** (2), 133–9.

85. **Schmitz, H.** (1998) Collective efficiency and increasing returns. *Cambridge Journal of Economics*, **23** (4), 465–83; **Nadvi, K. and H. Schmitz** (1994) *Industrial Clusters in Less Developed Countries: Review of Experiences and Research Agenda*, Institute of Development Studies, Brighton; **Keeble, D. and F. Williamson, eds** (2000) *High Technology Clusters, Networking and Collective Learning in Europe*, Aldershot, Ashgate.

86. **DTI/CBI** (2000) *Industry in Partnership*, Department of Trade and Industry/Confederation of British Industry, London.

87. **Bessant, J.** (1995) Networking as a mechanism for technology transfer: the case of continuous improvement. In R. Kaplinsky, F. den Hertog and B. Coriat, eds, *Europe's Next Step*, Frank Cass, London.

88. **Semlinger, K.** (1995) Public support for firm networking in Baden-Wurttemburg. In R. Kaplinsky, F. den Hertog and B. Coriat, eds, *Europe's Next Step*, Frank Cass, London; **Keeble, D. et al.** (1999) Institutional thickness in the Cambridge region. *Regional Studies*, **33** (4), 319–32.

89. **Piore, M. and C. Sabel** (1982) *The Second Industrial Divide*, Basic Books, New York; **Nadvi, K.** (1997) *The Cutting Edge: Collective Efficiency and International Competitiveness in Pakistan*, Institute of Development Studies, University of Sussex.

90. **Dodgson, M.** (1993) *Technological Collaboration in Industry*, Routledge, London; **Hobday, M.** (1996) *Complex Systems vs Mass Production Industries: A New Innovation Research Agenda*, Complex Product Systems Research Centre, Brighton; **Marceau, J.** (1994) Clusters, chains and complexes: three approaches to innovation with a public policy perspective. In R. Rothwell and M. Dodgson, eds, *The Handbook of Industrial Innovation*, Edward Elgar, Aldershot.

91. **Oliver, N. and M. Blakeborough** (1998) Innovation networks: the view from the inside. In J. Grieve Smith and J. Michie, eds, *Innovation, Cooperation and Growth*, Oxford University Press, Oxford; **Tidd, J.** (1997) Complexity, networks and learning: integrative themes for research on innovation management. *International Journal of Innovation Management*, **1** (1), 1–22.

92. **Hines, P.** (1999) *Value Stream Management: The Development of Lean Supply Chains*, Financial Times Management, London.

93. **Dyer, J. and K. Nobeoka** (2000) Creating and managing a high-performance knowledge-sharing network: the Toyota case. *Strategic Management Journal*, **21** (3), 345–67.

94. **Dell, M.** (1999) *Direct from Dell*, HarperCollins, New York;

95. **Best, M.** (2001) *The New Competitive Advantage*, Oxford University Press, Oxford.

96. **Hines, P.** (1994) *Creating World Class Suppliers: Unlocking Mutual Competitive Advantage*, Pitman, London; **Cusumano, M.** (1985) *The Japanese Automobile Industry: Technology and Management at Nissan and Toyota*, Harvard University Press, Boston, MA.

97. **Imai, K.** (1987) *Kaizen*, Random House, New York.

98. **Lamming, R.** (1993) *Beyond Partnership*, Prentice-Hall, London

99. **AFFA** (1998) *Chains of Success*, Department of Agriculture, Fisheries and Forestry – Australia (AFFA), Canberra; **Marsh, I. and B. Shaw** (2000) Australia's wine industry: collaboration and learning as causes of competitive success. In Working Paper, Australian Graduate School of Management, Melbourne

100. **AFFA** (2000) *Supply Chain Learning: Chain Reversal and Shared Learning for Global Competitiveness*, Department of Agriculture, Fisheries and Forestry – Australia (AFFA), Canberra; **Fearne, A. and D. Hughes** (1999) Success factors in the fresh produce supply chain: insights from the UK. *Supply Management*, **4** (3); **Dent, R.** (2001) *Collective Knowledge Development, Organisational Learning and Learning Networks: An Integrated Framework*, Economic and Social Research Council, Swindon.

101. **Humphrey, J., R. Kaplinsky and P. Saraph** (1998) *Corporate Restructuring: Crompton Greaves and the Challenge of Globalization*, Sage Publications, New Delhi.

102. **Bessant, J. and G. Tsekouras** (2001) Developing learning networks. *AI and Society*, **15** (2), 82–98.

103. **Barnes, J. and M. Morris** (1999) *Improving Operational Competitiveness through Firm-level Clustering: A Case Study of the KwaZulu-Natal Benchmarking Club*, School of Development Studies, University of Natal, Durban, South Africa.

104. **Kaplinsky, R., M. Morris and J. Readman** (2003) The globalization of product markets and immiserising growth: lessons from the South African furniture industry. *World Development*, **30** (7), 1159–78; **Gereffi, G.** (1994) The organisation of buyer-driven global commodity chains: how US retailers shape overseas production networks. In G. Gereffi and P. Korzeniewicz, eds, *Commodity Chains and Global Capitalism*, Praeger, London.

105. **Bessant, J., R. Kaplinsky and R. Lamming** (2003) Putting supply chain learning into practice. *International Journal of Operations and Production Management*, **23** (2), 167–84.

106. **Francis, D., J. Bessant and M. Hobday** (2003) Managing radical organisational transformation. *Management Decision*, **41** (1), 18–31.

107. **Foster, R. and S. Kaplan** (2002) *Creative Destruction*, Harvard University Press, Boston, MA.

108. **Welch, J.** (2001) *Jack! What I've Learned From Leading a Great Company and Great People*, Headline, New York.

109. **Groves, A.** (1999) *Only the Paranoid Survive*, Bantam Books, New York.

110. **Buckland, W., A. Hatcher and J. Birkinshaw** (2003) *Inventuring: Why Big Companies Must Think Small*, McGraw Hill, London.

111. **Birkinshaw, J. J. Bessant and R. Delbridge** (2007) Finding, forming and performing: creating networks for discontinuous innovation. *California Management Review*, **49** (3), 67–84.

112. **Day, G. and P. Schoemaker** (2004) Driving through the fog: managing at the edge. *Long Range Planning*, **37**, 127–42.

113. **Rich, B. and L. Janos** (1994) *Skunk Works*, Warner Books, London.

114. **Kingdon, M.** (2002) *Sticky Wisdom – How to Start a Creative Revolution at Work*, Capstone, London.

115. **Winter, S.** (2004) Specialized perception, selection and strategic surprise: learning from the moths and the bees. *Long Range Planning*, **37**, 163–9.

116. **Prahalad, C.** (2004) The blinders of dominant logic. *Long Range Planning*, **37** (2), 171–9.

117. **Huston, L.** (2004) Mining the periphery for new products. *Long Range Planning*, **37** (2), 191–6.

118. **Seely Brown, J.** (2004) Minding and mining the periphery. *Long Range Planning*, **37** (2), 143–51.

119. **Allen, P.** (2001) A complex systems approach to learning, adaptive networks. *International Journal of Innovation Management*, **5**, 149–80.

120. **Ayas, K.** (1997) *Design for Learning for Innovation*, Eburon, Delft.

121. **Mariotti, F. and R. Delbridge** A portfolio of ties: managing knowledge transfer and learning within network firms. *Academy of Management Review* (forthcoming).

CHAPTER 4

Developing an innovation strategy[*]

A great deal of business success depends on generating new knowledge and on having the capabilities to react quickly and intelligently to this new knowledge . . . I believe that strategic thinking is a necessary but overrated element of business success. If you know how to design great motorcycle engines, I can teach you all you need to know about strategy in a few days. If you have a Ph.D. in strategy, years of labor are unlikely to give you the ability to design great new motorcycle engines.

(Richard Rumelt (1996) *California Management Review,* **38**, 110, on the continuing debate about the causes of Honda's success in the US motorcycle market)

The above quotation from a distinguished professor of strategy appears on the surface not to be a strong endorsement of his particular trade. In fact, it offers indirect support for the central propositions of this chapter:

1. Firm-specific knowledge – including the capacity to exploit it – is an essential feature of competitive success.
2. An essential feature of corporate strategy should therefore be an innovation strategy, the purpose of which is deliberately to accumulate such firm-specific knowledge.
3. An innovation strategy must cope with an external environment that is complex and ever-changing, with considerable uncertainties about present and future developments in technology, competitive threats and market (and non-market) demands.
4. Internal structures and processes must continuously balance potentially conflicting requirements:
   (a) to identify and develop specialized knowledge within technological fields, business functions and product divisions;
   (b) to exploit this knowledge through integration across technological fields, business functions and product divisions.

Given complexity, continuous change and consequent uncertainty, we believe that the so-called rational approach to innovation strategy, still dominant in practice and in the teaching at many business schools, is less likely to be effective than an incremental approach that stresses continuous adjustment in the light of new knowledge and learning. We also argue that the approach pioneered by Michael Porter correctly identifies the nature of the competitive threats and opportunities that emerge from advances in technology, and rightly stresses the importance of developing and protecting firm-specific technology in order to enable firms to position themselves against the competition. But it underestimates the power of technology to change the rules of the competitive game by modifying industry

---

[*] The approach we adopt in this chapter is based upon the pioneering work of our former colleague, friend and co-author, the late Keith Pavitt. The more specific topic of *technology* strategy is explored in greater detail in earlier editions of this book, in particular the first (1997) and second (2001).

boundaries, developing new products and shifting barriers to entry. It also overestimates the capacity of senior management to identify and predict the important changes outside the firm, and to implement radical changes in competencies and organizational practices within the firm.

In this chapter, we develop what we think is the most useful framework for defining and implementing innovation strategy. We propose that such a framework is the one developed by David Teece and Gary Pisano. It gives central importance to the dynamic capabilities of firms, and distinguishes three elements of corporate innovation strategy: (i) competitive and national positions; (ii) technological paths; (iii) organizational and managerial processes. We begin by summarizing the fundamental debate in corporate strategy between 'rationalist' and 'incrementalist' approaches, and argue that the latter approach is more realistic, given the inevitable complexities and uncertainties in the innovation process.

4.1 'Rationalist' or 'incrementalist' strategies for innovation?

The long-standing debate between 'rational' and 'incremental' strategies is of central importance to the mobilization of technology and to the purposes of corporate strategy. We begin by reviewing the main terms of the debate, and conclude that the supposedly clear distinction between strategies based on 'choice' or on 'implementation' breaks down when firms are making decisions in complex and fast-changing competitive environments. Under such circumstances, formal strategies must be seen as part of a wider process of continuous learning from experience and from others to cope with complexity and change.

Notions of corporate strategy first emerged in the 1960s. A lively debate has continued since then amongst the various 'schools' or theories. Here we discuss the two most influential: the 'rationalist' and the 'incrementalist'. The main protagonists are Ansoff[1] of the rationalist school and Mintzberg[2] amongst the incrementalists. A face-to-face debate between the two in the 1990s can be found in the *Strategic Management Journal* and an excellent summary of the terms of the debate can be found in Whittington.[3]

Rationalist strategy

'Rationalist' strategy has been heavily influenced by military experience, where strategy (in principle) consists of the following steps: (i) describe, understand and analyze the environment; (ii) determine a course of action in the light of the analysis; (iii) carry out the decided course of action. This is a 'linear model' of rational action: appraise, determine and act. The corporate equivalent is SWOT: the analysis of corporate strengths and weaknesses in the light of external opportunities and threats. This approach is intended to help the firm to:

- Be conscious of trends in the competitive environment.
- Prepare for a changing future.
- Ensure that sufficient attention is focused on the longer term, given the pressures to concentrate on the day to day.
- Ensure coherence in objectives and actions in large, functionally specialized and geographically dispersed organizations.

However, as John Kay has pointed out, the military metaphor can be misleading.[4] Corporate objectives are different from military ones: namely, to establish a distinctive competence enabling them to satisfy customers better than the competition – and not to mobilize sufficient resources to destroy the enemy (with perhaps the exception of some Internet companies). Excessive concentration on the 'enemy' (i.e. corporate competitors) can result in strategies emphasizing large commitments of resources for the establishment of monopoly power, at the expense of profitable niche markets and of a commitment to satisfying customer needs (see Research Note).

More important, professional experts, including managers, have difficulties in appraising accurately their real situation, essentially for two reasons. First, their external environment is both *complex*,

---

RESEARCH NOTE | **Innovation strategy in the real world**

Since 2005 the international management consultants Booz Allen Hamilton have conducted a survey of the spending on and performance of innovation in the world's 1000 largest firms. The most recent survey found that there remain significant differences between spending on innovation across different sectors and regions. For example, the R&D-intensity (R&D spending divided by sales, expressed as a %) was an average of 13% in the software and healthcare industries, 7% in electronics, but only 1–2% in more mature sectors. Of the 1000 companies studied, representing annual R&D expenditure of US $447 billion, 95% of this spending was in the USA, Europe and Japan.

However, like most studies of innovation and performance (see Chapter 11 for a review), they find no correlation between R&D spending, growth and financial or market performance. They argue that it is how the R&D is managed and translated into successful new processes, products and services which counts more. Overall they identify two factors that are common to those companies which consistently leverage their R&D spending: strong alignment between innovation and corporate strategies; and close attention to customer and market needs. This is not to suggest that there is any single optimum strategy for innovation, and instead they argue that three distinct clusters of good practice are observable:

- *Technology drivers*, which focus on scouting and developing new technologies and matching these to unmet needs, with strong project and risk management capabilities.
- *Need seekers*, which aim to be first to market, by identifying emerging customer needs, with strong design and product development capabilities.
- *Market readers*, which aim to be fast followers, and conduct detailed competitors analysis, with strong process innovation.

They conclude 'Is there a best innovation strategy? No . . . Is there a best innovation strategy for any given company? Yes . . . the key to innovation success has nothing to do with how much money you spend. It is directly related to the effort expended to align innovation with strategy and your customers, and to manage the entire process with discipline and transparency' (p. 16).

Source: Jaruzelski, B. and K. Dehoff (2008) Booz Allen Hamilton Annual Innovation Survey. *Strategy and Business*, issue 49.

involving competitors, customers, regulators and so on; and *fast-changing*, including technical, economic, social and political change. It is therefore difficult enough to understand the essential features of the present, let alone to predict the future (see Box 4.1). Second, managers in large firms disagree on their firms' strengths and weaknesses in part because their knowledge of what goes on *inside* the firm is imperfect.

BOX 4.1   ## 'Strategizing in the real world'

'The war in Vietnam is going well and will succeed.'

(R. MacNamara, 1963)

'I think there is a world market for about five computers.'

(T. Watson, 1948)

'Gaiety is the most outstanding feature of the Soviet Union.'

(J. Stalin, 1935)

'Prediction is very difficult, especially about the future.'

(N. Bohr)

'I cannot conceive of any vital disaster happening to this vessel.'

(Captain of *Titanic*, 1912)

The above quotes are from a paper by William Starbuck,[5] in which he criticizes formal strategic planning:

> First, formalization undercuts planning's contributions. Second, nearly all managers hold very inaccurate beliefs about their firms and market environments. Third, no-one can forecast accurately over the long term . . . However, planners can make strategic planning more realistic and can use it to build healthier, more alert and responsive firms. They can make sensible forecasts and use them to foster alertness; exploit distinctive competencies, entry barriers and proprietary information; broaden managers' horizons and help them develop more realistic beliefs; and plan in ways that make it easier to change strategy later (p. 77).

As a consequence, internal corporate strengths and weaknesses are often difficult to identify before the benefit of practical experience, especially in new and fast-changing technological fields. For example:

- In the 1960s, the oil company Gulf defined its distinctive competencies as producing energy, and so decided to purchase a nuclear energy firm. The venture was unsuccessful, in part because the strengths of an oil company in finding, extracting, refining and distributing oil-based products, i.e. geology and chemical-processing technologies, logistics, consumer marketing, were largely irrelevant to the design, construction and sale of nuclear reactors, where the key skills are in electromechanical technologies and in selling to relatively few, but often politicized, electrical utilities.[6]

- In the 1960s and 1970s, many firms in the electrical industry bet heavily on the future of nuclear technology as a revolutionary breakthrough that would provide virtually costless energy. Nuclear energy failed to fulfil its promise, and firms only recognized later that the main revolutionary opportunities and threats for them came from the virtually costless storage and manipulation of information provided by improvements in semiconductor and related technologies.[7]
- In the 1980s, analysts and practitioners predicted that the 'convergence' of computer and communications technologies through digitalization would lower the barriers to entry of mainframe computer firms into telecommunications equipment, and vice versa. Many firms tried to diversify into the other market, often through acquisitions or alliances, e.g. IBM bought Rohm, AT&T bought NCR. Most proved unsuccessful, in part because the software requirements in the telecommunications and office markets were so different.[8]
- The 1990s similarly saw commitments in the fast-moving fields of ICT (information and communication technology) where initial expectations about opportunities and complementarities have been disappointed. For example, the investments of major media companies in the Internet in the late 1990s took more than a decade to prove profitable: problems remain in delivering products to consumers and in getting paid for them, and advertising remains ineffective.[9] There have been similar disappointments so far in development of 'e-entertainment'.[10]
- The Internet bubble, which began in the late 1990s but had burst by 2000, placed wildly optimistic and unrealistic valuations on new ventures utilizing e-commerce. In particular, most of the new e-commerce businesses selling to consumers which floated on the US and UK stock exchanges between 1998 and 2000 subsequently lost around 90% of their value, or were made bankrupt. Notorious failures of that period include Boo.com in the UK, which attempted to sell sports clothing via the Internet, and Pets.com in the USA, which attempted to sell pet food and accessories.

Incrementalist strategy

Given the conditions of uncertainty, 'incrementalists' argue that the complete understanding of complexity and change is impossible: our ability both to comprehend the present and to predict the future is therefore inevitably limited. As a consequence, successful practitioners – engineers, doctors and politicians, as well as business managers – do not, in general, follow strategies advocated by the rationalists, but incremental strategies which explicitly recognize that the firm has only very imperfect knowledge of its environment, of its own strengths and weaknesses, and of the likely rates and directions of change in the future. It must therefore be ready to adapt its strategy in the light of new information and understanding, which it must consciously seek to obtain. In such circumstances the most efficient procedure is to:

1. Make deliberate steps (or changes) towards the stated objective.
2. Measure and evaluate the effects of the steps (changes).
3. Adjust (if necessary) the objective and decide on the next step (change).

This sequence of behaviour goes by many names, such as incrementalism, trial and error, 'suck it and see', muddling through and learning. When undertaken deliberately, and based on strong background knowledge, it has a more respectable veneer, such as:

- Symptom → diagnosis → treatment → diagnosis → adjust treatment → cure (for medical doctors dealing with patients).
- Design → development → test → adjust design → retest → operate (for engineers making product and process innovations).

Corporate strategies that do not recognize the complexities of the present, and the uncertainties associated with change and the future, will certainly be rigid, will probably be wrong, and will potentially be disastrous if they are fully implemented (Case study 4.1). But this is not a reason for rejecting analysis and rationality in innovation management. On the contrary, under conditions of complexity and continuous change, it can be argued that 'incrementalist' strategies are more rational (that is, more efficient) than 'rationalist' strategies. Nor is it a reason for rejecting all notions of strategic planning. The original objectives of the 'rationalists' for strategic planning – set out above – remain entirely valid. Corporations, and especially big ones, without any strategies will be ill-equipped to deal with emerging opportunities and threats: as Pasteur observed '. . . chance favours only the prepared mind'.[11]

CASE STUDY 4.1

The limits of rational strategizing

Jonathan Sapsed's thought-provoking analysis of corporate strategies of entry into new digital media[12] concludes that the rationalist approach to strategy in emerging industries is prone to failure. Because of the intrinsic uncertainty in such an area, it is impossible to forecast accurately and predict the circumstances on which rationalist strategy, e.g. as recommended by Porter, will be based. Sapsed's book includes case studies of companies that have followed the classical rational approach and subsequently found their strategies frustrated.

An example is Pearson, the large media conglomerate, which conducted a SWOT analysis in response to developments in digital media. The strategizing showed the group's strong assets in print publishing and broadcasting, but perceived weaknesses in new media. Having established its 'gaps' in capability Pearson then searched for an attractive multimedia firm to fill the gap. It expensively acquired Mindscape, a small Californian firm. The strategy failed with Mindscape being sold for a loss of £212 million four years later, and Pearson announcing exit from the emerging market of consumer multimedia.

The strategy failed for various reasons. First, unfamiliarity with the technology and market; second, a misjudged assessment of Mindscape's position; and third, a lack of awareness of the multimedia activities already within the group. The formal strategy exercises that preceded action were prone to misinterpretation and misinformation. The detachment from operations recommended by rationalist strategy exacerbated the information problems. The emphasis of rational strategy is not on assessing information arising from operations, but places great credence in detached, logical thought.

Sapsed argues that whilst formal strategizing is limited in what it can achieve, it may be viewed as a form of therapy for managers operating under uncertainty. It can enable disciplined thought on linking technologies to markets, and direct attention to new information and learning. It focuses minds on products, financial flows and anticipating options in the event of crisis or growth. Rather than determining future action, it can prepare the firm for unforeseen change.

Implications for management

This debate has two sets of implications for managers. The first concerns the practice of corporate strategy, which should be seen as a form of corporate *learning, from analysis and experience, how to cope more effectively with complexity and change*. The implications for the processes of strategy formation are the following:

- Given uncertainty, explore the implications of a *range* of possible future trends.
- Ensure broad participation and informal channels of communication.
- Encourage the use of multiple sources of information, debate and scepticism.
- Expect to change strategies in the light of new (and often unexpected) evidence.

The second implication is that *successful management practice is never fully reproducible*. In a complex world, neither the most scrupulous practising manager nor the most rigorous management scholar can be sure of identifying – let alone evaluating – all the necessary ingredients in real examples of successful management practice. In addition, the conditions of any (inevitably imperfect) reproduction of successful management practice will differ from the original, whether in terms of firm, country, sector, physical conditions, state of technical knowledge, or organizational skills and cultural norms.

Thus, in conditions of complexity and change – in other words, the conditions for managing innovation – there are no easily applicable recipes for successful management practice. This is one of the reasons why there are continuous swings in management fashion (see Box 4.2). Useful learning from the experience and analysis of others necessarily requires the following:

1. *A critical reading of the evidence underlying any claims to have identified the factors associated with management success.* Compare, for example, the explanations for the success of Honda in penetrating the US motorcycle market in the 1960s, given (i) by the Boston Consulting Group: exploitation of cost reductions through manufacturing investment and production learning in deliberately targeted and specific market segments;[13] and (ii) by Richard Pascale: flexibility in product–market strategy in response to unplanned market signals, high-quality product design, manufacturing investment in response to market success.[14] The debate has recently been revived, although not resolved, in the *California Management Review*.[15]

2. *A careful comparison of the context of successful management practice, with the context of the firm, industry, technology and country in which the practice might be reused.* For example, one robust conclusion from management research and experience is that major ingredients in the successful implementation of innovation are effective linkages amongst functions within the firm, and with outside sources of relevant scientific and marketing knowledge. Although very useful to management, this knowledge has its limits. Conclusions from a drug firm that the key linkages are between university research and product development are profoundly misleading for an automobile firm, where the key linkages are amongst product development, manufacturing and the supply chain. And even within each of these industries, important linkages may change over time. In the drug industry, the key academic disciplines are shifting from chemistry to include more biology. And in automobiles, computing and associated skills have become important for the development of 'virtual prototypes', and for linkages between product development, manufacturing and the supply chain.[16]

BOX 4.2   **Swings in management fashion**

'*Upsizing. After a decade of telling companies to shrink, management theorists have started to sing the praises of corporate growth.*'

(Feature title from *The Economist*, 10 February 1996, p. 81)

'*Fire and forget? Having spent the 1990s in the throes of restructuring, reengineering and downsizing, American companies are worrying about corporate amnesia.*'

(Feature title from *The Economist*, 20 April 1996, pp. 69–70)

Above are two not untypical examples of swings in management fashion and practice that reflect the inability of any recipe for good management to reflect the complexities of the real thing, and to put successful experiences in the past in the context of the function, firm, country, technology, etc. More recently, a survey of 475 global firms by Bain and Co. showed that the proportion of companies using management tools associated with *business process reengineering, core competencies* and *total quality management* has been declining since the mid-1990s. But they still remain higher than the more recently developed tools associated with *knowledge management*, which have been less successful, especially outside North America (Management fashion: fading fads. *The Economist*, 22 April 2000, pp. 72–3).

The management tools and techniques described in this book represent only very imperfectly the complexities and changes of the real world. As such, they can be no more than aids to systematic thinking, and to collective learning based on analysis and experience. Especially in conditions of complexity and change, *tacit knowledge* of individuals and groups (i.e. know-how that is based on experience, and that cannot easily be codified and reproduced) is of central importance, whether in the design of automobiles and drugs, or in the strategic management of innovation.

RESEARCH NOTE   Blue Ocean innovation strategies

For the past decade INSEAD professors W. Chan Kim and Renée Mauborgne have researched innovation strategies, including work on new market spaces and value innovation. Their most recent contribution is the idea of Blue Ocean Strategies.

By definition, Blue Ocean represents all potential markets which currently do not exist and must be created. In a few cases whole new industries are created, such as those spawned by the Internet, but in most cases they are created by challenging the boundaries of existing industries and markets. Therefore both incumbents and new entrants can play a role.

They distinguish Blue Ocean strategies by comparing them to traditional strategic thinking, which they refer to as Red Ocean strategies:

1. Create uncontested market space, rather than compete in existing market space.
2. Make the competition irrelevant, rather than beat competitors.
3. Create and capture new demand, rather than fight for existing markets and customers.

4. Break the traditional value/cost trade-off: Align the whole system of a company's activities in pursuit of both differentiation *and* low cost.

In many cases a Blue Ocean is created where a company creates value by simultaneously reducing costs and offering something new or different. In their study of 108 company strategies they found that only 14% of innovations created new markets, whereas 86% were incremental line extensions. However, the 14% of Blue Ocean innovations accounted for 38% of revenues and 61% of profits.

The key to creating successful Blue Oceans is to identify and serve uncontested markets, and therefore benchmarking or imitating competitors is counter-productive. It often involves a radically different business model, offering a different value proposition at lower cost. It may be facilitated by technological or other radical innovations, but in most cases this is not the driver.

Sources: W. Chan Kim and R. Mauborgne (2005) Blue Ocean strategy: from theory to practice. *California Management Review*, **47** (3), Spring, 105–21; (2005) *Blue Ocean Strategy: How to Create Uncontested Market Space and Make the Competition Irrelevant*, Harvard Business School, Boston, MA; (2004) Blue Ocean strategy, *Harvard Business Review*, **82** (10), October, 76–84.

Innovation 'leadership' versus 'followership'

Finally, according to Porter, firms must also decide between two market strategies:[17]

1. Innovation 'leadership' – where firms aim at being first to market, based on technological leadership. This requires a strong corporate commitment to creativity and risk taking, with close linkages both to major sources of relevant new knowledge, and to the needs and responses of customers.
2. Innovation 'followership' – where firms aim at being late to market, based on imitating (learning) from the experience of technological leaders. This requires a strong commitment to competitor analysis and intelligence, to reverse engineering (i.e. testing, evaluating and taking to pieces competitors' products, in order to understand how they work, how they are made and why they appeal to customers), and to cost cutting and learning in manufacturing.

However, in practice the distinction between 'innovator' and 'follower' is much less clear. For example, a study of the product strategies of 2273 firms found that market pioneers continue to have high expenditures on R&D, but that this subsequent R&D is most likely to be aimed at minor, incremental innovations. A pattern emerges where pioneer firms do not maintain their historical strategy of innovation leadership, but instead focus on leveraging their competencies in minor incremental innovations. Conversely, late entrant firms appear to pursue one of two very different strategies. The first is based on competencies other than R&D and new product development – for example, superior distribution or greater promotion or support. The second, more interesting strategy, is to focus on major new product development projects in an effort to compete with the pioneer firm.[18]

However, this example also reveals the essential weaknesses of Porter's framework for analysis and action. As Martin Fransman has pointed out, technical personnel in firms like IBM in the 1970s were well aware of trends in semiconductor technology, and their possible effects on the competitive position of mainframe producers.[19] IBM in fact made at least one major contribution to developments in the revolutionary new technology: RISC microprocessors. Yet, in spite of this knowledge, none of the established firms proved capable over the next 20 years of achieving the primary objective of strategy, as defined by Porter: '. . . to find a position . . . where a company can best defend itself against these competitive forces or can influence them in its favour'.

Like many mainstream industrial economics, Porter's framework underestimates the power of technological change to transform industrial structures, and overestimates the power of managers to decide and implement innovation strategies. Or, to put it another way, it underestimates the importance of *technological trajectories*, and of the firm-specific *technological and organizational competencies* to exploit them. Large firms in mainframe computers could not control the semiconductor trajectory. Although they had the necessary technological competencies, their organizational competencies were geared to selling expensive products in a focused market, rather than a proliferating range of cheap products in an increasing range of (as yet) unfocused markets.

These shortcomings of Porter's framework in its treatment of corporate technology and organization led it to underestimate the constraints on individual firms in choosing their innovation strategies. In particular:

- Firm *size* influences the choice between 'broad front' and 'focused' technological strategies. Large firms typically have 'broad front' strategies whilst small firms are 'focused'.
- The firm's *established product base* and related technological competencies will influence the range of technological fields and industrial sectors in which it can hope to compete in future. Chemical-based firms do not diversify into making electronic products, and vice versa. It is very difficult (but not impossible, see, for example, the case of Nokia in Box 9.2) for a firm manufacturing traditional textiles to have an innovation strategy to develop and make computers.[20]
- The *nature of its products and customers* will strongly influence its degree of choice between quality and cost. Compare food products, where there are typically a wide range of qualities and prices, with ethical drugs and passenger aeroplanes where product quality (i.e. safety) is rigidly regulated by government legislation (e.g. FAR, JAR) or agencies (e.g. FAA, CAA). Food firms therefore have a relatively wide range of potential innovation strategies to be chosen from amongst those described by Porter. The innovation strategies of drug and aircraft firms, on the other hand, inevitably require large-scale expenditures on product development and rigorous testing.

In addition, technological opportunities are always emerging from advances in knowledge, so that:

- Firms and technologies do not fit tidily into preordained and static industrial structures. In particular, firms in the chemical and electrical-electronic industries are typically active in a number of product markets, and also create new ones, like personal computers. Really new innovations (as distinct from radical or incremental), which involve some discontinuity in the technological or marketing base of a firm are actually very common, and often evolve into new businesses and product lines, such as the Sony Walkman or Canon Laserjet printers.[21]
- Technological advances can increase opportunities for profitable innovation in so-called mature sectors. See, for example, the opportunities generated over the past 15 years by applications of IT in marketing, distribution and coordination in such firms as Benetton and Hotpoint.[22] See also the increasing opportunities for technology-based innovation in traditional service activities like banking, following massive investments in IT equipment and related software competencies.[23]
- Firms do not become stuck in the middle as Porter predicted. John Kay has shown that firms with medium costs and medium quality compared to the competition achieve higher returns on investment

than those with either low–low or high–high strategies.[24] Furthermore, some firms achieve a combination of high quality and low cost compared to competitors and this reaps high financial returns. These and related issues of product strategy will be discussed in Chapter 9.

There is also little place in Porter's framework for the problems of *implementing* a strategy:

- Organizations which are large and specialized must be capable of learning and changing in response to new and often unforeseen opportunities and threats. This does not happen automatically, but must be consciously managed. In particular, the continuous transfer of knowledge and information across functional and divisional boundaries is essential for successful innovation. Studies confirm that the explicit management of competencies across different business divisions can help to create radical innovations, but that such interactions demand attention to leadership roles, team composition and informal networks.[25]
- Elements of Porter's framework have been contradicted as a result of organizational and related technological changes. The benefits of non-adversarial relations with both suppliers and customers have become apparent. Instead of bargaining in what appears to be a zero-sum game, cooperative links with customers and suppliers can increase competitiveness, by improving both the value of innovations to customers, and the efficiency with which they are supplied.[26]

According to a survey of innovation strategies in Europe's largest firms, just over 35% replied that the technical knowledge they obtain from their suppliers and customers is very important for their own innovative activities.[27]

Christensen and Raynor provide a recent and balanced summary of the relative merits of the rational versus incremental approaches to strategy:

. . . core competence, as used by many managers, is a dangerously inward-looking notion. Competitiveness is far more about doing what customers value, than doing what you think you're good at . . . the problem with the core competence/not your core competence categorization is that what might seem to be a non-core activity today might become an absolutely critical competence to have mastered in a proprietary way in the future, and vice versa . . . emergent processes should dominate in circumstances in which the future is hard to read and it is not clear what the right strategy should be . . . the deliberate strategy process should dominate once a winning strategy has become clear, because in those circumstances effective execution often spells the difference between success and failure.[28]

4.2 The dynamic capabilities of firms

Teece and Pisano[29] integrate the various dimensions of innovation strategy identified above into what they call the 'dynamic capabilities' approach to corporate strategy, which underlines the importance of dynamic change and corporate learning:

This source of competitive advantage, dynamic capabilities, emphasizes two aspects. First, it refers to the shifting character of the environment; second, it emphasizes the key role of strategic management in appropriately adapting, integrating and re-configuring internal and external organizational skills, resources and functional competencies towards a changing environment (p. 537).

To be strategic, a capability must be honed to a user need (so that there are customers), unique (so that the products/services can be priced without too much regard for the competition), and difficult to replicate (so that profits will not be competed away) (p. 539).

We advance the argument that the strategic dimensions of the firm are its managerial and organizational *processes*, its present *position*, and the *paths* available to it. By managerial *processes* we refer to the way things are done in the firm, or what might be referred to as its 'routines', or patterns of current practice and learning. By *position*, we refer to its current endowment of technology and intellectual property, as well as its customer base and upstream relations with suppliers. By *paths* we refer to the strategic alternatives available to the firm, and the attractiveness of the opportunities which lie ahead (pp. 537–41, our italics).

Institutions: finance, management and corporate governance

Firms' innovative behaviours are strongly influenced by the competencies of their managers and the ways in which their performance is judged and rewarded (and punished). Methods of judgement and reward vary considerably amongst countries, according to their national systems of corporate governance: in other words, the systems for exercising and changing corporate ownership and control. In broad terms, we can distinguish two systems: one practised in the USA and UK; and the other in Japan, Germany and its neighbours, such as Sweden and Switzerland. In his book *Capitalism against Capitalism*, Michel Albert calls the first the 'Anglo-Saxon' and the second the 'Nippon–Rhineland' variety.[30] A lively debate continues about the essential characteristics and performance of the two systems, in terms of innovation and other performance variables. Table 4.1 is based on a variety of sources, and tries to identify the main differences that affect innovative performance.

In the UK and the USA, corporate ownership (shareholders) is separated from corporate control (managers), and the two are mediated through an active stock market. Investors can be persuaded to hold shares only if there is an expectation of increasing profits and share values. They can shift their investments relatively easily. On the other hand, in countries with governance structures like those of Germany or Japan, banks, suppliers and customers are more heavily locked into the firms in which they invest. Until the 1990s, countries strongly influenced by German and Japanese traditions persisted in investing heavily in R&D in established firms and technologies, whilst the US system has since been more effective in generating resources to exploit radically new opportunities in IT and biotechnology.

During the 1980s, the Nippon–Rhineland model seemed to be performing better. Aggregate R&D expenditures were on a healthy upward trend, and so were indicators of aggregate economic performance. Since then, there have been growing doubts. The aggregate technological and economic indicators have been performing less well. Japanese firms have proved unable to repeat in telecommunications, software, microprocessors and computing their technological and competitive successes in consumer electronics.[31] German firms have been slow to exploit radically new possibilities in IT and biotechnology,[32] and there have been criticisms of expensive and unrewarding choices in corporate strategy, like the entry of Daimler-Benz into aerospace.[33] At the same time, US firms appear to have learned important lessons, especially from the Japanese in manufacturing technology, and to have reasserted their eminence in IT and biotechnology. The 1990s also saw sustained increases in productivity in US industry. According to *The Economist* in 1995, in a report entitled 'Back on top?', one professor at the Harvard Business School believed that people will look back at this period as 'a golden age of entrepreneurial management in the USA'.[34]

TABLE 4.1	**The effects of corporate governance on innovation**	
Characteristics	**Anglo-Saxon**	**Nippon–Rhineland**
Ownership	Individuals, pension funds, insurers	Companies, individuals, banks
Control management	Dispersed, arm's length	Concentrated, close and direct
	Business schools (USA), accountants (UK)	Engineers with business training
Evaluation of R&D investments	Published information	Insider knowledge
Strengths	Responsive to radically new technological opportunities	Higher priority to R&D than to dividends for shareholders
	Efficient use of capital	Remedial investment in failing firms
Weakness	Short-termism	Slow to deal with poor investment choices
	Inability to evaluate firm-specific intangible assets	Slow to exploit radically new technologies

However, some observers have concluded that the strong US performance in innovation cannot be satisfactorily explained simply by the combination of entrepreneurial management, a flexible labour force and a well-developed stock exchange. The US experience has not been repeated in the other Anglo-Saxon country with apparently similar characteristics – the UK. They argue that the groundwork for US corporate success in exploiting IT and biotechnology was laid initially by the US Federal Government, with the large-scale investments by the Defense Department in California in electronics, and by the National Institutes of Health in the scientific fields underlying biotechnology.[35] In addition, we should not write off Germany and Japan too soon. The former is now dealing with the dirt and inefficiency of the former East Germany[36] (the inclusion of which in official statistics is one reason for the German decline in the 1990s in business R&D as a share of GDP). Japanese firms like Sony are world leaders in exploiting in home electronics the opportunities opened up by advances in digital technology. And Scandinavian countries are now well ahead of the rest of the world (including the USA) in mobile telephony,[37] as well as in more general indicators of skills and knowledge. The jury is still out.

Learning from foreign systems of innovation

National systems of innovation clearly influence the rate and direction of innovation of domestic firms, and vice versa, but larger firms also learn and exploit innovation from other countries (Table 4.2). Firms

TABLE 4.2   **Relative importance of national and overseas sources of technical knowledge (% firms judging source as being 'very important')**

	Home country	Other Europe	North America	Japan
Affiliated firms	48.9	42.9	48.2	33.6
Joint ventures	36.6	35.0	39.7	29.4
Independent suppliers	45.7	40.3	30.8	24.1
Independent customers	51.2	42.2	34.8	27.5
Public research	51.1	26.3	28.3	12.9
Reverse engineering	45.3	45.9	40.0	40.0

Source: Arundel, A., G. van der Paal and L. Soete (1995) *Innovation Strategies of Europe's Largest Industrial Firms*, PACE Report, MERIT, University of Limbourg, Maastricht. Reproduced by permission of Anthony Arundel.

have at least three reasons for monitoring and learning from the development of technological, production and organizational competencies of national systems of innovation, and especially from those that are growing and strong:

1. They will be the sources of firms with a strong capacity to compete through innovation. For example, beyond Japan, other East Asian countries are developing strong innovation systems. In particular, business firms in South Korea and Taiwan now spend more than 2% of GDP on R&D, which puts them up with the advanced OECD countries. By the early 1990s, Taiwan was granted more patents in the USA than Sweden, and together with South Korea, is catching up fast with Italy, the Netherlands and Switzerland. Other Asian countries like Malaysia are also developing strong technological competencies. Following the collapse of the Russian Empire, we can also anticipate the re-emergence of strong systems of innovation in the Czech Republic and Hungary.

2. They are also potential sources of improvement in the corporate management of innovation, and in national systems of innovation. However, as we shall see below, understanding, interpreting and learning general lessons from foreign systems of innovation is a difficult task. Effectiveness in innovation has become bound up with wider national and ideological interests, which makes it more difficult to separate fact from belief. Both the business press and business education are dominated by the English language and Anglo-Saxon examples: very little is available in English on the management of innovation in Germany; and much of the information about the management of innovation in Japan has been via interpretations of researchers from North America.

3. Finally, firms can benefit more specifically from the technology generated in foreign systems of innovation. A high proportion of large European firms attach great importance to foreign sources of technical

knowledge, whether obtained through affiliated firms (i.e. direct foreign investment) and joint ventures, links with suppliers and customers, or reverse engineering. In general, they find it is more difficult to learn from Japan than from North America and elsewhere in Europe, probably because of greater distances – physical, linguistic and cultural. Perhaps more surprising, European firms find it most difficult to learn from foreign publicly funded research. This is because effective learning involves more subtle linkages than straightforward market transactions: for example, the membership of informal professional networks. This public knowledge is often seen as a source of potential world innovative advantage, and we shall see that firms are increasingly active in trying to access foreign sources. In contrast, knowledge obtained through market transactions and reverse engineering enables firms to catch up, and keep up, with competitors. East Asian firms have been very effective over the past 25 years in making these channels an essential feature of their rapid technological learning (see Case study 4.2).

The slow but significant internationalization of R&D is also a means by which firms can learn from foreign systems of innovation. There are many reasons why multinational companies choose to locate R&D outside their home country, including regulatory regime and incentives, lower cost or more specialist human resources, proximity to lead suppliers or customers, but in many cases a significant motive is to gain access to national or regional innovation networks. Overall, the proportion of R&D expenditure made outside the home nation has grown from less than 15% in 1995, to more than 25% by 2009. However, some countries are more advanced in internationalizing their R&D than others (Figure 4.1). In this respect European firms are the most internationalized, and the Japanese the least.

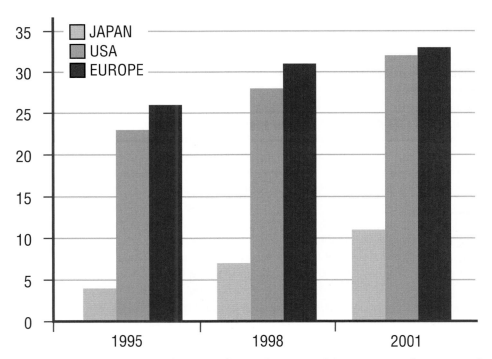

FIGURE 4.1: Internationalization of R&D by region (% R&D expenditure outside home region)

Source: Derived from J. Edler, F. Meyer-Krahmer and G. Reger (2002). Changes in the strategic management of technology: results of a global benchmarking study, *R & D Management* **32** (2): 149–164.

CASE STUDY 4.2

Technology strategies of latecomer firms in East Asia

The spectacular modernization over the past 25 years of the East Asian 'dragon' countries – Hong Kong, South Korea, Singapore and Taiwan – has led to lively debate about its causes. Michael Hobday has provided important new insights into how business firms in these countries succeeded in rapid learning and technological catch up, in spite of underdeveloped domestic systems of science and technology, and a lack of technologically sophisticated domestic customers.

Government policies provided the favourable general economic climate: export orientation; basic and vocational education, with strong emphasis on industrial needs; and a stable economy, with low inflation and high savings. However, of major importance were the strategies and policies of specific business firms for the effective assimilation of foreign technology.

The main mechanism for catching up was the same in electronics, footwear, bicycles, sewing machines and automobiles, namely the 'OEM' (original equipment manufacture) system. OEM is a specific form of subcontracting, where firms in catching-up countries produce goods to the exact specification of a foreign trans-national company (TNC) normally based in a richer and technologically more advanced country. For the TNC, the purpose is to cut costs, and to this end it offers assistance to the latecomer firms in quality control, choice of equipment, and engineering and management training.

OEM began in the 1960s, and became more sophisticated in the 1970s. The next stage in the mid-1980s was ODM (own design and manufacture), where the latecomer firms learned to design products for the buyer. The last stage was OBM (own brand manufacture) where latecomer firms market their own products under their own brand name (e.g. Samsung, Acer) and compete head on with the leaders.

For each stage of catching up, the company's technology position must be matched with a corresponding market position, as shown below:

Stage	Technology position	Market position
1. Assembly skills Basic production Mature products	Passive importer pull Cheap labour Distribution by buyers	
2. Incremental process change Reverse engineering	Active sales to foreign buyer Quality and cost-based	
3. Full production skills Process innovation Product design	Advanced production sales International marketing department Markets own design	
4. R&D Product innovation	Product marketing push Own-brand product range and sales	

Stage	Technology position	Market position

5. Frontier R&D
   R&D linked to market needs
   Advanced innovation
	Own-brand push
	In-house market research
	Independent distribution	

Source: Hobday, M. (1995) *Innovation in East Asia: The Challenge to Japan*, Edward Elgar, Cheltenham.

Learning and imitating

Whilst information on competitors' innovations is relatively cheap and easy to obtain, corporate experience shows that knowledge of how to replicate competitors' product and process innovations is much more costly and time consuming to acquire. Such imitation typically costs between 60 and 70% of the original, and typically takes three years to achieve.[38]

These conclusions are illustrated by the examples of Japanese and Korean firms, where very effective imitation has been sustained by heavy and firm-specific investments in education, training and R&D.[39] They are confirmed by a large-scale survey of R&D managers in US firms in the 1980s. As Table 4.3 shows, these managers reported that the most important methods of learning about competitors'

TABLE 4.3   Effectiveness of methods of learning about competitors

Method of learning	Overall sample means* Processes	Products
Independent R&D	4.76	5.00
Reverse engineering	4.07	4.83
Licensing	4.58	4.62
Hiring employees from innovating firm	4.02	4.08
Publications or open technical meetings	4.07	4.07
Patent disclosures	3.88	4.01
Consultations with employees of the innovating firm	3.64	3.64

*Range: 1 = not at all effective; 7 = very effective.
Source: Levin, R., A. Klevorick, R., Nelson and S. Winter (1987) Appropriating the returns from industrial research and development. *Brookings Papers on Economic Activity*, **3**, 783–820. Reproduced by permission of The Brookings Institution.

innovations were independent R&D, reverse engineering and licensing, all of which are expensive compared to reading publications and the patent literature. Useful and usable knowledge does not come cheap. A similar and more recent survey of innovation strategy in more than 500 large European firms also found that nearly half reported the great importance, for their own innovative activities, of the technical knowledge they accumulated through the reverse engineering of competitors' products.[27] For example, Bookpages, a UK Internet business, was developed in response to the success of Amazon in the USA, and based upon the founder's previous experience of the book trade in the UK and deep knowledge of IT systems. However, in order to raise sufficient resources to continue to grow, the business was later sold to Amazon.

More formal approaches to technology intelligence gathering are less widespread, and the use of different approaches varies by company and sector (Figure 4.2). For example, in the pharmaceutical sector, where much of the knowledge is highly codified in publications and patents, these sources of information are scanned routinely, and the proximity to the science base is reflected in the widespread use of expert panels. In electronics, product technology roadmaps are commonly used, along with the lead users (see the 3M video for a discussion of lead users). Surprisingly (according to this study of 26 large firms), long-established and proven methods such as Delphi studies, S-curve analysis and patent citations are not in widespread use.

4.3 Appropriating the benefits from innovation

Technological leadership in firms does not necessarily translate itself into economic benefits.[40] Teece argues that the capacity of the firm to appropriate the benefits of its investment in technology depends on two factors: (i) the firm's capacity to translate its technological advantage into commercially viable products or processes; (ii) the firm's capacity to defend its advantage against imitators. Thus, effective patent protection enabled Pilkington to defend its technological breakthrough in glass making, and stopped Kodak imitating Polaroid's instant photography. Lack of commitment of complementary assets in production and marketing resulted in the failure of EMI and Xerox to reap commercial benefits from their breakthroughs in medical scanning and personal computing technologies. De Havilland's pioneering Comet jet passenger aircraft paid the price of revealing the effects of metal fatigue on high-altitude flight, the details of which were immediately available to all competitors. In video recorders, Matsushita succeeded against the more innovative Sony in imposing its standard, in part because of a more liberal licensing policy towards competitors.

Some of the factors that enable a firm to benefit commercially from its own technological lead can be strongly shaped by its management: for example, the provision of complementary assets to exploit the lead. Other factors can be influenced only slightly by the firm's management, and depend much more on the general nature of the technology, the product market and the regime of intellectual property rights: for example, the strength of patent protection. We identify below nine factors which influence the firm's capacity to benefit commercially from its technology:

1. Secrecy
2. Accumulated tacit knowledge
3. Lead times and after-sales service
4. The learning curve

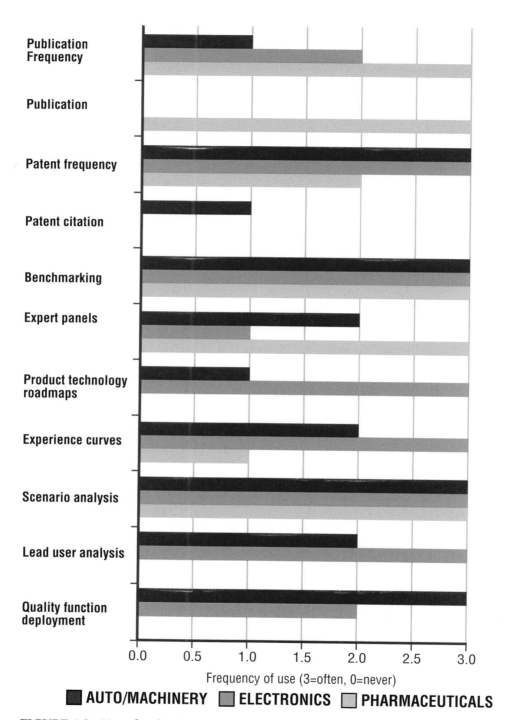

FIGURE 4.2: Use of technology intelligence methods by sector

Source: Derived from E. Lichtenthaler (2004). Technology intelligence processers in leading European and North American multinationals, *R & D Management,* **34**(2): 121–134.

5. Complementary assets
6. Product complexity
7. Standards
8. Pioneering radical new products
9. Strength of patent protection

We begin with those over which management has some degree of discretion for action, and move on to those where its range of choices is more limited.

1. *Secrecy* is considered an effective form of protection by industrial managers, especially for process innovations. However, it is unlikely to provide absolute protection, because some process characteristics can be identified from an analysis of the final product, and because process engineers are a professional community, who talk to each other and move from one firm to another, so that information and knowledge inevitably leak out.[41] Moreover, there is evidence that, in some sectors, firms that share their knowledge with their national system of innovation outperform those that do not, and that those that interact most with global innovation systems have the highest innovative performance.[42] Specifically, firms that regularly have their research (publications and patents) cited by foreign competitors are rated more innovative than others, after controlling for the level of R&D. In some cases this is because sharing knowledge with the global system of innovation may influence standards and dominant designs (see below), and can help attract and maintain research staff, alliance partners and other critical resources.

2. *Accumulated tacit knowledge* can be long and difficult to imitate, especially when it is closely integrated in specific firms and regions. Examples include product design skills, ranging from those of Benetton and similar Italian firms in clothing design, to those of Rolls-Royce in aircraft engines.

3. *Lead times and after-sales service* are considered by practitioners as major sources of protection against imitation, especially for product innovations. Taken together with a strong commitment to product development, they can establish brand loyalty and credibility, accelerate the feedback from customer use to product improvement, generate learning-curve cost advantages (see below) and therefore increase the costs of entry for imitators. Based on the survey of large European firms, Table 4.4 shows that there are considerable differences amongst sectors in product development lead times, reflecting differences both in the strength of patent protection and in product complexity.

4. *The learning curve* in production generates both lower costs and a particular and powerful form of accumulated and largely tacit knowledge that is well recognized by practitioners. In certain industries and technologies (e.g. semiconductors, continuous processes), the first-comer advantages are potentially large, given the major possibilities for reducing unit costs with increasing cumulative production. However, such 'experience curves' are not automatic, and require continuous investment in training, and learning.

5. *Complementary assets*. The effective commercialization of an innovation very often depends on assets (or competencies) in production, marketing and after-sales to complement those in technology. For example, EMI did not invest in them to exploit its advances in electronic scanning. On the other hand, Teece argues that strong complementary assets enabled IBM to catch up in the personal computer market.[40]

6. *Product complexity*. However, Teece was writing in the mid-1980s, and IBM's performance in personal computers has been less than impressive since then. Previously, IBM could rely on the size and

TABLE 4.4	Inter-industry differences in product development lead time
Industry	**% of firms noting > 5 years for development and marketing of alternative to a significant product innovation**
All	11.0
Pharmaceuticals	57.5
Aerospace	26.3
Chemicals	17.2
Petroleum products	13.6
Instruments	10.0
Automobiles	7.3
Machinery	5.7
Electrical equipment	5.3
Basic metals	4.2
Utilities	3.7
Glass, cement and ceramics	0
Plastics and rubber	0
Food	0
Telecommunications equipment	0
Computers	0
Fabricated metals	0

Source: Arundel, A., G. van der Paal and L. Soete (1995) *Innovation Strategies of Europe's Largest Industrial Firms*, PACE Report, MERIT, University of Limbourg, Maastricht. Reproduced by permission of Anthony Arundel.

complexity of its mainframe computers as an effective barrier against imitation, given the long lead times required to design and build copy products. With the advent of the microprocessor and standard software, these technological barriers to imitation disappeared and IBM was faced in the late 1980s with strong competition from IBM 'clones', made in the USA and in East Asia. Boeing and Airbus have faced no such threat to their positions in large civilian aircraft, since the costs and lead times for imitation remain very high. Product complexity is recognized by managers as an effective barrier to imitation.

7. *Standards*. The widespread acceptance of a company's product standard widens its own market and raises barriers against competitors. Carl Shapiro and Hal Varian have written the standard (so far) text on the competitive dynamics of the Internet economy,[43] where standards compatibility is an essential feature of market growth, and 'standards wars' an essential feature of the competitive process (see Box 4.3). However, they point out that such wars have been fought in the adoption

---

BOX 4.3   Standards and 'winner takes all' industries

Charles Hill has gone so far as to argue that standards competition creates 'winner takes all' industries.[44] This results from so-called 'increasing returns to adoption', where the incentive for customers to adopt a standard increases with the number of users who have already adopted it, because of the greater availability of complementary and compatible goods and services (e.g. content programmes for video recorders, and computer application programs for operating systems). While the experiences of Microsoft and Intel in personal computers give credence to this conclusion, it does not always hold. The complete victory of the VHS standard has not stopped the loser (Sony) from developing a successful business in the video market, based on its rival's standard.[45] Similarly, IBM has not benefited massively (some would say at all), compared to its competitors, from the success of its own personal computer standard.[46] In both cases, rival producers have been able to copy the standard and to prevent 'winner takes all', because the costs to producers of changing to other standards have been relatively small. This can happen when the technology of a standard is licensed to rivals, in order to encourage adoption. It can also happen when technical differences between rival standards are relatively small. When this is the case (e.g. in TV and mobile phones) the same firms will often be active in many standards.

A recent review by Fernando Suarez of the literature on standards criticized much of the research as being 'ex-post', and therefore offering few insights into the 'ex-ante' dynamics of standards formation most relevant to managers.[47] It identifies that both firm-level and environmental factors influence standards setting:

- *Firm-level factors*: technological superiority, complementary assets, installed base, credibility, strategic manoeuvering, including entry timing, licensing, alliances, managing market expectations.
- *Environmental factors*: regulation, network effects, switching costs, appropriability regime, number of actors and level of competition versus cooperation. The appropriability regime refers to the legal and technological features of the environment which allow the owner of a technology to benefit from the technology. A strong or tight regime makes it more difficult for a rival firm to imitate or acquire the technology.

of each new generation of radically new technology: for example, over the width of railway gauges in the nineteenth century, the provision of electricity through direct or alternating current early in the twentieth, and rival technical systems for colour television more recently. Amongst other things they conclude that the market leader normally has the advantage in a standards war, but this can be overturned through radical technological change, or a superior response to customers' needs. Competing firms can adopt either 'evolutionary' strategies minimizing switching costs for customers (e.g. backward compatibility with earlier generations of the product), or 'revolutionary' strategies based on greatly superior performance–price characteristics, such that customers are willing to accept higher switching costs. Standards wars are made less bitter and dramatic when the costs to the losers of adapting to the winning standard are relatively small (see Box 4.5).

Different factors will have an influence at different phases of the standards process. In the early phases, aimed at demonstrating technical feasibility, factors such as the technological superiority, complementary assets and credibility of the firm are most important, combined with the number and nature of other firms and appropriability regime. In the next phase, creating a market, strategic manoeuvering and regulation are most important. In the decisive phase, the most significant factors are the installed base, complementary assets, credibility and influence of switching costs and network effects. However, in practice it is not always easy to trace such ex-ante factors to ex-post success in successfully establishing a standard (Table 4.5). This is one reason that increasingly collaboration is occurring earlier in the standards process, rather than the more historical 'winner takes all' standards battles in the later stages.[48] Research in the telecommunications and other complex technological environments where system-wide compatibility is necessary, confirms that early advocates of standards via alliances are more likely to create standards and achieve dominant positions in the industry network (see also Case study 4.3 on Ericsson and the GSM standard).[49] Contrast the failure of Philips and Sony to establish their respective analogue video standards, and subsequent recordable digital media standards, compared to the success of VHS, CD and DVD standards, which were the result of early alliances. Where strong appropriability regimes exist, compatibility standards may be less important than customer interface standards, which help to 'lock-in' customers.[50] Apple's graphic user interface is a good example of this trade-off.

8. *Pioneering radical new products.* It is not necessarily a great advantage to be a technological leader in the early stages of the development of radically new products, when the product performance characteristics, and features valued by users, are not always clear, either to the producers or to the users themselves. Especially for consumer products, valued features emerge only gradually through a process of dynamic competition, which involves a considerable amount of trial, error and learning by both producers and users. New features valued by users in one product can easily be recognized by competitors and incorporated in subsequent products. This is why market leadership in the early stages of the development of personal computers was so volatile, and why pioneers are often displaced by new entrants.[51] In such circumstances, product development must be closely coupled with the ability to monitor competitors' products and to learn from customers. According to research by Tellis and Golder, pioneers in radical consumer innovations rarely succeed in establishing long-term market positions. Success goes to so-called 'early entrants' with the vision, patience and flexibility to establish a mass consumer market.[52] As a result, studies suggest that the success of product pioneers ranges between 25% (for consumer products) and 53% (for higher technology products), depending on the technological

TABLE 4.5   **Cases of standardization and innovation success and failure**

Standard	Outcome	Key actors and technology
Betamax	Failure	Sony, pioneering technology
VHS	Success	Matsushita and JVC alliance, follower technology
CD	Success	Sony and Philips alliance for hardware, Columbia and Polygram for content
DCC	Failure	Philips, digital evolution of analogue cassette
Minidisc	Failure	Sony competitor to DCC, relaunched after DCC withdrawn, limited subsequent success
MS-DOS	Success	Microsoft and IBM
Navigator	Mixed	Netscape was a pioneer and early standard for Internet browsers, but Microsoft's Explorer overtook this position

Source: Derived from Chiesa, V. and G. Toletti (2003) Standards-setting in the multimedia sector. *International Journal of Innovation Management*, **7** (3), 281–308.

and market conditions. For example, studies of the PIMS (Profit Impact of Market Strategy) database indicate that (surviving) product pioneers tend to have higher quality and a broader product line than followers, whereas followers tend to compete on price, despite having a cost disadvantage. A pioneer strategy appears more successful in markets where the purchasing frequency is high, or distribution important (e.g. fast-moving consumer goods), but confers no advantage where there are frequent product changes or high advertising expenditure (e.g. consumer durables).[53]

9. Strength of patent protection can, as we have already seen in the examples described above, be a strong determinant of the relative commercial benefits to innovators and imitators. Table 4.6 summarizes the results of the surveys of the judgements of managers in large European and US firms about the strength of patent protection. The firms' sectors are ordered according to the first column of figures, showing the strength of patent protection for product innovations for European firms. On the whole, European firms value patent protection more than their US counterparts. However, with one exception (cosmetics), the variations across industry in the strength of patent protection are very similar in Europe and the USA. Patents are judged to be more effective in protecting product innovations than process innovations in all sectors except petroleum refining, probably reflecting the importance of improvements in chemical catalysts for increasing process efficiency. It also shows that patent

| TABLE 4.6 | Inter-industry differences in the effectiveness of patenting | | | |

Industry	Products Europe	Processes USA	Europe	USA
Drugs	4.8	4.6	4.3	3.5
Plastic materials	4.8	4.6	3.4	3.3
Cosmetics	4.6	2.9	3.9	2.1
Plastic products	3.9	3.5	2.9	2.3
Motor vehicle parts	3.9	3.2	3.0	2.6
Medical instruments	3.8	3.4	2.1	2.3
Semiconductors	3.8	3.2	3.7	2.3
Aircraft and parts	3.8	2.7	2.8	2.2
Communications equipment	3.6	2.6	2.4	2.2
Steel mill products	3.5	3.6	3.5	2.5
Measuring devices	3.3	2.8	2.2	2.6
Petroleum refining	3.1	3.1	3.6	3.5
Pulp and paper	2.6	2.4	3.1	1.9

Range: 1 = not at all effective; 5 = very effective.
Note: Some industries omitted because of lack of Europe – USA comparability.

Sources: Arundel, A., G. van de Paal and L. Soete (1995) *Innovation Strategies of Europe's Largest Industrial Firms*, PACE Report, MERIT, University of Limbourg, Maastricht and Levin, R. *et al.* (1987) Appropriating the returns from industrial research and development. *Brookings Papers on Economic Activity*, **3**, 783–820. Reproduced by permission of Anthony Arundel.

protection is rated more highly in chemical-related sectors (especially drugs) than in other sectors. This is because it is more difficult in general to 'invent round' a clearly specified chemical formula than round other forms of invention.

Radically new technologies are now posing new problems for the protection of intellectual property, including the patenting system. The number of patents granted to protect software technology is growing in the USA, and so are the numbers of financial institutions getting involved in patenting for the first time.[54] Debate and controversy surround important issues, such as the possible effects of digital technology on copyright protection,[55] the validity of patents to protect living organisms, and the appropriate breadth of patent protection in biotechnology.[56]

Finally, we should note that firms can use more than one of the above nine factors to defend their innovative lead. For example, in the pharmaceutical industry secrecy is paramount during the early phases of research, but in the later stages of research patents become critical. Complementary assets such as global sales and distribution become more important at the later stages. Despite all the mergers and acquisitions in this sector, these factors, combined with the need for a significant critical mass of R&D, have resulted in relatively stable international positions of countries in pharmaceutical innovation over a period of some 70 years (Figure 4.3). By any measure, firms in the USA have dominated the industry since the 1940s, followed by a second division consisting of Switzerland, Germany, France and the UK. Some of the methods are mutually exclusive: for example, secrecy precludes patenting, which requires disclosure of information, although it can precede patenting. However, firms typically deploy all the useful means available to them to defend their innovations against imitation.[57]

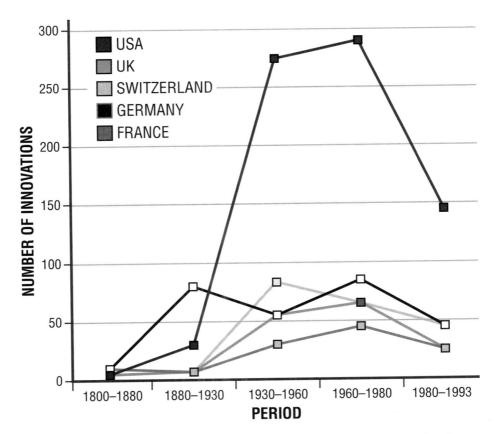

FIGURE 4.3: Innovative positions of countries in the pharmaceutical industry across time

Source: Derived from B. Achilladelis and N. Antonakis (2001). The dynamics of technological innovation: the case of the pharmaceutical industry, *Research Policy*, **30**, 535–588.

CASE STUDY 4.3

Standards, intellectual property and first-mover advantages: the case of GSM

The development of the global system for mobile communications (GSM) standard began around 1982. Around 140 patents formed the essential intellectual property behind the GSM standard. In terms of numbers of patents, Motorola dominated with 27, followed by Nokia (19) and Alcatel (14). Philips also had an initial strong position with 13 essential patents, but later made a strategic decision to exit the mobile telephony business. Ericsson was unusual in that it held only four essential patents for GSM, but later became the market leader. One reason for this was that Ericsson wrote the original proposal for GSM. Another reason is that it was second only to Philips in its position in the network of alliances between relevant firms. Motorola continued to patent after the basic technical decisions had been agreed, whereas the other firms did not. This allowed Motorola greater control over which markets GSM would be made available, and also enabled it to influence licensing conditions and to gain access to others' technology. Subsequently, virtually all the GSM equipment was supplied by companies which participated in the cross-licensing of this essential intellectual property: Ericsson, Nokia, Siemens, Alcatel and Motorola, together accounting for around 85% of the market for switching systems and stations, a market worth US $100 billion.

As the GSM standard moved beyond Europe, North American suppliers such as Nortel and Lucent began to license the technology to offer such systems, but never achieved the success of the five pioneers. Most recently Japanese firms have licensed the technology to provide GSM-based systems. Royalties for such technology can be high, representing up to 29% of the cost of a GSM handset.

Source: Bekkers R., G. Duysters and B. Verspagen (2002) Intellectual property rights, strategic technology agreements and market structure. *Research Policy*, **31**, 1141–61.

In some cases the advantages of pioneering technology, intellectual property and standards combine to create a sustainable market position (Case study 4.3).

4.4 Technological trajectories

In this section we focus on firms and broad technological trajectories.[58] This is because firms and industrial sectors differ greatly in their underlying technologies. For example, designing and making an automobile is not the same as designing and making a therapeutic drug, or a personal computer. We are dealing not with one *technology*, but with several *technologies*, each with its historical pattern of development, skill requirements and strategic implications. It is therefore a major challenge to develop a framework, for integrating changing technology into strategic analysis, that deals effectively with corporate and sectoral diversity. We describe below the framework that one of us has developed over the past 10 or more years to encompass diversity.[59] It has been strongly influenced by the analyses of the emergence of the major new technologies over the past 150 years by Chris Freeman and his colleagues,[60] and by David Mowery and Nathan Rosenberg.[61]

A number of studies have shown marked, similar and persistent differences amongst industrial sectors in the sources and directions of technological change. They can be summarized as follows:

- *Size of innovating firms*: typically *big* in chemicals, road vehicles, materials processing, aircraft and electronic products; and *small* in machinery, instruments and software.
- *Type of product made*: typically *price sensitive* in bulk materials and consumer products; and *performance sensitive* in ethical drugs and machinery.
- *Objectives of innovation*: typically *product* innovation in ethical drugs and machinery; *process* innovation in steel; and *both* in automobiles.
- *Sources of innovation*: *suppliers* of equipment and other production inputs in agriculture and traditional manufacture (like textiles); *customers* in instrument, machinery and software; *in-house* technological activities in chemicals, electronics, transport, machinery, instruments and software; and *basic research* in ethical drugs.
- *Locus of own innovation*: *R&D laboratories* in chemicals and electronics; *production engineering departments* in automobiles and bulk materials; *design offices* in machine building; and *systems departments* in service industries (e.g. banks and supermarket chains).

In the face of such diversity there are two opposite dangers. One is to generalize about the nature, source, directions and strategic implications of innovation on the basis of experience in one firm or in one sector. In this case, there is a strong probability that many of the conclusions will be misleading or plain wrong. The other danger is to say that all firms and sectors are different, and that no generalizations can be made. In this case, there can be no cumulative development of useful knowledge. In order to avoid these twin dangers, one of us distinguished five major technological trajectories, each with its distinctive nature and sources of innovation, and with its distinctive implications for technology strategy and innovation management. This was done on the basis of systematic information on more than 2000 significant innovations in the UK, and of a reading of historical and case material. In Table 4.7 we identify for each trajectory its typical core sectors, its major sources of technological accumulation and its main strategic management tasks.

Thus, in *supplier-dominated firms*, technical change comes almost exclusively from suppliers of machinery and other production inputs. This is typically the case in both agriculture and textiles, where most new techniques originate in firms in the machinery and chemical industries. Firms' technical choices reflect input costs, and the opportunities for firm-specific technological accumulation are relatively modest, being focused on improvements and modifications in production methods and associated inputs. *The main task of innovation strategy* is therefore to use technology from elsewhere to reinforce other competitive advantages. Over the past 10 years, advances elsewhere in IT have opened up radical new applications in design, distribution, logistics and transactions, thereby making production more responsive to customer demands. But, since these revolutionary changes are available from specialized suppliers to all firms, it is still far from clear whether they can be of a lasting advantage for firms competing in supplier-dominated sectors.

In *scale-intensive firms*, technological accumulation is generated by the design, building and operation of complex production systems and/or products. Typical core sectors include the extraction and processing of bulk materials, automobiles and large-scale civil engineering projects. Given the potential economic advantages of increased scale, combined with the complexity of products and/or

TABLE 4.7	**Five major technological trajectories**				
	Supplier-dominated	**Scale-intensive**	**Science-based**	**Information-intensive**	**Specialized suppliers**
Typical core products	Agriculture Services Traditional manufacture	Bulk materials Consumer durables Automobiles Civil engineering	Electronics Chemicals	Finance Retailing Publishing Travel	Machinery Instruments Software
Main sources of technology	Suppliers Production learning	Production engineering Production learning Suppliers Design offices	R&D Basic research	Software and systems departments Suppliers	Design Advanced users
Main tasks of innovation strategy					
Positions	Based on non-technological advantages	Cost-effective and safe complex products and processes	Develop technically related products	New products and services	Monitor and respond to user needs
Paths	Use of IT in finance and distribution	Incremental integration of new knowledge (e.g. virtual prototypes, new materials, B2B*)	Exploit basic science (e.g. molecular biology)	Design and operation of complex information processing systems	Matching changing technologies to user needs
Processes	Flexible response to user	Diffusion of best practice in design, production and distribution	Obtain complementary assets. Redefine divisional boundaries	To match IT-based opportunities with user needs	Strong links with lead users

*B2B = business to business.

production systems, the risks of failure associated with radical but untested changes are potentially very costly. Process and product technologies therefore develop incrementally on the basis of earlier operating experience, and improvements in components, machinery and subsystems. The main sources of technology are in-house design and production engineering departments, operating experience, and specialized suppliers of equipment and components. In these circumstances, *the main tasks of innovation strategy* are the incremental improvement of technological improvements in complex products or production systems, and the diffusion throughout the firm of best-practice methods in design and production. Recent advances in the techniques of large-scale computer simulation and modelling now offer considerable opportunities for saving time and money in the building and testing of prototypes and pilot plant.

In *science-based firms*, technological accumulation emerges mainly from corporate R&D laboratories, and is heavily dependent on knowledge, skills and techniques emerging from academic research. Typical core sectors are chemicals and electronics; fundamental discoveries (electromagnetism, radio waves, transistor effect, synthetic chemicals, molecular biology) open up major new product markets over a wide range of potential applications. The major directions of technological accumulation in the firm are horizontal searches for new and technologically related product markets. As a consequence, *the main tasks of innovation strategy* are to monitor and exploit advances emerging from basic research, to develop technologically related products and acquire the complementary assets (e.g. production and marketing) to exploit them, and to reconfigure the operating divisions and business units in the light of changing technological and market opportunities.

Information-intensive firms have begun to emerge only in the past 10 to 15 years, particularly in the service sector: finance, retailing, publishing, telecommunications and travel. The main sources of technology are in-house software and systems departments, and suppliers of IT hardware and of systems and applications software. The main purpose is to design and operate complex systems for processing information, particularly in distribution systems that make the provision of a service or a good more sensitive to customer demands. *The main tasks of innovation strategy* are the development and operation of complex information-processing systems, and the development of related and often radically new services.

Specialized supplier firms are generally small, and provide high-performance inputs into complex systems of production, of information processing and of product development, in the form of machinery, components, instruments and (increasingly) software. Technological accumulation takes place through the design, building and operational use of these specialized inputs. Specialized supplier firms benefit from the operating experience of advanced users, in the form of information, skills and the identification of possible modifications and improvements. Specialized supplier firms accumulate the skills to match advances in technology with user requirements, which – given the cost, complexity and interdependence of production processes – put a premium on reliability and performance, rather than on price. *The main tasks of innovation strategy* are keeping up with users' needs, learning from advanced users and matching new technologies to users' needs.

Knowledge of these major technological trajectories can improve analysis of particular companies' technological strategies, by helping answer the following questions:

- Where do the company's technologies come from?
- How do they contribute to competitive advantage?
- What are the major tasks of innovation strategy?
- Where are the likely opportunities and threats, and how can they be dealt with?

Although the above taxonomy has held up reasonably well to subsequent empirical tests,[62] it inevitably simplifies.[63] For example, we can find 'supplier-dominated' firms in electronics and chemicals, but they are unlikely to be technological pacesetters. In addition, firms can belong in more than one trajectory. In particular, large firms in all sectors have capacities in *scale-intensive* (mainly mechanical and instrumentation) technologies, in order to ensure efficient production. Software technology is beginning to play a similarly pervasive role across all sectors. We have recently extended this taxonomy based on survey and interview data on the innovative activities of almost 1000 firms (see Research Note and Table 4.8).

TABLE 4.8   Patterns of innovation in the 'new' and 'old' economies

Variable	New economy	Old economy
R&D sets strategic vision of firm	5.14	3.56
R&D active participant in making corporate strategy	5.87	4.82
R&D responsible for developing new business	5.05	3.76
Transforming academic research into products	4.64	3.09
Accelerating regulatory approval	4.62	3.02
Reliability and systems engineering	5.49	4.79
Making products de facto standard	3.56	2.71
Anticipating complex client needs	4.95	3.94
Exploration with potential customers and lead users	5.25	4.41
Probing user needs with preliminary designs	4.72	3.59
Using roadmaps of product generations	4.51	3.26
Planned replacement of current products	3.56	2.53
Build coalition with commercialization partners	4.18	3.38
Working with suppliers to create complementary offers	4.32	3.61

Scale: 1 (low) – 7 (high); only statistically significant differences shown, $n = 75$ firms).

Source: Derived from S. Floricel and R. Miller (2003) An exploratory comparison of the management of innovation in the new and old economies. *R&D Management*, **33** (5), 501–25.

| RESEARCH NOTE | Diversity of strategic games for innovation |

The MINE (Managing Innovation in the New Economy) research program at Ecole Polytechnique in Montreal, Canada, together with SPRU, University of Sussex, UK, conducted qualitative and quantitative studies to gain an understanding of the diversity of strategies for innovation. Almost 925 chief technology officers (CTOs) and senior managers of R&D (from Asia, North and South America, and Europe) across all industrial sectors of the economy responded to a global survey. The survey tool is available at www.minesurvey.polymtl.ca. Respondents come from firms such as Intel, Synopsys, Motorola, IBM Global Services, Novartis and Boeing. Executives were asked what competitive forces impact on innovation, what value-creation and -capture activities are pursued in innovating, and what strategies and practices are used.

Games of innovation involve many interdependent players, persist over time, and are strategically complex. Games are distinct, coherent scenarios of value creation and capture involving activities of collaboration and rivalry:

- Each involves a distinct logic of innovative activities that is largely contingent on product architectures and market lifecycle stage.
- They follow persistent trajectories, bound by some basic economic and technical forces and thus tend to fall into a small number of natural trajectories.
- They result in differing levels of performance. Market-creation games involve radical innovations, grow fast, and display high variations in profitability. By contrast, market-evolution games are characterized by process innovations, a slower pace of growth, but good profitability.
- However, games are not fully determined by their contexts, but allow degrees of strategic freedom to interact with members of relevant ecosystems and to adopt collaborative and competitive moves to expand markets.

Clustering analyses led to the identification of seven distinct and stable groups each containing at least 100 firms that create and capture value in similar ways. Each game is characterized by statistically different value-creation and -capture activities:

- Patent-driven discovery
- Cost-based competition
- Systems integration
- Systems engineering and consulting
- Platform orchestration
- Customized mass-production
- Innovation support and services.

Source: Miller, R. and S. Floricel (2007) Special Issue, *International Journal of Innovation Management*, **11** (1), vii–xvi.

4.5 Developing firm-specific competencies

The ability of firms to track and exploit the technological trajectories described above depends on their specific technological and organizational competencies, and on the difficulties that competitors have in imitating them. The notion of firm-specific competencies has become increasingly influential amongst economists, trying to explain why firms are different, and how they change over time,[20] and also amongst business practitioners and consultants, trying to identify the causes of competitive success. In the 1990s, management began to shift interest from improvements in short-term operational efficiency and flexibility (through 'de-layering', 'downsizing', 'outsourcing' and 'business process reengineering', etc.), to a concern that – if taken too far – the 'lean corporation' could become the 'anorexic corporation', without any capacities for longer-term change and survival.

Hamel and Prahalad on competencies

The most influential business analysts promoting and developing the notion of 'core competencies' have been Gary Hamel and C. K. Prahalad.[64] Their basic ideas can be summarized as follows:

1. The sustainable competitive advantage of firms resides not in their products but in their *core competencies*: 'The real sources of advantage are to be found in management's ability to consolidate corporate-wide technologies and production skills into competencies that empower individual businesses to adapt quickly to changing opportunities' (p. 81).
2. Core competencies feed into more than one core product, which in turn feed into more than one business unit. They use the metaphor of the tree:

    End products = Leaves, flowers and fruit
    Business units = Smaller branches
    Core products = Trunk and major limbs
    Core competencies = Root systems

    Examples of core competencies include Sony in miniaturization, Philips in optical media, 3M in coatings and adhesives and Canon in the combination of the precision mechanics, fine optics and microelectronics technologies that underlie all their products (see Case study 4.4). Examples of core products include Honda in lightweight, high-compression engines and Matsushita in key components in video cassette recorders.

3. The importance of associated organizational competencies is also recognized: 'Core competence is communication, involvement, and a deep commitment to working across organizational boundaries' (1990, p. 82).
4. Core competencies require focus: 'Few companies are likely to build world leadership in more than five or six fundamental competencies. A company that compiles a list of 20 to 30 capabilities has probably not produced a list of core competencies' (1990, p. 84).
5. As Table 4.9 shows, the notion of core competencies suggests that large and multidivisional firms should be viewed not only as a collection of strategic business units (SBUs), but also as bundles of competencies that do not necessarily fit tidily in one business unit.

CASE STUDY 4.4

Core competencies at Canon

Product	Competencies Precision mechanics	Fine optics	Microelectronics
Basic camera	X	X	
Compact fashion camera	X	X	
Electronic camera	X	X	
EOS autofocus camera	X	X	X
Video still camera	X	X	X
Laser beam printer	X	X	X
Colour video printer	X		X
Bubble jet printer	X		X
Basic fax	X		X
Laser fax	X		X
Calculator			X
Plain paper copier	X	X	X
Colour copier	X	X	X
Laser copier	X	X	X
Colour laser copier	X	X	X
Still video system	X	X	X
Laser imager	X	X	X
Cell analyzer	X	X	X
Mask aligners	X		X
Stepper aligners	X		X
Excimer laser aligners	X	X	X

Source: Prahalad, C. and G. Hamel (1990) The core competencies of the corporation. *Harvard Business Review*, May–June, 79–91.

According to Christer Oskarsson:[65]

In the late 1950s . . . the time had come for Canon to apply its precision mechanical and optical technologies to other areas [than cameras] . . . such as business machines. By 1964 Canon had begun by developing the world's first 10-key fully electronic calculator . . . followed by entry into the coated paper copier market with the development of an electrofax copier model in 1965, and then into . . . the revolutionary Canon plain paper copier technology unveiled in 1968 . . . Following these successes of product diversification, Canon's product lines were built on a foundation of precision optics, precision engineering and electronics . . .

The main factors behind . . . increases in the numbers of products, technologies and markets . . . seem to be the rapid growth of information technology and electronics, technological transitions from analogue to digital technologies, technological fusion of audio and video technologies, and the technological fusion of electronics and physics to optronics (pp. 24–6).

According to Hamel and Prahalad, the concept of the corporation based on core competencies should not replace the traditional one, but a commitment to it 'will inevitably influence patterns of diversification, skill deployment, resources allocation priorities, and approaches to alliances and outsourcing' (1990, p. 86). More specifically, the conventional multidivisional structure may facilitate efficient innovation

TABLE 4.9   **Two views of corporate structure: strategic business units and core competencies**

	Strategic business unit	Core competencies
Basis for competition	Competitiveness of today's products	Inter-firm competition to build competencies
Corporate structure	Portfolio of businesses in related product markets	Portfolio of competencies, core products and business
Status of business unit	Autonomy: SBU 'owns' all resources other than cash	SBU is a potential reservoir of core competencies
Resource allocation	SBUs are unit of analysis. Capital allocated to SBUs	SBUs and competencies are unit of analysis. Top management allocates capital and talent
Value added of top management	Optimizing returns through trade-offs among SBUs	Enunciating strategic architecture, and building future competencies

within specific product markets, but may limit the scope for learning new competencies: firms with fewer divisional boundaries are associated with a strategy based on capabilities broadening, whereas firms with many divisional boundaries are associated with a strategy based on capabilities deepening.[66]

6. The identification and development of a firm's core competencies depend on its *strategic architecture*, defined as:

> . . . a road map of the future that identifies which core competencies to build and their constituent technologies . . . should make resource allocation priorities transparent to the whole organization . . . Top management must add value by enunciating the strategic architecture that guides the competence acquisition process (1990, p. 89).

Examples given include:

- NEC = convergence of computing and communication technologies
- Vickers, USA = being the best power and motion control company in the world
- Honda = lightweight, high-compression engines
- 3M = coatings and adhesives

Assessment of the core competencies approach

The great strength of the approach proposed by Hamel and Prahalad is that it places the cumulative development of firm-specific technological competencies at the centre of the agenda of corporate strategy. Although they have done so by highlighting practice in contemporary firms, their descriptions reflect what has been happening in successful firms in science-based industries since the beginning of the twentieth century. For example, Gottfried Plumpe has shown that the world's leading company in the exploitation of the revolution in organic chemistry in the 1920s – IG Farben in Germany – had already established numerous 'technical committees' at the corporate level, in order to exploit emerging technological opportunities that cut across divisional boundaries.[67] These enabled the firm to diversify progressively out of dyestuffs into plastics, pharmaceutical and other related chemical products.

Other histories of businesses in chemicals and electrical products tell similar stories.[68] In particular, they show that the competence-based view of the corporation has major implications for the organization of R&D, for methods of resource allocation and for strategy determination, to which we shall return later. In the meantime, their approach does have limitations and leaves at least three key questions unanswered.

(a) *Differing potentials for technology-based diversification?* It is not clear whether the corporate core competencies in all industries offer a basis for product diversification. Compare the recent historical experience of most large chemical and electronics firms, where product diversification based on technology has been the norm, with that of most steel and textile firms, where technology-related product diversification has proved very difficult (see, for example, the unsuccessful attempts to diversify by the Japanese steel industry in the 1980s).[69]

(b) *Multi-technology firms?* Recommendations that firms should concentrate resources on a few fundamental (or 'distinctive') world-beating technological competencies are potentially misleading. Large firms are typically active in a wide range of technologies, in only a few of which do they achieve a 'distinctive' world-beating position.[70] In other technological fields, a background technological competence is necessary to enable the firm to coordinate and benefit from outside linkages, especially with suppliers of components, subsystems, materials and production machinery. In industries with complex products or production processes, a high proportion of a firm's technological competencies is deployed in such background competencies.[71] In addition, firms are constrained to develop competencies in an increasing range of technological fields (e.g. IT, new materials, nanotechnology) in order to remain competitive as products become even more 'multi-technological'.

Thus, as is shown in Table 4.10, a firm's innovation strategy will involve more than its distinctive core (or critical) competencies. In-house competencies in background (enabling) technologies are necessary for the effective coordination of changes in production and distribution systems, and in supply chains. In industries with complex product systems (like automobiles), background technologies can account for a sizable proportion of corporate innovative activities. Background technologies can also be the sources of revolutionary and disruptive change. For example, given the major opportunities for improved performance that they offer, all businesses today have no choice but to adopt advances in IT technology, just as all factories in the past had no choice but to convert to electricity as a power source. However, in terms of innovation strategy, it is important to distinguish firms where IT is a core technology and a source of distinctive competitive advantage (e.g. Cisco, the supplier of Internet equipment) from firms where it is a background technology, requiring major changes but available to all competitors from specialized suppliers, and therefore unlikely to be a source of distinctive and sustainable competitive advantage (e.g. Tesco, the UK supermarket chain).

In all industries, emerging (key) technologies can end up having pervasive and major impacts on firms' strategies and operations (e.g. software). A good example of how an emerging/key technology can transform a company is provided by the Swedish telecommunications firm Ericsson. Table 4.11 traces the accumulation of technological competencies, with successive generations of mobile cellular phones

---

TABLE 4.10   **The strategic function of corporate technologies**

Strategic functions	Definition	Typical examples
Core or critical functions	Central to corporate competitiveness. Distinctive and difficult to imitate	Technologies for product design and development. Key elements of process technologies
Background or enabling	Broadly available to all competitors, but essential for efficient design, manufacture and delivery of corporate products	Production machinery, instruments, materials, components (software)
Emerging or key	Rapidly developing fields of knowledge presenting potential opportunities or threats, when combined with existing core and background technologies	Materials, biotechnology, ICT-software

and telecommunication cables. In both cases, each new generation required competencies in a wider range of technological fields, and very few established competencies were made obsolete. The process of accumulation involved both increasing links with outside sources of knowledge, and greater expenditures on R&D, given greater product complexity. This was certainly not a process of concentration, but of diversification in both technology and product.

For these reasons, the notion of 'core competencies' should perhaps be replaced for technology by the notion of 'distributed competencies', given that, in large firms, they are distributed:

- over a large number of technical fields
- over a variety of organizational and physical locations within the corporation – in the R&D, production engineering and purchasing departments of the various divisions, and in the corporate laboratory

| TABLE 4.11 | **Technological accumulation across product generations** |

Product and generation	No. of important technologies			R&D costs		% of technologies acquired externally	Main technological fields (d)	No. of patent classes (e)
	(a)	(b)	Total	(c)	(base = 100)			
Cellular phones								
1. NMT-450	n.a.	n.a.	5	n.a.	100	12	E	17
2. NMT-900	5	5	10	0	200	28	EPM	25
3. GSM	9	5	14	1	500	29	EPMC	29
Telecomm cables								
1. Coaxial	n.a.	n.a.	5	n.a.	100	30	EPM	14
2. Optical	4	6	10	1	500	47	EPCM	17

n.a. = not applicable.
Notes:
(a) No. of technologies from the previous generation.
(b) No. of new technologies, compared to previous generation.
(c) No. of technologies now obsolete from previous generation.
(d) 'Main' = >15% of total engineering stock. Categories are: E = electrical; P = physics; K = chemistry; M = mechanical; C = computers.
(e) Number of international patent classes (IPC) at four-digit level.

Source: Derived from Granstrand, O., E. Bohlin, C. Oskarsson and N. Sjorberg (1992) External technology acquisition in large multi-technology corporations. *R&D Management*, **22** (2), 111–133.

• amongst different strategic objectives of the corporation, which include not only the establishment of a distinctive advantage in existing businesses (involving both core and background technologies), but also the exploration and establishment of new ones (involving emerging technologies).

(c) *Core rigidities?* As Dorothy Leonard-Barton has pointed out, 'core competencies' can also become 'core rigidities' in the firm, when established competencies become too dominant.[72] In addition to sheer habit, this can happen because established competencies are central to today's products, and because large numbers of top managers may be trained in them. As a consequence, important new competencies may be neglected or underestimated (e.g. the threat to mainframes from mini- and micro-computers by management in mainframe companies). In addition, established innovation strengths may overshoot the target. In Box 4.4, Leonard-Barton gives a fascinating example from the Japanese automobile industry: how the highly successful 'heavyweight' product managers of the 1980s (see Chapter 9) overdid it in the 1990s. Many examples show that, when 'core rigidities' become firmly entrenched, their removal often requires changes in top management.

Developing and sustaining competencies

The final question about the notion of core competencies is very practical: how can management identify and develop them?

   Definition and measurement. There is no widely accepted definition or method of measurement of competencies, whether technological or otherwise. One possible measure is the level of *functional*

BOX 4.4   Heavyweight product managers and fat product designs

Some of the most admired features . . . identified . . . as conveying a competitive advantage [to Japanese automobile companies] were: (1) overlapping problem solving among the engineering and manufacturing functions, leading to shorter model change cycles; (2) small teams with broad task assignments, leading to high development productivity and shorter lead times; and (3) using a 'heavyweight' product manager – a competent individual with extensive project influence . . . who led a cohesive team with autonomy over product design decisions. By the early 1990s, many of these features had been emulated . . . by US automobile manufacturers, and the gap between US and Japanese companies in development lead time and productivity had virtually disappeared.

   However . . . there was another reason for the loss of the Japanese competitive edge – 'fat product designs' . . . an excess in product variety, speed of model change, and unnecessary options . . . 'overuse' of the same capability that created competitive advantages in the 1980s has been the source of the new problem in the 1990s. The formerly 'lean' Japanese producers such as Toyota had overshot their targets of customer satisfaction and overspecified their products, catering to a long 'laundry list' of features and carrying their quest for quality to an extreme that could not be cost-justified when the yen appreciated in 1993 . . . Moreover, the practice of using heavyweight managers to guide important projects led to excessive complexity of parts because these powerful individuals disliked sharing common parts with other car models.

Source: Leonard-Barton, D. (1995) *Wellsprings of Knowledge*, Harvard Business School Press, Boston, MA, p. 33.

---

performance in a generic product, component or subsystem: in, for example, performance in the design, development, manufacture and performance of compact, high-performance combustion engines. As a strategic technological *target* for a firm like Honda, this obviously makes sense. But its achievement requires the combination of technological competencies from a wide variety of *fields* of knowledge, the composition of which changes (and increases) over time. Twenty years ago, they included mechanics (statics and dynamics), materials, heat transfer, combustion, fluid flow. Today they also include ceramics, electronics, computer-aided design, simulation techniques and software.

Thus, the functional definition of technological competencies bypasses two central tasks of corporate technology strategy: first, to identify and develop the range of disciplines or fields that must be combined into a functioning technology; second (and perhaps more important) to identify and explore the new competencies that must be added if the functional capability is not to become obsolete. This is why a definition based on the measurement of the combination of competencies in different technological fields is more useful for formulating innovation strategy, and is in fact widely practised in business.[73]

Richard Hall goes some way towards identifying and measuring core competencies.[74] He distinguishes between intangible assets and intangible competencies. Assets include intellectual property rights and reputation. Competencies include the skills and know-how of employees, suppliers and distributors, and the collective attributes which constitute organizational culture. His empirical work, based on a survey and case studies, indicates that managers believe that the most significant of these intangible resources are company reputation and employee know-how, both of which may be a function of organizational culture. Thus organizational culture, defined as the shared values and beliefs of members of an organizational unit, and the associated artefacts, becomes central to organizational learning. This framework provides a useful way to assess the competencies of an organization, and to identify how these contribute to performance.

Sidney Winter links the idea of competencies with his own notion of organizational 'routines', in an effort to contrast capabilities from other generic formulas for sustainable competitive advantage or managing change.[75] A routine is an organizational behaviour that is highly patterned, is learned, derived in part from tacit knowledge and with specific goals, and is repetitive. He argues that an organizational capability is a high-level routine, or collection of routines, and distinguishes between 'zero-level' capabilities as the 'how we earn our living now', and true 'dynamic' capabilities which change the product, process, scale or markets; for example, new product development. These dynamic capabilities are not the only or even the most common way organizations can change. He uses the term '*ad hoc* problem solving' to describe these other ways to manage change. In contrast, dynamic capabilities typically involve long-term commitments to specialized resources, and consist of patterned activity to relatively specific objectives. Therefore dynamic capabilities involve both the exploitation of existing competencies and the development of new ones. For example, leveraging existing competencies through new product development can consist of de-linking existing technological or commercial competencies from one set of current products, and linking them in a different way to create new products. However, new product development can also help to develop new competencies. For example, an existing technological competence may demand new commercial competencies to reach a new market, or conversely a new technological competence might be necessary to service an existing customer.[76]

The trick is to get the right balance between exploitation of existing competencies and the exploitation and development of new competencies. Research suggests that over time some firms are more successful at this than others, and that a significant reason for this variation in performance is due to difference in the ability of managers to build, integrate and reconfigure organizational competencies and resources.[77] These 'dynamic' managerial capabilities are influenced by managerial cognition, human capital and social capital. Cognition refers to the beliefs and mental models which influence decision making. These affect the knowledge and assumptions about future events, available alternatives and association between cause and effect. This will restrict a manager's field of vision, and influence perceptions and interpretations. Case study 4.5 discusses the role of (limited) cognition in the case of Polaroid and digital imaging. Human capital refers to the learned skills that require some investment in education, training experience and socialization, and these can be generic, industry- or firm-specific. It is the firm-specific factors that appear to be the most significant in dynamic managerial capability, which can lead to different decisions when faced with the same environment. Social capital refers to the internal

CASE STUDY 4.5

Capabilities and cognition at Polaroid

Polaroid was a pioneer in the development of instant photography. It developed the first instant camera in 1948, the first instant colour camera in 1963, and introduced sonar automatic focusing in 1978. In addition to its competencies in silver halide chemistry, it had technological competencies in optics and electronics, and mass manufacturing, marketing and distribution expertise. The company was technology-driven from its foundation in 1937, and the founder Edwin Land had 500 personal patents. When Kodak entered the instant photography market in 1976, Polaroid sued the company for patent infringement, and was awarded $924.5 million in damages. Polaroid consistently and successfully pursued a strategy of introducing new cameras, but made almost all its profits from the sale of the film (the so-called razor-blade marketing strategy also used by Gillette), and between 1948 and 1978 the average annual sales growth was 23%, and profit growth 17% per year.

Polaroid established an electronic imaging group as early as 1981, as it recognized the potential of the technology. However, digital technology was perceived as a potential technological shift, rather than as a market or business disruption. By 1986 the group had an annual research budget of $10 million, and by 1989 42% of the R&D budget was devoted to digital imaging technologies. By 1990 28% of the firm's patents related to digital technologies. Polaroid was therefore well positioned at that time to develop a digital camera business. However, it failed to translate prototypes into a commercial digital camera until 1996, by which time there were 40 other companies in the market, including many strong Japanese camera and electronics firms. Part of the problem was adapting the product development and marketing channels to the new product needs. However, other more fundamental problems related to long-held cognitions: a continued commitment to the razor-blade business model, and pursuit of image quality. Profits from the new market for digital cameras were derived from the cameras rather than the consumables (film). Ironically, Polaroid had rejected the development of ink-jet printers, which rely on consumables for profits, because of the relatively low quality of their (early) outputs. Polaroid had a long tradition of improving its print quality to compete with conventional 35mm film. A more detailed case study is available.

Source: Tripsas, M. and G. Gavetti (2000) Capabilities, cognition, and inertia: evidence from digital imaging. *Strategic Management Journal*, **21**, 1147–61.

and external relationships which affect managers' access to information, their influence, control and power.

Top management and 'strategic architecture' for the future. The importance given by Hamel and Prahalad to top management in determining the 'strategic architecture' for the development of future technological competencies is debatable. As *The Economist* has argued:[78]

> It is hardly surprising that companies which predict the future accurately make more money than those who do not. In fact, what firms want to know is what Mr Hamel and Mr Prahalad steadfastly fail to tell them: how to guess correctly. As if to compound their worries, the authors are oddly reticent about those who have gambled and lost.

The evidence in fact suggests that the successful development and exploitation of core competencies does not depend on management's ability to forecast accurately long-term technological and product developments: as Box 4.5 illustrates, the record here is not at all impressive.[79] Instead, the importance of new technological opportunities and their commercial potential emerge not through a flash of genius (or a throw of the dice) from senior management, but gradually through an incremental corporate-wide process of learning in knowledge building and strategic positioning. New core competencies cannot be

---

| BOX 4.5 | **The overvaluation of technological wonders** |

In 1986, Schnaars and Berenson published an assessment of the accuracy of forecasts of future growth markets since the 1960s, with the benefit of 20 or more years of hindsight.[80] The list of failures is as long as the list of successes. Below are some of the failures.

> The 1960s were a time of great economic prosperity and technological advancement in the United States . . . One of the most extensive and widely publicized studies of future growth markets was TRW Inc.'s 'Probe of the Future'. The results . . . appeared in many business publications in the late 1960s . . . Not all . . . were released. Of the ones that were released, nearly all were wrong! Nuclear-powered underwater recreation centers, a 500-kilowatt nuclear power plant on the moon, 3-D color TV, robot soldiers, automatic vehicle control on the interstate system, and plastic germproof houses were amongst some of the growth markets identified by this study.
>
> . . . In 1966, industry experts predicted, 'The shipping industry appears ready to enter the jet age.' By 1968, large cargo ships powered by gas turbine engines were expected to penetrate the commercial market. The benefits of this innovation were greater reliability, quicker engine starts and shorter docking times.
>
> . . . Even dentistry foresaw technological wonders . . . in 1968, the Director of the National Institute of Dental Research, a division of the US Public Health Service, predicted that 'in the next decade, both tooth decay and the most prevalent form of gum disease will come to a virtual end'. According to experts at this agency, by the late 1970s false teeth and dentures would be 'anachronisms' replaced by plastic teeth implant technology. A vaccine against tooth decay would also be widely available and there would be little need for dental drilling.

CASE STUDY 4.6

Learning about optoelectronics in Japanese companies

Using a mixture of bibliometric and interview data, Kumiko Miyazaki traced the development and exploitation of optoelectronics technologies in Japanese firms. Her main conclusions were as follows:

. . . Competence building is strongly related to a firm's past accomplishments. The notions of path dependency and cumulativeness have a strong foundation. Competence building centres in key areas to enhance a firm's core capabilities. . . . by examining the different types of papers related to semiconductor lasers over a 13-year period, it was found that in most firms there was a decrease in experimental type papers accompanied by a rise in papers marking 'new developments' or 'practical applications'. The existence of a wedge pattern for most firms confirmed . . . that competence building is a cumulative and long process resulting from trial and error and experimentation, which may eventually lead to fruitful outcomes. The notion of search trajectories was tested using . . . INSPEC and patent data. Firms search over a broad range in basic and applied research and a narrower range in technology development . . . In other words, in the early phases of competence building, firms explore a broad range of technical possibilities, since they are not sure how the technology might be useful for them. As they gradually learn and accumulate their knowledge bases, firms are able to narrow the search process to find fruitful applications.

Source: Miyazaki, K. (1994) Search, learning and accumulation of technological competencies: the case of optoelectronics. *Industrial and Corporate Change*, **3**, 653.

identified immediately and without trial and error: thus Case study 4.6 neglects to show that Canon failed in electronic calculators and in recording products.[81]

It was through a long process of trial and error that Ericsson's new competence in mobile telephones first emerged.[82] As Case study 4.6 shows, it is also how Japanese firms developed and exploited their competencies in optoelectronics.

A study of radical technological innovations found how visions can influence the development or acquisition of competencies, and identified three related mechanisms through which firms link emerging technologies to markets that do not yet exist: motivation, insight and elaboration.[83] Motivation serves to focus attention and to direct energy, and encourages the concentration of resources. It requires senior management to communicate the importance of radical innovation, and to establish and enforce challenging goals to influence the *direction* of innovative efforts. Insight represents the critical connection between technology and potential application. For radical technological innovations, such insight is rarely from the marketing function, customers or competitors, but is driven by those with extensive technical knowledge and expertise with a sense of both market needs and opportunities. Elaboration involves the demonstration of technical feasibility, validating

> ## RESEARCH NOTE   Business model innovation
>
> For many years, Costas Markides at London Business School has been researching the links between strategy, innovation and firm performance. In recent work he argues for the need to make a clearer distinction between the technological and market aspects of disruptive innovations, and to pay greater attention to business model innovation.
>
> By definition, business model innovation enlarges the existing value of a market, either by attracting new customers or by encouraging existing customers to consume more. Business model innovation does not require the discovery of new products or services, or new technology, but rather the redefinition of existing products and services and how these are used to create value.
>
> For example, Amazon did not invent book selling, and low-cost airlines such as Southwest and easyJet are not pioneers of air travel. Such innovators tend to offer different product or service attributes to existing firms, which emphasize different value propositions. As a result, business model innovation typically requires different and often conflicting systems, structures, processes and value chains to existing offerings.
>
> However, unlike the claims made for disruptive innovations, new business models can coexist with more mainstream approaches. For example, Internet banking and low-cost airlines have not displaced the more mainstream approaches, but have captured around 20% of the total demand for these services. Also, while many business model innovations are introduced by new entrants, which have none of the legacy systems and products of incumbent firms, the more mainstream firms may simply choose not to adopt the new business models as they make little sense for them. Alternatively, they may make other innovations to create or recapture customers.
>
> *Sources*: C. Markides (2006) Disruptive innovation: in need of a better theory. *Journal of Product Innovation Management,* **23**, 19–25; (2004) *Fast Second: How Smart Companies Bypass Radical Innovation to Enter and Dominate New Markets,* Jossey Bass, San Francisco.

the idea within the organization, prototyping and the building and testing of different business models (see Research note).

At this point the concept is sufficiently well elaborated to work with the marketing function and potential customers. Market visioning for radical technologies is necessarily the result of individual or technological leadership. '*There were multiple ways for a vision to take hold of an organization . . . our expectation was that a single individual would create a vision of the future and drive it across the organization. But just as we discovered that breakthrough innovations don't necessarily arise simply because of a critical scientific discovery, neither do we find that visions are necessarily born of singular prophetic individuals*' (pp. 239–44).[83] Case study 4.7 illustrates how Corning developed its ceramic technologies and deep process competencies to develop products for the emerging demand for catalytic converters in the car industry, and for glass fibre for telecommunications. Case study 4.8 shows the limited role of technology in the Internet search engine business, and the central role of an integrated approach to process, product and business innovation.

CASE STUDY 4.7

Market visions and technological innovation at Corning

Corning has a long tradition of developing radical technologies to help create emerging markets. It was one of the first companies in the USA to establish a corporate research laboratory in 1908. The facility was originally set up to help solve some fundamental process problems in the manufacture of glass, and resulted in improved glass for railroad lanterns. This led to the development of Pyrex in 1912, which was Corning's version of the German-invented borosilicate glass. In turn, this led to new markets in medical supplies and consumer products.

In the 1940s the company began to develop television tubes for the emerging market for colour television sets, drawing upon its technology competencies developed for radar during the war. Corning did not have a strong position in black-and-white television tubes, but the tubes for colour television followed a different and more challenging technological trajectory, demanding a deep understanding of the fundamental phenomena to achieve the alignment of millions of photorescent dots to a similar pattern of holes.

In 1966, in response to a joint enquiry from the British Post Office and British Ministry of Defence, Corning supplied a sample of high-quality glass rods to determine the performance in transmitting light. Based on the current performance of copper wire, a maximum loss of 20db/km was the goal. However, at that time the loss of the optical fibre (waveguide) was 10 times this: 200db/km. The target was theoretically possible given the properties of silica, and Corning began research on optical fibre. Corning pursued a different approach to others, using pure silica, which demanded very high temperatures, making it difficult to work with. The company had developed this tacit knowledge in earlier projects, and this would take time for others to acquire. In 1970 the research group developed a composition and fibre design that exceeded the target performance. Excluded from the US market by an agreement with AT&T, Corning formed a five-year joint development agreement with five companies from the UK, Germany, France, Italy and Japan. Corning subsequently developed key technologies for waveguides, filed the 12 key patents in the field, and after a number of high-profile but successful patent infringement actions against European, Japanese and Canadian firms, it came to dominate what would become $10 million annual sales by 1982.

Corning also had close relationships with the main automobile manufacturers as a supplier of headlights, but it had failed to convince these companies to adopt its safety glass for windscreens (windshields) due to the high cost and low importance of safety at that time. Corning had also developed a ceramic heat exchanger for petrol (gasoline) turbine engines, but the automobile manufacturers were not willing to reverse their huge investments for the production of internal combustion engines. However, discussion with GM, Ford and Chrysler indicated that future legislation would demand reduced vehicle emissions, and therefore some form of catalytic converter would become standard for all cars in the USA. However, no one knew how to make these at that time. The passing of the Clean Air Act in 1970 required reductions in emissions by 1975, and accelerated development. Competitors included 3M and GM. However, Corning had the advantage of having already developed the new ceramic for its (failed) heat exchanger project, and its competencies in R&D organization and production processes. Unlike its competitors, which organized development along divisional lines, Corning was able to apply as many researchers as it had to tackle the project, what became known as 'flexible critical mass'. In 1974 it filed a patent for its new extrusion production technology, and in 1975 for a new development of its ceramic material. The competitors' technologies proved unable to match the increasing reduction in emissions needed, and by 1994 catalytic converters generated annual sales of $1 billion for Corning. A more detailed case study is available.

Source: Graham, M. and A. Shuldiner (2001) *Corning and the Craft of Innovation*, Oxford University Press, Oxford.

CASE STUDY 4.8

Innovation in Internet search engines

Internet search engines demonstrate the need for an integrated approach to innovation, which includes process, product and business innovation. Perhaps surprisingly, the leading companies such as Google and Yahoo! have not based their innovation strategies on technological research and development, but rather on the novel combinations of technological, process, product and business innovations.

For example, of the 126 search engine patents granted in the USA between 1999 and 2001, the market leaders Yahoo! and Google each only had a single patent, whereas IBM led the technology race with 16 patents, but no significant search business. However, over the same period Yahoo! published more than 1000 new feature releases and Google over 300. These new releases included: new configurations of search engine; new components for existing search engines; new functions and improved usability.

Moreover, this strategy of a broad range of type of innovations, rather than a narrow focus on technological innovations, did not follow the classic product–process life cycle. A strong consistent emphasis on process innovation throughout the company histories was punctuated with multiple episodes of significant product and business innovation, in particular new offerings which integrated core search functions and other services. This pattern confirms that even in so-called high-tech sectors, other competencies are equally or even more important for continued business success.

Source: Lan, P., G.A. Hutcheson, Y. Markov and N.W. Runyan (2007) An examination of the integration of technological and business innovation: cases of Yahoo! and Google. *International Journal of Technology Marketing*, **2** (4), 295–316.

4.6 Globalization of innovation

Many analysts and practitioners have argued that, following the 'globalization' of product markets, financial transactions and direct investment, large firms' R&D activities should also be globalized – not only in their traditional role of supporting local production, but also in order to create interfaces with specialized skills and innovative opportunities at a world level.[84] This is consistent with more recent notions of 'open innovation', rather than 'closed innovation' which relies on internal development. However, although striking examples of the internationalization of R&D can be found (e.g. the large Dutch firms, particularly Philips[85]), more comprehensive evidence casts doubt on the strength of such a trend (Table 4.12). This evidence is based on the countries of origin of the inventors cited on the front page of patents granted in the USA, to nearly 359 of the world's largest, technologically active firms (and which account for about half of all patenting in the USA). This information turns out to be an accurate guide to the international spread of large firms' R&D activities.

Taken together, the evidence shows that:[86]

- Twenty years ago the world's large firms perform about 12% of their innovative activities outside their home country. The equivalent share of production is now about 25%.
- The most important factor explaining each firm's share of foreign innovative activities is its share of foreign production. Firms from smaller countries in general have higher shares of foreign innovative activities. On average, foreign production is less innovation-intensive than home production.
- Most of the foreign innovative activities are performed in the USA and Europe (in fact, Germany). They are *not* 'globalized'.
- Since the late 1980s, European firms – and especially those from France, Germany and Switzerland – have been performing an increasing share of their innovative activities in the USA, in large part in order to tap into local skills and knowledge in such fields as biotechnology and IT.

Controversy remains both in the interpretation of this general picture, and in the identification of implications for the future. Our own views are as follows:

1. There are major efficiency advantages in the geographical concentration in one place of strategic R&D for *launching major new products and processes* (first model and production line). These are:
   (a) dealing with unforeseen problems, since proximity allows quick, adaptive decisions;
   (b) integrating R&D, production and marketing, since proximity allows integration of tacit knowledge through close personal contacts.
2. The nature and degree of international dispersion of R&D will also depend on the company's major technological trajectory, and the strategically important points for integration and learning that relate to it. Thus, whereas automobile firms find it difficult to separate their R&D geographically from production when launching a major new product, drug firms can do so, and instead locate their R&D close to strategically important basic research and testing procedures.
3. In deciding about the internationalization of their R&D, managers must distinguish between:
   (a) becoming part of global *knowledge networks* – in other words, being aware of, and able to absorb – the results of R&D being carried out globally. Practising scientists and engineers have always done this, and it is now easier with modern IT. However, business firms are finding it increasingly useful to establish relatively small laboratories in foreign countries in order to become strong members of local research networks and thereby benefit from the person-embodied knowledge behind the published papers;
   (b) the *launching of major innovations*, which remains complex, costly, and depends crucially on the integration of tacit knowledge. This remains difficult to achieve across national boundaries. Firms therefore still tend to concentrate major product or process developments in one country. They will sometimes choose a foreign country when it offers identifiable advantages in the skills and resources required for such developments, and/or access to a lead market.[87]

Matching global knowledge networks with the localized launching of major innovations will require increasing international mobility amongst technical personnel, and the increasing use of multinational teams in launching innovations.

4. Advances in IT will enable spectacular increases in the international flow of codified knowledge in the form of operating instructions, manuals and software. They may also have some positive impact on international exchanges of tacit knowledge through teleconferencing, but not anywhere near to the same extent. The main impact will therefore be at the second stage of the 'product cycle',[88] when product design has stabilized, and production methods are standardized and documented, thereby facilitating the internationalization of production. Product development and the first stage of the product cycle will still require frequent and intense personal exchanges, and be facilitated by physical proximity. Advances in IT are therefore more likely to favour the internationalization of production than of the process of innovation.

The two polar extremes of organizing innovation globally are the specialization-based and integration-based, or network structure.[89] In the specialization-based structure the firm develops global centres of excellence in different fields, which are responsible globally for the development of a specific technology or product or process capability. The advantage of such global specialization is that it helps to achieve a critical mass of resources and makes coordination easier. As one R&D director notes:

> . . . the centre of excellence structure is the most preferable. Competencies related to a certain field are concentrated, coordination is easier, and economies of scale can be achieved. Any R&D director has the dream to structure R&D in such a way. However, the appropriate conditions seldom occur.[90]

RESEARCH NOTE   Globalization strategies for Innovation

It is possible to distinguish between two conflicting strategies for the globalization of innovation: augmenting, in which firms locate innovation activities overseas primarily in order to learn from foreign systems of innovation, public and private; and exploiting, the exact opposite, where the main motive is to gain competitive advantage from existing corporate-specific capabilities in an environment overseas. In practice firms will adopt a combination of these two different approaches, and need to manage the trade-offs on a technology- and market-specific basis.

Christian Le Bas and Pari Patel analysed the patenting behaviour of 297 multinational firms over a period of eight years. They found that overall the augmenting strategy was the most common, but this varied by nationality of the firm and technical field. Consistent with other studies, they confirm that the strategy of augmenting was strongest for European firms and weakest for Japanese firms. The Japanese firms were more likely to adopt a strategy of exploiting home technology overseas. By technological field, the ranking for the importance of augmenting was (augmenting strategy most common in the first): instrumentation, consumer goods, civil engineering, industrial processes, engineering and machinery, chemicals and pharmaceuticals and electronics. Moreover, they argue that these different strategies are persistent over time, and are not the result of changes in the internationalization of innovation.

Source: Le Bas, C. and P. Patel (2007) The determinants of homebase-augmenting and homebase-exploiting technological activities: some new results on multinationals' locational strategies. *SPRU Electronic Working Paper Series (SEWPS)*, www.sussex.ac.uk/spru/publications.

| TABLE 4.12 | **Indicators of the geographic location of the innovative activities of firms** |

Nationality of large firms (no.)	% share of origin of US patents in 1992–96		% share of foreign-performed R&D expenditure (year)	% share of foreign origin of US patents in 1992–96				% change in foreign origin of US patents, since 1980–84
	Home	Foreign		US	Europe	Japan	Other	
Japan (95)	97.4	2.6	2.1 (1993)	1.9	0.6	0.0	0.1	−0.7
USA (128)	92.0	8.0	11.9 (1994)	0.0	5.3	1.1	1.6	2.2
Europe (136)	77.3	22.7		21.1	0.0	0.6	0.9	3.3
Belgium	33.2	66.8		14.0	52.6	0.0	0.2	4.9
Finland	71.2	28.8	24.0 (1992)	5.2	23.5	0.0	0.2	6.0
France	65.4	34.6		18.9	14.2	0.4	1.2	12.9
Germany	78.2	21.8	18.0 (1995)	14.1	6.5	0.7	0.5	6.4
Italy	77.9	22.1		12.0	9.5	0.0	0.6	7.4
Netherlands	40.1	59.9		30.9	27.4	0.9	0.6	6.6
Sweden	64.0	36.0	21.8 (1995)	19.4	14.2	0.2	2.2	−5.7
Switzerland	42.0	58.0		31.2	25.0	0.9	0.8	8.2
UK	47.6	52.4		38.1	12.0	0.5	1.9	7.6
All firms (359)	87.4	12.6	11.0 (1997)	5.5	5.5	0.6	0.9	2.4

Sources: Derived from Patel, P. and K. Pavitt (2000) National systems of innovation under strain: the internationalization of corporate R&D. In R. Barrell, G. Mason and M. O'Mahoney, eds, *Productivity, Innovation and Economic Performance*, Cambridge University Press, Cambridge; and Patel, P. and M. Vega (1998) *Technology Strategies of Large European Firms*, P.R.Patel@sussex.ac.uk

In addition, it may allow location close to a global innovation cluster. The main disadvantages of global specialization are the potential isolation of the centre of excellence from global needs, and the subsequent transfer of technologies to subsidiaries worldwide. In contrast, in the integration-based structure different units around the world each contribute to the development of technology projects. The advantage of this approach is that it draws upon a more diverse range of capabilities and international perspectives. In addition, it can encourage competition amongst different units. However, the integrated approach suffers from very high costs of coordination, and commonly suffers from duplication of efforts and inefficient use of scarce resources. In practice, hybrids of these two extreme structures are common, often as a result of practical compromises and trade-offs necessary to accommodate history, acquisitions and politics. For example, specialization by centre of excellence may include contributions from other units, and integrated structures may include the contribution of specialized units. The main factors influencing the decision where to locate R&D globally are, in order of importance:[90]

1. The availability of critical competencies for the project.
2. The international credibility (within the organization) of the R&D manager responsible for the project.
3. The importance of external sources of technical and market knowledge, e.g. sources of technology, suppliers and customers.
4. The importance and costs of internal transactions, e.g. between engineering and production.
5. Cost and disruption of relocating key personnel to the chosen site.

VIEWS FROM THE FRONT LINE

Location of global innovation

Large companies swing between 'distributed R&D' where researchers are based in small business units (SBU), and centralized R&D. The reason for this is that there are merits in both approaches. Centralized R&D improves recruitment and development of world-class specialists, whereas distributed R&D improves researchers' understanding of business strategy. Anyone working in centralized R&D must make the most of the advantages and work to overcome the disadvantages. The biggest challenge for centralized R&D is connectivity with the SBU.

In Sharp Laboratories of Europe we have found that the probability of success of our projects is the probability of technical success multiplied by the probability of commercial success. Technical success is fundamentally easier to manage because so many of the parameters are within our control. It is easy for us to increase the effort, bring in outside expertise or try different routes. Commercial success is much harder for us to manage and we have learnt that the quality of relationships is fundamental to success. There are well-understood motivational and cultural differences between R&D and other company functions such as manufacturing or marketing. Manufacturing is measured by quality, yield, availability, low inventory and low cost. Parameters all disrupted by the introduction of new products. Marketing is seeking to provide customers with exactly what they want, but those goals may not be technically achievable. Researchers are measured by the strength of the technology and are always looking for a better solution.

Inability to bridge these different motivations and cultures is a major barrier to delivering innovation in products. Engaging in short-term R&D projects is the most useful way to build a bridge between a centralized R&D centre and SBU. It creates understanding on both sides and in our experience is a vital precursor to a major technology transfer. There is risk associated with it that vital long-term R&D resource will be diverted into fire-fighting activities and this needs to be managed. It is our experience that managing commercial risk through strong relationships is vital to project success.

Source: Dr Stephen Bold FREng, Managing Director, Sharp Laboratories of Europe Ltd, **www.sle.sharp.co.uk.**

4.7 Enabling strategy making

Scanning and searching the environment identifies a wide range of potential targets for innovation and effectively answers the question, 'What could we do?' But even the best-resourced organization will need to balance this with some difficult choices about *which* options it will explore – and which it will leave aside. This process should not simply be about responding to what competitors do or what customers ask for in the marketplace. Nor should it simply be a case of following the latest technological fashion. Successful innovation strategy requires understanding the key parameters of the competitive game (markets, competitors, external forces, etc.) and also the role which technological knowledge can play as a resource in this game. How can it be accumulated and shared, how can it be deployed in new products/services and processes, how can complementary knowledge be acquired or brought to bear, etc.? Such questions are as much about the management of the learning process within the firm as about investments or acquisitions – and building effective routines for supporting this process is critical to success.

Building a strategic framework to guide selection of possible innovation projects is not easy. In a complex and uncertain world it is a nonsense to think that we can make detailed plans ahead of the game and then follow them through in systematic fashion. Life – and certainly organizational life – isn't like that; as John Lennon said, it's what happens when you're busy making other plans!

Equally organizations cannot afford to innovate at random – they need some kind of framework which articulates how they think innovation can help them survive and grow, and they need to be able to allocate scarce resources to a portfolio of innovation projects based on this view. It should be flexible enough to help monitor and adapt projects over time as ideas move towards more concrete innovations – and rigid enough to justify continuation or termination as uncertainties and risky guesswork become replaced by actual knowledge.

Although developing such a framework is complex we can identify a number of key routines which organizations use to create and deploy such frameworks. These help provide answers to three key questions:

- Strategic analysis – what, realistically, could we do?
- Strategic choice – what are we going to do (and in choosing to commit our resources to that, what will we leave out)?
- Strategic monitoring – over time reviewing to check is this still what we want to do?

Routines to help strategic analysis

Research has repeatedly shown that organizations which simply innovate on impulse are poor performers. For example, a number of studies cite firms that have adopted expensive and complex innovations to upgrade their processes but which have failed to obtain competitive advantage from process innovation.[91] By contrast, those which understand the overall business, including their technological competence and their desired development trajectory are more likely to succeed. In similar fashion, studies of product/service innovation regularly point to lack of strategic underpinning as a key problem.[92] For this reason many organizations take time – often off-site and away from the day-to-day pressures of their 'normal' operations – to reflect and develop a shared strategic framework for innovation.

The underlying question this framework has to answer is about balancing fit with business strategy – does the innovation we are considering help us reach the strategic goals we have set ourselves (for growth, market share, profit margin, etc.)? – with the underlying competencies – do we know enough about this to pull it off (or if not do we have a clear idea of how we would get hold of and integrate such knowledge)? Much can be gained through taking a systematic approach to answering these questions – a typical approach might be to carry out some form of competitive analysis which looks at the positioning of the organization in terms of its environment and the key forces acting upon competition. Within this picture questions can then be asked about how a proposed innovation might help shift the competitive positioning favourably – by lowering or raising entry barriers, by introducing substitutes to rewrite the rules of the game, etc. A wide range of tools is available to help this process. For example, the 4Ps framework introduced in Chapter 2 can be used to review the potential for innovation.

Many structured methodologies exist to help organizations work through these questions and these are often used to help smaller and less experienced players build management capability. Examples include the SWORD approach to help SMEs find and develop appropriate new product opportunities or the 'Making IT Pay' framework offered by the UK DTI to help firms make strategic process innovation decisions.[93] Increasing emphasis is being placed on the role of intermediaries – innovation consultants and advisors – who can provide a degree of assistance in thinking through innovation strategy – and a number of regional and national government support programmes include this element. Examples include the IRAP programme (developed in Canada but widely used by other countries such as Thailand), the European Union's MINT programme, the TEKES counselling scheme in Finland, the Manufacturing Advisory Service in the UK (modelled in part on the US Manufacturing Extension Service in the USA) and the AMT programme in Ireland.[94]

In carrying out such a systematic analysis it is important to build on multiple perspectives. Reviews can take an 'outside-in' approach, using tools for competitor and market analysis, or they can adopt an 'inside-out' model, looking for ways of deploying competencies. They can build on explorations of the future such as the scenarios described earlier in this chapter, and they can make use of techniques like 'technology road-mapping' to help identify courses of action which will deliver broad strategic objectives.[95] But in the process of carrying out such reviews it is critical to remember that strategy is not an exact science so much as a process of building shared perspectives and developing a framework within which risky decisions can be located.

It is also important not to neglect the need to communicate and share this strategic analysis. Unless people within the organization understand and commit to the analysis it will be hard for them to use it to frame their actions. The issue of strategy *deployment* – communicating and enabling people to use the framework – is essential if the organization is to avoid the risk of having 'know-how' but not 'know-why' in its innovation process. Deployment of this kind comes to the fore in the case of focused incremental improvement activities common to implementations of the 'lean' philosophy or of *kaizen*. In principle it is possible to mobilize most of the people in an organization to contribute their ideas and creativity towards continuous improvement, but in practice this often fails. A key issue is the presence – or absence – of some strategic focus within which they can locate their multiple small-scale innovation activities. This requires two key enablers – the creation of a clear and coherent strategy for the business and the deployment of it through a cascade process which builds understanding and ownership of the goals and sub-goals.

This is a characteristic feature of many Japanese *kaizen* systems and may help explain why there is such a strong 'track record' of strategic gains through continuous improvement. In such plants overall business strategy is broken down into focused three-year mid-term plans (MTPs); typically the plan is given a slogan or motto to help identify it. This forms the basis of banners and other illustrations, but its real effect is to provide a backdrop against which efforts over the next three years can be focused. The MTP is specified not just in vague terms but with specific and measurable objectives – often described as pillars. These are, in turn, decomposed into manageable projects which have clear targets and measurable achievement milestones, and it is to these that workplace innovation activities are systematically applied.

Policy deployment of this kind requires suitable tools and techniques and examples include *hoshin* (participative) planning, how–why charts, 'bowling charts' and briefing groups. Chapter 9 picks up this theme in more detail.

Portfolio management approaches

There are a variety of approaches which have developed to deal with the question of what is broadly termed 'portfolio management'. These range from simple judgements about risk and reward to complex quantitative tools based on probability theory.[96] But the underlying purpose is the same – to provide a coherent basis on which to judge which projects should be undertaken, and to ensure a good balance across the portfolio of risk and potential reward. Failure to make such judgements can lead to a number of problem issues, as Table 4.13 indicates.

In general we can identify three approaches to this problem of building a strategic portfolio – benefit measurement techniques, economic models and portfolio models. Benefit measurement approaches are usually based on relatively simple subjective judgements – for example, checklists which ask whether certain criteria are met or not. More advanced versions attempt some kind of scoring or weighting so that projects can be compared in terms of their overall attractiveness. The main weakness here is that they consider each project in relative isolation.

Economic models attempt to put some financial or other quantitative data into the equation – for example, by calculating a payback time or discounted cash flow arising from the project. Once again these suffer from only treating single projects rather than reviewing a bundle, and they are also heavily dependent on the availability of good financial data – not always the case at the outset of a risky project. The third group – portfolio methods – tries to deal with the issue of reviewing across a set of projects and looks for balance. Chapter 8 discusses portfolio methods and tools in more detail.

TABLE 4.13 Criteria for evaluating different types of research project

Objective	Technical activity	Evaluation criteria (% of all R&D)	Decision-takers	Market analysis	Nature of risk	Higher volatility	Longer time horizons	Nature of external alliances
Knowledge building	Basic research, monitoring	Overhead cost allocation (2–10%)	R&D	None	Small = cost of R&D	Reflects wide potential	Increases search potential	Research grant
Strategic positioning	Focused applied research, exploratory development	'Options' evaluation (10–25%)	Chief executive R&D division	Broad	Small = cost of R&D	Reflects wide potential	Increases search potential	R&D contract Equity
Business Investment	Development and production engineering	'Net present value' analysis (70–99%)	Division	Specific	Large = total cost of launching	Uncertainty reduces net present value	Reduces present value	Joint venture Majority control

RESEARCH NOTE   Strategy making in practice

We examined how strategy develops and evolves over time, and how different tools and processes are used in practice. Unlike most studies, which rely on surveys or interviews after the event, in this study we collected data from two case study companies by direct observation over many months, in *real* time. The data we generated included:

(a) 1392 digital photographs – the photographs we had taken of activities in the two settings included pictures taken during project and client meetings, interactions with visual materials, individual working and office conversations.
(b) Field notebooks – the notebooks had been used by each researcher to keep a diary of their time in the field, jotting down observations alongside the date and time, and at times relinquishing control to engineers and designers who took the notebooks and drew directly into them.
(c) 34 hours of audio material – taped during the project meetings attended as part of the observational work and follow-up interviews. This was also transcribed.
(d) Digital and physical files – additional documentation relating to the new product development project was archived in both digital and hard-copy formats.

The more useful practices we observed included:

- **Business strategy charts and roadmaps**: These timeline charts are generated in PowerPoint, and used by the general managers to disseminate corporate strategy, showing gross margin and the competitive roadmap. They were used in a meeting called by the general manager and attended by everybody in the division. Copies were then published on the server.
- **Technology development roadmap**: This is a sector-level roadmap for silicon implant technology; which also shows R&D and product release schedules. It shows the lifetime of product models, with quarterly figures for spending on R&D and continuous improvement. A printed version sits on the desk of the assistant to the product manager. A PowerPoint version was published on the server.
- **Financial forecast spreadsheets**: These are used to manage cost reduction and projections of revenue flow; the charts have a time dimension. For example, versions of cost reduction spreadsheets, generated by senior management, are used in a frozen way in cross-function team meetings between representatives of the engineering and procurement departments to negotiate and coordinate around delivery of targets and responsibilities for cost.
- **Strategic project timelines**: These are timelines showing the goals of the project; the different streams of business and relationships with clients that relate to it. The general manager used a whiteboard to sketch the first version, which was then converted over a number of weeks into a proliferation of more formalized and detailed versions.
- **Gantt charts**: These are timelines for scheduling activities. As the project progressed, versions of this timeline were widely used by the project team to keep present understanding of the activities involved in achieving production against a tight deadline. An example is posted up on the office wall of the assistant to the product manager. Hard copies and PowerPoint versions were used in cross-function product development team meetings.

- **Progress charts**: These are timelines for progress towards phase exit (and hence, revenue generation) shown in a standardized format with 'smileys' used to represent the project manager's assessment of risks. It is used by the quality manager for generic product development process, in a fortnightly cross-function meeting to review progress across the entire portfolio of new product development activity.

Source: Whyte, J., B. Ewenstein, M. Hales and J. Tidd (2008) How to visualize knowledge in project-based work. *Long Range Planning*, **41**(1), 74–92. Reproduced by permission of Elsevier.

Summary and further reading

In formulating and executing their innovation strategies, organizations cannot ignore the national systems of innovation and international value chains in which they are embedded. Through their strong influences on demand and competitive conditions, the provision of human resources, and forms of corporate governance, national systems of innovation both open opportunities and impose constraints on what firms can do.

However, although firms' strategies are influenced by their own national systems of innovation, and their position in international value chains, they are not determined by them. Learning (i.e. assimilating knowledge) from competitors and external sources of innovation is essential for developing capabilities, but does require costly investments in R&D, training and skills development in order to develop the necessary absorptive capacity. This depends in part on what management itself does, by way of investing in complementary assets in production, marketing, service and support, and its position in local and international systems of innovation. It also depends on a variety of factors that make it more or less difficult to appropriate the benefits from innovation, such as intellectual property and international trading regimes, and over which management can sometimes have very little influence.

There are a number of texts which describe and compare different systems of national innovation policy, including *National Innovation Systems* (Oxford University Press, 1993), edited by Richard Nelson; *National Systems of Innovation* (Pinter, 1992), edited by B.-A. Lundvall; and *Systems of Innovation: Technologies, Institutions and Organisations* (Pinter, 1997), edited by Charles Edquist. The former is stronger on US policy, the other two on European, but all have an emphasis on public policy rather than corporate strategy. Michael Porter's *The Competitive Advantage of Nations* (Macmillan, 1990) provides a useful framework in which to examine the direct impact on corporate behaviour of innovation systems. At the other extreme, David Landes' *Wealth and Poverty of Nations* (Little Brown, 1998) takes a broad (and stimulating) historical and cultural perspective. The best overview is provided by the anthology of Chris Freeman's work in *Systems of Innovation* (Edward Elgar, 2008).

Comprehensive and balanced reviews of the arguments and evidence for product leadership versus follower positions are provided by G.J. Tellis and P.N. Golder: *Will and Vision: How Latecomers Grow to Dominate Markets* (McGraw-Hill, 2002) and *Fast Second: How Smart Companies Bypass Radical Innovation to Enter and Dominate New Markets* (Jossey Bass, 2004) by Costas Markides. More relevant to firms from emerging economies, and our favourite text on the subject, is Naushad Forbes and David Wield's *From Followers to Leaders: Managing Technology and Innovation* (Routledge, 2002), which includes numerous case examples. For recent reviews of the core competence and dynamic capability perspectives see David Teece's *Essays in Technology Management and Policy: Selected Papers* (World Scientific Press, 2004) and Connie Helfat's *Dynamic Capabilities: Understanding Strategic Change in Organizations* (Blackwell, 2006). Davenport, Leibold and Voelpel provide an recent edited compilation of leading strategy writers in *Strategic Management in the Innovation Economy* (2nd edition, Wiley, 2006), and the review edited by Robert Galavan, John Murray and Costas Markides, *Strategy, Innovation and Change* (Oxford University Press, 2008), is excellent. On the more specific issue of technology strategy Vittorio Chiesa's *R&D Strategy and Organization* (Imperial College Press, 2001) is a good place to start. The renewed interest in business model innovation, that is how value is created and captured, is discussed in *Strategic Market Creation: A New Perspective on Marketing and Innovation Management*, a review of research at Copenhagen Business School and Bocconi University, edited by Karin Tollin and Antonella Carù (Wiley, 2008).

Web links

Here are the full details of the resources available on the website flagged throughout the text:

Case studies:
Polaroid
Corning

Interactive exercises:
4Ps analysis

Tools:
Identifying innovative capabilities

Video podcast:
3M: Identifying lead users

Audio podcast:
Simon Murdoch: Bookpages

References

1. **Ansoff, I.** (1965) The firm of the future. *Harvard Business Review*, Sept–Oct, 162–178.
2. **Mintzberg, H.** (1987) Crafting strategy., *Harvard Business Review*, July–August, 66–75. See also the interview with Mintzberg in *The Academy of Management Executive* (2000) **14** (3), 31–42.
3. **Whittington, R.** (1994) *What is Strategy and Does it Matter?* Routledge, London.
4. **Kay, J.** (1993) *Foundations of Corporate Success: How Business Strategies Add Value*, Oxford University Press, Oxford.
5. **Starbuck, W.H.** (1992) Strategizing in the real world. *International Journal of Technology Management*, special publication on 'Technological Foundations of Strategic Management'. **8** (1/2), 77–85.
6. **Howard, N.** (1983) A novel approach to nuclear fusion. *Dun's Business Month*, **123**, 72, 76.

7. **Berton, L.** (1974) Nuclear energy stocks, set to explode. *Financial World*, **141** (16 Jan), 8–11; **Freeman, C.** (1984) Prometheus unbound. *Futures*, **16**, 495–507.

8. **Duysters, G.** (1995) The Evolution of Complex Industrial Systems: The Dynamics of Major IT Sectors, MERIT, University of Maastricht, Maastricht; ***The Economist*** (1996) Fatal attraction: why AT&T was led astray by the lure of computers. *Management Brief*, 23 March; **Von Tunzelmann, N.** (1999) Technological accumulation and corporate change in the electronics industry. In A. Gambardella and F. Malerba (eds), *The Organization of Scientific and Technological Research in Europe*, Cambridge University Press, Cambridge.

9. ***The Economist*** (2000) The failure of new media. 19 August, 59–60.

10. ***The Economist*** (2000) A survey of e-entertainment. 7 October.

11. **Pasteur, L.** (1854) Address given on the inauguration of the Faculty of Science, University of Lille, 7 December. Reproduced in *Oxford Dictionary of Quotations*, Oxford University Press, Oxford.

12. **Sapsed, J.** (2001) *Restricted Vision: Strategizing Under Uncertainty*, Imperial College Press, London.

13. **Boston Consulting Group** (1975) *Strategy Alternatives for the British Motorcycle Industry*, HMSO, London.

14. **Pascale, R.** (1984) Perspectives on strategy: the real story behind Honda's success. *California Management Review*, **26**, 47–72.

15. **Mintzberg, H., R.T. Pascale, M. Goold, and R.P. Rumelt** (1996) The 'Honda effect' revisited. *California Management Review*, **38** (4), 78–117.

16. **Lee, G.** (1995) Virtual prototyping on personal computers. *Mechanical Engineering*, **117** (July), 70–3.

17. **Porter, M.** (1980) *Competitive Strategy*, Free Press, New York.

18. **Robinson, W. and Chiang, J.** (2002) Product development strategies for established market pioneers, early followers, and late entrants. *Strategic Management Journal*, **23**, 855–66.

19. **Fransman, M.** (1994) Information, knowledge, vision and theories of the firm. *Industrial and Corporate Change*, **3**, 713–57.

20. **Patel, P. and K. Pavitt** (1998) The wide (and increasing) spread of technological competencies in the world's largest firms: a challenge to conventional wisdom. In A. Chandler, P. Hagstrom and O. Solvell (eds), *The Dynamic Firm*, Oxford University Press, Oxford.

21. **Garcia, R. and R. Calantone** (2002) A critical look at technological innovation typology and innovativeness terminology: a literature review. *Journal of Product Innovation Management*, **19**, 110–32.

22. **Baden-Fuller, C. and J. Stopford** (1994) *Rejuvenating the Mature Business: The Competitive Challenge*, Harvard Business School Press, Boston, MA; **Belussi, F.** (1989) Benetton – a case study of corporate strategy for innovation in traditional sectors. In M. Dodgson (ed.), *Technology Strategy and the Firm: Management and Public Policy* (pp. 116–33), Longman, London.

23. **Barras, R.** (1990) Interactive innovation in financial and business services: the vanguard of the service revolution. *Research Policy*, **19**, 215–38.

24. **Kay, J.** (1996) Oh Professor Porter, whatever did you do? *Financial Times*, 10 May, 17.

25. **Tidd, J.** (1995) The development of novel products through intra- and inter-organizational networks: the case of home automation. *Journal of Product Innovation Management*, **12** (4), 307–22; **McDermott, C. and G. O'Connor** (2002) Managing radical innovation: an overview of emergent strategy issues. *Journal of Product Innovation Management*, **19**, 424–38.

26. **Lamming, R.** (1993) *Beyond Partnership*, Prentice-Hall, Hemel Hempstead.

27. **Arundel, A., G. van de Paal and L. Soete** (1995) *Innovation Strategies of Europe's Largest Industrial Firms*, PACE Report, MERIT, University of Limbourg, Maastricht.

28. **Christensen, C. and M. Raynor** (2003) *The Innovator's Solution: Creating and Sustaining Successful Growth*, Harvard Business School Press, Boston, MA.

29. **Teece, D. and G. Pisano** (1994) The dynamic capabilities of firms: an introduction. *Industrial and Corporate Change*, **3**, 537–56.

30. **Albert, M.** (1992) *Capitalism against Capitalism*, Whurr, London.

31. **Fransman, M.** (1995) *Japan's Computer and Communications Industry*, Oxford University Press, Oxford.

32. **Albach, H.** (1996) Global competitive strategies for scienceware products. In G. Koopmann and H. Scharrer (eds), *The Economics of High Technology Competition and Cooperation in Global Markets* (pp. 203–17), Nomos, Baden-Baden.

33. *The Economist* (1995) Dismantling Daimler-Benz. 18 November, 99–100.

34. *The Economist* (1995) Back on top. A survey of American business. 16 September.

35. **Gordon, P.** (1996) Industrial districts and the globalization of innovation: regions and networks in the new economic space. In X. Vence-Deza and J. Metcalfe (eds), *Wealth from Diversity* (pp. 103–34), Kluwer, Dordrecht; **Computer Science and Telecommunications Board** (1999) *Funding a Revolution: Government Support for Computing Research*, National Research Council, Washington, DC.

36. **Harding, R. and W. Paterson** (eds) (2000) *The Future of the German Economy: An End to the Miracle?* Manchester University Press, Manchester.

37. *The Economist* (1999) The world in your pocket: a survey of telecommunications, 9 October.

38. **Levin, R., A. Klevoric, R. Nelson, and S. Winter** (1987) Appropriating the returns from industrial research and development. *Brookings Papers on Economic Activity*, **3**, 783–820; **Mansfield, E., M. Schwartz and S. Wagner** (1981) Imitation costs and patents: an empirical study. *Economic Journal*, **91**, 907–18.

39. **Kim, L.** (1993) National system of industrial innovation: dynamics of capability building in Korea and **Odagiri, H. and A. Goto** (1993) The Japanese system of innovation: past, present and future. In R. Nelson (ed.), *National Innovation Systems* (pp. 357–83, 76–114), Oxford University Press, Oxford.

40. **Teece, D.** (1986) Profiting from technological innovation: implications for integration, collaboration, licensing and public policy. *Research Policy*, **15**, 285–305.

41. **Von Hippel, E.** (1987) Cooperation between rivals: informal know-how training. *Research Policy*, **16**, 291–302.

42. **Spencer, J.** (2003) Firms' knowledge-sharing strategies in the global innovation system: empirical evidence from the flat panel display industry. *Strategic Management Journal*, **24**, 217–33.

43. **Shapiro, C. and H. Varian** (1998) *Information Rules: A Strategic Guide to the Network Economy*, Harvard Business School Press, Boston, MA.

44. **Hill, C.** (1997) Establishing a standard: competitive strategy and technological standards in winner-take-all industries. *Academy of Management Executive*, **11**, 7–25.

45. **Rosenbloom, R. and M. Cusumano** (1987) Technological pioneering and competitive advantage: the birth of the VCR industry. *California Management Review*, **24**, 51–76.

46. **Chesbrough, H. and D. Teece** (1996) When is virtual virtuous? Organizing for innovation. *Harvard Business Review*, Jan–Feb, 65–73.

47. **Suarez, F.** (2004) Battles for technological dominance: an integrative framework. *Research Policy*, **33**, 271–86.

48. **Chiesa, V., R. Manzini, and G. Toletti** (2002) Standards-setting processes: evidence from two case studies. *R&D Management*, **32** (5), 431–50.

49. **Soh, P. and E. Roberts** (2003) Networks of innovators: a longitudinal perspective. *Research Policy*, 32, 1569–88.

50. **Sahay, A. and D. Riley** (2003) The role of resource access, market conditions, and the nature of innovation in the pursuit of standards in the new product development process. *Journal of Product Innovation Management*, **20**, 338–55.

51. **Steffens, J.** (1994) *Newgames: Strategic Competition in the PC Revolution*, Pergamon Press, Oxford.

52. **Tellis, G. and P. Golder** (1996) First to market, first to fail? Real causes of enduring market leadership. *Sloan Management Review*, Winter, 65–75; **Tellis, G. and P. Golder** (2002) *Will and Vision: How Latecomers Grow to Dominate Markets*, McGraw-Hill, New York.

53. **Lambkin, M.** (1992) Pioneering new markets. A comparison of market share winners and losers. *International Journal of Research on Marketing*, 5–22; **Robinson, W.** (2002) Product development strategies for established market pioneers, early followers and late entrants. *Strategic Management Journal*, **23**, 855–66.

54. *The Economist* (2000) The knowledge monopolies: patent wars. 8 April, 95–9; (1996) A dose of patent medicine. 10 February, 93–4.

55. *The Economist* (1999) Digital rights and wrongs. 17 July, 99–100.

56. **Mazzolini, R. and R. Nelson** (1998) The benefits and costs of strong patent protection: a contribution to the current debate. *Research Policy*, **26**, 405.

57. **Bertin, G. and S. Wyatt** (1988) *Multinationals and Industrial Property: The Control of the World's Technology*, Harvester-Wheatsheaf, Hemel Hempstead.

58. **Dosi, G.** (1982) Technological paradigms and technological trajectories. *Research Policy*, **11**, 147–62.

59. **Pavitt, K.** (1984) Sectoral patterns of technical change: towards a taxonomy and a theory. *Research Policy*, **13**, 343–73; **Pavitt, K.** (1990) What we know about the strategic management of technology. *California Management Review*, **32**, 17–26.

60. **Freeman, C., J. Clark and L. Soete** (1982) *Unemployment and Technical Innovation: A Study of Long Waves and Economic Development*, Frances Pinter, London.

61. **Mowery, D. and N. Rosenberg** (1989) *Technology and the Pursuit of Economic Growth*, Cambridge University Press, Cambridge.

62. **Arundel, A., G. van de Paal and L. Soete** (1995) *Innovation Strategies of Europe's Largest Industrial Firms*, PACE Report, MERIT, University of Limbourg, Maastricht; **Cesaretto, S. and S. Mangano** (1992) Technological profiles and economic performance in the Italian manufacturing sector. *Economics of Innovation and New Technology*, **2**, 237–56.

63. **Coombs, R. and A. Richards** (1991) Technologies, products and firms' strategies. *Technology Analysis and Strategic Management*, **3**, 77–86, 157–75.

64. **Prahalad, C. and G. Hamel** (1990) The core competencies of the corporation. *Harvard Business Review*, May–June, 79–91; **Prahalad, C. and Hamel, G.** (1994) *Competing for the Future*, Harvard Business School Press, Cambridge, MA.

65. **Oskarsson, C.** (1993) *Technology Diversification: The Phenomenon, its Causes and Effects*, Department of Industrial Management and Economics, Chalmers University, Gothenburg.

66. **Argyres, N.** (1996) Capabilities, technological diversification and divisionalization. *Strategic Management Journal*, **17**, 395–410.

67. **Plumpe, G.** (1995) Innovation and the structure of IG Farben. In F. Caron, P. Erker and W. Fischer (eds), *Innovations in the European Economy between the Wars*, De Gruyter, Berlin.

68. **Graham, M.** (1986) *RCA and the Videodisc: The Business of Research*, Cambridge University Press, Cambridge; **Hounshell, D. and J. Smith** (1988) *Science and Corporate Strategy: Du Pont R&D, 1902–1980*, Cambridge University Press, New York; **Reader, W.** (1975) *Imperial Chemical Industries,*

a History, Oxford University Press, Oxford; **Reich, L.** (1985) *The Making of American Industrial Research: Science and Business at GE and Bell*, Cambridge University, Cambridge. For a discussion of the implications for innovation strategy of these and related studies, see **Pavitt, K. and W. Steinmueller** (2001) Technology in corporate strategy: change, continuity and the information revolution. In A. Pettigrew, H. Thomas and R. Whittington (eds), *Handbook of Strategy and Management*, Sage, London.

69. ***The Economist*** (1989) Japan's smokestack fire-sale. 19 August, 63–4.

70. **Granstrand, O., P. Patel and K. Pavitt** (1997) Multi-technology corporations: why they have 'distributed' rather than 'distinctive core' competencies. *California Management Review*, **39**, 8–25; **Patel, P. and K. Pavitt** (1998) The wide (and increasing) spread of technological competencies in the world's largest firms: a challenge to conventional wisdom. In A. Chandler, P. Hagstrom and O. Solvell (eds), *The Dynamic Firm*, Oxford University Press, Oxford.

71. **Prencipe, A.** (1997) Technological competencies and product's evolutionary dynamics: a case study from the aero-engine industry. *Research Policy*, **25**, 1261.

72. **Leonard-Barton, D.** (1995) *Wellsprings of Knowledge*, Harvard Business School Press, Boston, MA.

73. **Capon, N. and R. Glazer** (1987) Marketing and technology: a strategic coalignment. *Journal of Marketing*, **51**, 1–14.

74. **Hall, R.** (2006) What are competencies. In J. Tidd (ed), *From Knowledge Management to Strategic Competence*, Second Edition, Imperial College Press, London; (1994) A framework for identifying the intangible sources of sustainable competitive advantage. In G. Hamel and A. Heene (eds), *Competence-Based Competition* (pp. 149–69), John Wiley & Sons, Ltd, Chichester.

75. **Winter, S.G.** (2003) Understanding dynamic capabilities. *Strategic Management Journal*, **24**, 991–5

76. **Danneels, E.** (2002) The dynamic effects of product innovation and firm competencies. *Strategic Management Journal*, **23**, 1095–21.

77. **Adner, R. and Helfat, C.** (2003) Corporate effects and dynamic managerial capabilities. *Strategic Management Journal*, 1011–25.

78. ***The Economist*** (1994) The vision thing. 3 September, 77.

79. For more detail, see **Schnaars, S.** (1989) *Megamistakes: Forecasting and the Myth of Rapid Technological Change*, Free Press, New York.

80. **Schnaars, S. and C. Berenson** (1986) Growth market forecasting revisited: a look back at a look forward. *California Management Review*, **28**, 71–88.

81. **Sandoz, P.** (1997) *Canon*, Penguin, London.

82. **Granstrand, O., E. Bohlin, C. Oskarsson, and N. Sjoberg** (1992) External technology acquisition in large multi-technology corporations. *R&D Management*, **22** (2), 111–33.

83. **O'Connor, G. and R. Veryzer** (2001) The nature of market visioning for the technology-based radical innovation. *Journal of Product Innovation Management*, **18**, 231–46.

84. **Ohmae, K.** (1990) *The Borderless World: Power and Strategy in the Interlinked Economy*, Collins, London; **Friedman, T.** (2006) *The World is Flat: The Globalized World in the 21st Century*, Penguin, London.

85. **Ghoshal, S. and C. Bartlett** (1987) Innovation processes in multinational corporations. *Strategic Management Journal*, **8**, 425–39.

86. **Cantwell, J. and J. Molero** (2003) *Multinational Enterprises, Innovative Systems and Systems of Innovation*, Edward Elgar, Cheltenham; **Cantwell, J.** (1992) The internationalisation of technological activity and its implications for competitiveness. In O. Granstrand, L. Hakanson and S. Sjolander (eds), *Technology Management and International Business*, John Wiley & Sons, Ltd,

Chichester; **Patel, P.** (1996) Are large firms internationalising the generation of technology? Some new evidence. *IEEE Transactions on Engineering Management*, **43**, 41–47; **Ariffin, L. and M. Bell** (1999) Firms, politics and political economy: patterns of subsidiary–parent linkages and technological capability-building in electronics TNC subsidiaries in Malaysia. In K. Jomo, G. Felker and R. Rasiah (eds), *Industrial Technology Development in Malaysia: Industry and Firm Studies*, Routledge, London; **Hu, Y-S.** (1995) The international transferability of competitive advantage. *California Management Review*, **37**, 73–88; **Senker, J.** (1995) Tacit knowledge and models of innovation. *Industrial and Corporate Change*, **4**, 425–47; **Senker, J., P. Benoit-Joly and M. Reinhard** (1996) *Overseas Biotechnology Research by Europe's Chemical-Pharmaceuticals Multinationals: Rationale and Implications*, STEEP Discussion Paper No. 33, Science Policy Research Unit, University of Sussex, Brighton; **Niosi, J.** (1999) The internationalization of industrial R&D. *Research Policy*, **29**, 107.

87. **Gerybadze, A. and G. Reger** (1999) Globalisation of R&D: recent changes in the management of innovation in transnational corporations. *Research Policy*, **28**, 251–74.

88. **Vernon, R.** (1966) International investment and international trade in the product cycle. *Quarterly Journal of Economics*, **80**, 190–207.

89. **Chiesa, V.** (2001) *R&D Strategy and Organization*, Imperial College Press, London.

90. **Chiesa, V.** (2000) Global R&D project management and organization: a taxonomy. *Journal of Product Innovation Management*, **17**, 341–59.

91. **Ettlie, J.** (1988) *Taking Charge of Manufacturing*, Jossey-Bass, San Francisco; **Bessant, J.** (1991) *Managing Advanced Manufacturing Technology: The Challenge of the Fifth Wave*, NCC-Blackwell, Oxford.

92. **Griffin, A., M.D. Rosenau, G.A. Castellion, and N.F. Anschuetz** (1996) *The PDMA Handbook of New Product Development*, John Wiley & Sons, Inc., New York; **Ernst, H.** (2002) Success factors of new product development: a review of the empirical literature. *International Journal of Management Reviews*, **4** (1), 1–40.

93. **Carson, J.** (1989) *Innovation: A Battleplan for the 1990s*, Gower, Aldershot. **Bessant, J.** (1997) Developing technology capability through manufacturing strategy. *International Journal of Technology Management*, **14** (2/3/4), 177–95; **DTI** (1998) *Making IT Fit: Guide to Developing Strategic Manufacturing*, UK Department of Trade and Industry, London.

94. **Mills, J., K. Platts, A. Neely, A. Richards, and M. Bourne** (2002) *Creating a Winning Business Formula*, Cambridge University Press, Cambridge; **Mills, J., K. Platts, M. Bourne, and A. Richards** (2002) *Competing through Competencies*, Cambridge University Press, Cambridge.

95. **Crawford, M. and C. Di Benedetto** (1999) *New Products Management*, McGraw-Hill/Irwin, New York; **Floyd, C.** (1997) *Managing Technology for Corporate Success*, Gower, Aldershot.

96. **Cooper, R.** (1988) The new product process: a decision guide for management. *Journal of Marketing Management*, **3** (3), 238–55.

PART 3

Search

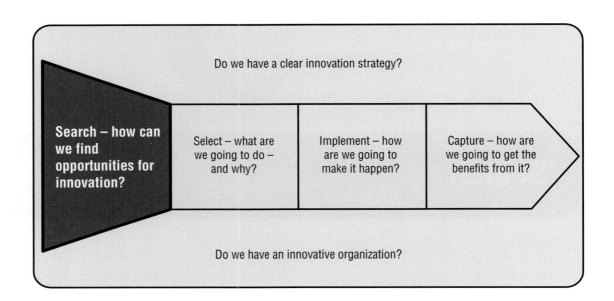

Do we have a clear innovation strategy?

Search – how can we find opportunities for innovation?

Select – what are we going to do – and why?

Implement – how are we going to make it happen?

Capture – how are we going to get the benefits from it?

Do we have an innovative organization?

In this section we move on to the first of the core elements in our process model – the 'search' question. Chapter 5 explores the issues around the question of what triggers the innovation process – the multiple sources which we need to be aware of and the challenges involved in searching for and picking up signals from them. And it looks at the complementary question – *how* do we carry out this search activity? Which structures, tools and techniques are appropriate under what conditions? How do we balance search around exploration of completely new territory with exploiting what we already know in new forms? In Chapter 6 we look at the major challenge of building and sustaining rich networks to enable what has become labelled 'open innovation'.

CHAPTER 5

Sources of innovation

5.1 Where do innovations come from?

Where do innovations come from? There's a good chance that asking that question will conjure images like that of Archimedes, jumping up from his bath and running down the street, so enthused by the desire to tell the world his discovery that he forgot to get dressed. Or Newton, dozing under the apple tree until a falling apple helped kick his brain into thinking about the science of gravity. Or James Watt, also asleep, until woken by the noise of a boiling kettle. Such 'Eureka' moments are certainly a part of innovation folklore – and they underline the importance of flashes of insight which make new connections. They form the basis of the cartoon model of innovation which usually involves thinking bubbles and flashing light bulbs. And from time to time they do happen, for example, Percy Shaw's observation of the reflection in a cat's eye at night led to the development of one of the most widely used road-safety innovations in the world. Or George de Mestral, returning home from a walk in the Swiss Alps noticing the way plant burrs became attached to his dog's fur, found the inspiration behind Velcro fasteners. The myths of innovation – Scott Berkun, who worked on developing Internet Explorer, discusses Eureka moments and the reality of how innovations happen. Video of a lecture to Carnegie Mellon University (http://www.youtube.com/watch?v=m6gaj6huCp0).

But of course there is much more to it than that – as we saw in Chapter 2. Innovation is a process of taking ideas forward, revising and refining them, weaving the different strands of 'knowledge spaghetti' together towards a useful product, process or service. Triggering that process is not just about occasional flashes of inspiration – innovation comes from many other directions, and if we are to manage it effectively we need to remind ourselves of this diversity. Figure 5.1 indicates the wide range of stimuli which could be relevant to kick-starting the innovation journey, and we will explore some of the important triggers in this chapter.

5.2 Knowledge push . . .

One obvious source of innovation is the possibilities which emerge as a result of scientific research. From the earliest days curious men and women have experimented and explored the world around them and various Greek philosophers, Roman engineers, Egyptian astronomers, Persian mathematicians, Chinese doctors and a host of others laid the foundations of what we loosely call 'science'. Although some of the earliest work was something of a solo act we should remember that from a very early stage this process of exploring and codifying at the frontiers of knowledge became a systematic activity – and one which involved a wide network of people sharing their ideas. We sometimes think that organized science is a child of the twentieth century but a quick look at the ways in which the medieval Guilds managed the processes of knowledge acquisition, extension and diffusion reminds us that this is

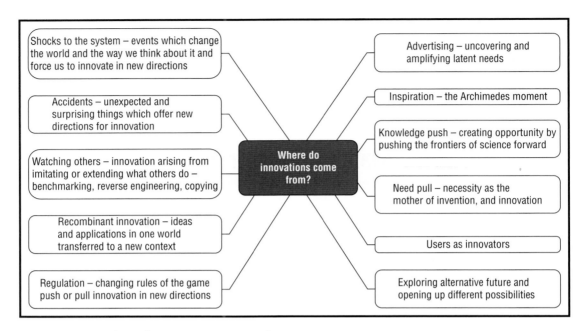

FIGURE 5.1: Where do innovations come from?

a well-established pattern. The fame of key cities like Venice or regions like Flanders owed as much to the organized scientific knowledge in fields like gun-making or textile manufacture as to the entrepreneurial activities of traders and merchants.

In the twentieth century the rise of the modern large corporation brought with it the emergence of the research laboratory as a key instrument of progress. Bell Labs, ICI, Bayer, BASF, Philips, Ford, Western Electric, Du Pont – all were founded in the 1900s as powerhouses of ideas.[1] They produced a steady stream of innovations which fed rapidly growing markets for automobiles, consumer electrical products, synthetic materials, industrial chemicals – and the vast industrial complexes needed to fight two major wars. Their output wasn't simply around product innovation – many of the key technologies underpinning process innovations, especially around the growing field of automation and information/communications technology also came from such organized R&D effort. Table 5.1 gives some examples of science-push innovations. The Corning case study provides an example of a long-term knowledge-push innovator.

TABLE 5.1	Some examples of knowledge-push innovations	
Nylon	Radar	Antibiotics
Microwave	Synthetic rubber	Cellular telephony
Medical scanners	Photocopiers	Hovercraft
Fibre optic cable	Digital imaging	Transistor/integrated circuits

It's important to see the pattern which such activity established in terms of innovation. Organized R&D became a systematic commitment of specialist staff, equipment, facilities and resources targeted at key technological problems or challenges. The aim was to explore, but much of that exploration was elaborating and stretching trajectories that were established as a result of occasional breakthroughs. So the leap in technology, which the invention of synthetic materials like nylon or polyethylene represented, was followed by innumerable small-scale developments around and along that path. The rise of 'big Pharma' – the huge global pharmaceutical industry – was essentially about large R&D expenditure, but much of it spent on development and elaboration punctuated by the occasional breakthrough into 'blockbuster' drug territory. The computer and other industries that depend on semiconductors have become linked to a long-term trajectory, which followed from the early 'breakthrough' years of the industry. Moore's law (named after one of the founders of Intel) essentially sets up a trajectory which shapes and guides innovation based on the idea that the size will shrink and the power will increase by a factor of two every two years.[2] This affects memory, processor speed, display drivers and various other components, which in turn drives the rate of innovation in computers, digital cameras, mobile phones and thousands of other applications.

This can apply to products or processes: in both cases the key characteristics become stabilized and experimentation moves to getting the bugs out and refining the dominant design. For example, the nineteenth-century chemical industry moved from producing soda ash (an essential ingredient in making soap, glass and a host of other products) from the earliest days where it was produced by burning vegetable matter through to a sophisticated chemical reaction which was carried out in a batch process (the Leblanc process), which was one of the drivers of the Industrial Revolution. This process dominated for nearly a century but was in turn replaced by a new generation of continuous processes that used electrolytic techniques and which originated in Belgium where they were developed by the Solvay brothers. Moving to the Leblanc process or the Solvay process did not happen overnight – it took decades of work to refine and improve the process, and to fully understand the chemistry and engineering required to get consistent high quality and output.

The same pattern can be seen in products. For example, the original design for a camera is something which goes back to the early nineteenth century and – as a visit to any science museum will show – involved all sorts of ingenious solutions. The dominant design gradually emerged with an architecture which we would recognize today – shutter and lens arrangement, focusing principles, back plate for film or plates, etc. But this design was then modified still further, for example, with different lenses, motorized drives, flash technology, and, in the case of George Eastman's work, to creating a simple and relatively

---

BOX 5.1   ## The ubiquitous tale of polyethylene

Like it or loathe it, polythene is one of the key material innovations to come out of the twentieth century. It is the world's 'favourite' plastic measured in terms of consumption – 60 million tonnes/year find their way into films, plastic bags, packaging, cosmetics and a host of other applications. Discovered by accident by chemists working at ICI in the UK in 1933 the original low-density polyethylene product has gone through a classic pattern of incremental and occasional breakthrough innovation giving rise to new products like high-density polyethylene and film and to process innovations such as the Phillips catalysis process which enabled better yields in production.

'idiot-proof' model camera (the Box Brownie) which opened up photography to a mass market. More recent development has seen a similar fluid phase around digital imaging devices.

This idea of occasional breakthroughs followed by extended periods of exploring and elaboration along those paths has been studied and mapped by a number of writers.[3,4] It's a common pattern and one which helps us deal with the key management question of how and where to direct our search activity for innovation – a theme we will return to shortly.

5.3 Need pull . . .

Knowledge creation provides a push, creates an 'opportunity field' which sets up possibilities for innovation. But – as we saw in Chapter 2 – we know from innumerable examples that simply having a bright idea is no guarantee of adoption. The American writer Ralph Waldo Emerson is supposed to have said *'build a better mousetrap and the world will beat a path to your door'** – but the reality is that there are plenty of bankrupt mousetrap salesmen around! Knowledge push creates a field of possibilities – but not every idea finds successful application and one of the key lessons is that innovation requires some form of demand if it is to take root. Bright ideas are not, in themselves, enough – they may not meet a real or perceived need and people may not feel motivated to change.

We should recognize that another key driver of innovation is need – the complementary pull to the knowledge push. In its simplest form it is captured in the saying that 'necessity is the Mother of invention' – innovation is often the response to a real or perceived need for change. Basic needs – for shelter, food, clothing, security – led early innovation as societies evolved and we are now at a stage where the need pull operates on more sophisticated higher level needs but via the same process. In innovation management the emphasis moves to ensuring we develop a clear understanding of needs and finding

VIEWS FROM THE FRONT LINE

Two hundred years ago Churchill Potteries began life in the UK making a range of crockery and tableware. That they are still able to do so today, despite a turbulent and highly competitive global market, says much for the approach that they have taken to ensure a steady stream of innovation. Chief Executive Andrew Roper highlights the way in which listening to users and understanding their needs has changed the business. *'We have taken on a lot of service disciplines, so you could think of us as less of a pure manufacturer and more as a service company with a manufacturing arm'*. Staff spend a significant proportion of their time talking to chefs, hoteliers and others. *'. . . sales, marketing and technical people spend far more of their time than I could ever have imagined checking out what happens to the product in use and asking the customer, professional or otherwise, what they really want next'*.

Source: Peter Marsh (2008) Ingredients for success on a plate. *Financial Times*, 26 March, p. 16.

---

* Ralph Waldo Emerson, *'If a man has good corn, or wood, or boards, or pigs to sell, or can make better chairs or knives, crucibles or church organs than anybody else, you will find a broad-beaten road to his home, though it be in the woods.'*

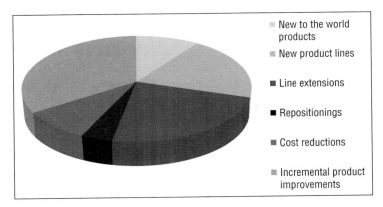

FIGURE 5.2: Types of new product
Source: Based on **Griffin, A.** (1997) PDMA research on new product development practices. *Journal of Product Innovation Management,* **14,** 429

ways to meet those needs. For example, Henry Ford was able to turn the luxury plaything that was the early automobile into something which became 'a car for Everyman', whilst Procter & Gamble began a business meeting needs for domestic lighting (via candles) and moved across into an ever-widening range of household requirements from soap to nappies to cleaners, toothpaste and beyond.

Just as the knowledge-push model involves a mixture of occasional breakthrough followed by extensive elaboration on the basic theme, searching around the core trajectory, so the same is true of need. Occasionally it involves a new-to-the-world idea which offers an innovative way of meeting a need – but mostly it is elaboration and differentiation. Various attempts have been made to classify product innovations in terms of their degree of novelty, and whilst the numbers and percentages vary slightly, the underlying picture is clear – there are very few 'new-to-the-world' products and very many extensions, variations and adaptations around those core ideas.[5,6] Figure 5.2 indicates a typical breakdown – and we could construct a similar picture for process innovations.

Understanding buyer/adopter behaviour has become a key theme in marketing studies since it provides us with frameworks and tools for identifying and understanding user needs[7] (we return to this theme in Chapter 9). Advertising and branding play a key role in this process – essentially using psychology to tune into – or even stimulate and create – basic human needs.[8,9] Much recent research has focused on detailed ethnographic studies of what people actually do and how they really use products and services – using the same approaches which anthropologists use to study strange new tribes to uncover hidden and latent needs[10] (see Case study 5.1 for an example). An example of using ethnographic methods to help get closer to user needs can be found in the case study of Tesco's Fresh & Easy store design in the USA.

Need-pull innovation is particularly important at mature stages in industry or product life cycles when there is more than one offering to choose from – competing depends on differentiating on the basis of needs and attributes, and/or segmenting the offering to suit different adopter types. There are differences between business-to-business markets (where emphasis is on needs amongst a shared group, e.g. along a supply chain) and consumer markets (where the underlying need may be much more basic, e.g. food, shelter, mobility, and appeals to a much greater number of people). Importantly there is also a 'bandwagon' effect – as more people adopt so the innovation becomes modified to take on board their needs – and the process accelerates.[11]

CASE STUDY 5.1

Understanding user needs in Hyundai

One of the problems facing global manufacturers is how to tailor their products to suit the needs of local markets. For Hyundai this has meant paying considerable attention to getting deep insights into customer needs and aspirations – an approach which they used to good effect in developing the 'Santa Fe', reintroduced to the US market in 2007. The headline for their development programme was 'touch the market' and they deployed a number of tools and techniques to enable it. For example, they visited an ice rink and watched an Olympic medallist skate around to help them gain an insight into the ideas of grace and speed which they wanted to embed in the car. This provided a metaphor – 'assertive grace' – which the development teams in Korea and the US were able to use.

Analysis of existing vehicles suggested some aspects of design were not being covered, for example, many sport/utility vehicles (SUVs) were rather 'boxy' so there was scope to enhance the image of the car. Market research suggested a target segment of 'glamour mums' who would find this attractive and the teams then began an intensive study of how this group lived their lives. Ethnographic methods looked at their homes, their activities and their lifestyles – for example, team members spent a day shopping with some target women to gain an understanding of their purchases and what motivated them. The list of key motivators that emerged from this shopping study included durability, versatility, uniqueness, child-friendly and good customer service from knowledgeable staff. Another approach was to make all members of the team experience driving routes around southern California, making journeys similar to those popular with the target segment and in the process getting first-hand experience of comfort, features and fixtures inside the car.

Source: Kluter, H. and D. Mottram (2007) Hyundai uses 'Touch the market' to create clarity in product concepts. *PDMA Visions*, **31**, 16–19.

Of course needs aren't just about external markets for products and services – we can see the same phenomenon of need pull working inside the business, as a driver of process innovation. 'Squeaking wheels' and other sources of frustration provide rich signals for change – and this kind of innovation is often something that can engage a high proportion of the workforce who experience these needs first hand. (The successful model of '*kaizen*' which underpins the success of firms like Toyota is fundamentally about sustained, high-involvement incremental process innovation along these lines.[12]). *Kaizen* provided the basic philosophy behind the 'total quality management' movement in the 1980s, the 'business process re-engineering' ideas of the 1990s and the current widespread application of concepts based on the idea of 'lean thinking' – essentially taking waste out of existing processes.[13–15]

Once again we can see the pattern – most of the time such innovation is about 'doing what we do better' but occasionally it involves a major leap. The example of glassmaking (Case study 5.2) provides a good illustration – for decades the need to produce smooth flat glass for windows had been met by a steady stream of innovations around the basic trajectory of grinding and polishing. There is plenty of scope for innovation in machinery, equipment, working practices, etc. – but such innovation tends to meet with diminishing returns as some of the fundamental bottlenecks emerge – the limits of how much you can improve an existing process. Eventually the stage is set for a breakthrough – like the emergence

CASE STUDY 5.2

Innovation in the glass industry

It's particularly important to understand that change doesn't come in standard sized jumps. For much of the time it is essentially incremental, a process of gradual improvement over time on dimensions like price, quality, choice, etc. For long periods of time nothing much shifts in either product offering or the way in which this is delivered (product and process innovation is incremental). But sooner or later someone somewhere will come up with a radical change which upsets the apple cart.

For example, the glass window business has been around for at least 600 years and is – since most houses, offices, hotels and shops have plenty of windows – a very profitable business to be in. But for most of those 600 years the basic process for making window glass hasn't changed. Glass is made in approximately flat sheets which are then ground down to a state where they are flat enough for people to see through them. The ways in which the grinding takes place have improved – what used to be a labour-intensive process became increasingly mechanized and even automated, and the tools and abrasives became progressively more sophisticated and effective. But underneath the same core process of grinding down to flatness was going on.

Then in 1952 Alastair Pilkington working in the UK firm of the same name began working on a process which revolutionized glass making for the next 50 years. He got the idea whilst washing up when he noticed that the fat and grease from the plates floated on the top of the water – and he began thinking about producing glass in such a way that it could be cast to float on the surface of some other liquid and then allowed to set. If this could be accomplished it might be possible to create a perfectly flat surface without the need for grinding and polishing.

Five years, millions of pounds and over 100 000 tonnes of scrapped glass later the company achieved a working pilot plant and a further two years on began selling glass made by the float glass process. The process advantages included around 80% labour and 50% energy savings plus those which came about because of the lack of need for abrasives, grinding equipment, etc. Factories could be made smaller and the overall time to produce glass dramatically cut. So successful was the process that it became – and still is – the dominant method for making flat glass around the world.

of float glass – which then creates new space within which incremental innovation along a new trajectory can take place.

Sometimes the increase in the urgency of a need or the extent of demand can have a forcing effect on innovation – the example of wartime and other crises supports this view. For example, the demand for iron and iron products increased hugely in the Industrial Revolution and exposed the limitations of the old methods of smelting with charcoal – it created the pull which led to developments like the Bessemer converter. In similar fashion the emerging energy crisis with oil prices reaching unprecedented levels has created a significant pull for innovation around alternative energy sources – and an investment boom for such work.

It's also important to recognize that innovation is not always about commercial markets or consumer needs. There is also a strong tradition of social need providing the pull for new products, processes and

CASE STUDY 5.3

The emergence of microfinance

One of the biggest problems facing people living below the poverty line is the difficulty of getting access to banking and financial services. As a result they are often dependent on moneylenders and other unofficial sources – and are often charged at exorbitant rates if they do borrow. This makes it hard to save and invest – and puts a major barrier in the way of breaking out of this spiral. Awareness of this problem led Muhammad Yunus, Head of the Rural Economics Program at the University of Chittagong, to launch a project to examine the possibility of designing a credit delivery system to provide banking services targeted at the rural poor. In 1976 the Grameen Bank Project (Grameen means 'rural' or 'village' in Bangla language) was established, aiming to

- extend banking facilities to the poor;
- eliminate the exploitation of the poor by money lenders;
- create opportunities for self-employment for unemployed people in rural Bangladesh;
- offer the disadvantaged an organizational format which they can understand and manage by themselves;
- reverse the age-old vicious circle of 'low income, low saving and low investment', into virtuous circle of 'low income, injection of credit, investment, more income, more savings, more investment, more income'.

The original project was set up in Jobra (a village adjacent to Chittagong University) and some neighbouring villages and ran during 1976–79. The core concept was of 'micro-finance' – enabling people (and a major success was with women) to take tiny loans to start and grow tiny businesses. With the sponsorship of the central bank of the country and support of the nationalized commercial banks, the project was extended to Tangail district (a district north of Dhaka, the capital city of Bangladesh) in 1979. Its further success there led to the model being extended to several other districts in the country and in 1983 it became an independent bank as a result of government legislation. Today Grameen Bank is owned by the rural poor whom it serves. Borrowers of the bank own 90% of its shares, while the remaining 10% is owned by the government. It now serves over 5 million clients, and has enabled 10 000 families to escape the poverty trap every month. Younis received the Nobel Peace Prize for this innovation in 2006.

services. A recent example was the development of innovations around the concept of 'micro-finance' – see Case study 5.3.

5.4 Whose needs?

When considering need pull as a source of innovation we should remember that one size doesn't fit all. Differences amongst potential users can also provide rich triggers for innovation in new directions. Disruptive innovation – a theme to which we will return later – is often associated with entrepreneurs working at the fringes of a mainstream market and finding groups whose needs are not being met. It poses a problem for existing incumbents because the needs of such fringe groups are not seen as relevant to their 'mainstream' activities – and so they tend to ignore them or to dismiss them as not being important. But working with these users and their different needs creates different innovation options – and sometimes what has relevance for the fringe begins to be of interest to the mainstream. Clayton Christensen, in his many studies of such 'disruptive innovation', shows this has been the pattern across industries as diverse as computer disk drives, earth-moving equipment, steel making and low-cost air travel.[16]

For much of the time there is stability around markets where innovation of the 'do better' variety takes place and is well managed. Close relationships with existing customers are fostered and the system is configured to deliver a steady stream of what the market wants – and often a great deal more! (What he terms 'technology overshoot' is often a characteristic of this, where markets are offered more and more features which they may not ever use or place much value on but which come as part of the package.)

But somewhere else there is another group of potential users who have very different needs – usually for something much simpler and cheaper – which will help them get something done. For example the emergent home computer industry began amongst a small group of hobbyists who wanted simple computing capabilities at a much lower price than was available from the mini-computer suppliers. In turn the builders of those early PCs wanted disk drives which were much simpler technologically but – importantly – much cheaper and so were not really interested in what the existing disk drive industry had to offer. It was too high tech, massively overengineered for their needs and, most important, much too expensive.

Although they approached the existing drive makers none of them was interested in making such a device – not surprisingly since they were doing very comfortably supplying expensive high-performance equipment to an established mini-computer industry. Why should they worry about a fringe group of hobbyists as a market? Steve Jobs described in an interview their attempts to engage interest, '. . . *So we went to Atari and said, "Hey, we've got this amazing thing, even built with some of your parts, and what do you think about funding us? Or we'll give it to you. We just want to do it. Pay our salary, we'll come work for you." And they said, "No." So then we went to Hewlett-Packard, and they said, "Hey, we don't need you. You haven't got through college yet."'*

Consequently the early PC makers had to look elsewhere – and found entrepreneurs willing to take the risks, and experiment with trying to come up with a product which met their needs. It didn't happen overnight and there were plenty of failures on the way – and certainly the early drives were very poor performers in comparison with what was on offer in the mainstream industry. But gradually the PC market grew, moving from hobbyists to widespread home use and from there – helped by the emergence and standardization of the IBM PC – to the office and business environment. And as it grew and matured so it learned and the performance of the machines became much more impressive and reliable – but coming from a much lower cost base than mini-computers. The same thing happened to the disk drives within them – the small entrepreneurial firms who began in the game grew and learned and became large suppliers of reliable products which did the job – but at a massively lower price.

Eventually the fringe market, which the original disk drive makers had ignored because it didn't seem relevant or important enough to worry about, grew to dominate – and by the time they realized this it was too late for many of them. The best they could hope for would be to be late entrant imitators, coming from behind and hoping to catch up.

This pattern is essentially one of disruption – the rules of the game changed dramatically in the marketplace with some new winners and losers. Figure 5.3 shows the transition where the new market and

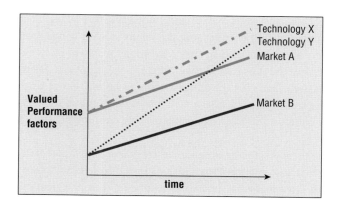

FIGURE 5.3: The pattern of disruptive innovation

suppliers gradually take over from the existing players. It can be seen in many industries – think about the low-cost airlines, for example. Here the original low-cost players didn't go head to head with the national flag carriers who offered the best routes, high levels of service and prime airport slots – all for a high price. Instead they sought new markets at the fringe – users who would accept a much lower level of service (no food, no seat allocation, no lounges, no frills at all), but for a basic safe flight would pay a much lower price. As these new users began to use the service and talk about it, so the industry grew and came to the attention of existing private and business travellers who were interested in lower cost flights at least for short-haul, because it met their needs for a 'good enough' solution to their travel problem. Eventually the challenge hit the major airlines who found it difficult to respond because of their inherently much higher cost structure – even those like BA and KLM, which set up low-cost subsidiaries, found they were unable to manage with the very different business model that low-cost flying involved.

Low-end market disruption of this kind is a potent threat – think what a producer in China might do to an industry like pump manufacturing if it began to offer a simple, low-cost 'good enough' household pump for $10 instead of the high-tech high-performance variants available from today's industry at prices 10 to 50 times as high. Or how manufacturers of medical devices like asthma inhalers will need to respond once they have come off patent – a challenge already being posed in markets such as generic pharmaceuticals.

But it is also important to recognize that similar challenges to existing market structures can happen through 'high-end' disruption – as Utterback points out.[17] Where a group of users requires something at a higher level than the current performance this can create new products or services which then migrate to mainstream expectations – for example, in the domestic broadband or mobile telephone markets.

Disruptive innovation examples of this kind focus attention on the requirement to look for needs which are not being met, or poorly met or sometimes where there is an overshoot.[18] Each of these can provide a trigger for innovation – and often involve disruption because existing players don't see the different patterns of needs. This thinking is behind, for example, the concept of 'Blue Ocean strategy'[19] which argues for firms to define and explore uncontested market space by spotting latent needs that are not well served. See Case study 5.4.

Over-served markets might include those for office software or computer operating systems where the continuing trend towards adding more and more features and functionality has possibly outstripped

CASE STUDY 5.4

Gaining competitive edge through meeting unserved needs

An example of the 'Blue Ocean' approach is the Nintendo Wii which has carved a major foothold in the lucrative computer games market – a business which is in fact bigger than Hollywood in terms of overall market value. The Wii console is not a particularly sophisticated piece of technology – compared to rivals Sony PS3 or Microsoft Xbox it has less computing power, storage or other features and the games graphics are much lower resolution than major sellers like *Grand Theft Auto*. But the key to the phenomenal success of the Wii has been its appeal to an under-served market. Where computer games were traditionally targeted at boys the Wii extends – by means of a simple interface wand – interest to all members of the family. Add-ons to the platform like the Wii board for keep fit and other applications mean that market reach extends further, for example to include the elderly or patients suffering the after-effects of stroke.

Nintendo has performed a similar act of opening up the marketplace with its DS handheld device – again by targeting unmet needs across a different segment of the population. Many DS users are middle-aged or retired and the best-selling games are for brain training and puzzles.

user needs for or ability to use them all. Linux and open office applications such as 'Star Office' represent simpler, 'good enough' solutions to the basic needs of users – and are potential disruptive innovations for players like Microsoft.

The role of 'emerging markets'

On a global scale there is growing interesting in what have been termed the 'bottom of the pyramid' (BoP) markets.[20] This term comes from a book by C.K. Prahalad who argued that 80% of the world's population lived on incomes below the poverty line – around $2 a day – and therefore did not represent markets in the traditional sense. But seeing them as a vast reservoir of under-served needs opens up a significant challenge and opportunity for innovation (see Table 5.2 for some examples of the challenge and opportunities).

Solutions to meeting these needs will have to be highly innovative but the prize is equally high – access to a high-volume low-margin marketplace. For example Unilever realized the potential of selling its shampoos and other cosmetic products not in 250ml bottles (which were beyond the price range of most BoP customers) but in single sachets. As G. Gilbert Cloyd, Chief Technology Officer, Procter & Gamble commented in a *Business Week* interview, '. . . *We've put more emphasis on serving an even broader base of consumers. We have the goal of serving the majority of the world's consumers someday. Today, we probably serve about 2 billion-plus consumers around the globe, but there are 6 billion consumers out there. That has led us to put increased emphasis on low-end markets and in mid- and low-level pricing tiers in developed geographies. That has caused us to put a lot more attention on the cost aspects of our products . . .*'

Prahalad's original book contains a wide range of case examples where this is beginning to happen in fields as diverse as healthcare, agriculture, consumer white goods and home improvements.[19] Subsequently there has been significant expansion of innovative activity in these emerging market areas – driven in part by a realization that the major growth in global markets will come from regions with a

| TABLE 5.2 | **Challenging assumptions about the bottom of the pyramid** |

Assumption	Reality – and innovation opportunity
The poor have no purchasing power and do not represent a viable market	Although low income, the sheer scale of this market makes it interesting. Additionally the poor often pay a premium for access to many goods and services, e.g. borrowing money, clean water, telecommunications and basic medicines, because they cannot address 'mainstream' channels like shops and banks. The innovation challenge is to offer low-cost, low-margin but high-quality goods and services across a potential market of 4 billion people
The poor are not brand-conscious	Evidence suggests a high degree of brand and value consciousness – so if an entrepreneur can come up with a high-quality low cost solution it will be subject to hard testing in this market. Learning to deal with this can help migrate to other markets – essentially the classic pattern of 'disruptive innovation'
The poor are hard to reach	By 2015 there are likely to be nearly 400 cities in the developing world with populations over 1 million and 23 with over 10 million. 30–40% of these will be poor – so the potential market access is considerable. Innovative thinking around distribution – via new networks or agents (such as the women village entrepreneurs used by Hindustan Lever in India or the 'Avon ladies' in rural Brazil) – can open up untapped markets
The poor are unable to use and not interested in advanced technology	Experience with PC kiosks, low-cost mobile phone sharing and access to the Internet suggests that rates of take-up and sophistication of use are extremely fast amongst this group. In India the e-choupal (e-meeting place) set up by software company ITC enabled farmers to check prices for their products at the local markets and auction houses. Very shortly after that the same farmers were using the web to access prices of their soybeans at the Chicago Board of Trade and strengthen their negotiating hand!

Source: Prahalad, C.K. (2006) *The Fortune at the Bottom of the Pyramid*, Wharton School Publishing, New Jersey.

high BoP profile. There are several video clips accompanying the book *The Fortune at the Bottom of the Pyramid,* available from Wharton School Publishing. In addition there is a YouTube clip of an interview with C.K. Prahalad at http://www.youtube.com/watch?v=ew2zQnUh_uw.

Importantly many companies are actively using 'bottom of pyramid' markets as places to search for weak signals of potentially interesting new developments. For example, Nokia has been sending scouts to study how people in rural Africa and India are using mobile phones and the potential for new services which this might offer, whilst the pharmaceutical firm Novo Nordisk has been learning about low-cost provision of diabetes care in Tanzania as an input to a better understanding of how such models might be developed for different regions.[21,22]. We'll return to this theme when we look at the idea of 'extreme users' as sources of innovation.

CASE STUDY 5.5

Learning from extreme conditions

The Aravind Eye Care System has become the largest eye care facility in the world with its headquarters in Madurai, India. Its doctors perform over 200 000 cataract operations – and with such experience have developed state-of-the art techniques to match their excellent facilities. Yet the cost of these operations runs from $50 to $300, with over 60% of patients being treated free. Despite only 40% paying customers the company is highly profitable and the average cost per operation (across free and paying patients) at $25 is the envy of most hospitals around the world.

Aravind was founded by Dr G. Venkataswamy in 1976 on his retirement from the Government Medical College and represents the result of a passionate concern to eradicate needless blindness in the population. Within India there are an estimated 9 million (and worldwide 45 million) people who suffer from blindness which could be cured via corrective glasses and simple cataract or other surgery. Building on his experience in organizing rural eye camps to deal with diagnosis and treatment he set about developing a low-cost high-quality solution to the problem, originally aiming its treatment in his home state of Tamil Nadu.

One of the key building blocks in developing the Aravind system has been transferring the ideas of another industry concerned with low-cost, high and consistent quality provision – the hamburger business pioneered by the Croc brothers and underpinning McDonald's. By applying the same process innovation approaches to standardization, workflow and tailoring tasks to skills he created a system which not only delivered high quality but was also reproducible. The model has now diffused widely – there are now five hospitals within Tamil Nadu offering nearly 4000 beds, the majority of which are free. It has moved beyond cataract surgery to education, lens manufacturing, research and development and other linked activities around the theme of improving sight and access to treatment.

In making this vision come alive Dr Venkataswamy has not only demonstrated considerable entrepreneurial flair – he has also created a template which others, including health providers in the advanced industrial economies, are now looking at very closely. It has provided both the trigger and some of the trajectory for innovative approaches in health care – not just in eye surgery but across a growing range of operations.

5.5 Towards mass customization

Arguably Henry Ford's plant, based on principles of mass production, represented the most efficient response to the market environment of its time. But that environment changed rapidly during the 1920s, so that what had begun as a winning formula for manufacturing began gradually to represent a major obstacle to change. Production of the Model T began in 1909 and for 15 years or so it was the market leader. Despite falling margins the company managed to exploit its blueprint for factory technology and organization to ensure continuing profits. But growing competition (particularly from General Motors with its strategy of product differentiation) was shifting away from trying to offer the customer low-cost personal transportation and towards other design features – such as the closed body – and Ford was increasingly forced to add features to the Model T. Eventually it was clear that a new model was needed and production of the Model T stopped in 1927. See detailed case study of Model T on web.

The trouble is that markets are not made up of people wanting the same thing – and there is an underlying challenge to meet their demands for variety and increasing customization. This represents a powerful driver for innovation – as we move from conditions where products are in short supply to one of mass production so the demand for differentiation increases. There has always been a market for personalized custom-made goods – and similarly custom-configured services, for example, personal shoppers, personal travel agents, personal physicians. But until recently there was an acceptance that this customization carried a high price tag and that mass markets could only be served with relatively standard product and service offerings.[23]

However a combination of enabling technologies and rising expectations has begun to shift this balance and resolve the trade-off between price and customization. 'Mass customization' is a widely used term which captures some elements of this.[24] Mass customization is the ability to offer highly configured bundles of non-price factors configured to suit different market segments (with the ideal target of total customization, i.e. a market size of one), but to do this without incurring cost penalties and the setting up of a trade-off of agility versus prices.

Of course there are different levels of customizing – from simply putting a label 'specially made for . . . (insert your name here)' on a standard product right through to sitting down with a designer and co-creating something truly unique. Table 5.3 gives some examples of this range of options.

This trend has important implications for services, in part because of the difficulty of sustaining an entry barrier for long. Service innovations are often much easier to imitate and the competitive advantages which they offer can quickly be competed away because there are fewer barriers to entry or options for protecting intellectual property. The pattern of airline innovation on the transatlantic route provides a good example of this – there is a fast pace of innovation but as soon as one airline introduces something like a flat bed, others will quickly emulate it. Arguably the drive to personalization of the service experience will be strong because it is only through such customized experiences that a degree of customer 'lock on' takes place.[25] Certainly the experience of Internet banking and insurance suggests that, despite attempts to customize the experience via sophisticated web technologies, there is little customer loyalty and a high rate of churn. However, the lower capital cost of creating and delivering services and their relative simplicity make co-creation more of an option. Where manufacturing may require sophisticated tools like computer-aided design and rapid prototyping, services lend themselves to shared experimentation at relatively lower cost. There is growing interest in such models involving active users in design of services, for example in the open source movement around software or in the digital entertainment and communication fields where community and social networking sites like

TABLE 5.3	Options in customization	
Type of customization	**Characteristics**	**Examples**
Distribution customization	Customers may customize product/service packaging, delivery schedule and delivery location but the actual product/service is standardized	Sending a book to a friend from Amazon.com. They will receive an individually wrapped gift with a personalized message from you – but it's actually all been done online and in their distribution warehouses. iTunes appears to offer personalization of a music experience but in fact it does so right at the end of the production and distribution chain
Assembly customization	Customers are offered a number of predefined options. Products/services are made to order using standardized components	Buying a computer from Dell or another online retailer. Customers choose and configure to suit their exact requirements from a rich menu of options – but Dell only start to assemble this (from standard modules and components) when their order is finalized. Banks offering tailor-made insurance and financial products are actually configuring these from a relatively standard set of options
Fabrication customization	Customers are offered a number of predefined designs. Products/services are manufactured to order	Buying a luxury car like a BMW, where the customers are involved in choosing ('designing') the configuration which best meets their needs and wishes, e.g. engine size, trim levels, colour, fixtures and extras. Only when they are satisfied with their virtual model does the manufacturing process begin – and customers can even visit the factory to watch their car being built
		Services allow a much higher level of such customization since there is less of an asset base needed to set up for 'manufacturing' the service – examples here would include made to measure tailoring, personal planning for holidays and pensions

(continued)

TABLE 5.3 (Continued)		
Type of customization	**Characteristics**	**Examples**
Design customization	Customer input stretches to the start of the production process. Products do not exist until initiated by a customer order	Co-creation, where end users may not even be sure what it is they want but where – sitting down with a designer – they co-create the concept and elaborate it. It's a little like having some clothes made but rather than choosing from a pattern book they actually have a designer with them and create the concept together. Only when it exists as a firm design idea does it then get made. Co-creation of services can be found in fields like entertainment (where user-led models like YouTube are posing significant challenges to mainstream providers) and in healthcare where experiments towards radical alternatives for healthcare delivery are being explored – see for example, the Design Council RED project

Source: After Lampel, J. and H. Mintzberg (1996) Customizing, customization. *Sloan Management Review*, **38** (1), 21–30.

MySpace, Flickr and YouTube have had a major impact. See video links of Stan Davies discussing the future of mass customization and an interview with Frank Piller talking about mass customization and configurators.

5.6 Users as innovators

Although need pull represents a powerful trigger for innovation it is easy to fall into the trap of thinking about the process as a serial one in which user needs are identified and then something is created to meet those needs. The assumption underpinning this is that users are passive recipients – but this is often not the case. Indeed history suggests that users are sometimes ahead of the game – their ideas plus their frustrations with existing solutions lead to experiment and prototyping and create early versions of what eventually become mainstream innovations. Eric von Hippel of Massachusetts Institute of Technology has made a lifelong study of this phenomenon and gives the example of the pickup truck – a long-time staple of the world automobile industry. This major category did not begin life on the drawing boards of Detroit but rather on the farms and homesteads of a wide range of users who wanted more than a family saloon. They adapted their cars by removing seats, welding new pieces on and cutting off the roof – in the process prototyping and developing the early model of the pickup. Only later did Detroit pick up on the idea and then begin the incremental innovation process to refine and mass produce the vehicle.[26] A host of other examples support the view that user-led innovation matters, for example petroleum refining, medical devices, semiconductor equipment, scientific instruments, the Polaroid camera and a wide range of sports goods.

Importantly active and interested users – 'lead users' – are often well ahead of the market in terms of innovation needs. In Mansfield's detailed studies of diffusion of a range of capital goods into major firms in the bituminous coal, iron and steel, brewing and railroad industries, he found that in 75% of the cases it took over 20 years for complete diffusion of these innovations to major firms.[27] As von Hippel points out some users of these innovations could be found far in advance of the general market.[28]

One of the fields where this has played a major role is in medical devices where active users amongst medical professionals have provided a rich source of innovations for decades. Central to their role in the innovation process is that they are very early on the adoption curve for new ideas – they are concerned with getting solutions to particular needs and prepared to experiment and tolerate failure in their search for a better solution. One strategy – which we will explore later – around managing innovation is thus to identify and engage with such 'lead users' to co-create innovative solutions. Tim Craft, a practising anaesthetist, developed a range of connectors and other equipment as a response to frustrations and concerns about the safety aspects of the equipment he was using in operating theatres. He describes the birth of the company, Anaesthetic Medical Systems, and the underlying philosophy in the podcast interview on the website.

CASE STUDY 5.6

User involvement in innovation – the Coloplast example

One of the key lessons about successful innovation is the need to get close to the customer. At the limit (and as Eric Von Hippel and other innovation scholars have noted), the user can become a key part of the innovation process, feeding in ideas and improvements to help define and shape the innovation. The Danish medical devices company, Coloplast, was founded in 1954 on these principles when nurse Elise Sorensen developed the first self-adhering ostomy bag as a way of helping her sister, a stomach cancer patient. She took her idea to a various plastics manufacturers, but none showed interest at first. Eventually Aage Louis-Hansen discussed the concept with his wife, also a nurse, who saw the potential of such a device and persuaded her husband to give the product a chance. Hansen's company, Dansk Plastic Emballage, produced the world's first disposable ostomy bag in 1955. Sales exceeded expectations and in 1957, after having taken out a patent for the bag in several countries, the Coloplast company was established. Today the company has subsidiaries in 20 countries and factories in five countries around the world, with specialist divisions dealing with incontinence care, wound care, skin care, mastectomy care, consumer products (specialist clothing etc.) as well as the original ostomy care division.

Keeping close to users in a field like this is crucial and Coloplast have developed novel ways of building in such insights by making use of panels of users, specialist nurses and other healthcare professionals located in different countries. This has the advantage of getting an informed perspective from those involved in post-operative care and treatment and who can articulate needs which might for the individual patient be difficult or embarrassing to express. By setting up panels in different countries the varying cultural attitudes and concerns could also be built into product design and development.

An example is the Coloplast Ostomy Forum (COF) board approach. The core objective within the COF Boards is to try and create a sense of partnership with key players, either as key customers or key influencers. Selection is based on an assessment of their technical experience and competence but also on the degree to which they will act as opinion leaders and gatekeepers, for example by influencing colleagues, authorities, hospitals and patients. They are also a key link in the clinical trials process. Over the years Coloplast has become quite skilled in identifying relevant people who would be good COF board members, for example by tracking people who author clinical articles or who have a wide range of experience across different operation types. Their specific role is particularly to help with two elements in innovation:

- to identify, discuss and prioritize user needs
- to evaluate product development projects from idea generation right through to international marketing.

Importantly COF Boards are seen as integrated with the company's product development system and they provide valuable market and technical information into the stage-gate decision process. This input is mainly associated with early stages around concept formulation (where the input is helpful in testing and refining perceptions about real user needs and fit with new concepts). There is also significant involvement around project development where Board members are concerned with evaluating and responding to prototypes, suggesting detailed design improvements, design for usability, etc.

Sometimes user-led innovation involves a community which creates and uses innovative solutions on a continuing basis. Good examples of this include the Linux community around computer operating systems or the Apache server community around web server development applications, where communities have grown up and the resulting range of applications is constantly growing – a state which has been called 'perpetual beta' referring to the old idea of testing new software modules across a community to get feedback and development ideas.[29] A growing range of Internet-based applications make use of communities – for example Mozilla and its Firefox and other products, Propellerhead and other music software communities and the emergent group around Apple's i-platform devices like the iPhone.[30]

Increasing interest is being shown in such 'crowd-sourcing' approaches to co-creating innovations – and to finding new ways of creating and working with such communities. The principle extends beyond software and virtual applications – for example, LEGO makes extensive use of communities of developers in its LEGO factory and other online activities linked to its manufactured products.[31] Adidas has taken the model and developed its 'mi Adidas' concept where users are encouraged to co-create their own shoes using a combination of website (where designs can be explored and uploaded) and in-store mini-factories where user-created and customized ideas can then be produced. See LEGO and Threadless case studies on the web.

5.7 Extreme users

An important variant that picks up on both the lead user and the fringe needs concepts lies in the idea of extreme environments as a source of innovation. The argument here is that the users in the toughest environments may have needs which by definition are at the edge – so any innovative solution which meets those needs has possible applications back into the mainstream. An example would be antilock braking systems (ABS) which are now a commonplace feature of cars but which began life as a special add-on for premium high-performance cars. The origins of this innovation came from a more extreme case, though – the need to stop aircraft safely under difficult conditions where traditional braking might lead to skidding or other loss of control. ABS was developed for this extreme environment and then migrated across to the (comparatively) easier world of automobiles.[29] A set of videos outlining the experience of 3M working with Eric von Hippel as it tries to make use of lead user methods in its innovation development work can be found via links on the website.

Looking for extreme environments or users can be a powerful source of stretch in terms of innovation – meeting challenges which can then provide new opportunity space. As Roy Rothwell put it in the title of a famous paper, 'tough customers mean good designs'.[32] For example, stealth technology arose out of a very specific and extreme need for creating an invisible aeroplane – essentially something which did not have a radar signature. It provided a powerful pull for some radical innovation which challenged fundamental assumptions about aircraft design, materials, power sources etc. and opened up a wide frontier for changes in aerospace and related fields.[33] The 'bottom of the pyramid' concept mentioned earlier also offers some powerful extreme environments in which very different patterns of innovation are emerging.

For example in the Philippines there is little in the way of a formal banking system for the majority of people – and this has led to users creating very different applications for their mobile phones where pay as you go credits become a unit of currency to be transferred between people and used as currency for various goods and services. In Kenya the mobile phone is used to increase security – if a traveller wishes to move between cities he or she will not take money but instead forward it via mobile phone in the form of credits, which can then be collected from the phone recipient at the other end. This is only one of hundreds of new applications being developed in extreme conditions and by under-served users – and represents a powerful laboratory for new concepts which companies like Nokia and Vodafone are working closely to explore.[20] The potential exists to use this kind of extreme environment as a laboratory to test and develop concepts for wider application – for example, Citicorp has been experimenting with a design of ATM based on biometrics for use with the illiterate population in rural India. The pilot involves some 50 000 people but as a spokesman for the company explained, *we see this as having the potential for global application*.

Such experiments can open up significant new innovation space by bringing new rules to the game. For example, India's giant Tata Corporation has been developing the '1-lakh car' – essentially a car for the Indian market which would retail for around $2000. Despite considerable cynicism from the industry the Nano has been launched at close to this price and represents the first response to a classic extreme environment challenge. Producing something at this target cost, which also meets emission controls and provides a level of features to satisfy the growing Indian middle class, is already a significant innovation achievement given that the closest competitor cars retail for nearly twice that price. Creating the wider system for service and support, for insurance, for financing purchase, for driver training and so on implies a very different approach to bringing driving within the reach of a large population. As low-cost airlines and other disruptive innovators found, the learning effects across large volumes of rapidly growing markets mean that many innovative solutions are developed and create a business model which has significant challenges for established incumbents. Arguably this is not simply a local innovation but an experiment towards the kind of industry-changing system that Henry Ford pioneered a century ago.

5.8 Watching others

Innovation is essentially a competitive search for new or different solutions – whether in the sense of commercial enterprises competing with each other for market share or in the wider sense of public service, where the competition is for doing more with limited resources, or between law and order and crime, or education and illiteracy. In such a contest one important strategy involves learning from others – imitation is not only the sincerest form of flattery but also a viable and successful strategy for sourcing innovation. For example, reverse engineering of products and processes and development of imitations – even around impregnable patents – is a well-known route to find ideas. Much of the rapid progress of Asian economies in the post-war years was based on a strategy of 'copy and develop', taking Western ideas and improving on them.[34] For example much of the early growth in Korean manufacturing industries in fields like machine tools came from adopting a strategy of 'copy and develop' – essentially learning (often as a result of taking licenses or becoming service agents) by working with established products and understanding how they might be adapted or developed for the local market. Subsequently this learning could be used to develop new generations of products or services.[35]

 A wide range of tools for competitor product and process profiling has been developed which provide structured ways of learning from what others do or offer.[36] See web for examples of competitiveness profiling.

One powerful variation on this theme is the concept of benchmarking.[37] In this process enterprises make structured comparisons with others to try and identify new ways of carrying out particular processes or to explore new product or service concepts. The learning triggered by benchmarking may arise from comparing between similar organizations (same firm, same sector, etc.), or it may come from looking outside the sector but at similar products or processes. For example, Southwest Airlines became the most successful carrier in the USA by dramatically reducing the turnaround times at airports – an innovation which it learned from studying pit-stop techniques in the Formula 1 Grand Prix events. Similarly the Karolinska Hospital in Stockholm made significant improvements to its cost and time performance through studying inventory management techniques in advanced factories.[38]

Benchmarking of this kind is increasingly being used to drive change across the public sector, both via 'league tables' linked to performance metrics which aim to encourage fast transfer of good practice between schools or hospitals and also via secondment, visits and other mechanisms designed to facilitate learning from other sectors managing similar process issues such as logistics and distribution. One of the most successful applications of benchmarking has been in the development of the concept of 'lean thinking', now widely applied to a many public- and private-sector organizations.[39] The origins were in a detailed benchmarking study of car manufacturing plants during the 1980s, which identified significant performance differences and triggered a search for the underlying process innovations that were driving the differences.[40]

5.9 Recombinant innovation

Another easy assumption to make about innovation is that it always has to involve something new to world. The reality is that there is plenty of scope for crossover – ideas and applications which are commonplace in one world may be perceived as new and exciting in another. This is an important principle in sourcing innovation where transferring or combining old ideas in new contexts – a process called 'recombinant innovation' by Andrew Hargadon – can be a powerful resource.[41] The Reebok pump running shoe, for example, was a significant product innovation in the highly competitive world of sports equipment – yet although this represented a breakthrough in that field it drew on core ideas which were widely used in a different world. Design Works – the agency which came up with the design – brought together a team that included people with prior experience in fields like paramedic equipment (from which they took the idea of an inflatable splint providing support and minimizing shock to bones) and operating theatre equipment (from which they took the micro-bladder valve at the heart of the pump mechanisms). Many businesses – as Hargadon points out – are able to offer rich innovation possibilities primarily because they have deliberately recruited teams with diverse industrial and professional backgrounds and thus bring very different perspectives to the problem in hand. His studies of the design company, IDEO, show the potential for such recombinant innovation work.[42]

Nor is this a new idea. Thomas Edison's famous 'Invention Factory' in New Jersey was founded in 1876 with the grand promise of 'a minor invention every ten days and a big thing every six months or so'. They were able to deliver on that promise not because of the lone genius of Edison himself but rather from taking on board the recombinant lesson – Edison hired scientists and engineers (he called them 'muckers') from all the emerging new industries of early twentieth-century USA. In doing so he brought experience in technologies and applications like mass production and precision machining (gun industry), telegraphy and telecommunications, food processing and canning and automobile manufacture. Some of the early innovations that built the reputation of the business – for example the teleprinter for the NYSE – were really simple cross-over applications of well-known innovations in other sectors.[41]

In many ways recombinant innovation involves a core principle understood by researchers on human creativity. Very often original – breakthrough – ideas come about through a process of what Arthur Koestler called 'bisociation' – the bringing together of apparently unrelated things which can somehow be connected and yield an interesting insight.[43] The key message here for managing innovation is to look to diversity to provide the raw material which might be combined in interesting ways – and realizing this makes the search for unlikely bedfellows a useful strategy.

5.10 Regulation

Photographs of the pottery towns around Stoke on Trent in the Midlands of the UK taken in the early part of the twentieth century would not be much use in tracing landmarks or spotting key geographical features. The images in fact would reveal very little at all – not because of a limitation in the photographic equipment or processing but because the subject matter itself – the urban landscape – was rendered largely invisible by the thick smog which regularly enveloped the area. Yet 60 years later the same images would show up crystal clear – not because the factories had closed (although there are fewer of them) but because of the continuing effects of the Clean Air Act and other legislation. They provide a clear reminder of another important source of innovation – the stimulus given by changes in the rules and regulations which define the various 'games' for business and society. The Clean Air Act didn't specify how but only what had to change – achieving the reduction in pollutants emitted to the atmosphere involved extensive innovation in materials, processes and even in product design made by the factories.

Regulation in this way provides a two-edged sword – it both restricts certain things (and closes off avenues along which innovation had been taking place) and opens up new ones along which change is mandated to happen.[44] And it works the other way – deregulation – the slackening off of controls – may open up new innovation space. The liberalization and then privatization of telecommunications in many countries led to rapid growth in competition and high rates of innovation, for example.

Given the pervasiveness of legal frameworks in our lives we shouldn't be surprised to see this source of innovation. From the moment we get up and turn the radio on (regulation of broadcasting shaping the range and availability of the programmes we listen to), to eating our breakfast (food and drink is highly regulated in terms of what can and can't be included in ingredients, how foods are tested before being allowed for sale, etc.), to climbing into our cars and buckling on our safety belt whilst switching on our hands-free phone devices (both the result of safety legislation), the role of regulation in shaping innovation can be seen.[45]

Regulation can also trigger counter innovation – solutions designed to get round existing rules or at least bend them to advantage. The rapid growth in speed cameras as a means of enforcing safety legislation on roads throughout Europe has led to the healthy growth of an industry providing products or services for detecting and avoiding cameras. And at the limit changes in the regulatory environment can create radical new space and opportunity. Although Enron ended its days as a corporation in disgrace due to financial impropriety it is worth asking how a small gas pipeline services company rose to become such a powerful beast in the first place. The answer was its rapid and entrepreneurial take up of the opportunities opened up by deregulation of markets for utilities like gas and electricity.[46]

5.11 Futures and forecasting

Another source of stimuli for innovation comes through imagining and exploring alternative trajectories to the dominant version in everyday use. Various tools and techniques for forecasting and imagining alternative futures are used to help strategy making – but can also be used to stimulate imagination around new possibilities in innovation. For example, Shell has a long history of exploring future options and driving innovations, most recently through its GameChanger programme.[47] Sometimes various 'transitional objects' are used, like concept models and prototypes in the context of product development, to explore reactions and provide a focus for various different kinds of input which might shape and co-create future products and services.[48,49]

Chapter 8 explores this theme and the related toolkits in detail.

5.12 Accidents

Accidents and unexpected events happen – and in the course of a carefully planned R&D project they could be seen as annoying disruptions. But on occasions accidents can also trigger innovation, opening up surprisingly new lines of attack. The famous example of Fleming's discovery of penicillin is but

CASE STUDY 5.7

Cleaning up by accident

Audley Williamson is not a household name of the Thomas Edison variety but he was a successful innovator whose UK business sold for £135 million in 2004. The core product which he invented was called 'Swarfega' and offered a widely used and dermatologically safe cleaner for skin. It is a greenish gel which has achieved widespread use in households as a simple and robust aid with the advertising slogan 'clean hands in a flash!' But the original product was not designed for this market at all – it was developed in 1941 as a mild detergent to wash silk stockings. Unfortunately the invention of nylon and its rapid application in stockings meant that the market quickly disappeared and he was forced to find an alternative. Watching workers in a factory trying to clean their hands with an abrasive mixture of petrol, paraffin and sand which left their hands cracked and sore led him to rethink the use of his gel as a safer alternative.

Source: The Independent, 28 February 2006, p.7.

one of many stories in which mistakes and accidents turned out to trigger important innovation directions. For example, the famous story of 3M's 'Post-it' notes began when a polymer chemist mixed an experimental batch of what should have been a good adhesive but which turned out to have rather weak properties – sticky but not very sticky. This failure in terms of the original project provided the impetus for what has become a billion dollar product platform for the company. Henry Chesbrough calls this process 'managing the false negatives' and draws attention to a number of cases.[50] For example, in the late 1980s, scientists working for Pfizer began testing what was then known as compound UK-92,480 for the treatment of angina. Although promising in the lab and in animal tests, the compound showed little benefit in clinical trials in humans. Despite these initial negative results the team pursued what was an interesting side effect which eventually led to UK-92,480 becoming the blockbuster drug Viagra.

The secret is not so much recognizing that such stimuli are available but rather in creating the conditions under which they can be noticed and acted upon. As Pasteur is reputed to have said, '*chance favours the prepared mind!*' Using mistakes as a source of ideas only happens if the conditions exist to help it emerge. For example Xerox developed many technologies in its laboratories in Palo Alto which did not easily fit their image of being 'the document company'. These included Ethernet (later successfully commercialized by 3Com and others) and PostScript language (taken forward by Adobe Systems). Chesbrough reports that 11 of 35 rejected projects from Xerox's labs were later commercialized with the resulting businesses having a market capitalization of twice that of Xerox itself.[50]

In similar fashion shocks to the system which fundamentally change the rules provide not only a threat to the existing status quo but a powerful stimulus to find and develop something new. The tragedy of the 9/11 bombing of the Twin Towers served to change fundamentally the public sense of security – but it has also provided a huge stimulus to innovate in areas like security, alternative transportation, fire safety and evacuation.[45]

RESEARCH NOTE

In a major research project around 'ideation' – where do innovation ideas come from? – Robert Cooper and Scott Edgett looked at 18 possible sources in the field of product innovation. Their sample covered 160 firms in the business-to-business and business-to-consumer markets, split approximately 70%/30% and covering a wide size range. They looked at how extensively each method was used but also asked managers to report on how effective they felt each technique to be. Their results are summarized below:

Approach	How extensively used (% of sample using)	Rank	How effective (scale of 1–10)	Rank
Ethnography	12.9	13	6.8	1
Customer visit teams	30.6	4	6.6	2
Customer focus groups for problem detection	25.5	5	6.4	3
Lead user methods	24	6	6.4	4
User design	17.4	11	6.0	5
Customer brainstorming	17.4	11	5.9	6
Peripheral vision tools	33.1	2	5.9	7
Customer advisory board	17.6	10	5.8	8
Community of enthusiasts	8	15	5.7	9
Disruptive technologies	22	8	5.7	10
Internal idea capture	38	1	5.5	11
Partners and vendors	22.1	7	5.5	12
Patent mining	33	3	5.5	13
Accessing external technical community	19.5	9	4.9	14
Scanning small businesses and start-ups	13	13	4.9	15
External product design/ crowdsourcing	2	18	4.8	16
External submitted ideas	7.9	16	4.5	17
External idea contest	4.1	17	4.3	18

Source: Cooper, R. and S. Edgett (2008) Ideation for product innovation: What are the best methods? *PDMA Visions*, Product Development Management Association, March, 12–16.

5.13 A framework for looking at innovation sources

It's clear that opportunities for innovation are not in short supply – and they arise from many different directions. The key challenge for innovation management is how to make sense of the potential input – and to do so with often limited resources. No organization can hope to cover all the bases so there needs to be some underlying strategy to how the search process is undertaken. One way is to impose some dimensions on the search space to help us frame where and why we might search for innovation triggers.

One important question is the relative importance of the push or pull forces outlined above. This has been the subject of many innovation studies over the years, using a variety of different methods to try and establish which is more important (and therefore where organizations might best place their resources). The reality is that innovation is never a simple matter of push or pull but rather their interaction; as Chris Freeman said 'necessity may be the mother of invention but procreation needs a partner'! Innovations tend to resolve into vectors – combinations of the two core principles. And these direct our attention in two complementary directions – creating possibilities (or at least keeping track of what others are doing along the R&D frontier) and identifying and working with needs. Importantly the role of needs in innovation is often to translate or select from the range of knowledge-push possibilities the variant which becomes the dominant strain. Out of all the possible bicycle ideas we eventually get to the dominant design – which is with us today.[51] The iPod wasn't the first MP3 player but it somehow clicked as the one which resonated best with user needs.

In fact most of the sources of innovation we mentioned above involve both push and pull components – for example, 'applied R&D' involves directing the push search in areas of particular need. Regulation both pushes in key directions and pulls innovations through in response to changed conditions. User-led innovation may be triggered by user needs but it often involves creating new solutions to old problems – essentially pushing the frontier of possibility in new directions.

There is a risk in focusing on either of the 'pure' forms of push or pull sources. If we put all our eggs in one basket we risk being excellent at invention but without turning our ideas into successful innovations – a fate shared by too many would-be entrepreneurs. But equally too close an ear to the market may limit us in our search – as Henry Ford is reputed to have said, 'if I had asked the market they would have said they wanted faster horses!' The limits of even the best market research lie in the fact that they represent sophisticated ways of asking people's reactions to something which is already there – rather than allowing for something completely outside their experience so far.

Another key dimension is around incremental or radical innovation. We've seen that there is a pattern of what could be termed 'punctuated equilibrium' with innovation – most of the time innovation is about exploiting and elaborating, creating variations on a theme within an established technical, market or regulatory trajectory. But occasionally there is a breakthrough which creates a new trajectory – and the cycle repeats itself. This suggests that much of our attention in searching for innovation triggers will be around incremental improvement innovation – the different versions of a piece of software, the mark 2, 3, 4 of a product or the continuing improvement of a business process to make it closer to lean. But we will need to have some element of our portfolio focused on the longer range, higher risk, which might lead to the breakthrough and set up a new trajectory.

A third issue is around timing – at different stages in the product or industry life cycle the emphasis may be more or less on push or pull. For example, mature industries will tend to focus on pull, responding to different market needs and differentiating by incremental innovation in key directions of user need. By contrast a new industry – for example the emergent industries based on genetics or nano materials

---

RESEARCH NOTE — **Where do innovations come from? Transformations in the US National Innovation System, 1970–2006**

Using an innovative research method, UC Davis scholars Fred Block and Mathew Keller analysed a sample of innovations recognized by *R&D Magazine* as being among the top 100 innovations of the year over the last four decades. They found that while in the 1970s almost all winners came from corporations acting on their own, more recently over two-thirds of the winners have come from partnerships involving business and government, including federal labs and federally funded university research. Moreover, in 2006 77 of the 88 US entities that produced award-winning innovations were beneficiaries of federal funding.

Source: http://www.itif.org.

---

technology – is often about solutions looking for a problem. So we would expect a different balance of resources committed to push or pull within these different stages.

This kind of thinking is reflected in the Abernathy/Utterback model of innovation life cycle which we covered in Chapter 1.[52] This sees innovation at the early fluid stage being characterized by extensive experimentation and with emphasis on product – creating a radical new offering. As the dominant design emerges attention shifts towards more incremental variation around the core trajectory – and as the industry matures so emphasis shifts to process innovation aimed at improving parameters like cost and quality. Once again this helps allocate scarce search resources in particular ways.

A fourth and related issue is around diffusion – the adoption and elaboration of innovation over time. Innovation adoption is not a binary process but rather one which takes place gradually over time, following some version of an S-curve.[53] At the early stages innovative users with high tolerance for failure will explore, to be followed by early adopters. This gives way to the majority following their lead until finally the remnant of a potential adopting population – the laggards in Roger's terms – adopt or remain stubbornly resistant. Understanding diffusion processes and the influential factors (which we will explore in more detail in Chapter 8) is important because it helps us understand where and when different kinds of triggers are picked up. Lead users and early adopters are likely to be important sources of ideas and variations, which can help shape an innovation in its early life, whereas the early and late majority will be more a source of incremental improvement ideas.[54]

5.14 How to search

Of course the challenge in managing innovation is not one of classifying different sources but rather how to seek out and find the relevant triggers early and well enough to do something about them. In developing search strategies we can make use of some of the broad dimensions highlighted above – for example by ensuring we have a balance between push and pull, and between incremental and radical. A good place to start understanding broad strategies is to look at what firms actually do in searching for innovation triggers. There are many large-scale innovation surveys which ask around this theme, for example, the European Community Innovation Survey (`www.cordis.europa.eu/cip/index.html`) which looks at the innovative behaviour of firms across 27 EU states (Table 5.4).

TABLE 5.4	Innovation activity and cooperation during 2002–2004

	Enterprises with innovation activity, % of all enterprises	All types of co-operation with other enterprises or institutions	Co-operation partners:			
			Suppliers	Clients or customers	Universities or other higher education institutes	Government or public research institutes
			% of all innovative enterprises			
EU27	42	42	17	14	9	6
Belgium	51	36	26	21	13	9
Bulgaria	16	22	16	13	6	4
Czech Republic*	38	38	31	26	13	7
Denmark	52	43	28	28	14	7
Germany	65	16	7	8	8	4
Estonia	49	35	23	23	9	6
Ireland	52	32	23	25	10	6
Greece	36	24	11	8	6	2
Spain	35	18	9	4	5	5
France	33	40	26	20	10	7
Italy	36	13	7	5	5	1
Cyprus	46	37	24	4	2	2
Latvia	18	39	33	29	14	12
Lithuania	29	56	45	35	12	10
Luxembourg	52	30	24	22	10	8
Hungary	21	37	26	20	14	5
Malta	21	32	22	17	4	4
Netherlands	34	39	30	22	12	9
Austria	53	17	7	8	10	5
Poland	25	42	28	16	6	9
Portugal	41	19	14	12	8	5
Romania	20	17	14	10	4	4
Slovenia	27	47	38	33	19	13
Slovakia	23	38	32	30	15	11
Finland	43	44	41	41	33	26
Sweden	50	43	32	28	17	6
United Kingdom	43	31	23	22	10	8
Iceland	52	29	20	20	5	13
Norway	37	33	23	22	15	16

*Data for Czech Republic correspond to the reference period 2003–2005

Source: fourth Community Innovation Survey

Source: Fourth European Community Innovation Survey. (c) European Communities 2007. Reproduced with permission.

Data from studies like the Community Innovation Survey gives us one picture – and it reinforces the view that successful innovation is about spreading the net as widely as possible, mobilizing multiple channels. Although surveys of this kind tell us a lot they also miss important elements in the sources of innovation picture. A lot of incremental innovation and how it is triggered lies beneath the radar screen, and there is a bias towards product innovation where we know that a great deal of incremental process improvement goes on. And surveys don't capture position or business model innovation so well, again

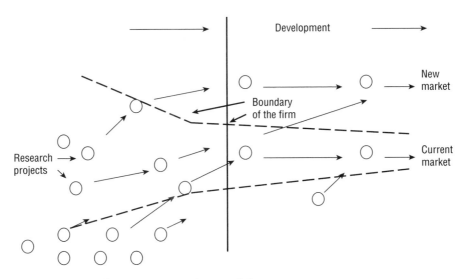

FIGURE 5.4: The open innovation model

Source: Chesbrough, H. (2003) *Open Innovation: The New Imperative for Creating and Profiting from Technology,* Harvard Business School Press, Boston, MA.

especially at the incremental end. It tends to focus on the 'obvious' search agents like R&D or market research departments – but others are involved, e.g. purchasing, and within the business the idea of suggestion schemes and high-involvement innovation. But surveys give us a broad picture – and underline the need for an extensive net.

Building rich and extensive linkages with potential sources of innovation has always been important – for example studies by Carter and Williams in the UK in the 1950s identified one key differentiator between successful and less successful innovating firms as the degree to which they were 'cosmopolitan' as opposed to 'parochial' in their approach towards sources of innovation.[55] There are, of course, arguments for keeping a relatively closed approach, for example there is a value in doing your own R&D and market research because the information collected is then available to be exploited in ways that the business can control. It can choose to push certain lines, hold back on others, keeping things essentially within a closed system. But as we've seen the reality is that innovation is triggered in all sorts of ways and a sensible strategy is to cast the net as widely as possible. In what is termed 'open innovation' organizations move to a more permeable view of knowledge in which they recognize the importance of external sources and also make their own knowledge more widely available.[56] Figure 5.4 illustrates this principle.

This is not without its difficulties – on the one hand it makes sense to recognize that in a knowledge-rich world 'not all the smart guys work for us'. Even large R&D spenders like Procter & Gamble (annual R&D budget around $3 billion and about 7000 scientists and engineers working globally in R&D) are fundamentally rethinking their models – in its case switching from 'Research and Develop' to 'Connect and Develop' as the dominant slogan, with the strategic aim of moving from closed innovation to sourcing 50% of its innovations from outside the business.[57] But on the other we should recognize the tensions that arise around intellectual property (how do we protect and hold on to knowledge when it is now much more mobile – and how do we access other people's knowledge?), around appropriability (how do we ensure a return on our investment in creating knowledge?) and around the mechanisms to make sure we can find and use relevant knowledge (when we are now effectively sourcing it from across the globe and in all sorts of unlikely locations). In this context innovation management emphasis shifts from knowledge creation to knowledge trading and managing knowledge flows.[58]

We will return to this theme of 'open innovation' and how to enable it, shortly.

5.15 Balancing exploitation and exploration

A core theme in discussion of innovation relates to the tensions in search behaviour between 'exploitation' and 'exploration' activities.[59,60] On the one hand firms need to deploy knowledge resources and other assets to secure returns and a 'safe' way of doing so is to harvest a steady flow of benefits derived from 'doing what we do better'. This has been termed 'exploitation' by innovation researchers, and it essentially involves *'the use and development of things already known'*.[61] It builds strongly through *'knowledge leveraging activities'*[62] on what is already well established – but in the process leads to a high degree of path dependency – *'firms' accumulated exploitation experience reinforces established routines within domains'*.[63]

The trouble is that in an uncertain environment the potential to secure and defend a competitive position depends on 'doing something different', i.e. radical product or process innovation rather than imitations and variants of what others are also offering.[64] This kind of search had been term 'exploration' and is the kind which involves *'long jumps'* or *'re-orientations that enable a firm to adopt new attributes and attain new knowledge outside its domain'*[65,66]

The above-mentioned tension comes because the organizational routines needed to support these activities differ. Incremental exploitation innovation is about highly structured processes and often high-frequency small-scale innovation carried out within operating units. Radical innovation, by contrast, is occasional and high risk, often requiring a specific and cross-functional combination of resources and a looser approach to organization and management.[67]

There is no easy prescription for doing these two activities but most organizations manage a degree of 'ambidexterity' through the use of a combination of approaches across a portfolio.[68,69] So, for example, technological search activity is managed by investment in a range of R&D projects with a few 'blue-sky'/high-risk outside bets and a concentration of projects around core technological trajectories.[70] Market research is similarly structured to develop deep and responsive understanding of key market segments but also allowing some search around peripheral and emergent constituencies.[71]

5.16 Absorptive capacity

One more broad strategic point concerns the question of where, when and how organizations make use of external knowledge to grow. It's easy to make the assumption that because there is a rich environment full of potential sources of innovation that every organization will find and make use of these. The reality is, of course, that they differ widely in their ability to make use of such trigger signals – and the measure of this ability to find and use new knowledge has been termed 'absorptive capacity'.

The concept was first introduced by Cohen and Levinthal who described it as *'the ability of a firm to recognize the value of new, external information, assimilate it, and apply it to commercial ends'* and saw it as *'largely a function of the firm's level of prior related knowledge'*.[72] It is an important construct because it shifts our attention to how well firms are equipped to search out, select and implement knowledge.

The underlying construct of absorptive capacity is not new – discussion of firm learning forms the basis of a number of studies going back to the work of Arrow,[73] Simon and March[74] and others. In the area of innovation studies the ideas behind 'technological learning' – the processes whereby firms acquire and use new technological knowledge and the underlying organizational and managerial processes which are involved – were extensively discussed by, inter alia, Freeman,[1] Bell and Pavitt[75] and Lall.[76] Cohen and Levinthal's original work was based on exploring (via mathematical modelling) the premise that firms might incur substantial long-run costs for learning a new 'stock' of information and that R&D needed to be viewed as an investment in today's and tomorrow's technology.[72] In later work they broadened and refined the model and definition of absorptive capacity to include more than just the R&D function and also explored the role of technological opportunity and appropriability in determining the firm's incentive to build absorptive capacity.

Absorptive capacity is clearly not evenly distributed across a population. For various reasons firms may find difficulties in growing through acquiring and using new knowledge. Some may simply be unaware of the need to change never mind having the capability to manage such change. Such firms – a classic problem of SME growth for example – differ from those which recognize in some strategic way the need to change, to acquire and use new knowledge but lack the capability to target their search or to assimilate and make effective use of new knowledge once identified. Others may be clear what they need but lack capability in finding and acquiring it. And others may have well-developed routines for dealing with all of these issues and represent resources on which less experienced firms might draw – as is the case with some major supply chains focused around a core central player.[77]

Reviewing the literature on why and when firms take in external knowledge suggests that this is not – as is sometimes assumed – a function of firm size or age. It appears instead that the process is more one of transitions via crisis-turning points. Some firms do not make the transition, others learn up to a

---

RESEARCH NOTE   **Absorptive capacity**

Research by Zahra and George (2002) noted that carrying out studies of absorptive capacity (AC) has become fraught with difficulty owing to the diversity and ambiguity surrounding its definition and components. Zahra and George decided to review and extend the absorptive capacity construct and suggested that several different processes were involved – rather than a simple absorption of new knowledge there were discrete activities linked to search, acquisition, assimilation and exploitation. Potential AC relates to Cohen and Levinthal's (1990) research on how a firm may value and acquire knowledge, although not necessarily exploit it. The firm's ability to transform and exploit the knowledge is captured by Realized AC. In short, absorptive capacity is a set of organizational routines and processes which are used to create a dynamic organizational capability. The authors state that firms need to build both types of absorptive capacity in order to maintain a competitive advantage.

Zahra and George discuss how Potential and Realized AC are separate but complementary, and why the distinction is useful. By distinguishing between Potential and Realized absorptive capacity we are able to ascertain which firms are unable to leverage and exploit external information. This can provide useful implications for managerial competences in developing both aspects of AC. They use the Potential and Realized absorptive capacity constructs to build a model of the antecedents, moderators and outcomes of the construct. For instance, they propose that a firm's experience and exposure to external knowledge will influence the development of Potential AC. Activation triggers, such as a change in dominant design may also play a moderating influence in determining the locus of search for external sources of knowledge. Finally they introduce the role of the social integration mechanism in reducing the gap between Potential and Realized AC. These mechanisms can help distribute information throughout the firm and provide an environment whereby information can be exploited.

Their work spawned extensive discussion and application – but the resulting proliferation of use of the term led to problems highlighted by Lane *et al.* (2006), who tried to evaluate how much divergence there has been in the field. These authors analysed 289 absorptive capacity papers from 14 journals to understand how the construct had been used and to identify the contributions to the broader literature of absorptive capacity. From their analysis, the authors concluded that the construct had become reified. '*Reification is the outcome of the process by which we forget the authorship of ideas and theories, objectify them (turn them into things), and then forget that we have done so*' (p. 835). They identified only six papers which extended the understanding of absorptive capacity in any meaningful way.

Todorova and Durisin (2007) also focus on the dynamic characteristics of the absorptive capacity construct, by examining the relationship between identification and acquisition of relevant knowledge, and the ability to apply that knowledge to commercial ends. In particular they claim that 'transformation' should be regarded not as a consequence but as an alternative process to 'assimilation' suggesting a more complex relationship between the components of absorptive capacity. In addition, they highlight the role of power relationships and socialization mechanisms within the dynamic model of absorptive capacity.

Sources: Zahra, S.A. and G. George (2002) Absorptive capacity: a review, reconceptualization and extension. *Academy of Management Review*, **27**, 185–94; Cohen, W. and D. Levinthal (1990) Absorptive capacity: a new perspective on learning and innovation. *Administrative Science Quarterly*, **35** (1), 128–52; Lane, P., B. Koka and S. Pathar (2006) The reification of absorptive capacity: a critical review and rejuvenation of the construct. *Academy of Management Review*, **31** (4), 833–63; Todorova, G. and B. Durisin (2007) Absorptive capacity: valuing a reconceptualization. *Academy of Management Review*, **32** (3), 774–96.

limited level. Equally the ability to move forwards depends on the past – a point made forcibly by Cohen and Levinthal in their original studies.

The key message from research on AC is that this complex construct – acquiring and using new knowledge – involves multiple and different activities around search, acquisition, assimilation and implementation. Connectivity between these is important – the ability to search and acquire (Potential AC in Zahra and George's model) may not lead to innovation. To complete the process further capabilities around assimilation and exploitation (Realized AC) are also needed. Importantly AC is associated with various kinds of search and subsequent activities, not just large firm formal R&D; mechanisms whereby SMEs explore and develop their process innovation, for example are also relevant.

AC is essentially about accumulated learning and embedding of capabilities – search, acquire, assimilate, etc. – in the form of routines (structures, processes, policies and procedures) which allow organizations to repeat the trick. Firms differ in their levels of AC and this places emphasis on how they develop, establish and reinforce these routines. In other words their ability to learn. Developing AC involves two complementary kinds of learning. Type 1 – adaptive learning – is about reinforcing and establishing relevant routines for dealing with a particular level of environmental complexity; and type 2 – generative learning – for taking on new levels of complexity.[78,79]

5.17 Tools and mechanisms to enable search

Within this broad framework firms deploy a range of approaches to organizing and managing the search process. For example, much experience has been gained in how R&D units can be structured to enable a balance between applied research (supporting the 'exploit' type of search) and more wide-ranging 'blue-sky' activities (which facilitate the 'explore' side of the equation.[70] These approaches have been refined further along 'open innovation' lines where the R&D work of others is brought into play, and by ways of dealing with the increasingly global production of knowledge – for example the pharmaceutical giant GSK deliberately pursues a policy of R&D competition across several major facilities distributed around the world. In similar fashion market research has evolved to produce a rich portfolio of tools for building a deep understanding of user needs – and which continues to develop new and further refined techniques – for example, empathic design, lead-user methods and increasing use of ethnography.

Choice of techniques and structures depends on a variety of strategic factors like those explored above – balancing their costs and risks against the quality and quantity of knowledge they bring in. Throughout the book we have stressed the idea that managing innovation is a dynamic capability – something which needs to be updated and extended on a continuing basis to deal with the 'moving frontier' problem. As markets, technologies, competitors, regulations and all sorts of other elements in a complex environment shift so we need to learn new tricks and sometimes let go of older ones which are no longer appropriate. In the following section we'll look at some examples of tools and mechanisms for innovation search, which are emerging in response to a context that sees very high levels of knowledge production, global distribution of such production and of the marketplaces providing the demand signals, increasing virtualization of those markets, growing involvement of users in shaping and 'co-creating' innovation, etc.

Managing internal knowledge connections

One area which has seen growing activity addresses a fundamental knowledge management issue which is well expressed in the statement – '*if only xxx (insert the name of any large organization) knew what it knows!*' In other words how can organizations tap into the rich knowledge (and potential innovation triggers) within its existing structures and amongst its workforce?

This has led to renewed efforts to deal with what is an old problem, for example, Procter & Gamble's successes with 'connect and develop' owe much to their mobilizing rich linkages between people who know things within their giant global operations and increasingly outside it. They use 'communities of practice'[80] – Internet-enabled 'clubs' where people with different knowledge sets can converge around core themes – and they deploy a small army of innovation 'scouts' who are licensed to act as prospectors, brokers and gatekeepers for knowledge to flow across the organization's boundaries (we discuss this in more detail in Chapter 6). Intranet technology links around 10 000 people in an internal 'ideas market' – and some of their significant successes have come from making better internal connections.

3M – another firm with a strong innovation pedigree dating back over a century – similarly put much of its success down to making and managing connections. Larry Wendling, Vice President for Corporate Research, talks of 3M's 'secret weapon' – the rich formal and informal networking which links the thousands of R&D and market-facing people across the organization. Their long-history of breakthrough innovations – from masking tape, through Scotchgard, Scotch tape, magnetic recording tape to Post-its and their myriad derivatives – arise primarily out of people making connections.

It's important to recognize that much of the knowledge lies in the experience and ideas of 'ordinary' employees rather than solely with specialists in formal innovation departments like R&D or market research. Increasingly organizations are trying to tap into such knowledge as a source of innovation via various forms of what can be termed 'high-involvement innovation' systems such as suggestion schemes, problem-solving groups and innovation 'jams'.

VIEWS FROM THE FRONT LINE

Sources of innovation

We look in the usual places for our industry. We look at our customers. We look at our suppliers. We go to trade bodies. We go to trade fairs. We present technical papers. We have an input coming from our customers. What we also try to do is develop inputs from other areas. We've done that in a number of ways. Where we're recruiting, we try to bring in people who can bring a different perspective. We don't necessarily want people who've worked in the type of instruments we have in the same industry . . . certainly in the past we've brought in people who bring a completely different perspective, almost like introducing greensand into the oyster. We deliberately look outside. We will look in other areas. We will look in areas that are perhaps different technology. We will look in areas that are adjacent to what we do, where we haven't normally looked. And we also do encourage the employees themselves to come forward with ideas.

Some of our product ideas have come from an individual who was sitting as a peripheral part of a little project team that was looking at different project ideas, different products for the future of the business. He had an idea. He created something in his garage. He brought it into me and says, what about this? And we looked at it. We had a quick discussion about it, talked to the management team and initiated a development that we did for one of our suppliers. That came right from outside the area we normally operate in. It came through one of our employees, a long-service employee, so not someone who was recent to the business. But it was triggered by him thinking in a different way. An idea came that he has married up to a potential market need because of the job he worked in when he was working in the service and repair area. He said, right, there's an opportunity for this product. He created a prototype out of a piece of drainpipe and some pieces he had taken from the repair area and made a functional model. And from that, we actually created a product that has spawned a product range of small manual instruments, which traditionally the business hasn't been involved with for probably 20 years. So, that's an idea that came from within the business. It came from an existing employee, but it's not something that we would have thought of as part of our normal pipeline.

We didn't immediately see, oh, there's a demand for this, let's do that. This came from him having some local knowledge and talking to customers at lower levels and saying, there's actually a demand for this small product. It's small, it's relatively niche, it's not going to set the world alight, but it enhances our product range and it puts us into an area where we've never been before. So, we're very receptive to those ideas coming forward. We create an environment where we encourage people to question and challenge. We've actually got an appraisal system where we look at people's competencies rather than performance, and one of the competencies we want is, is that person going to question and challenge? Are they willing to say, how can we do this better, how can we do this more effectively? So, continuous improvement is something we look for. But we also want people to hold up hands and say, hang on a minute, why are you doing it that way? What about this? I've seen this because of something I've done, one of my hobbies or in some of the social activities, and we encourage people to bring those ideas in and work with us to develop that into a product idea. We've actually set up a mechanism where we run a project team where we take people from all areas of the business . . . this is no longer just a product development area. We then put them in a room with all the resources they need for three or four days and say, what we want out of this is a number of product ideas that are different to what we do. Where can we go in the future? Where can you take this little business? Working within the limits of what we're capable of they will come up with product ideas, and the last one that we ran, we had seven or eight product ideas came out.

A full video version and transcript of the interview with Patrick is on the website.

Source: Patrick McLaughlin, Managing Director, Cerulean

One rich source of internal innovation lies in the entrepreneurial ideas of employees – projects which are not formally sanctioned by the business but which build on the energy, enthusiasm and inspiration of people passionate enough to want to try out new ideas. Encouraging internal entrepreneurship – 'intrapreneurship' as it has been termed[81] – is increasingly popular and organizations like 3M and Google make attempts to manage it in a semi-formal fashion, allocating a certain amount of time/space to employees to explore their own ideas.[82] Managing this is a delicate balancing act – on the one hand there is a need to give both permission and resources to enable employee-led ideas to flourish, but on the other there is the risk of these resources being dissipated with nothing to show for them. In many cases there is an attempt to create a culture of what can be termed 'bootlegging' in which there is tacit support for projects which go against the grain.[83] An example in BMW – where these are called 'U-boat projects' –was the Series 3 Estate version which the mainstream company thought was not wanted and would conflict with the image of BMW as a high-quality, high-performance and somewhat 'sporty' car. A small group of staff worked on a U-boat project, even using parts cannibalized from an old VW Rabbit to make a prototype – and the model has gone on to be a great success and opened up new market space.[84]

Extending external connections

The principle of spreading the net widely is well established in innovation studies as a success factor – and places emphasis on building strong relationships with key stakeholders. In a recent IBM survey of 750 CEOs around the world 76% ranked business partner and customer collaboration as top sources for new ideas whilst internal R&D ranked only eighth. The study also indicated that 'outperformers' – in terms of revenue growth – used external sources 30% more than underperformers. It's not hard to see why – the managers interviewed listed the clear benefits from collaboration with partners as things like reduced costs, higher quality and customer satisfaction, access to skills and products, increased revenue, and access to new markets and customers. As one CEO put it, '*We have at our disposal today a lot more capability and innovation in the marketplace of competitive dynamic suppliers than if we were to try to create on our own*' while another stated simply, '*If you think you have all of the answers internally, you are wrong.*'

This emphasizes the need both for better use of existing mainstream innovation agents – for example sales or purchasing as channels to monitor and bring back potential sources of innovation – and for establishing new roles and structures. In the former case there is already strong evidence of the importance of customers and suppliers as sources of innovation and the key role which relevant staff have in managing these knowledge sources. In the field of process innovation, for example, where the 'lean' agenda of improving on cost, quality and delivery is a key theme, there is strong evidence that diffusion can be accelerated through supply-chain learning initiatives like the UK Industry Forum in the auto components, aerospace, textiles and other sectors.[85,86]

But the 'open innovation' challenge also points us to where further experimentation is needed to make new connections. Table 5.5 identifies these and the following section explores some approaches which represent this 'frontier' in terms of search behaviour.[84]

Table 5.5	**Extending search strategies for innovation**
Search strategy	**Mode of operation**
Sending out scouts	Dispatch idea hunters to track down new innovation triggers
Exploring multiple futures	Use futures techniques to explore alternative possible futures; and then develop innovation options
Using the web	Harness the power of the web, through online communities, and virtual worlds, for example, to detect new trends
Working with active users	Team up with product and service users to see the ways in which they change and develop existing offerings
Deep diving	Study what people actually do, rather than what they say they do
Probe and learn	Use prototyping as a mechanism to explore emergent phenomena and act as boundary object to bring key stakeholders into the innovation process
Mobilize the mainstream	Bring mainstream actors into the product and service development process
Corporate venturing	Create and deploy venture units
Corporate entrepreneurship and intrapreneuring	Stimulate and nurture the entrepreneurial talent inside the organization
Use brokers and bridges	Cast the ideas net far and wide and connect with other industries
Deliberate diversity	Create diverse teams and a diverse workforce
Idea generators	Use creativity tools

Sending out scouts

This is a widely used strategy which involves sending out people (full or part time) to search actively for new ideas to trigger the innovation process. (In German they are called *Ideen-Jäger* – idea-hunters – a term which captures the concept well.) They could be searching for technological triggers, emerging markets or trends, competitor behaviour, etc., but what they have in common is a remit to seek things out, often in unexpected places. Search is not restricted to the organization's particular industry; on the contrary, the fringes of an industry or even currently entirely unrelated fields can be of interest.

For example, the mobile phone company O2 has a trend-scouting group of about 10 people who interpret externally identified trends into their specific business context whilst BT has a scouting unit in Silicon Valley which assesses some 3000 technology opportunities a year in California. The four-person operation was established in 1999 to make venture investments in promising telecom start-ups, but after the dotcom bubble burst it shifted its mission towards identifying partners and technologies that BT was interested in. The small team looks at more than 1000 companies per year and then, based on their deep knowledge of the issues facing the R&D operations back in England, they target the small number of cases where there is a direct match between BT's needs and the Silicon Valley company's technology. While the number of successful partnerships that result from this activity is small – typically four or fve per year – the unit serves an invaluable role in keeping BT abreast of the latest developments in its technology domain.[84]

Exploring multiple futures

Futures studies of various kinds can provide a powerful source of ideas about possible innovation triggers, especially those which do not necessarily follow the current trajectory. Shell's 'GameChanger' programme is a typical example which makes extensive use of alternative futures as a way of identifying domains of interest for future business which may lie outside the 'mainstream' of their current activities. Increasingly these rich 'science fiction' views of how the world might develop (and the threats and opportunities which it might pose in terms of discontinuous innovations) are being constructed by using a wide and deliberately diverse set of inputs rather than using the relatively narrow frame of reference that company staff might bring. One consequence has been the growth of specialist service companies, which offer help in building and exploring models of alternative futures. See interview with Helen King from Bord Bia about the Irish food industry futures project.

For example, Novo Nordisk, a major Danish pharmaceuticals business makes use of a company-wide scenario-based programme to explore radical futures around their core business. Its 'Diabetes 2020' process involved exploring radical alternative scenarios for chronic disease treatment and the roles which a player like Novo Nordisk could play. As part of the follow-up from this initiative, in 2003 the company helped set up the Oxford Health Alliance, a non-profit collaborative entity which brought together key stakeholders – medical scientists, doctors, patients and government officials – with views and perspectives which were sometimes quite widely separated. To make it happen, Novo Nordisk made clear that its goal was nothing less than the prevention or cure of diabetes – a goal which if it were achieved would potentially kill off the company's main line of business. As Lars Rebien Sørensen, the CEO of Novo Nordisk, explained:

> In moving from intervention to prevention – that's challenging the business model where the pharmaceuticals industry is deriving its revenues! . . .We believe that we can focus on some major global health issue – mainly diabetes – and at the same time create business opportunities for our company.

Another related approach is to build 'concept' models and prototypes to explore reactions and provide a focus for various different kinds of input which might shape/co-create future products and services. Concept cars are commonly used in the automotive industry not as production models but as stepping stones to help understand and shape the products of the future. Similarly Airbus and other aerospace firms have concept aircraft whilst Toyota is working on concept projects around housing, transportation and energy systems.

More recently companies have started to see value in developing such scenarios jointly with other organizations and discover exciting opportunities for cross-industry collaboration (which often means the creation of an entirely new market).

Using the web

At one level the Internet offers a vast library – and the mechanisms to make new connections to and amongst the information it contains. This is, naturally, a widely used approach but it is interesting to look a little more deeply at how particular forms are developing and shaping this powerful tool.

In its simplest form the web is a passive information resource to be searched – an additional space into which the firm might send its scouts. Increasingly there are professional organizations who offer focused search capabilities to help with this hunting, for example, in trying to pick up on emerging 'cool' trends among particular market segments. High-velocity environments like mobile telecommunications, gaming and entertainment depend on picking up early warning signals and often make extensive use of these search approaches across the web.

Developments in communications technology also make it possible to provide links across extranets and intranets to speed up the process of bringing signals into where they are needed. Firms like Zara and Benetton have sophisticated IT systems giving them early warning of emergent fashion trends, which can be used to drive a high-speed flexible response on a global basis.

This rich information source aspect can quickly be amplified in its potential if it is seen as a two-way or multi-way information marketplace. One of the first companies to take advantage of this was Eli Lilly who set up InnoCentive.com as a match-making tool, connecting those with scientific problems to those being able to offer solutions. As InnoCentive CEO Darrel Carroll says, '*Lilly hires a large number of extremely talented scientists from around the world, but like every company in its position, it can never hire all the scientists it needs. No company can.*' There are now multiple sites offering a brokering service, linking needs and means and essentially creating a global marketplace for ideas – in the process providing a rich source of early warning signals.

A further extension of this is to use websites in a more open-ended fashion, as laboratories in which experiments can be conducted or prototypes tested. For example, a site which is growing in popularity is www.secondlife.com – essentially a role-playing game with over six million users. In this alternative world people can create different characters for themselves and interact with each other – in the process creating a powerful laboratory for testing out ideas. Since, by definition, Second Life is the result of people projecting their aspirations and interests in a different space, it offers significant scope for early warning about or even creating new trends. The potential of 'advergaming' is being explored, for example, by US clothing retailer American Apparel which opened a virtual store in Second Life in 2006. In similar fashion social networking sites such as MySpace (with 120 million members) have become a powerful channel for finding and developing music and other entertainment ideas, challenging 'traditional' marketing approaches.

The largest network of web-based communities for innovation is organized by CommuniSpace, a Boston-based company that organizes and hosts communities around products and brands for major manufacturers around the world. At the beginning of 2007, CommuniSpace operated more than 300 parallel communities. In each of these communities, members discuss either concrete product concepts posted by companies, or develop in a more open discussion new ideas and trends. Each community contains between 50 and 200 members, who are screened, selected and invited by CommuniSpace to participate.

Beyond these uses come those which bring users into the equation as 'co-creators' – a theme we discussed earlier. For example, BMW makes use of the web to enable a 'Virtual Innovation Agency' – a forum where suppliers from outside the normal range of BMW players can offer ideas that BMW may be able to use. These can be both product related and also process related, for example a recent suggestion was for carbon recycling out of factory waste. Although this carries the risk that many 'cranks' will offer ideas, suggestions may also provide stepping stones to new domains of interest.

Working with active users

As we saw earlier, an increasingly significant strategy involves seeing users not as passive consumers of innovations created elsewhere but rather as active players in the process. Their ideas and insights can provide the starting point for very new directions and create new markets, products and services. The challenge now is to find ways of identifying and working with such lead users.

One of the clues is that active users are often at the fringes of the mainstream – in diffusion theory they are not even early adopters but rather active innovators. They are tolerant of failure, prepared to accept that things go wrong but through mistakes they can get to something better – hence the growing interest in participating in 'perpetual beta testing' and development of software and other online products. More often than not active users love to get involved because they feel strongly about the product or service in question; they really want to help and improve things. LEGO found that the prime motivator amongst its communities of user-developers was the recognition which came with having their products actually made and distributed. Microsoft maintains a group of so-called 'Microsoft buddies' – about 1500 power users of their products such as web masters, programmers, software vendors, etc. Strong ties to these customers support Microsoft. They participate in beta testing, help to improve existing products, and submit ideas for new functionalities. The users get no monetary rewards, but receive free software and are invited to bi-annual meetings. To prevent a 'not-invented-here' problem within Microsoft's internal development teams, special liaison officers act as bridges between the 'buddies' and the development teams of the company.

The German firm Webasto makes a wide range of roofing systems for cars including the sophisticated cabriolet features on luxury cars like the Porsche, Volvo, Saab and Ferrari. They went through a systematic approach to understand what lead users are and how to identify them. Building on existing literature they identified four aspects that really drive people's propensity to innovate (cognitive complexity, team expertise, general knowledge, willingness to help). Based on those aspects they developed a questionnaire that they sent out, depending on the project in question, to up to 5000 people from their database. About 20% returned the questionnaires, there were several selection steps (e.g. age bracket, innovation potential) before they arrived at a lead-user group of between 10 and 30. The lead users committed to come for an entire weekend, and without pay.

'Deep diving'

Most market research has become adept at hearing the 'voice of the customer' via interviews, focus groups, panels, etc. But sometimes what people say and what they actually do is different. In recent years there has been an upsurge in the use of anthropological-style techniques to get closer to what people need/want in the context in which they operate. 'Deep diving' is one of many terms used to describe the approach – 'empathic design' and 'ethnographic methods' are others.

Much of the research toolkit here originates from the field of anthropology where the researcher aims to gain insights primarily through observation and immersing herself in the day-to-day life of the object of study – rather than through questioning only. For example, to ensure their new terminal at Heathrow would address user needs well into the future, BAA commissioned some research into what users in 2020 might look like, and what their needs might be. Of course the ageing population came up as an issue; focusing on the behaviour of older people at the airport they noticed that they tend to go to the toilet rather frequently. So, the conclusion was to plan for more toilets at Terminal 5. However, when someone really followed people around they noted that many people going to the restrooms did not actually go to the toilet – but went there because it was quiet, and they could actually hear the announcements!

Probe and learn

One of the problems about a radically different future is that it is hard to imagine it and to predict how things will play out. Sometimes a powerful approach is to try something out – probe – and learn from the results, even if they represent a 'failure'. In this way emergent trends, potential designs, etc. can be explored and refined in a continuing learning process.

There are two complementary dimensions here – the concept of 'prototyping' as a means of learning and refining an idea; and the concept of pilot-scale testing before moving across to a mainstream market. In both cases the underlying theme is essentially one of 'learning as you go', trying things out, making mistakes but using the experience to get closer to what is needed and will work. As Geoff Penney, Chief Information Officer of the US-based investment house Charles Schwab, once said, '*To avoid running too much risk we run pilots, and everyone knows it is "just" a pilot and is not afraid of making suggestions for improvement – or killing it.*'

Not surprisingly prototyping is particularly relevant in product-based firms. For example, Bang & Olufsen has revitalized its prototyping department and made it refer directly to the innovation hub of the company. The prototyping department is engaged in new ideas as early as possible and the experiences are that this strongly supports the process. And, after a period with disappointing results in applying electronics in toys, LEGO made a change in their development approach towards more intensive use of prototypes. Prototypes were created within days – often within hours – after the ideas matured. The result was a much more precise dialogue both within the organization and with the main customers. Eventually, this led to more simple technology – and more success in terms of sales.

But the principles also apply in services – for example the UK National Health Service and the Design Council have been prototyping new options for dealing with chronic diseases like diabetes, heart conditions and Alzheimer's disease. The aim is to learn by doing and also by engaging with the multiple stakeholders who will be part of whatever new system co-evolves. See case study of NHS/RED and podcast interview with Lynn Maher.

Corporate venturing

One widely used approach involves setting up of special units with the remit – and more importantly the budget – to explore new diversification options. Loosely termed 'corporate venture' (CV) units they actually cover a spectrum ranging from simple venture capital funds (for internal and externally generated ideas) through to active search and implementation teams, acquisition and spin-out specialists. For example, Nokia has a very interesting corporate venturing approach for finding innovation. It has moved beyond 'not invented here' and is embracing 'let's find the best ideas wherever they are'. Nokia

Venturing Organization is focused on corporate venturing activities that include identifying and developing new businesses, or as they put it 'the renewal of Nokia'. Nokia Venture Partners invests exclusively in mobile and Internet Protocol (I/P) related start-up businesses. They have a very interesting third group called Innovent that directly supports and nurtures nascent innovators with the hope of growing future opportunities for Nokia.

SAP has set up a venture unit called SAP Inspire to fund start-ups with interesting technologies. The mission of the group is to 'be a world-class corporate venturing group that will contribute, through business and technical innovation, to SAP's long-term growth and leadership'. It does so by:

- seeking entrepreneurial talent within SAP and providing an environment where ideas are evaluated on an open and objective basis
- actively soliciting and cultivating ideas from the SAP community as well as effectively managing the innovation process from idea generation to commercialization
- looking for growth opportunities that are beyond the existing portfolio but within SAP's overall vision and strategy.

The purpose of corporate venturing is to provide some ring-fenced funds to invest in new directions for the business. Such models vary from being tightly controlled (by the parent organization) to being fully autonomous. (Chapter 10 discusses this approach in detail.)

Use brokers and bridges

As we saw earlier, innovation can often take a 'recombinant' form – and the famous saying of William Gibson is relevant here – 'the future is already here, it's just unevenly distributed'. Much recent research work on networks and broking suggests that a powerful search strategy involves making or facilitating connections – 'bridging small worlds'. Increasingly organizations are looking outside their 'normal' knowledge zones as they begin to pursue 'open innovation' strategies. But sending out scouts or mobilizing the Internet can result simply in a vast increase in the amount of information coming at the firm – without necessarily making new or helpful connections. There is a clear message that networking – whether internally across different knowledge groups – or externally – is one of the big management challenges in the twenty-first century. Increasingly organizations are making use of social networking tools and techniques to map their networks and spot where and how bridges might be built – and this is a source of a growing professional service sector activity. Firms like IDEO specialize in being experts in nothing except the innovation process itself – their key skill lies in making and facilitating connections.

A number of new brokers today use the Internet to facilitate innovation. We have already mentioned InnoCentive and CommuniSpace above. Other web-based brokers are companies like YET2.com, who provide bridging capabilities for (external) inventors with ideas or concepts to corporate development units.

Learning to search at the frontier

As we saw earlier there is a long-standing discussion in innovation literature around 'exploration' and 'exploitation' – both are search behaviours but one is essentially incremental, doing what we do better, adaptive learning; whilst the second is radical, do different, generative learning. A key issue is how organizations can operationalize these different behaviours – what 'routines' (structures, processes, behaviours) can they embed to enable effective exploration and exploitation? Whilst the literature is fairly clear about routines for exploitation – essentially innovation approaches to enable continuous incremental extension and adaptation – there is less about exploration.

Striking a suitable balance is tricky enough under what might be called 'steady-state' innovation conditions, but the work of Christensen and others on disruptive innovation suggests that under certain conditions (for example the emergence of completely new markets) established incumbents get into difficulties. They are too focused in their search routines (both explore and exploit) for dealing with what they perceive as a relevant part of the environment (their market 'value network') and they fail to respond to a new emerging challenge until it is often too late. This is partly because their search behaviour is so routinized, embedded in reward structures and other reinforcement mechanisms, that it blinds the organization to other signals.[87–89]

Importantly this is not a failure in innovation management per se – the firms described are in fact very successful innovators under the 'steady-state' conditions of their traditional marketplace, deploying textbook routines and developing close and productive networks with customers and suppliers. The problem arises at the edge of their 'normal' search space and under the discontinuous conditions of new market emergence.

In similar fashion incumbent organizations often suffer when technologies shift in discontinuous fashion. Again their established repertoire of search routines tends towards exploitation and bounds their search space – with the risk that developments outside can achieve considerable momentum and by the time they are visible the organization has little reaction time.[90] This is further complicated by the issue of sunk costs which commit the incumbent to the earlier generation of technology, and the 'sailing ship' effect whereby their exploitation routines continue to bring a stream of improvements to the old technology and sustain that pathway while the new technology matures.[91] (The 'sailing ship' effect refers to the fact that when steamships were first invented it gave a spur to an intensive sequence of innovation in sailing ship technology which meant the two could compete for an extended period before the underlying superiority of steamship technology worked through.)

It is also clear that another key issue is how to integrate these different approaches within the same organization – how (or even if it is possible) to develop what Tushman and O'Reilly call 'ambidextrous' capability around innovation management.[92] Much recent literature on disruptive, radical, discontinuous innovation highlights the tensions which are set up and the fundamental conflicts between certain sets of routines – for example, Christensen's theory suggests that by being too good at 'exploit' routines to listen to and work with the market, incumbent firms fail to pick up or respond to other signals from new fringe markets until it is too late.

5.18 Two dimensions of innovation search

The problem is not just that such firms fail to get the balance between exploit and explore right but also because there are choices to be made about the overall direction of search. Characteristic of many of these businesses is that they continue to commit to 'explore' search behaviour – but in directions which reinforce the boundaries between them and emergent new innovation space. For example, in many of the industries that Christensen studied high rates of R&D investment were going on to push technological frontiers even further – resulting in many cases in 'technology overshoot'. This is not a lack of search activity but rather a problem of direction.

The issue is that the search space is not one-dimensional. As Henderson and Clark point out it is not just a question of searching near or far from core knowledge concepts but also across configurations – the 'component/architecture challenge'. They argue that innovation rarely involves dealing with a single technology or market but rather a bundle of knowledge, which is brought together into a configuration.

Successful innovation management requires that we can get hold of and use knowledge about components but also about how those components can be put together – what they termed the architecture of an innovation.[93]

One way of looking at the search problem is in terms of the ways in which 'innovation space' is framed by the organization. Just as human beings need to develop cognitive schemas to simplify the 'blooming, buzzing confusion' which the myriad stimuli in their environment offer them, so organizations make use of simplifying frames. They 'look' at the environment and take note of elements which they consider relevant – threats to watch out for, opportunities to take advantage of, competitors and collaborators, etc. The construction of such frames helps give the organization some stability and – amongst other things – defines the space within which it will search for innovation possibility. Whilst there is scope for organizations to develop their own individual ways of seeing the world – their business models – in practice there is often commonality within a sector. So most firms in a particular field will adopt similar ways of framing – assuming certain 'rules of the game', following certain trajectories in common.

These frames correspond to accepted 'architectures' – the ways in which players see the configuration within which they innovate. The dominant architecture emerges over time but once established becomes the 'box' within which further innovation takes place. We are reminded of the difficulties in thinking and working outside this box because it is reinforced by the structures, processes and toolkit – the core routines – which the organization (and its key reference points in a wider network of competitors, customers and suppliers) has learned and embedded.

This perspective highlights the challenge of moving between knowledge sets. Firms can be radical innovators but still be 'upstaged' by developments outside their search trajectory. The problem is that search behaviour is essentially bounded exploration and raises a number of challenges:

- When there is a shift to a new mindset – cognitive frame – established players may have problems because of the reorganization of their thinking which is required. It is not simply adding new information but changing the structure of the frame through which they see and interpret that information. They need to 'think outside the box' within which their *bounded* exploration takes place – and this is difficult because it is highly structured and reinforced.[94]
- This is not simply a change of personal or even group mindset – the consequence of following a particular mindset is that artefacts and routines come into place which block further change and reinforce the status quo. Christensen points out, for example, the difficulty of seeing and accepting the relevance of different signals about emerging markets because the reward systems around sales and marketing are biased towards reinforcing the established market.[87] Henderson and Clark highlight the problems of social and knowledge networks which need to be abandoned and new ones set up in the move to new architectures in photolithography equipment.[93] Day and Shoemaker show how organizations develop particular ways of seeing and not seeing.[95] These are all part of the bounding process – essentially they create the box we need to get out of.
- Architectural – as opposed to component innovation – requires letting go of existing networks and building new ones.[96] This is easier for new players to do, hard for established players, because the inertial tendency is to revert to established pathways for knowledge and other exchange – the finding, forming and performing problem.
- The new frame may not necessarily involve radical change in technology or markets but rather a rearrangement of the existing elements. Low-cost airlines did not, for example, involve major technological shifts in aircraft or airport technology but rather problem solving to make flying available to

an under-served market segment. Similarly the 'bottom of the pyramid' development is not about radical new technologies but about applying existing concepts to under-served markets with different characteristics and challenges.[20] There may be incremental innovation to make the new configuration work but this is not usually new to the world but rather problem solving.

5.19 A map of innovation search space

In summarizing the different sources of innovation and how we might organize and manage the process of searching for them we can use a simple map – see Figure 5.5. The vertical axis refers to the familiar 'incremental/radical' dimension in innovation whilst the second relates to environmental complexity – the number of elements and their potential interactions. Rising complexity means that it becomes increasingly difficult to predict a particular state because of the increasing number of potential configurations of these elements. In this way we capture the 'component/architecture' challenge outlined above. Firms can innovate at component level – the left-hand side – in both incremental and radical fashion but such changes take place within an assumed core configuration of technological and market elements – the dominant architecture. Moving to the right introduces the problem of new and emergent architectures arising out of alternative ways of framing amongst complex elements.

Organizations simplify their perceptions of complex environments, choosing to pay attention to certain key features which they interpret via a shared mental model. They learn to manage innovation within this space and construct routines – embedding structures and processes and building networks to support and enable work within it. In mature sectors a characteristic is the dominance of a particular logic which gives rise to business models of high similarity, for example, industries like pharmaceuticals or integrated circuit design and manufacture are characterized by a small number of actors playing to a similar set of rules involving R&D spend, sales and marketing, etc.

But whilst such models represent a 'dominant logic' or trajectory for a sector they are not the only possible way of framing things. In high-complexity environments with multiple sources of variety it becomes possible to configure alternative models – to 'reframe' the game and arrive at an alternative architecture. Whilst many attempts at reframing may fail, from time to time alternatives do emerge which better deal with the environmental complexity and become the new dominant model.

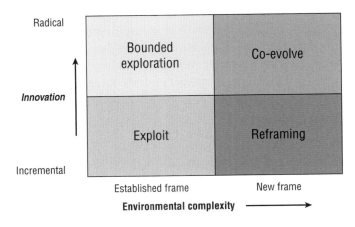

FIGURE 5.5: Innovation search space

Using this idea of different 'frames' we can explore four zones in Figure 5.5 which have different implications for the ways in which innovation is managed. Whilst those approaches for dealing with the left-hand side – zones 1 and 2 – are well developed we argue that there is still much to learn about the right-hand side challenges and how to approach them in practical terms – via methods and tools.

Zone 1 corresponds to the 'exploit' field discussed earlier and assumes a stable and shared frame within which adaptive and incremental development takes place. Search routines here are associated with refining tools and methods for technological and market research, deepening relationships with established key players. Examples would be working with key suppliers, getting closer to customers and building key strategic alliances to help deliver established innovations more efficiently.

The structures for carrying out this kind of search behaviour are clearly defined with relevant actors – department or functions responsible for market research, product (service) development, etc. They involve strong ties in external networks with customers, suppliers and other relevant actors in their wider environment. The work of core groups like R&D is augmented by high levels of participation across the organization – because the search questions are clearly defined and widely understood high involvement of nonspecialists is possible. So procurement and purchasing can provide a valuable channel as can sales and marketing – since these involve contact with external players.[97] Process innovation can be enabled by inviting suggestions for incremental improvement across the organization – a high-involvement *kaizen* model.[15]

Zone 2 involves search into new territory, pushing the frontiers of what is known and deploying different search techniques for doing so. But this still takes place within an established framework – a shared mental model which we could term 'business model as usual'. R&D investments here are on big bets with high strategic potential, patenting and IP strategies aimed at marking out and defending territory, riding key technological trajectories (such as Moore's law in semiconductors). Market research similarly aims to get close to customers but to push the frontiers via empathic design, latent needs analysis, etc. Although the activity is risky and exploratory it is still governed strongly by the frame for the sector – as Pavitt observed there are certain sectoral patterns which shape the behaviour of all the players in terms of their innovation strategies.[98]

The structures involved in such exploration are, of necessity, highly specialized. Formal R&D and within that sophisticated specialization is the pattern on the science/technology frontier, often involving separate facilities. Here too there is mobilization of a network of external but similarly specialized researchers – in university, public and commercial laboratories – and the formation of specific strategic alliances and joint ventures around a particular area of deep technology exploration. The highly specialized nature of the work makes it difficult for others in the organization to participate – and indeed this gap between worlds can often lead to tensions between the 'operating' and the 'exploring' units and the boardroom battles between these two camps for resources are often tense. In similar fashion market research is highly specialized and may include external professional agencies in its network with the task of providing sophisticated business intelligence around a focused frontier.

These two zones represent familiar territory in discussion of exploit/explore in innovation search. But arguably they take place within an accepted frame, a way of seeing the world which essentially filters and shapes perceptions of what is relevant and important. This corresponds to Henderson and Clark's architecture and, as we have argued, defines the 'box' within which innovative activity is expected to occur. Such framing is, however, a construct and open to alternatives – and Zone 3 is essentially associated with reframing. It involves searching a space where alternative architectures are generated, exploring different permutations and combinations of elements in the environment. Importantly this often happens by working with elements in the environment not embraced by established business models – for example, Christensen's work on fringe markets,[87] Prahalad's bottom of the pyramid[20] or von Hippel's extreme users.[26]

For example, the low-cost airline industry was not a development of new product or process – it still involves airports, aircraft, etc. Instead the innovation was in position and paradigm, reframing the business model by identifying new elements in the markets – students, pensioners, etc. – who did not yet fly but might if the costs could be brought down. Rethinking the business model required extensive product and process innovation to realize it – for example in online booking, fast turnaround times at airports, multi-skilling of staff, etc. – but the end result was a reframing and creation of new innovation space.

Zone 4 represents the 'edge of chaos' complex environment where innovation emerges as a product of a process of co-evolution. This is not the product of a predefined trajectory so much as the result of complex interactions between many independent elements.[99,100] Processes of amplification and feedback reinforce what begin as small shifts in direction and gradually define a trajectory. This is the pattern – the 'fluid state' – before a dominant design emerges and sets the standard.[52] As a result it is characterized by very high levels of experimentation.

Search strategies here are difficult since it is impossible to predict what is going to be important or where the initial emergence will start and around which feedback and amplification will happen. The best an organization can do is to try and place itself within that part of its environment where something might emerge and then develop fast reactions to weak signals. 'Strategy' here can be distilled down to three elements – be in there, be in there early and be in there actively (i.e. in a position to be part of the feedback and amplification mechanisms).

With these four zones we have a simple map on which to explore innovation routines. Our concern in this chapter is with search routines – how do organizations manage the process of recognizing and acquiring key new knowledge to enable the innovation process? There are also implications for how they assimilate and transform (select) and how they exploit and implement but we will not focus on those at this stage. As we have suggested each zone represents a different kind of challenge and leads to the use of different methods and tools. And whilst the toolbox is well stocked for zones 1 and 2 there is value in experimentation and experience sharing around zones 3 and 4.

Table 5.6 summarizes the challenge.

Table 5.6

Zone	Search challenges	Tools and methods	Enabling structures
1 'Business as usual' – innovation but under 'steady-state' conditions, little disturbance around core business model	Exploit – extend in incremental fashion boundaries of technology and market. Refine and improve. Close links/strong ties with key players	'Good practice' new product/service development Close to customer Technology platforms and systematic exploitation tools	Formal and mainstream structures High involvement across organization Established roles and functions (including production, purchasing, etc.)

(continued)

Table 5.6	(Continued)		
Zone	**Search challenges**	**Tools and methods**	**Enabling structures**
2 'Business model as usual' – bounded exploration within this frame	Exploration – pushing frontiers of technology and market via advanced techniques. Close links with key strategic knowledge sources	Advanced tools in R&D, market research. Increasing 'open innovation' approaches to amplify strategic knowledge search resources	Formal investment in specialized search functions – R&D, market research, etc.
3 Alternative frame – taking in new/different elements in environment Variety matching, alternative architectures	Reframe – explore alternative options, introduce new elements Experimentation and open-ended search Breadth and periphery important	Alternative futures Weak signal detection User-led innovation Extreme and fringe users Prototyping – probe and learn Creativity techniques Bootlegging, etc.	Peripheral/ad hoc Challenging – 'licensed fools' CV units Internal entrepreneurs Scouts Futures groups, brokers, boundary spanning and consulting agencies
4 Radical – new to the world – possibilities. New architecture around as yet unknown and established elements	Emergence – need to co-evolve with stakeholders Be in there Be in there early Be in there actively	Complexity theory – feedback and amplification, probe and learn, prototyping and use of boundary objects	Far from mainstream 'Licensed dreamers' Outside agents and facilitators

Summary and further reading

In this chapter we've looked at the many ways in which the innovation process can be triggered – and the need for multiple approaches to the problem of searching for them. The management challenge lies in recognizing the rich variety of sources and configuring search mechanisms which balance the 'exploit' and 'explore' domains, providing a steady stream of both incremental (do what we do better) ideas and more radical (do different) stimuli – and doing so with limited resources.

The long-running debate about which sources – demand pull or knowledge push – are most important is well covered in Freeman and Soete's work (*The Economics of Industrial Innovation*, MIT Press, 1997), which brings in the ideas of Schumpeter, Schmookler and other key writers. Particular discussion of fringe markets and unmet or poorly met needs as a source of innovation is covered by Christensen and colleagues (*Seeing What's Next*, Harvard Business School Press, 2007) and by Ulnwick (*What Customers Want*, McGraw-Hill, 2005), whilst the 'bottom of the pyramid' and extreme user potential is explored in Prahalad's work (*The Fortune at the Bottom of the Pyramid*, Wharton School Publishing, 2006). Next Billion (`www.nextbillion.net`) provides a wide range of resources and information about 'bottom of the pyramid' and extreme user activity including video and case studies. User-led innovation has been researched extensively by Eric von Hippel and his books (*The Sources of Innovation*, 1988; *The Democratization of Innovation*, 2005, MIT Press) and website (`http://web.mit.edu/evhippel/www/`) provide an excellent starting point for further exploration of this approach. Frank Piller, Professor at Aachen University in Germany, has a rich website around the theme of mass customization with extensive case examples and other resources (`http://www.mass-customization.de/`); the original work on the topic is covered in Joseph Pine's book (*Mass Customization*, Harvard University Press, 1993). Andrew Hargadon has done extensive work on 'recombinant innovation' (*How Breakthroughs Happen*, Harvard Business School Press, 2003) and Mohammed Zairi provides a good overview of benchmarking (*Effective Benchmarking*, Chapman and Hall, 1996). The 'future of the automobile' project offers a good example of this approach in practice (Womack *et al.*, *The Machine that Changed the World*, Rawson Associates, 1991).

The model of 'punctuated equilibrium' and the different phases of innovation activity linked to search is explored by Tushman and Anderson (Technological discontinuities and organizational environments. *Administrative Science Quarterly*, **31** (3), 439–65, 1987) and Utterback (*Mastering the Dynamics of Innovation*, Harvard Business School Press, 1994) amongst others. The concept of open innovation was originated by Henry Chesbrough (*Open Innovation: The New Imperative for Creating and Profiting from Technology*, Harvard Business School Press, 2003) but has been elaborated in a number of other studies (Chesborough *et al.*, *Open Innovation: Researching a New Paradigm*, Oxford University Press, 2006). Case examples include the Procter & Gamble story and Alan Lafley's book (Lafley, A. and R. Charan, *The Game Changer*, Profile, 2008) provides a readable account from the perspective of the CEO. The concept of absorptive capacity was originated by Cohen and Levinthal (Absorptive capacity: a new perspective on learning and innovation. *Administrative Science Quarterly*, **35** (1), 128–52, 1990) and developed by Zahra and George (Absorptive capacity: a review, reconceptualization and extension. *Academy of Management Review*, **27**, 185–94, 2002); Lane *et al.* provide an extensive review of developments and models (The reification of absorptive capacity: a critical review and rejuvenation of the construct. *Academy of Management Review*, **31** (4), 833–63, 2006). Searching at the frontier is one of the questions being addressed by the Discontinuous Innovation Laboratory, a network of around 30 academic institutions and 150 companies – see www.innovation-lab.org for more details. A report on their work is available at www.aim-research.org.

Web links

Here are the full details of the resources available on the website flagged throughout the text:

 Case studies:
Corning
Tesco Fresh & Easy

Aravind
Novo Nordisk
Model T
RED/NHS
LEGO
Threadless
3M
Cerulean

 Interactive exercises:
Do better/do different search strategy

 Tools:
Benchmarking

 Video podcast:
Scott Berkun
C.K. Prahalad – BoP cases
Stan Davis
Frank Piller
Patrick McLaughlin – Cerulean

 Audio podcast:
Tim Craft – AMS Ltd
Helen King – Bord Bia
Lynne Maher – NHS

References

1. **Freeman, C. and L. Soete** (1997) *The Economics of Industrial Innovation*, third edition, MIT Press, Cambridge, MA.
2. **Moore, Gordon E.** (1965) Cramming more components onto integrated circuits. *Electronics Magazine*.
3. **Dosi, G.** (1982) Technological paradigms and technological trajectories. *Research Policy*, **11**, 147–62.
4. **Tushman, M. and P. Anderson** (1987) Technological discontinuities and organizational environments. *Administrative Science Quarterly*, **31** (3), 439–65.
5. **Trott, P.** (2004) *Innovation Management and New Product Development*, second edition, Prentice-Hall, London.
6. **Booz, Allen and Hamilton Consultants** (1982) *New Product Management for the 1980s*, Booz, Allen and Hamilton Consultants.
7. **Griffin, A.** (1997) PDMA research on new product development practices. *Journal of Product Innovation Management*, **14**, 429.
8. **Kotler, P.** (2003) *Marketing Management, Analysis, Planning and Control*, 11th edition, Prentice Hall, Englewood Cliffs, NJ.

9. **Goffin, K. and R. Mitchell** (2005) *Innovation Management*, Pearson, London.

10. **Kelley, T., J. Littman, and T. Peters** (2001) *The Art of Innovation: Lessons in Creativity from IDEO, America's Leading Design Firm*, Currency, New York.

11. **Rosenberg, N.** (1982) *Inside the Black Box: Technology and Economics*, Cambridge University Press, Cambridge.

12. **Imai, K.** (1987) *Kaizen*, Random House, New York.

13. **Davenport, T.** (1992) *Process Innovation: Re-Engineering Work through Information Technology*, Harvard University Press, Boston, MA.

14. **Womack, J. and D. Jones** (2005) *Lean Solutions*, Free Press, New York.

15. **Bessant, J.** (2003) *High-Involvement Innovation*, John Wiley & Sons, Ltd, Chichester.

16. **Christensen, C., S. Anthony, and E. Roth** (2007) *Seeing What's Next*, Harvard Business School Press, Boston, MA.

17. **Utterback, J. and H. Acee** (2005) Disruptive technologies – an expanded view. *International Journal of Innovation Management*, **9**(1), 1–17.

18. **Ulnwick, A.** (2005) *What Customers Want: Using Outcome-Driven Innovation to Create Breakthrough Products and Services*, McGraw-Hill, New York.

19. **Kim, W. and R. Mauborgne** (2005) *Blue Ocean Strategy: How to Create Uncontested Market Space and Make the Competition Irrelevant*, Harvard Business School Press, Boston, MA.

20. **Prahalad, C.K.** (2006) *The Fortune at the Bottom of the Pyramid*, Wharton School Publishing, New Jersey.

21. **Corbett, S.** (2008) Can the cellphone help end global poverty? *New York Times*.

22. **Bessant, J. and B. Wesley** (2009) Radical service development. *Creativity and Innovation Management, forthcoming*.

23. **Brown, S., J. Bessant, R. Lamming, and P. Jones** (2004) *Strategic Operations Management*, second edition, Butterworth Heinemann, Oxford.

24. **Pine, B.J.** (1993) *Mass Customization: The New Frontier in Business Competition*, Harvard University Press, Cambridge, MA.

25. **Vandermerwe, S.** (2004) *Breaking Through: Implementing Customer Focus in Enterprises*, Palgrave Macmillan, London.

26. **Von Hippel, E.** (1988) *The Sources of Innovation*, MIT Press, Cambridge, MA.

27. **Mansfield, E.** (1968) *Industrial Research and Technological Innovation: An Econometric Analysis*, Norton, New York.

28. **Von Hippel, E.** (1986) Lead users: a source of novel product concepts. *Management Science*, **32** (7), 791–805.

29. **Von Hippel, E.** (2005) *The Democratization of Innovation*, MIT Press, Cambridge, MA.

30. **Piller, F.** (2006) *Mass Customization: Ein wettbewerbsstrategisches Konzept im Informationszeitalter*, fourth edition, Gabler Verlag, Frankfurt.

31. **Moser, K. and F. Piller** (2006) Special issue on mass customization case studies: cases from the international mass customization case collection. *International Journal of Mass Customization*, **1** (4).

32. **Rothwell, R. and P. Gardiner** (1983) Tough customers, good design. *Design Studies*, **4** (3), 161–9.

33. **Rich, B. and L. Janos** (1994) *Skunk Works*, Warner Books, London.

34. **Hobday, M.** (1995) *Innovation in East Asia – The Challenge to Japan*, Edward Elgar, Cheltenham.

35. **Kim, L.** (1997) *Imitation to Innovation: The Dynamics of Korea's Technological Learning*, Harvard Business School Press, Boston, MA.

36. **Belliveau, P., A. Griffin, and S. Somermeyer** (2002) *The PDMA Toolbook for New Product Development: Expert Techniques and Effective Practices in Product Development*, John Wiley & Sons, Inc., New York.

37. **Camp, R.** (1989) *Benchmarking – The Search for Industry Best Practices that Lead to Superior Performance*, Quality Press, Milwaukee, WI.

38. **Kaplinsky, R., F. den Hertog, and B. Coriat** (1995) *Europe's Next Step*, Frank Cass, London.

39. **Womack, J. and D. Jones** (1996) *Lean Thinking*, Simon and Schuster, New York.

40. **Womack, J., D. Jones and D. Roos** (1991) *The Machine that Changed the World*, Rawson Associates, New York.

41. **Hargadon, A.** (2003) *How Breakthroughs Happen*, Harvard Business School Press, Boston, MA.

42. **Hargadon, A. and R. Sutton** (1997) Technology brokering and innovation in a product development firm. *Administrative Science Quarterly*, **42**, 716–49.

43. **Koestler, A.** (1964) *The Act of Creation*, Hutchinson, London.

44. **Blind, K.** (2007) Special issue on innovation and regulation. *International Journal of Public Policy*, **2** (1).

45. **Dodgson, M., D. Gann, and A. Salter** (2007) In case of fire, please take the elevator. *Organization Science*, **18**(5), 849–64.

46. **Hamel, G.** (2000) *Leading the Revolution*, Harvard Business School Press, Boston. MA.

47. **de Geus, A.** (1996) *The Living Company*, Harvard Business School Press, Boston, MA.

48. **Schwartz, P.** (1991) *The Art of the Long View*, Doubleday, New York.

49. **Fahey, L. and R. Randall** (1998) *Learning from the Future*, John Wiley & Sons, Ltd, Chichester.

50. **Chesborough, H.** (2003) Managing your false negatives. *Harvard Management Updates*, **8** (8).

51. **Walsh, V., R. Roy, S. Potter, and M. Bruce** (1992) *Winning by Design: Technology, Product Design and International Competitiveness*, Basil Blackwell, Oxford.

52. **Utterback, J.** (1994) *Mastering the Dynamics of Innovation*, Harvard Business School Press, Boston, MA.

53. **Rogers, E.** (1995) *Diffusion of Innovations*, Free Press, New York.

54. **Moore, G.** (1999) *Crossing the Chasm; Marketing and Selling High-Tech Products to Mainstream Customers*, Harper Business, New York.

55. **Carter, C. and B. Williams** (1957) *Industry and Technical Progress*, Oxford University Press, Oxford.

56. **Chesbrough, H.** (2003) *Open Innovation: The New Imperative for Creating and Profiting from Technology*, Harvard Business School Press, Boston, MA.

57. **Huston, L. and N. Sakkab** (2006) Connect and develop: inside Procter & Gamble's new model for innovation. *Harvard Business Review*, March, 58–66.

58. **Bessant, J. and T. Venables** (2008) *Creating Wealth from Knowledge: Meeting the Innovation Challenge*, Edward Elgar, Cheltenham.

59. **March, J.** (1991) Exploration and exploitation in organizational learning. *Organization Science*, **2** (1), 71–87.

60. **Benner, M.J. and M.L. Tushman** (2003) Exploitation, exploration, and process management: the productivity dilemma revisited. *Academy of Management Review*, **28** (2), 238.

61. **Levinthal, D. and J. March** (1993) The myopia of learning. *Strategic Management Journal*, **14**, 95.

62. **Rothaermel, F.T.** (2001) Technological discontinuities and interfirm cooperation: What determines a startup's attractiveness as alliance partner? *IEEE Transactions on Engineering Management*, **49** (4), 388.

63. **Lavie, D. and L. Rosenkopf** (2006) Balancing exploration and exploitation in alliance formation. *Academy of Management Journal*, **49** (4), 797–818.

64. **McGrath, R.G.** (2001) Exploratory learning, innovative capacity, and managerial oversight. *Academy of Management Journal*, **44** (1), 118.

65. **Levinthal, D.** (1997) Adaptation on rugged landscapes. *Management Science*, **43**, 934–50.

66. **Rosenkop, L. and A. Nerkar** (2001) Beyond local search: boundary-spanning, exploration, and impact in the optical disk industry. *Strategic Management Journal*, **22** (4), 287–306.

67. **Leifer, R., C. McDermott, G. O'Conner, L. Peters, M. Rice, and R. Veryzer** (2000) *Radical Innovation*, Harvard Business School Press, Boston MA.

68. **Tushman, M. and C. O'Reilly** (1996) Ambidextrous organizations: managing evolutionary and revolutionary change. *California Management Review*, **38** (4), 8–30.

69. **Birkinshaw, J. and C. Gibson** (2004) Building ambidexterity into an organization. *Sloan Management Review*, **45** (4), 47–55.

70. **Roussel, P., K. Saad and T. Erickson** (1991) *Third Generation R&D: Matching R&D Projects with Corporate Strategy*, Harvard Business School Press, Cambridge, MA.

71. **Baker, M.** (1983) *Market Development*, Penguin, Harmondsworth.

72. **Cohen, W. and D. Levinthal** (1990) Absorptive capacity: a new perspective on learning and innovation. *Administrative Science Quarterly*, **35** (1), 128–52.

73. **Arrow, K.** (1962) The economic implications of learning by doing. *Review of Economic Studies*, **29** (2), 155–73.

74. **Simon, H. and J. March** (1992) *Organizations*, second edition, Basil Blackwell, Oxford.

75. **Bell, M. and K. Pavitt** (1993) Technological accumulation and industrial growth. *Industrial and Corporate Change*, **2** (2), 157–211.

76. **Lall, S.** (1992) Technological capabilities and industrialization. *World Development*, **20** (2), 165–86.

77. **Hobday, M., H. Rush and J. Bessant** (2005) Reaching the innovation frontier in Korea: a new corporate strategy dilemma. *Research Policy*, **33**, 1433–57.

78. **Senge, P.** (1990) *The Fifth Discipline*, Doubleday, New York.

79. **Argyris, C. and D. Schon** (1970) *Organizational Learning*, Addison Wesley, Reading, MA.

80. **Wenger, E.** (1999) *Communities of Practice: Learning, Meaning, and Identity*, Cambridge University Press, Cambridge.

81. **Pinchot, G.** (1999) *Intrapreneuring in Action – Why You Don't Have to Leave a Corporation to Become an Entrepreneur*, Berrett-Koehler, New York.

82. **Gundling, E.** (2000) *The 3M Way to Innovation: Balancing People and Profit*, Kodansha International, New York.

83. **Augsdorfer, P.** (1996) *Forbidden Fruit*, Avebury, Aldershot.

84. **Bessant, J. and B. Von Stamm** (2007) *Twelve Search Strategies Which Might Save Your Organization*, AIM Executive Briefing, London.

85. **AFFA** (2000) *Supply Chain Learning: Chain Reversal and Shared Learning for Global Competitiveness*, Department of Agriculture, Fisheries and Forestry – Australia (AFFA), Canberra.

86. **Bessant, J., R. Kaplinsky and R. Lamming** (2003) Putting supply chain learning into practice. *International Journal of Operations and Production Management*, **23** (2), 167–84.

87. **Christensen, C.** (1997) *The Innovator's Dilemma*, Harvard Business School Press, Boston, MA.

88. **Foster, R. and S. Kaplan** (2002) *Creative Destruction*, Harvard University Press, Boston, MA.

89. **March, J. and J. Olsen** (1981) Ambiguity and choice in organizations. In W. Starbuck and H. Nystrom, eds., *Handbook of Organization Design*, Oxford University Press, Oxford.

90. **Tripsas, M. and G. Gavetti** (2000) Capabilities, cognition and inertia: evidence from digital imaging. *Strategic Management Journal*, **21**, 1147–61.

91. **Gilfillan, S.** (1935) *Inventing the Ship*, Follett, Chicago.

92. **Tushman, M. and C. O'Reilly** (1996) *Winning through Innovation*, Harvard Business School Press, Boston, MA.

93. **Henderson, R. and K. Clark** (1990) Architectural innovation: the reconfiguration of existing product technologies and the failure of established firms. *Administrative Science Quarterly*, **35**, 9–30.

94. **Hodgkinson, G. and P. Sparrow** (2002) *The Competent Organization*, Open University Press, Buckingham.

95. **Day, G. and P. Schoemaker** (2006) *Peripheral Vision: Detecting the Weak Signals That Will Make or Break Your Company*, Harvard Business School Press, Boston, MA.

96. **Prahalad, C.** (2004) The blinders of dominant logic. *Long Range Planning*, **37** (2), 171–9.

97. **Lamming, R.** (1993) *Beyond Partnership*, Prentice-Hall, London.

98. **Pavitt, K.** (1984) Sectoral patterns of technical change: towards a taxonomy and a theory. *Research Policy*, **13**, 343–73.

99. **McKelvey, B.** (2004) Simple rules for improving corporate IQ: basic lessons from complexity science. In P. Andirani and G. Passiante, eds, *Complexity Theory and the Management of Networks*, Imperial College Press, London.

100. **Allen, P.** (2001) A complex systems approach to learning, adaptive networks. *International Journal of Innovation Management*, **5**, 149–80.

CHAPTER 6

Innovation networks

6.1 No man is an island . . .

Eating out in the days of living in caves was not quite the simple matter it has become today. For a start there was the minor difficulty of finding and gathering the roots and berries – or, being more adventurous, hunting and (hopefully) catching your mammoth. And raw meat isn't necessarily an appetizing or digestible dish so cooking it helps – but for that you need fire and for that you need wood, not to mention cooking pots and utensils. If any single individual tried to accomplish all of these tasks alone they would quickly die of exhaustion, never mind starvation! We could elaborate but the point is clear – like almost all human activity, it is dependent on others. But it's not simply about spreading the workload – for most of our contemporary activities the key is shared creativity – solving problems together, and exploiting the fact that different people have different skills and experiences that they can bring to the party.

It's easy to think of innovation as a solo act – the lone genius, slaving away in his or her garret or lying, Archimedes-like, in the bath before that moment of inspiration when they run through the streets proclaiming their 'Eureka!' moment. But although that's a common image, it lies a long way from the reality. In reality taking any good idea forward relies on all sorts of inputs from different people and perspectives.

BOX 6.1   The power of group creativity

Take any group of people and ask them to think of different uses for an everyday item – a cup, a brick, a ball, etc. Working alone they will usually develop an extensive list – but then ask them to share the ideas they have generated. The resulting list will not only be much longer but will also contain much greater diversity of possible classes of solution to the problem. For example, uses for a cup might include using it as a container (vase, pencil holder, drinking vessel, etc.), a mould (for sandcastles, cakes, etc.), a musical instrument, a measure, a template around which one can draw, a device for eavesdropping (when pressed against a wall) and even, when thrown, a weapon!

The psychologist J.P. Guilford classed these two traits as: 'fluency' – the ability to produce ideas; and 'flexibility' – the ability to come up with different types of idea.[1] The above experiment will quickly show that when working as a group people are usually much more fluent and flexible than any single individual. When working together people spark each other off, jump on and develop each other's ideas, encourage and support each other through positive emotional mechanisms like laughter and agreement – and in a variety of ways stimulate a high level of shared creativity.

(This is the basis of 'brainstorming' and a wide range of creativity enhancement techniques which have been developed over many years. Chapter 3 gives more detail on this.)

For example, the technological breakthrough that makes a better mousetrap is only going to mean something if people can be made aware of it and persuaded that this is something they cannot live without – and this requires all kinds of inputs from the marketing skill set. Making it happen is going to need skills in manufacturing, in procurement of the bits and pieces to make it, in controlling the quality of the final product. None of this will happen without some funding so other skills round getting access to finance – and the understanding of how to spend the money wisely – become important. And coordinating the diverse inputs needed to turn the mousetrap into a successful reality rather than as a gleam in the eye will require project management skills, balancing resources against the clock and facilitating a team of people to find and solve the thousand and one little problems which crop up as you make the journey.

As we saw in the last chapter, innovation is not a solo act but a multi-player game. Whether it is the entrepreneur who spots an opportunity or an established organization trying to renew its offerings or sharpen up its processes, making innovation happen depends on working with many different players. This raises questions about team working, bringing the different people together in productive and creative ways inside an organization – a theme we discussed in Chapter 3. But increasingly it's also about links between organizations, developing and making use of increasingly wide networks. Smart firms have always recognized the importance of linkages and connections – getting close to customers to understand their needs, working with suppliers to deliver innovative solutions, linking up with collaborators, research centres, even competitors, to build and operate innovation systems. In an era of global operations and high-speed technological infrastructures populated by people with highly mobile skills, building and managing networks and connections becomes the key requirement for innovation. It's not about knowledge creation so much as knowledge flows. Even major research and development players like Siemens or GlaxoSmithKline (GSK) are realizing that they can't cover all the knowledge bases they need and instead are looking to build extensive links and relationships with players around the globe.

This chapter explores some of the emerging themes around the question of innovation as a network-based activity. And of course, in the twenty-first century this game is being played out on a global stage but with an underlying networking technology – the Internet – which collapses distances, places geographically far-flung locations right alongside each other in time and enables increasingly exciting collaboration possibilities. However, just because we have the technology to make and live in a global village doesn't necessarily mean we'll be able to do so – much of the challenge, as we'll see, lies in organizing and managing networks so that they perform. Rather than simply being the coming together of different people and organizations, successful networks have what are called emergent properties – the whole is greater than the sum of the parts.

6.2 The 'spaghetti' model of innovation

As we showed in Chapter 2, innovation can be seen as a core process with a defined structure and a number of influences. This is helpful in terms of simplifying the picture into some clear stages and recognizing the key levers we might have to work with if we are going to manage the process successfully. But, like any simplification, the model isn't quite as complex as the reality. Whilst our model works as an aerial view of what goes on and has to be managed, the close-up picture is much more complicated. The ways knowledge actually flows around an innovation project are complex and interactive, woven together in a kind of 'social spaghetti' where different people talk to each other in different ways, more or less frequently, and about different things.

This complex interaction is all about *knowledge* and the ways it flows and is combined and deployed to make innovation happen. Whether it's our entrepreneur building a network to help him get his mousetrap to market or a company like Apple bringing out the latest generation iPod or phone the process will involve building and running knowledge networks. And, as the innovation becomes more complex, so the networks have to involve more different players, many of whom may lie outside the firm. By the time we get to big complex projects – like building a new aeroplane or hospital facility – the number of players and the management challenges the networks pose get pretty large. There is also the complication that increasingly the networks we have to learn to deal with are becoming more virtual, a rich and global set of human resources distributed and connected by the enabling technologies of the Internet, broadband and mobile communications and shared computer networks.

None of this is a new concept in innovation studies. Research going back to the work of Carter and Williams in the 1950s in the UK, for example, noted that 'technically progressive' – innovative – firms were far more 'cosmopolitan' than their 'parochial' and inward-looking counterparts.[2] Similar findings emerged from Project SAPPHO, from the 'Wealth from Knowledge' studies and from other work such as Allen's detailed study of innovation across the US space programme during the 1960s and 70s.[3–5] Andrew Hargadon's work on Thomas Edison and Henry Ford highlights the fact that they were not just solo geniuses, but rather that they understood the network dynamics of innovation and built teams around them capable of creating and sustaining rich innovation networks.[6] In fact studies of early industries – such as Flemish weavers or gun making in Italy or the UK – suggests that innovation networks have been long-established ways of creating a steady stream of successful new products and processes.[7,8]

We should not forget the importance of managing this 'knowledge spaghetti' within the organization. Recent years have seen an explosion of interest in 'knowledge management' and attention has

---

BOX 6.2   Why networks?

There are four major arguments pushing for greater levels of networking in innovation:

- Collective efficiency – in a complex environment requiring a high variety of responses it is hard for all but the largest firm to hold these competencies in-house. Networking offers a way of getting access to different resources through a shared exchange process – the kind of theme underlying the cluster model, which has proved so successful for small firms in Italy, Spain and many other countries.
- Collective learning – networking offers not only the opportunity to share scarce or expensive resources; it can also facilitate a shared learning process in which partners exchange experiences, challenge models and practices, bring new insights and ideas and support shared experimentation. 'Learning networks' have proved successful vehicles in industrial development in a variety of cases – see later in the chapter for some examples.
- Collective risk taking – building on the idea of collective activity networking also permits higher levels of risk to be considered than any single participant might be prepared to undertake. This is the rationale behind many pre-competitive consortia around high-risk R&D.
- Intersection of different knowledge sets – networking also allows for different relationships to be built across knowledge frontiers and opens up the participating organization to new stimuli and experiences.

---

focused on mechanisms to enable better flow, such as communities of practice, gatekeepers and, recently, social network analysis.[9]

Networking of this kind is something that Roy Rothwell foresaw in his pioneering work on models of innovation, which predicted a gradual move away from thinking about (and organizing) a linear science/technology push or demand pull process to one that saw increasing inter-activity. At first, this exists across the company with cross-functional teams and other boundary-spanning activities. Increasingly, it then moves outside it with links to external actors. Rothwell's vision of the 'fifth-generation' innovation is essentially the one in which we now need to operate, with rich and diverse network linkages accelerated and enabled by an intensive set of information and communication technologies.[10]

6.3 Innovation networks

A network can be defined as '*a complex, interconnected group or system*', and networking involves using that arrangement to accomplish particular tasks. As we've suggested innovation has always been a multi-player game and we can see a growing number of ways in which such networking takes place. The concept of innovation networks has become popular in recent years, as it appears to offer many of the benefits of internal development, but with few of the drawbacks of collaboration. (We explore the theme of collaboration in more detail in Chapter 10). Networks have been claimed by some to be a new hybrid form of organization that has the potential to replace both firms (hierarchies) and markets, in essence the 'virtual corporation', whereas others believe them to be simply a transitory form of organization, positioned somewhere between internal hierarchies and external market mechanisms. Whatever the case, there is little agreement on what constitutes a network, and the term and alternatives such as 'web' and 'cluster' have been criticized for being too vague and all-inclusive.[11]

Different authors adopt different meanings, levels of analysis and attribute networks with different characteristics. For example, academics on the continent have focused on social, geographical and institutional aspects of networks, and the opportunities and constraints these present for innovation.[12] In contrast, Anglo-Saxon studies have tended to take a systems perspective, and have attempted to identify how best to design, manage and exploit networks for innovation.[13] Figure 6.1 presents a framework for the analysis of different network perspectives in innovation studies.

Whilst there is little consensus in aims or means, there appears to be some agreement that a network is more than an aggregation of bilateral relationships or dyads, and therefore the configuration, nature and content of a network impose additional constraints and present additional opportunities. A network can be thought of as consisting of a number of positions or nodes, occupied by individuals, firms, business units, universities, governments, customers or other actors, and links or interactions between these nodes. By the same token, a network perspective is concerned with how these economic actors are influenced by the social context in which they are embedded and how actions can be influenced by the position of actors.

A network can influence the actions of its members in two ways.[14] First, through the flow and sharing of information within the network. Secondly, through differences in the position of actors in the network, which cause power and control imbalances. Therefore the position an organization occupies in a network is a matter of great strategic importance, and reflects its power and influence in that network. Sources of power include technology, expertise, trust, economic strength and legitimacy. Networks can

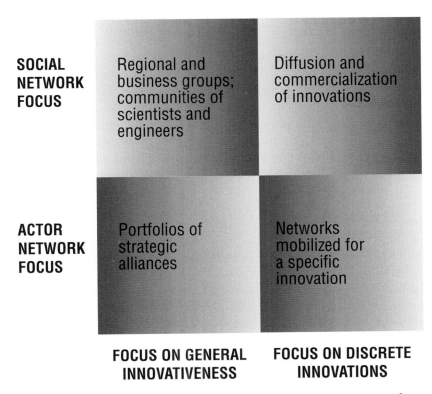

FIGURE 6.1: Different network perspectives in innovation research

Source: Derived from Conway, S. and Steward, F. (1998) Mapping innovation networks. *International Journal of Innovation Management*, **2** (2), 223–54

be tight or loose, depending on the quantity (number), quality (intensity) and type (closeness to core activities) of the interactions or links. Such links are more than individual transactions, and require significant investment in resources over time.

Networks are appropriate where the benefits of co-specialization, sharing of joint infrastructure and standards and other network externalities outweigh the costs of network governance and maintenance. Where there are high transaction costs involved in purchasing technology, a network approach may be more appropriate than a market model, and where uncertainty exists, a network may be superior to full integration or acquisition. Historically, networks have often evolved from long-standing business relationships. Any firm will have a group of partners that it does regular business with – universities, suppliers, distributors, customers and competitors. Over time mutual knowledge and social bonds develop through repeated dealings, increasing trust and reducing transaction costs. Therefore a firm is more likely to buy or sell technology from members of its network.[15]

Firms may be able to access the resources of a wide range of other organizations through direct and indirect relationships, involving different channels of communication and degrees of formalization. Typically, this begins with stronger relationships between a firm and a small number of primary suppliers, which share knowledge at the concept development stage. The role of the technology gatekeeper, or heavyweight project manager, is critical in this respect. In many cases organizational linkages can be traced to strong personal relationships between key individuals in each organization. These linkages may subsequently evolve into a full network of secondary and tertiary suppliers, each contributing to the development of a subsystem or component technology, but links with these organizations are weaker and filtered by the primary suppliers. However, links amongst the primary, secondary and tertiary supplier groups may be stronger to facilitate the exchange of information.

This process is path-dependent in the sense that past relationships between actors increase the likelihood of future relationships, which can lead to inertia and constrain innovation. Indeed much of the early research on networks concentrated on the constraints networks impose on members, for example preventing the introduction of 'superior' technologies or products by controlling supply and distribution networks. Organizational networks have two characteristics that affect the innovation process: activity cycles and instability.[16] The existence of activity cycles and transaction chains creates constraints within a network. Different activities are systematically related to each other and through repetition are combined to form transaction chains. This repetition of transactions is the basis of efficiency, but systemic interdependencies create constraints to change.

For example, the Swiss watch industry was based on long-established networks of small firms with expertise in precision mechanical movements, but as a result was slow to respond to the threat of electronic watches from Japan.

Similarly, Japan has a long tradition of formal business groups, originally the family-based *zaibatsu*, and more recently the more loosely connected *keiretsu*. The best-known groups are the three ex-*zaibatsu* – Mitsui, Mitsubishi and Sumitomo – and the three newer groups based around commercial banks – Fuji, Sanwa and Dai-Ichi Kangyo (DKB). There are two types of *keiretsu*, although the two overlap. The vertical type organizes suppliers and distribution outlets hierarchically beneath a large, industry-specific manufacturer, for example Toyota Motor.

These manufacturers are in turn members of *keiretsu*, which consist of a large bank, insurance company, trading company and representatives of all major industrial groups. These inter-industry *keiretsu* provide a significant internal market for intermediate products. In theory, benefits of membership of a *keiretsu* include access to low-cost, long-term capital, and access to the expertise of firms in related industries.

This is particularly important for high-technology firms. In practice, research suggests that membership of *keiretsu* is associated with below-average profitability and growth,[17] and independent firms like Honda and Sony are often cited as being more innovative than established members of *keiretsu*. However, the *keiretsu* may not be the most appropriate unit of analysis, as many newer, less formal clusters of companies have emerged in modern Japan. As the role of a network is different for all its members, there will always be reasons to change the network and possibilities to do so. A network can never be optimal in any generic sense, as there is no single reference point, but is inherently adaptable. This inherent instability and imperfection mean that networks can evolve over time. For example, Belussi and Arcangeli discuss the evolution of innovation networks in a range of traditional industries in Italy.[18]

More recent research has examined the opportunities networks might provide for innovation, and the potential to explicitly design or selectively participate in networks for the purpose of innovation, that is, a path-creating rather than path-dependent process.[19] A study of 53 research networks found two distinct dynamics of formation and growth. The first type of network emerges and develops as a result of environmental interdependence, and through common interests – an emergent network. However, the other type of network requires some triggering entity to form and develop – an engineered network.[20] In an engineered network a nodal firm actively recruits other members to form a network, without the rationale of environmental interdependence or similar interests.

TABLE 6.1 **Competitive dynamics in network industries**

	Type of network	
	Unconnected, closed	**Connected, open**
System attributes	Incompatible technologies	Compatible across vendors and products
	Custom components and interfaces	Standard components
Firm strategies	Control standards by protecting proprietary knowledge	Shape standards by sharing knowledge with rivals and complementary markets
Source of advantage	Economies of scale, customer lock-in	Economies of scope, multiple segments

Source: Adapted from Garud, R. and A. Kumaraswamy (1993) Changing competitive dynamics in network industries. *Strategic Management Journal,* **14**, 351–69.

Different types of network may present different opportunities for learning (Table 6.1). In a closed network a company seeks to develop proprietary standards through scale economies and other actions, and thereby lock customers and other related companies into its network.[21] Obvious examples include Microsoft in operating systems and Intel in microprocessors for PCs. In the case of open networks, complex products, services and businesses have to interface with others, and it is in everyone's interest to share information and to ensure compatibility.

Virtual innovation networks are beginning to emerge, based on firms that are connected via intranet/extranet/Internet and exchange information within a business relationship to create value. To date such virtual networks are most common in supply-chain and customer order automation, but recent examples include product development. For example, in supply-chain management Hervé Thermique, a French manufacturer of heating and air conditioners, uses an extranet to coordinate its 23 offices and 8000 suppliers; General Electric has an extranet bidding and trading system to manage its 1400 suppliers; Boeing has a web-based order system for its 700 customers worldwide, which features 410 000 spare parts; in product development, Caterpillar's customers can amend designs during assembly; and Adaptec coordinates design and production of microchips in Hong Kong, Taiwan and Japan.[22]

Innovation networks are more than just ways of assembling and deploying knowledge in a complex world. They can also have what are termed 'emergent properties' – that is, the potential for the whole to be greater than the sum of its parts. Being in an effective innovation network can deliver a wide range of benefits beyond the collective knowledge efficiency mentioned above. These include getting access to different and complementary knowledge sets, reducing risks by sharing them, accessing new markets and technologies and otherwise pooling complementary skills and assets. Without such networks it would be nearly impossible for the lone inventor to bring his or her idea successfully to market. And it's one of the main reasons why established businesses are increasingly turning to cooperation and alliances – to extend their access to these key innovation resources.

For example, participating in innovation networks can help companies bump into new ideas and creative combinations – even for mature businesses. It is well known in studies of creativity that the process involves making associations. And sometimes, the unexpected conjunction of different perspectives can lead to surprising results. The same seems to be true at the organizational level: studies of networks indicate that getting together in such a fashion can help open up new and productive territory.[23]

Another way in which networking can help innovation is in providing support for shared learning. A lot of process innovation is about configuring and adapting what has been developed elsewhere and applying it to your processes – for example, in the many efforts which organizations have been making to adopt world-class manufacturing (and increasingly, service) practice. While it is possible to go it alone in this process, an increasing number of companies are seeing the value in using networks to give them some extra traction on the learning process. Case study 6.1 gives some examples of building and operating learning networks and more examples can be found in Bessant *et al.*[24,25]

CASE STUDY 6.1

The potential of learning networks

Learning together has its advantages. For example, in the UK, the Society of Motor Manufacturers and Traders has run the successful Industry Forum for many years, helping a wide range of businesses adopt and implement process innovations around world-class manufacturing. This model has been rolled out (with support from the Department of Trade and Industry) to sectors as diverse as ceramics, aerospace, textiles and tourism. Many regional development agencies (such as Advantage West Midlands with its 'Innovation Networks' programme) now try and use networks and clusters as a key aid to stimulating economic growth through innovation. The same principles can help to diffuse innovative practices along supply chains: companies such as IBM and BAE Systems have made extensive efforts to make 'supply-chain learning' the next key thrust in their supplier development programmes.

CASE STUDY 6.2

Casting the innovation net wider

Cosworth is an automotive engineering company founded in London in 1958 specializing in engines for automobile racing. It supplies a wide range of motorsport series, including the World Rally Championship and, until the end of 2006, Formula One. They were seeking a source of aluminium castings which were cheap enough for volume use but of high enough precision and quality for their product; having searched throughout the world they were unable to find anyone suitable. Either they took the low-price route and used some form of die casting which often lacked the precision and accuracy, or they went along the investment-casting route which added significantly to the cost. Eventually they decided to go right back to basics and design their own manufacturing process. They set up a small pilot facility and employed a team of metallurgists and engineers with the brief to come up with an alternative approach that could meet their needs. After three years of work and a very wide and systematic exploration of the problem the team came up with a process which combined conventional casting approaches with new materials (especially a high grade of sand) and other improvements. The breakthrough was, however, the use of an electromagnetic pump which forced molten metal into a shell in such a way as to eliminate the air which normally led to problems of porosity in the final product. This innovation came from well outside the foundry industry, from the nuclear power field where it had been used to circulate the liquid sodium coolant used in the fast breeder reactor programme! The results were impressive; not only did Cosworth meet its own needs, it was also able to offer the service to other users of castings and to license the process to major manufacturers such as Ford and Daimler-Benz.

Consider another example from the motorsport industry: leading race-car makers are continually seeking innovation in support of enhanced performance and may take ideas, materials, technology or products from very different sectors. Indeed some have people (called 'technological antennae') whose sole responsibility is to search for new technologies that might be used. For instance, recent developments in the use of titanium components in Formula 1 engines have been significantly advanced by lessons learned about the moulding process from a company producing golf clubs.

Source: Birkinshaw, J., J. Bessant and R. Delbridge (2007) Finding, forming, and performing: creating networks for discontinuous innovation. *California Management Review*, **49** (3), 67–83.

Innovation is about taking risks and deploying what are often scarce resources on projects which may not succeed. So, another way in which networking can help is by helping to spread the risk and, in the process, extending the range of things which might be tried. This is particularly useful in the context of smaller businesses where resources are scarce and it is one of the key features behind the success of many industrial clusters. Case study 6.3 and the article on which it is based gives a good example of such cooperative effort in developing R&D capability.

Long-lasting innovation networks can create the capability to ride out major waves of change in the technological and economic environment. We think of places like Silicon Valley in the USA, Cambridge in the UK or the island of Singapore as powerhouses of innovation but they are just the latest in a long-running list of geographical regions which have grown and sustained themselves through a continuous stream of innovation.[26–28]

At its simplest networking happens in an informal way when people get together and share ideas as a by-product of their social and work interactions. But we'll concentrate our attention on more formal networks which are deliberately set up to help make innovation happen, whether it is creating a new product or service or learning to apply some new process thinking more effectively within organizations.

Table 6.2 gives an idea of the different ways in which such 'engineered' networks can be configured to help with the innovation process. In the following section we'll look a little more closely at some of these, how they operate and the benefits they can offer.

CASE STUDY 6.3

Networked-based R&D

The case of the Italian furniture industry is one in which a consistently strong export performance has been achieved by companies with an average size of fewer than 20 employees. Keeping their position at the frontier in terms of performance is the result of sustained innovation in design and quality enabled by a network-based approach. This isn't an isolated case – one of the most respected research institutes in the world for textiles is CITER, based in Emilia-Romagna. Unlike so many world-class institutions, this was not created in top-down fashion but evolved from the shared innovation concerns of a small group of textile producers who built on the network model to share risks and resources. Their initial problems with dyeing and with computer-aided design helped them to gain a foothold in terms of innovation in their processes. In the years since its founding in 1980, it has helped its 500 (mostly small business) members develop a strong innovation capability.

Source: Rush, H. *et al.* (1996) *Technology Institutes: Strategies for Best Practice*, International Thomson Business Press, London.

TABLE 6.2   **Types of innovation networks**

Network type	Examples
Entrepreneur-based	Bringing different complementary resources together to help take an opportunity forward. Often a combination of formal and informal, depends a lot on the entrepreneur's energy and enthusiasm in getting people interested to join – and stay in – the network
Internal project teams	Formal – and informal – networks of knowledge and key skills which can be brought together to help enable some opportunity to be taken forward – essentially like entrepreneur networks but on the inside of established organizations. May run into difficulties because of having to cross internal organizational boundaries
Communities of practice	These are networks which can involve players inside and across different organizations – what binds them together is a shared concern with a particular aspect or area of knowledge
Spatial clusters	Networks which form because of the players being close to each other, for example, in the same geographical region. Silicon Valley is a good example of a cluster which thrives on

(continued)

Network type	Examples
	proximity – knowledge flows amongst and across the members of the network but is hugely helped by the geographical closeness and the ability of key players to meet and talk
Sectoral networks	Networks which bring different players together because they share a common sector – and often have the purpose of shared innovation to preserve competitiveness. Often organized by sector or business associations on behalf of their members. Shared concern to adopt and develop innovative good practice across a sector or product market grouping, for example, the SMMT Industry Forum or Logic (Leading Oil and Gas Industry Competitiveness), a gas and oil industry forum
New product or process development consortium	Sharing knowledge and perspectives to create and market a new product or process concept, for example, the Symbian consortium (Sony, Ericsson, Motorola and others) working towards developing a new operating system for mobile phones and PDAs
Sectoral forum	Working together across a sector to improve competitiveness through product, process and service innovation
New technology development consortium	Sharing and learning around newly emerging technologies, for example, the pioneering semiconductor research programmes in the US and Japan
Emerging standards	Exploring and establishing standards around innovative technologies, for example, the Motion Picture Experts Group (MPEG) working on audio and video compression standards
Supply-chain learning	Developing and sharing innovative good practice and possibly shared product development across a value chain, for example, the SCRIA initiative in aerospace

TABLE 6.2 (Continued)

6.4 Networks at the start-up

The idea of the lone inventor pioneering his or her way through to market success is something of a myth – not least because of the huge efforts and different resources needed to make innovation happen. Say the name 'Thomas Edison' – and people instinctively imagine a great inventor, the lone genius who gave us so many twentieth-century products and services – the gramophone, the light bulb, electric power, etc. But he was actually a very smart networker. His 'Invention Factory' in Menlo Park, New Jersey employed a team of engineers in a single room filled with workbenches, shelves of chemicals, books and other resources. The key to their undoubted success was to bring together a group of young, entrepreneurial and enthusiastic men from very diverse backgrounds – and allow the emerging community to tackle a wide range of problems. Ideas flowed across the group and were combined and recombined into an astonishing array of innovations.[6]

Whilst individual ideas, energy and passion are key requirements, most successful entrepreneurs recognize the need to network extensively and to collect the resources they require via complex webs of relationships. They are essentially highly skilled at networking, both in building and in maintaining those networks to help build a sustainable business model.

Nowhere is this more clearly seen than in the case of social entrepreneurship where the challenge is to mobilize a wide range of supporting resources often at low or no cost – and to weave them into a network, which enables the launch of a new idea. As Case study 6.4 shows, this requires considerable network-building and managing skills. The Spanish textile and clothing company, Inditex, which includes famous brands like Zara, has grown through a mixture of entrepreneurial flair and extensive use of networking as a source of innovation.

CASE STUDY 6.4

Power to the people – Freeplay Energy

Trevor Baylis was quite a swimmer in his youth, representing Britain at the age of 15. So it wasn't entirely surprising that he ended up working for a swimming pool firm in Surrey before setting up his own company. He continued his swimming passion – working as a part-time TV stuntman doing underwater feats – but also followed an interest in inventing things. One of the projects he began work on in 1991 was to have widespread impact despite – or rather because of – being a 'low-tech' solution to a massive problem.

Having seen a documentary about AIDS in Africa he began to see the underlying need for something that could help communication. Much of the AIDS problem lies in the lack of awareness and knowledge across often isolated rural communities – people don't know about causes or prevention of this devastating disease. And this reflects a deeper problem – of *communication*. Experts estimate that less than 20% of the world's population has access to a telephone, while even fewer have a regular supply of electricity, much less television or Internet access. Very low literacy levels exclude most people from reading newspapers and other print media.

Radio is an obvious solution to the problem – but how can radio work when the receivers need power and in many places mains electricity is simply non-existent? An alternative is battery power – but batteries are equally problematic – even if they were of good quality and freely available via village stores people couldn't afford to buy them regularly. In countries where $1 a day is the standard wage, batteries can cost from a day's to a week's salary. The HIV/AIDS pandemic also means that household incomes are under increased pressure as earners become too ill to work while greater expenditure goes towards health care, leaving nothing for batteries.

What was needed was a radio which ran on some different source of electricity. In thinking about the problem Baylis remembered the old-fashioned telephones of pre-war days which had wind-up handles to generate power. He began experimenting, linking together odd items such as a hand brace, an electric motor and a small radio. He found that the brace turning the motor would act as a generator which would supply sufficient electricity to power the radio. By adding a clockwork mechanism he found that a spring could be wound up – and as it unwound the radio would play. This first working prototype ran for 14 minutes on a two-minute wind. Trevor had invented a clockwork (windup) radio! As a potential solution to the communication problem the idea had real merit. The trouble was that, like thousands of entrepreneurs before him, Trevor couldn't convince others of this. He spent nearly four years approaching major radio manufacturers like Philips and Marconi but to no avail. But luck often plays a significant part in the innovation story – and this was no exception. The idea came to the attention of some TV researchers and the product was featured in 1994 on a BBC TV programme, *Tomorrow's World*, which showcased interesting and exciting new inventions.

Amongst those who saw it and whose interest was taken by the wind-up radio were a corporate finance expert, Christopher Staines and a South African entrepreneur Rory Stear. They bought the rights from Baylis and received a UK government grant to help develop the product further, including the addition of solar panel options. In South Africa, the details of the invention were featured in a new broadcast and heard by Hylton Appelbaum, head of an organization called the Liberty Life Foundation, who saw the potential. Even in relatively rich South Africa, half the homes have no electricity, and elsewhere in Africa the problem is even more severe.

Liberty Life is a body set up by a major South African insurance company, and Anita and Gordon Roddick, the socially conscious owners of The Body Shop. Part of the work of the Foundation is in providing access to employment for the disabled and a third of the company's factory workers are blind, deaf, in wheelchairs, or mentally ill. Through Applebaum Liberty Life provided the $1.5 million in venture capital that founded the company. Baygen Power Industries was set up by Staines and Stear in 1995 in Cape Town. Sixty percent of the shares were held by a group of organizations for the disabled, a condition of Liberty's support. Technical development was provided by the Bristol University Electronics Engineering Department. Shortly thereafter production of the radio began in Cape Town by BayGen Products PTY South Africa. It came on the market at the beginning of 1996 and one year later around 160 000 units had been sold. Much of the early production was purchased by aid charities working in Rwanda and other African countries where relief efforts were underway.

This was not a glamorous product – as a *New York Times* article described it, 'It is no threat to a Sony Walkman. It weighs six pounds, it's built like an overstuffed lunch box, and it has a tinny speaker. But its wholesale price is only $40 and it gets AM, FM, and shortwave, meaning it can pick up the British Broadcasting Corporation or the Voice of America, so a circle of mud huts can zip back into the Information Age with a twist of the wrist' (*Source:* Donald G. McNeil Jr., *New York Times News Service*, 1996).

The impact was significant. In 1996 another BBC TV programme, *QED*, featured the radio and at one point showed footage of Baylis, Staines and Stear together with Nelson Mandela who commented that this was a '*fantastic product that can provide an opportunity for those people who have been despised by society*'.

A fuller version of this case is on the website.

6.5 Networks on the inside

'If only x knew what x knows. . .?' We can fill the x in with the name of almost any large contemporary organization – Siemens, Philips, GSK, Citibank – they all wrestle with the paradox that they have hundreds or thousands of people spread across their organizations with all sorts of knowledge. The trouble is that – apart from some formal project activities which bring them together – many of these knowledge elements remain unconnected, like a giant jigsaw puzzle in which only a small number of the pieces have so far been fitted together. This kind of thinking was behind the fashion for 'knowledge management' in the late 1990s, and one response, popular then, was to make extensive use of information technology to try and improve the connectivity. The trouble was that, whilst the computer and database systems were excellent at storage and transmission, they didn't necessarily help make the connections that turned data and information into useful – and used – knowledge. Increasingly firms are recognizing that, whilst advanced information and communications technology can support and enhance, the real need is for improved knowledge networks inside the organization.

It's back to the spaghetti model of innovation – how to ensure that people get to talk to others and share and build on each other's ideas. This might not be too hard in a three- or four-person business but it gets much harder across a typical sprawling multinational corporation. Although this is a long-standing problem there has been quite a lot of movement in recent years towards understanding how to build more effective innovation networks within such businesses. Research by Tom Allen during the US space programme highlighted the importance of social networks and coined the term 'technological gate-keeper' – and 30 years later we are seeing widespread interest in communities of practice, social networking and other mechanisms designed to build on these insights.[3,29]

CASE STUDY 6.5

Connect and develop at Procter & Gamble (P&G)

P&G's successes with 'connect and develop' owe much to their mobilizing rich linkages between people who know things within their giant global operations. Amongst their successes in internal networking was the 'Crest Whitestrips' product – essentially linking oral care experts with researchers working on film technology and others in the bleach and household cleaning groups. Another is 'Olay Daily Facials' which linked the surface active agents expertise in skin care with people from the tissue and towel areas and from the fabric property-enhancing skills developed in 'Bounce' a fabric-softening product.

Making it happen as part of daily life rather than as a special initiative is a big challenge. They use multiple methods including extensive networking via an intranet site called 'Ask me' which links 10 000 technical people across the globe. It acts as a signposting and web-market for ideas and problems across the company. They also operate 21 'communities of practice' built around key areas of expertise such as polymer chemists, biological scientists and people involved with fragrances. And they operate a global technology council, which is made up of representatives of all of their business units.

Enabling connect and develop

Roy Sandbach is a Research Fellow within P&G and his job is to enable connections within and across the business to create innovative new ideas. He has been responsible for a variety of innovations including the 'Tide to go' stain removal pen. The website contains an interview podcast which explores how networking on the inside of a large corporation can enable innovation.

6.6 Networks on the outside

Creating and combining different knowledge sets has always been the name of the game both inside and outside the firm. But there has been a dramatic acceleration in recent years led by major firms like P&G, GSK, 3M, Siemens and GE towards what has been termed 'open innovation'. The idea behind this – as we saw in Chapter 5 – is that even large-scale R&D in a closed system like an individual firm isn't going to be enough in the twenty-first-century environment.[30]

Knowledge production is taking place at an exponential rate and the OECD countries spend getting on for $1 trillion on R&D in the public and private sector – a figure which is probably an underestimate since it ignores the considerable amount of 'research' which is not captured in official statistics.[31] How can any single organization keep up with – or even keep tabs on – such a sea of knowledge? And this is happening in widely distributed fashion – R&D is no longer the province of the advanced industrial nations like USA, Germany or Japan but is increasing most rapidly in the newly growing economies like India and China. In this kind of context it's going to be impossible to pick up on every development and even smart firms are going to miss a trick or two.

BOX 6.3   **Open innovation**

Chesbrough's principles of open innovation can be summarized as:

- Not all the smart people work for you
- External ideas can help create value, but it takes internal R&D to claim a portion of that value for you
- It is better to build a better business model than to get to market first
- If you make the best use of internal and external ideas, you will win
- Not only should you profit from others' use of your intellectual property, you should also buy others' IP whenever it advances your own business model
- You should expand R&D's role to include not only knowledge generation, but knowledge brokering as well

Source: Chesbrough, H. (2003) *Open Innovation: The New Imperative for Creating and Profiting From Technology*, Harvard Business School Press, Boston, MA.

The case of Procter & Gamble provides a good example of this shift in approach. In the late 1990s there were concerns about their traditional inward-focused approach to innovation. Whilst it worked there were worries – not least the rapidly rising costs of carrying out R&D. Additionally there were many instances of innovations which they might have made but which they passed on – only to find someone else doing so and succeeding. As CEO Alan Lafley explained '*Our R&D productivity had levelled off, and our innovation success rate – the percentage of new products that met financial objectives – had stagnated at about 35%. Squeezed by nimble competitors, flattening sales, lacklustre new launches, and a quarterly earnings miss, we lost more than half our market cap when our stock slid from $118 to $52 a share. Talk about a wake-up call*'.[32]

They recognized that much important innovation was being carried out in small entrepreneurial firms, or by individuals, or in university labs and that other major players like IBM, Cisco, Eli Lilly and Microsoft were beginning to open up their innovation systems.

As a result they moved to what they have called 'connect and develop' – an innovation process based on the principles of 'open innovation'.

Lafley's original stretch goal was to get 50% of innovations coming from outside the company; by 2006 more than 35% of new products had elements which originated from outside, compared with 15% in 2000. Over 100 new products in the past two years came from outside the firm and 45% of innovations in the new product pipeline have key elements which were discovered or developed externally. They estimate that R&D productivity has increased by nearly 60% and their innovation success rate has more than doubled. One consequence is that they increased innovation whilst *reducing* their R&D spend, from 4.8% of turnover in 2000 to 3.4%.

Central to the model is the concept of mobilizing innovation networks. As chief technology officer Gilbert Cloyd explained, '*It has changed how we define the organization . . . We have 9000 people on our R&D staff and up to 1.5 million researchers working through our external networks. The line between the two is hard to draw. . . . We're . . . putting a lot more attention on what we call 360-degree innovation*'. But this is not simply a matter of outsourcing what used to happen internally. As vice president Larry Huston comments, '*People mistake this for outsourcing, which it most definitely is not . . . . . . Outsourcing is when I hire someone to perform a service and they do it and that's the end of the relationship. That's not much different from the way employment has worked throughout the ages. We're talking about bringing people in from outside and involving them in this broadly creative, collaborative process. That's a whole new paradigm*'.

Enabling external networking involves a number of mechanisms. One is a group of 80 'technology entrepreneurs' whose task is to roam the globe and find and make interesting connections. They visit conferences and exhibitions, talk with suppliers, visit universities, scour the Internet – essentially a no-holds-barred approach to searching for new possible connections.

They also make extensive use of the Internet. An example is their involvement as founder members of a site called InnoCentive (`www.innocentive.com`) originally set up by the pharmaceutical giant Eli Lilly in 2001. This is essentially a web-based marketplace where problem owners can link up with problem solvers – and it currently has around 90 000 solvers available around the world. The business model is simple – companies – like P&G, Boeing, DuPont – post their problems on the site and if any of the solvers can help they pay for the idea. Importantly the solvers are a very wide mix, from corporate and university laboratory staff through to lone inventors, retired scientists and engineers and professional design houses. Jill Panetta, InnoCentive's chief scientific officer, says more than 30% of the problems posted on the site have been cracked, '*which is 30% more than would have been solved using a traditional, in-house approach*'.

Other mechanisms include a website called YourEncore, which allows companies to find and hire retired scientists for one-off assignments. NineSigma is an online marketplace for innovations, matching seeker companies with solvers in a marketplace similar to InnoCentive. As chief technology officer Gil Cloyd comments, '*NineSigma can link us to solutions that are more cost efficient, give us early access to potentially disruptive technologies, and facilitate valuable collaborations much faster than we imagined*'. And Yet2.com looks for new technologies and markets across a broad frontier, involving around 40% of the world's major R&D players in their network.

The challenge in open innovation is less about understanding the concept than in developing mechanisms which can enable its operation in practice. Approaches like P&G's 'connect and develop' provide powerful templates but these are only relevant for certain kinds of organization – in other areas new models are being experimented with. For many this involves the construction of different kinds of shared platforms on which different partners can collaborate to create new products and services – such as the BBC Backstage project. This seeks to do with new media development what the open source community did with LINUX and other software development. The model is deceptively simple – developers are invited to make free use of various elements of the BBC's site (such as live news feeds, weather, TV listings, etc.) to integrate and shape innovative applications. The strap line of the Backstage website is '*use our stuff to build your stuff*'– and as soon as the site was launched in May 2005 it attracted the interest of hundreds of software developers and led to some high potential product ideas. Ben Metcalf, one of the programme's founders, summed up the approach:

Top line, we are looking to be seen promoting innovation and creativity on the Internet, and if we can be seen to be doing that, we will be very pleased. In terms of projects coming out of it, if we can see a few examples that offered real value to our end users to build something new, we would be happy with that as well. And if someone is doing something really innovative, we would like to invite them into the BBC and see if some of that value can be incorporated into the BBC's core propositions.

The UK's public-sector mapping organization, Ordnance Survey, has begun opening up its approach to sharing geographical information to a wide variety of partners and has extended an invitation to co-create similar to the BBC's – in part as they recognize the huge changes in their sector with the entry of players like Google. The website contains an interview podcast with David Overton of Ordnance Survey.

Others have gone further down the road towards creating open-source communities in which co-creation amongst different stakeholders takes place. Nokia, for example, is involved in trying to establish a platform for mobile computing/entertainment as well as communications and to enable this has created an open platform. To enable its effective development they recently bought out the other partners in the long-standing Symbian venture which was set up with Motorola, Sony Ericsson and others to develop operating systems for mobile devices. Their purpose in doing so was to allow them to open up Symbian as an open-source platform on which other actors can also develop applications.

The logic of open innovation is that organizations need to open up their innovation processes, searching widely outside their boundaries and working towards managing a rich set of network connections and relationships right across the board.[33] Their challenge becomes one of improving the knowledge *flows* in and out of the organization, trading in knowledge as much as goods and services. To assist in this process a new service sector of organizations offering various kinds of brokering and bridging

RESEARCH NOTE   Models for open innovation

A number of models are emerging around enabling open innovation, for example, Nambisan and Sawhney (2007) identify four. The 'orchestra' model is typified by a firm like Boeing which has created an active global network around the 787 Dreamliner with suppliers as both partners and investors and moving from 'build to print' to 'design and build to performance'. In this mode they retain considerable autonomy around their specialist tasks whilst Boeing retains the final integrating and decision making – analogous to professional musicians in an orchestra working under a conductor.

By contrast the 'creative bazaar' model involves more of a 'crowd-sourcing' approach in which a major firm goes shopping for innovation inputs – and then integrates and develops them further. Examples here would include aspects of the InnoCentive approach being used by P&G, Eli Lilly and others, or the Dial Corporation in the US which launched a 'Partners in Innovation' website where inventors could submit ideas. BMW's Virtual Innovation Agency operates a similar model.

A third model is what they term 'Jam central' which involves creating a central vision and then mobilizing a wide variety of players to contribute towards reaching it. It is the kind of approach found in many pre-competitive alliances and consortia where difficult technological or market challenges are used – such as the 5th Generation Computer project in Japan – to focus efforts of many different organizations. Once the challenges are met the process shifts to an exploitation mode, for example, in the 5th Generation programme the pre-competitive efforts by researchers from all the major electronics and IT firms led to generation of over 1000 patents which were then shared out amongst the players and exploited in 'traditional' competitive fashion. Philips deploys a similar model via its InnoHub, which selects a team from internal and external businesses and staff and covers technology, marketing and other elements. They deliberately encourage fusion of people with varied expertise in the hope that this will enhance the chances of 'breakthrough' thinking.

Their fourth model is called 'mod station', drawing on a term from the personal computer industry which allows users to make modifications to games and other soft and hardware. This is typified by many open source projects such as Sun Microsystems's OpenSPARC or Nokia's recent release of the Symbian operating system to the developer community in an attempt to establish an open platform for the development of mobile applications. It reflects models used by the BBC, by LEGO and many other organizations trying to mobilize external communities and amplify their own research efforts whilst retaining an ability to exploit the new and growing space.

Other models which might be added include Nasa's 'infusion' approach in which a major public agency uses its Innovative Partnerships Programme (IPP) to co-develop key technologies such as robotics. The model is essentially one of drawing in partners who work alongside NASA scientists – a process of 'infusion' in which ideas developed by NASA or by one or more of the partners are worked on. There is particular emphasis on spreading the net widely and seeking partnerships with 'unusual suspects' – companies, university departments and others which might not immediately recognize that they have something of value to offer (Cheeks, 2007).

Source: Nambisan, S. and M. Sawhney (2007) *The Global Brain: Your Roadmap for Innovating Smarter and Faster in a Networked World*, Philadelphia Wharton School Publishing; Cheeks, N. (2007) How NASA uses 'Infusion Partnerships'. *PDMA Visions*, Product Development Management Association: Mount Laurel, 9–12.

activity has begun to emerge. Examples include mainstream design houses like IDEO and ?Whatif!, which help to link clients with new ideas and connections on the technology and market side, technology brokers aiming at match-making between different needs and means (both web-enabled and on a face-to-face basis) and intellectual property transfer agents like the Innovation Exchange, which seek to identify, value and exploit internal IP which may be under-utilized.

6.7 Learning networks

In principle firms have a number of opportunities available to them to enable learning – through experiment (e.g. R&D), through transfer of ideas from outside, through working with different players (suppliers, partners, customers), through reflecting and reviewing previous projects and even from failure. Studies of organizational learning suggest that it can be supported by structures, procedures, etc. to facilitate the operation of the learning cycle, for example, through challenging reflection, facilitated sharing of experiences or planned experimentation.[34,35]

Experience and research suggests that shared learning can help deal with some of the barriers to learning that individual firms might face.[36] For example:

- In shared learning there is the potential for challenge and structured critical reflection from different perspectives.
- Different perspectives can bring in new concepts (or old concepts which are new to the learner).
- Shared experimentation can reduce perceived and actual costs risks in trying new things.
- Shared experiences can provide support and open new lines of enquiry or exploration.
- Shared learning helps explicate the systems principles, seeing the patterns – separating 'the wood from the trees'.
- Shared learning provides an environment for surfacing assumptions and exploring mental models outside of the normal experience of individual organizations – helps prevent 'not invented here' and other effects.
- Shared learning can reduce costs (for example, in drawing on consultancy services and learning about external markets) which can be particularly useful for SMEs and for developing country firms.

A key element in shared learning is the active participation of others in the process of challenge and support. Its potential as an aid to firms trying to cope with a challenging and continuing learning agenda has led to a number of attempts to establish formal arrangements for inter-organizational learning. For example, the experience of regional clusters of small firms, which have managed to share knowledge about product and process technology and to extend the capabilities of the sector as a whole, is recognized as central to their abilities to achieve export competitiveness. In work on supply-chain development there is a growing recognition that the next step after moving from confrontational to cooperative relationships within supply chains is to engage in a process of shared development and learning.[37,38]

Learning is often involved as a 'by-product' of network activities, for example, emerging through exchange of views or through shared attempts at problem solving. But it is also possible to see learning as the primary purpose around which a network is built: this concept of a 'learning network' can be expressed as '*a network formally set up for the primary purpose of increasing knowledge*'. Such networks share a number of characteristics:

- They are formally established and defined.
- They have a primary learning target – some specific learning/knowledge which the network is going to enable.
- They have a structure for operation, with boundaries defining participation.
- Processes which can be mapped on to the learning cycle.
- Measurement of learning outcomes which feeds back to operation of the network and which eventually decides whether or not to continue with the formal arrangement.

The website contains a framework for thinking about the design and operation of learning networks based on a variety of case examples in different sectors and locations.

Examples include 'best practice' clubs (whose members have formed together to try and understand and share experiences about new production concepts), 'co-laboratories' (shared pre-competitive R&D projects), supplier associations and sectoral research organizations (where the aim is to upgrade knowledge across a system of firms). Learning may also involve 'horizontal' collaboration (between like firms) or 'vertical' cooperation (as in supply-chain learning programmes), or a combination of the two.[39]

6.8 Networks into the unknown

Much of the time the challenge in innovation is one of 'doing what we do, but better' – continuously improving products and services and enhancing our processes. The scope here is enormous – both in terms of incremental modifications and additions of features and enhancements and in delivering on cost savings and quality improvements. Taken on their own these may not be as eye-catching as the launch of a radically new product but the historical evidence is that continuous incremental innovation of this kind has enormous economic impact. It's the glacier model rather than the violently fast-running stream – but in the long run the impact on the economic geography is significant.

But as we have seen when discontinuous events occur existing players often perform badly and it is the new entrant firms who succeed. Part of the problem is the commitment to existing networks by established players. Long-term relationships are recognized as powerful positive resources for incremental innovation, but under some circumstances '*the ties that bind may become the ties that blind*'.[6] For example, Christensen showed in his work on disruptive innovation, that when new markets emerge they do so at the fringe of existing ones and are often easy to ignore and dismiss as not being relevant. Under these conditions organizations need a different approach to managing innovation – much more exploratory, and engaged in developing new networks.[40]

Research suggests the challenge facing firms in building new networks can be broken down into two separate activities: identifying the relevant new partners; and learning how to work with them. Once the necessary relationships have been built, they can then be converted into high-performing partnerships. It's a little like the recipe for effective team working (forming, storming, norming and performing) except that here it is a three-stage process: finding, forming and performing.[23]

Finding refers essentially to the breadth of search that is conducted. How easy is it to identify the right organizations with which to interact? Finding is enabled by the scope and diversity of current

TABLE 6.3	**Barriers to new network formation**

Primary objective	Type of barrier	Description
Finding prospective partners	Geographical	Discontinuities often emerge in unexpected corners of the world. Geographical and cultural distance makes complex opportunities more difficult to assess, and as a result they typically get discounted
	Technological	Discontinuous opportunities often emerge at the intersection of two technological domains
	Institutional	Institutional barriers often arise because of the different objectives or origins of two groups, such as those dividing public sector from private sector
Forming relationships with prospective partners	Ideological	Many potential partners do not share the values and norms of the focal firm, which can blind it from seeing the threats or opportunities which might arise at the interfaces between the two world views
	Demographic	Barriers to building effective networks can arise from the different values and needs of different demographic groups
	Ethnic	Ethnic barriers arise from deep-rooted cultural differences between countries or regions of the world

Source: Based on Birkinshaw, J., J. Bessant and R. Delbridge (2007) Finding, forming, and performing: creating networks for discontinuous innovation. *California Management Review*, **49** (3), 67–83.

operations but also by capacity to move beyond the dominant mental models in the industry. It is also hindered by a combination of geographical, technological and institutional barriers (see Table 6.3). Forming refers to the attitude of prospective partners. How likely is a link-up and what are the advantages or barriers?

When these two aspects are set against each other, four separate approaches can be identified.[23] See Figure 6.2.

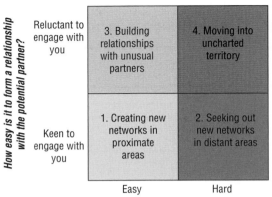

FIGURE 6.2: Four generic approaches to network building

Source: Based on Birkinshaw, J., J. Bessant, and R. Delbridge (2007) Finding, forming, and performing: creating networks for discontinuous innovation. *California Management Review,* **49** (3), 67–83.

Zone 1 represents the relatively straightforward challenge of creating new networks with potential partners that are both easy to find and keen to interact. Although this is where traditional business relationships are formed it also contains examples of uncertain projects even if the partners are known to each other.

For example LEGO's decision to develop its next-generation Mindstorms product involved using a network of lead users of the first-generation product. LEGO's experience after the first Mindstorms product had been that the enthusiastic user community was an asset, despite its approaches such as hacking into the old software and sharing this information on the web. As described by LEGO senior vice president Mads Nipper, '*We came to understand that this is a great way to make the product more exciting. It's a totally different business paradigm.*'

In zone 2 the emphasis is on new network partners. The barriers here are typically geographical, ethnic and institutional, and the challenge is to locate the appropriate organizations from among many prospective partners. It is here that scouts and other boundary-spanning agents can play a key role – as in P&G's connect and develop model.

Zone 3 is where the potential partners are easy to find but may be reluctant to engage. This might occur for ideological reasons, or because of institutional or demographic barriers. An illustration of this approach can be seen in the Danish pharmaceutical company, Novo Nordisk. Faced with long-term changes in the business environment towards greater obesity and rising healthcare costs associated with diabetes (its core market), Novo Nordisk realized that it needed to start exploring opportunities for discontinuous innovation in its products and offerings. Its 'Diabetes 2020' process involved exploring radical alternative scenarios for chronic disease treatment and the roles which a player like Novo Nordisk could play. As part of the follow-up from this initiative, in 2003 the company helped set up the Oxford Health Alliance, a nonprofit collaborative entity which brought together key stakeholders – medical scientists, doctors, patients and government officials – with views and perspectives which were sometimes quite widely separated. To make it happen, Novo Nordisk made clear that its goal was nothing less than the prevention or cure of diabetes – a goal which if it were achieved would potentially kill off the company's main line of business. As Lars Rebien Sørensen, the CEO of Novo Nordisk, explained: '*In moving from intervention to prevention – that's challenging the business model where the pharmaceuticals industry is deriving its revenues! . . . We believe that we can focus on some major global health issue – mainly diabetes – and at the same time create business opportunities for our company.'*

In zone 4 potential partners are neither easily identified nor necessarily keen to engage. One approach is gradually to reduce the reluctance of prospective partners by breaking down the institutional or demographic barriers that separate them – essentially pushing the prospective relationship into zone 2. The example of BBC Backstage offers a good illustration of this approach.

So far we have considered the 'finding' and 'forming' aspects of novel networks – the third question posed is how to make them effectively perform. Challenges in this connection include keeping the network up to date and engaged, building trust and reciprocity, positioning within the network and decoupling from existing networks.

6.9 Managing innovation networks

Throughout the book we have seen the growing importance of viewing innovation as something which needs to be managed at a system level and which is increasingly inter-organizational in nature. The rise of networking, the emergence of small-firm clusters, the growing use of 'open innovation' principles and the globalization of knowledge production and application are all indicators of the move to what Rothwell called a 'fifth-generation' innovation model. This has a number of implications for the ways in which we deal with the practical organization and management of the process.[5]

The basic model which we have been using throughout the chapter is still relevant but the ways in which the different phases are enabled now need to build on an increasing network orientation. For example, networking provides a powerful mechanism for extending and covering a richer selection environment and can bring into play a degree of collective efficiency in picking up relevant signals. Strategies like 'connect and develop' are predicated on the potential offered by increasing the range of connections available to an enterprise.

Configuring innovation networks

Whatever the purpose in setting it up, actually operating an innovation network is not easy – it needs a new set of management skills. Operating an innovation network depends heavily on the type of network and the purposes it is set up to achieve. For example, there is a big difference between the demands for an innovation network working at the frontier where issues of intellectual property management and risk are critical, and one where there is an established innovation agenda as might be the case in using supply chains to enhance product and process innovation. We can map some of these different types of innovation network on to a simple diagram (Figure 6.3) which positions them in terms of:

• how radical the innovation target is with respect to current innovative activity
• the similarity of the participating companies.

Different types of networks have different issues to resolve. For example, in zone 1 we have firms with a broadly similar orientation working on tactical innovation issues. Typically, this might be a cluster or sector forum concerned with adopting and configuring 'good practice' manufacturing. Issues here would involve enabling firms to share experiences, disclose information, develop trust and transparency and build a system-level sense of shared purpose around innovation.

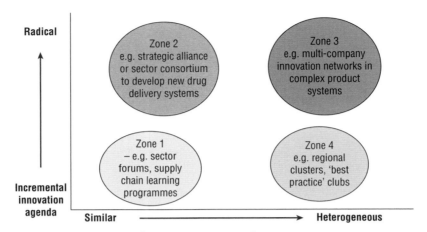

FIGURE 6.3: Types of innovation network

Zone 2 activities might involve players from a sector working to explore and create new product or process concepts, for example, the emerging biotechnology/pharmaceutical networking around frontier developments and the need to look for interesting connections and synthesis between these adjacent sectors. Here, the concern is exploratory and challenges existing boundaries, but will rely on a degree of information sharing and shared risk taking, often in the form of formal joint ventures and strategic alliances.

In Zones 3 and 4, the players are highly differentiated and bring different key pieces of knowledge to the party. Their risks in disclosing can be high so ensuring careful IP management and establishing ground rules will be crucial. At the same time, this kind of innovation is likely to involve considerable risk and so putting in place risk- and benefit-sharing arrangements will also be critical. For example, in a review of 'high-value innovation networks' in the UK, researchers from the Advanced Institute of Management Research (AIM) found the following characteristics were important success factors:[41]

- Highly diverse: network partners from a wide range of disciplines and backgrounds who encourage exchanges about ideas across systems.
- Third-party gatekeepers: science partners such as universities but also consultants and trade associations, who provide access to expertise and act as neutral knowledge brokers across the network.
- Financial leverage: access to investors via business angels, venture capitalists firms and corporate venturing, which spreads the risk of innovation and provides market intelligence.
- Proactively managed: participants regard the network as a valuable asset and actively manage it to reap the innovation benefits.

Learning to manage innovation networks

We have enough difficulties trying to manage within the boundaries of a typical business. So, the challenge of innovation networks takes us well beyond this. The challenges include:

- How to manage something we don't own or control.
- How to see system-level effects not narrow self-interests.
- How to build trust and shared risk taking without tying the process up in contractual red tape.
- How to avoid 'free riders' and information 'spillovers'.

www.managing-innovation.com

It's a new game and one in which a new set of management skills becomes important.

Innovation networks can be broken down into three stages of a life cycle. Table 6.4 looks at some of the key management questions associated with each stage. The website contains an exercise for thinking about the design and operation of learning networks based on a variety of case examples in different sectors and locations.

TABLE 6.4 | **Challenges in managing innovation networks**

Set-up stage	Operating stage	Sustaining (or closure) stage
Issues here are around providing the momentum for bringing the network together and clearly defining its purpose. It may be crisis triggered, e.g., perception of the urgent need to catch up via adoption of innovation. Equally, it may be driven by a shared perception of opportunity – the potential to enter new markets or exploit new technologies. Key roles here will often be played by third parties, e.g., network brokers, gatekeepers, policy agents and facilitators	The key issues here are about trying to establish some core operating processes about which there is support and agreement. These need to deal with: – network boundary management: how the membership of the network is defined and maintained – decision making: how (where, when, who) decisions get taken at the network level – conflict resolution: how conflicts are resolved effectively – information processing: how information flows among members and is managed – knowledge management: how knowledge is created, captured, shared and used across the network – motivation: how members are motivated to join/remain within the network – risk/benefit sharing: how the risks and rewards are allocated across members of the network – coordination: how the operations of the network are integrated and coordinated	Networks need not last forever – sometimes they are set up to achieve a highly specific purpose (e.g. development of a new product concept) and once this has been done the network can be disbanded. In other cases there is a case for sustaining the networking activities for as long as members see benefits. This may require periodic review and 're-targeting' to keep the motivation high. For example, CRINE, a successful development programme for the offshore oil and gas industry, was launched in 1992 by key players in the industry such as BP, Shell and major contractors with support from the UK government with the target of cost reduction. Using a network model, it delivered extensive innovation in product/services and processes. Having met its original cost-reduction targets, the programme moved to a second phase with a focus aimed more at capturing a bigger export share of the global industry through innovation.

Summary and further reading

In this chapter we have looked at the particular challenges in setting up and running networks designed to enable innovation. We have reviewed the different – and often confusing – types and models of networks and tried to focus on what can be termed 'engineered' networks, established and operated specifically to enable innovation. The chapter has looked at networks at the early stages of developing an entrepreneurial idea, at networks within organizations and at the increasingly important theme of external networks, which enable and facilitate the move to more open models of innovation. We have also looked at the particular case of finding, forming and getting new networks with unknown partners to support innovation. Finally we looked at the question of how networks are set up, operated and sustained.

The work of Andrew Hargadon has highlighted the importance of networks and brokers going back to the days of Edison and Ford (*How Breakthroughs Happen*, Harvard Business School Press, 2003). One of the strong examples of this approach today is IDEO, the design consultancy, which Kelley *et al.* have described in detail in *The Art of Innovation: Lessons in Creativity from IDEO, America's Leading Design Firm* (Currency, 2001). Conway and Steward (Mapping innovation networks, *International Journal of Innovation Management*, **2** (2), 165–96, 1998) look at the concept of innovation networks and this theme is also picked up by Swan *et al.* (Knowledge management and innovation: networks and networking, *Journal of Knowledge Management*, **3** (4), 262, 1999). Learning networks are discussed in Bessant and Tsekouras (Developing learning networks, *A.I. and Society*, **15** (2), 82–98, 2001) and their use in sectors, supply chains and regional clusters in Morris *et al.* (Using learning networks to enable industrial development, *International Journal of Operations and Production Management*, **26** (5), 557–68, 2006). High-value innovation networks are discussed in several reports from AIM – the Advanced Institute for Management Research (**www.aimresearch.org**).

Web links

Here are the full details of the resources available on the website flagged throughout the text:

 Case studies:
Learning networks
Zara
Freeplay radio
Lego

 Interactive exercises:
Setting up and running networks

 Tools:
Guide to building learning networks

 Audio podcast:
Roy Sandbach – P&G
David Overton – Ordnance Survey

References

1. **Guilford, J.** (1967) *The Nature of Human Intelligence*, McGraw-Hill, New York.
2. **Carter, C. and B. Williams** (1957) *Industry and Technical Progress*, Oxford University Press, Oxford.
3. **Allen, T.** (1977) *Managing the Flow of Technology*, MIT Press, Cambridge, MA.
4. **Langrish, J., M. Gibbons, W. Evans, and F. Jerons** (1972) *Wealth from Knowledge*, Macmillan, London.
5. **Rothwell, R.** (1977) The characteristics of successful innovators and technically progressive firms. *R&D Management*, 7 (3), 191–206.
6. **Hargadon, A.** (2003) *How Breakthroughs Happen*, Harvard Business School Press, Boston, MA.
7. **Jaikumar, R.** (1988) *From Filing and Fitting to Flexible Manufacturing*, Harvard Business School Press, Boston, MA.
8. **Williams, D.** (2004) *The Birmingham Gun Trade*, Tempus, Stroud.
9. **Wenger, E.** (1999) *Communities of Practice: Learning, Meaning, and Identity*, Cambridge University Press, Cambridge.
10. **Rothwell, R.** (1992) Successful industrial innovation: critical success factors for the 1990s. *R&D Management*, **22** (3), 221–39.
11. **DeBresson, C. and F. Amesse** (1991) Networks of innovators: a review and introduction. *Research Policy*, **20**, 363–79.
12. **Camagni, R.** (1991) *Innovation Networks: Spatial Perspectives*, Belhaven Press, London.
13. **Nohria, N. and R. Eccles** (1992) *Networks and Organizations: Structure, Form and Action*, Harvard Business School Press, Boston, MA.
14. **Gulati, R.** (1998) Alliances and networks. *Strategic Management Journal*, **19**, 293–317.
15. **Bidault, F. and W. Fischer** (1994) Technology transactions: networks over markets. *R&D Management*, **24** (4), 373–86.
16. **Hakansson, H.** (1995) Product development in networks. In D. Ford, ed., *Understanding Business Markets*, The Dryden Press, New York.
17. **Nakateni, L.** (1984) The economic role of financial corporate groupings. In M. Aoki, ed., *The Economic Analysis of the Japanese Firm*, North Holland, Amsterdam.
18. **Belussi, F. and F. Arcangeli** (1998) A typology of networks: flexible and evolutionary firms. *Research Policy*, **27**, 415–28.
19. **Galaskiewicz, J.** (1996) The 'new' network analysis. In D. Iacobucci, ed., *Networks in Marketing*, Sage, London.
20. **Conway, S. and F. Steward** (1998) Mapping innovation networks. *International Journal of Innovation Management*, **2** (2), 165–96.
21. **Hooi-Soh, P. and E. Roberts** (2003) Networks of innovators: a longitudinal perspective. *Research Policy*, **32**, 1569–88.
22. **Passiante, G. and P. Andriani** (2000) Modelling the learning environment of virtual knowledge networks. *International Journal of Innovation Management*, **4** (1), 1–31.
23. **Birkinshaw, J., J. Bessant and R. Delbridge** (2007) Finding, forming, and performing: creating networks for discontinuous innovation. *California Management Review*, **49** (3), 67–83.
24. **Bessant, J., M. Morris, and R. Kaplinski** (2003) Developing capability through learning networks. *International Journal of Technology Management and Sustainable Development*, **2** (1).

25. **Morris, M., J. Bessant, and J. Barnes** (2006) Using learning networks to enable industrial development: case studies from South Africa. *International Journal of Operations and Production Management*, **26** (5), 557–68.

26. **Garnsey, E. and E. Stam** (2008) Entrepreneurship in the knowledge economy. In J. Bessant and T. Venables, eds, *Creating Wealth from Knowledge*, Edward Elgar, Cheltenham.

27. **Best, M.** (2001) *The New Competitive Advantage*, Oxford University Press, Oxford.

28. **Humphrey, J. and H. Schmitz** (1996) The Triple C approach to local industrial policy. *World Development*, **24** (12), 1859–77.

29. **Brown, J.S. and P. Duguid** (2000) Knowledge and organization: a social-practice perspective. *Organization Science*, **12** (2),198.

30. **Chesbrough, H.** (2003) *Open Innovation: The New Imperative for Creating and Profiting From Technology*, Harvard Business School Press, Boston, MA.

31. **Bessant, J. and T. Venables, eds** (2008) *Creating Wealth from Knowledge: Meeting the Innovation Challenge*, Edward Elgar, Cheltenham.

32. **Lafley, A. and R. Charan** (2008) *The Game Changer*, Profile, New York.

33. **Dahlander, L. and D. Gann** (2008) How open is innovation? In J. Bessant and T. Venables, eds, *Creating Wealth from Knowledge: Meeting the Innovation Challenge*, Edward Elgar, Cheltenham.

34. **Argyris, C. and D. Schon** (1970) *Organizational Learning*, Addison Wesley, Reading, MA.

35. **Phelps, R., R.J. Adams and J. Bessant** (2007) Models of organizational growth: a review with implications for knowledge and learning. *International Journal of Management Reviews*, **9** (1), 53–80.

36. **Bessant, J. and G. Tsekouras** (2001) Developing learning networks. *A.I. and Society*, **15** (2), 82–98.

37. **Brown, J.E. and C. Hendry** (1998) Industrial districts and supply chains as vehicles for managerial and organizational learning. *International Studies of Management and Organization*, **27** (4), 127.

38 **Bessant, J.** (2004) Supply chain learning. In R. Westbrook and S. New, eds, *Understanding Supply Chains: Concepts, Critiques, Futures*, Oxford University Press, Oxford.

39. **Dyer, J. and K. Nobeoka** (2000) Creating and managing a high-performance knowledge-sharing network: the Toyota case. *Strategic Management Journal*, **21** (3), 345–67.

40. **Christensen, C.** (1997) *The Innovator's Dilemma*, Harvard Business School Press, Boston, MA.

41. **AIM** (2004) *i-works: How High-Value Innovation Networks can Boost UK Productivity*, ESRC/EPSRC Advanced Institute of Management Research, London.

PART 4

Select

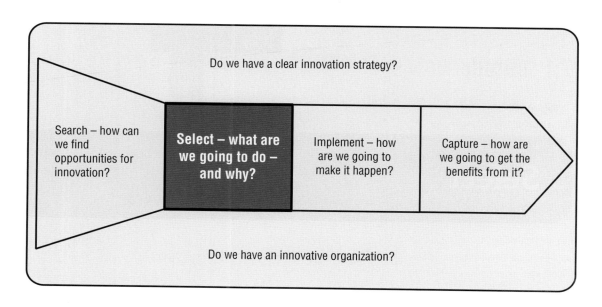

In this section we move into the area of selection in the core process model. Chapter 7 looks at how the innovation decision process works – of all the possible options generated by effective search which ones will we back – and why? Making decisions of this kind is not simple because of the underlying uncertainty involved – so which approaches, tools and techniques can we bring to bear? Chapter 8 picks up another core theme – how to build an innovation plan. This demands an understanding of the dynamics of forecasting and diffusion of market and technological innovations, and the assessment of the risk and resources involved.

CHAPTER 7

Decision making under uncertainty

7.1 Introduction

Triggers for innovation – as we saw in Chapter 5 – can be found all over the place. The world is full of interesting and challenging possibilities for change – the trouble is that even the wealthiest organization doesn't have deep enough pockets to do them all. Sooner or later it has to confront the issue of 'out of all the things we could do, what are we going to do?' This isn't easy – making decisions is about resource commitment and so choosing to go in one direction closes off opportunities elsewhere. Organizations cannot afford to innovate at random – they need some kind of framework which articulates how they think innovation can help them survive and grow, and they need to be able to allocate scarce resources to a portfolio of innovation projects based on this view. This underlines the importance of developing an innovation strategy – a theme we explored in Chapter 4.

But in a complex and uncertain world it is nonsense to think that we can make detailed plans ahead of the game and then follow them through in systematic fashion. Life – and certainly organizational life – isn't like that. So our strategic framework for innovation should be flexible enough to help monitor and adapt projects over time as ideas move towards more concrete solutions – and rigid enough to justify continuation or termination as uncertainties and risky guesswork become replaced by actual knowledge.

The challenge of innovation decision making is made more complex by the fact that it isn't a simple matter of selecting amongst clearly defined options. By its nature innovation is about the unknown, about possibilities and opportunities associated with doing something new and so the process involves dealing with *uncertainty*. The problem is that we don't know in advance if an innovation will work – will the technology actually do what we hope, will the market still be there and behave as we anticipated, will competitors move in a different and more successful direction, will the government change the rules of the game, and so on? All of these are uncertain variables which makes our act of decision making a little like driving in the fog. The only way we can get more certainty is by starting the project and learning as we go along. So making the initial decision – and the subsequent ones about whether to keep going or cut our losses and move in a different direction – becomes a matter of calculating as best we can the risks associated with different options. In this chapter we'll explore some of the ways in which organizations deal with this difficult area of decision making under uncertainty.

7.2 Meeting the challenge of uncertainty

What distinguishes innovation management from gambling? Both involve committing resources to something which (unless the game is rigged) has an uncertain outcome. But innovation *management* tries to convert that uncertainty at the outset to something closer to a calculated risk – there is still no guarantee of success but at least there is an attempt to review the options and assign some probabilities as to the chances of a successful outcome. This isn't simply a mechanical process – first the assessment of risk is still based on very limited information and secondly there is a balance between the risks involved and the potential rewards which might follow if the innovation project is successful.

Some 'bets' are safer than others because they carry lower risk – incremental innovation is about doing what we do – and therefore know about – better. We have some prior knowledge about markets, technologies, regulatory frameworks, etc. and so can make reasonably accurate assessments of risks using this information. But some bets are about radical innovation, doing something completely different and carrying a much higher level of risk because of the lack of information. These could pay off handsomely – but there are also many unforeseen ways in which they could run into trouble.

And we shouldn't forget that under such conditions decision making is often shaped by emotional forces as well as limited facts and figures. The economist John Maynard Keynes famously pointed out the important role which 'animal spirits' play in shaping decisions.* People can be persuaded to take a risk by a convincing argument, by expressions of energy or passion, by hooking into powerful emotions like fear (of not moving in the proposed direction) or reward (resulting from the success of the proposed innovation).[1]

7.3 The funnel of uncertainty

Central to this process is *knowledge* – this is what converts uncertainty to risk. The more we know about something, the more we can take calculated decisions about whether or not to proceed. And in a competitive environment this puts a premium on getting hold of knowledge as early as possible – this explains the value of an insider tip-off in horse racing or stock-market dealings. In innovation management the challenge is to invest in acquiring early knowledge – through technological R&D, market research, competitor analysis, trend spotting and a host of other mechanisms – to get early information to feed decision making. Robert Cooper uses the powerful metaphor of 'Russian roulette', suggesting that most people when faced with the uncertainty of pulling the trigger would be happy to 'buy a look' at the gun chamber to improve their knowledge of whether or not there is a bullet in it![2]

Thinking of innovation as a process of reducing uncertainty but increasing resource commitment gives us a classic graph (Figure 7.1) – the further into a project we go the more it costs but the more we know.

In practice this translates into what we can call the 'innovation funnel' – a roadmap which helps us make (and review) decisions about resource commitment. Figure 7.2 gives an illustration.

At the outset anything is possible, but increasing commitment of resources during the life of the project makes it increasingly difficult to change direction. Managing innovation is a fine balancing act, between the costs of continuing with projects which may not eventually succeed (and which represent opportunity costs in terms of other possibilities) and the danger of closing down too soon and eliminating potentially fruitful options. Taking these decisions can be done on an ad hoc basis but experience suggests some form of structured development system with clear decision points and agreed rules on which to base go/no-go decisions, is a more effective approach.[3]

---

* 'The thought of ultimate loss which often overtakes pioneers, as experience undoubtedly tells us and them, is put aside as a healthy man puts aside the expectation of death.'

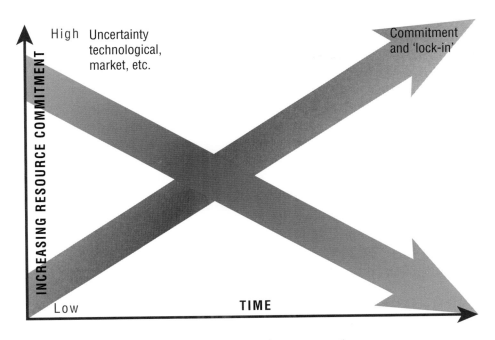

FIGURE 7.1: Increasing innovation commitment over time

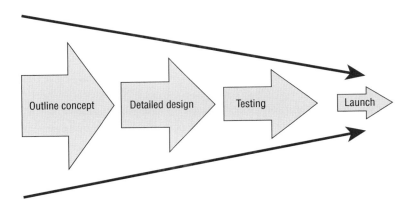

FIGURE 7.2: The innovation funnel

Given this model it makes sense not just to make one big decision to commit everything at the outset when uncertainty is very high but instead to make a series of stepwise decisions. Each of these involves committing more resources, but this only takes place if the risk/reward assessment justifies it – and the further into the project, the more information about technologies, markets, competitors, etc. we have to help with the assessment. We move from uncertainty to increasingly well-calculated risk management. Such a staged review process is particularly associated with the work of Robert Cooper, a Canadian researcher who studied thousands of new product development projects.[4]

This model essentially involves putting in a series of gates at key stages and reviewing the project's progress against clearly defined and accepted criteria. Only if it passes will the gate open – otherwise the project should be killed off or at least returned for further development work before proceeding. Many variations (e.g. 'fuzzy gates') on this approach exist; the important point is to ensure that there is a structure in place which reviews information about both technical and market aspects of the innovation as we move from high uncertainty to high resource commitment but gain a clearer picture of progress. We will explore this 'stage-gate' approach – and variations on that – in Chapter 9.

Models of this kind have been widely applied in different sectors, both in manufacturing and services.[5,6] We need to recognize the importance here of configuring the system to the particular contingencies of the organization, for example, a highly procedural system which works for a global multi-product company like Siemens or GM will be far too big and complex for many small organizations. And not every project needs the same degree of scrutiny – for some there will be a need to develop parallel 'fast tracks' where monitoring is kept to a light touch to ensure speed and flow in development.

We also need to recognize that the effectiveness of any stage-gate system will be limited by the extent to which it is accepted as a fair and helpful framework against which to monitor progress and continue to allocate resources.[7] This places emphasis on some form of shared design of the system – otherwise there is a risk of lack of commitment to decisions made and/or the development of resentment at the progress of some 'pet' projects and the holding back of others. The ABC Electronics case study shows how to design an effective stage-gate system which people buy into.

7.4 Decision making for incremental innovation

When we are deciding about incremental innovation – essentially doing what we already do but better – the process of deciding is (relatively) straightforward. Since this involves comparing something new with something that already exists we can set up criteria and measure against these – both at the outset and during progression of the project through our funnel. Systematic decision making of this kind is common in product development systems, which are discussed in detail in Chapter 9. Whilst risks are involved these can be calculated and relevant information collected to help guide judgement in a (relatively) mechanistic fashion. This is where stage-gate systems become powerful tools for innovation management – the Coloplast case gives an example. The website offers some examples of stage-gate tools and approaches to managing risk in innovation projects.

CASE STUDY 7.1

Accelerating ideas to market – the AIM process

Coloplast is a Danish company involved in the manufacture of a wide range of medical products. Its stage-gate process is called AIM and the basic structure is given in the diagram below:

AIM's purpose can be expressed as being:

- To provide common rules of the game for product development within Coloplast
- To make clear decisions at the right moment
- To clarify responsibility

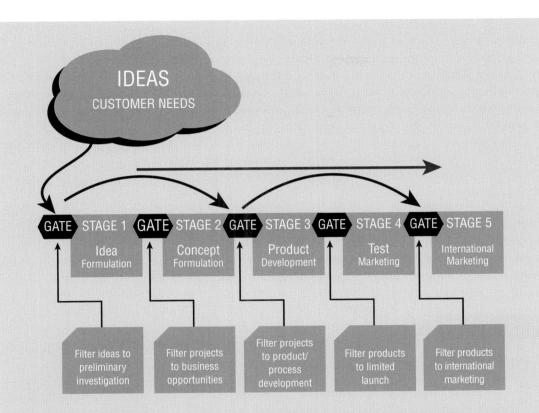

The objective of the AIM process is to ensure a high, uniform level of professionalism in product development yielding high quality products. It is based on the view that Coloplast must increase the success rate and reduce the development time for new products in order to become a 'world-class innovator'.

The stage-gate system

Much of the work in product development is carried out by project teams consisting of selected specialists from marketing (product divisions and subsidiaries), R&D, clinical affairs and manufacturing. Each project team works under the leadership of a skilled and enthusiastic project manager, and the AIM process defines the rules to be followed by the project team.

The AIM process divides the development of new products into five manageable 'stages'. Each stage contains a number of parallel and coordinated activities designed to refine the definitions of customer needs and to develop technological solutions and capacity for efficient manufacturing.

Each stage is followed by a 'gate', a decision point at which the project is reviewed by the 'gatekeepers', senior managers with authority to keep worthy projects moving ahead quickly. The gates serve as the critical quality control checkpoints between the stages. A 'go' decision is made when the gatekeepers decide that a project is likely, technically and economically, to meet the needs of the customers as well as to comply with Coloplast's high standards for return on investment, quality and environmental impact.

One area where systematic management of incremental innovation becomes important is in 'high-involvement' systems where a large proportion of the workforce becomes engaged in innovation.[8–10] Such *kaizen* or continuous improvement activities can have a significant cumulative effect – as we saw in Chapter 1. But there is a problem – if we are successful in persuading most of the workforce to make innovation proposals, then how will we manage the volume of ideas that result? (To put this in perspective many firms with a strong tradition of high-involvement innovation, for example Toyota, Kawasaki or Matsushita, receive several million suggestions per year.[11])

The solution to this is to make use of approaches which have been termed 'policy deployment' (sometimes called '*hoshin* planning') – essentially devolving the top-level innovation strategy to lower levels in the organization and allowing people at those levels to make the decisions. This provides a strategic focus within which they can locate their multiple small-scale innovation activities. But it requires two key enablers: the creation of a clear and coherent strategy for the business; and the deployment of it through a cascade process which builds understanding and ownership of the goals and subgoals.[10,12]

Policy deployment – which we discussed in Chapter 4 – is a characteristic feature of many Japanese *kaizen* systems. Case study 7.2 gives an example and there are other more detailed cases on the website.

CASE STUDY 7.2

Policy deployment in action

This is a major Japanese producer of fork-lift trucks and related machinery and at this particular plant they employ around 900 staff producing three main product lines: industrial trucks, construction equipment and other new products.

Strategy is now focused on the 'Aggressive 30' programme, reflecting the 30 years since the plant was set up, and total productive maintenance (TPM) and indirect cost reduction are the key themes. Typical targets within the plan are:

- 1.5 times increase in overall productivity
- Breakdown reduced to 10% of current levels
- Streamline production flow by 30%
- Reduction in new product development/introduction time of 50%

To deliver these they have a nine-pillar structure to the programme within which the total cost of waste is calculated and broken down into 46 areas, each of which becomes the target for improvement activity.

Kaizen operates in both top-down and bottom-up modes. Each work group studies its 'waste map' and identifies a series of projects which are led by section managers. Each section has specific targets to achieve, for example, increase machine availability from 49% to 86%, or cut work in progress from 100 to 20 vehicles.

Each waste theme is plotted on a matrix, with the other axis being a detailed description of the types and nature of waste arising. This matrix gives a picture of the project targets which are then indicated by a red (= unsolved) or a green (= solved) dot. Importantly projects completed in one year can be revisited and the targets increased in subsequent years to drive through continuous improvement.

Making things visible is a key theme – the use of the matrix charts with their red and green dots everywhere is a constant reminder of the overall continuous improvement programme. Also each project, as it is completed, is painted a shocking pink colour so that it is clear on walking through the factory where and what has been done – often sparking interest and application elsewhere but at least reminding on a continuing basis.

7.5 Building the business case

Even though projects may be incremental in nature and build on established experience, the development and presentation of a persuasive business case is important and much can be done with tools and techniques to explore and elaborate the core concept. The purpose here is to move an outline idea to something with clearer shape and form on which decisions about resource commitments can be made. As we move to more radical innovation projects – which are by definition higher risk – so the 'business case' needs to be more strongly made and to mobilize both emotional and factual components to secure 'buy-in' from decision makers.

Tools for helping here include simulation and prototyping, for example, in introducing new production management software a common practice is to 'walk through' the operation of core processes using computer and organizational simulation. Major developments in recent years have expanded the range of tools available for this exploration in ways which allow much higher levels of experimentation without incurring time or cost penalties. Dodgson, Gann and Salter use the phrase 'Think, Play, Do' to describe the innovation process and argue that, under intensifying pressure to improve efficiency and effectiveness, innovation practitioners have adopted a wide range of powerful tools to enable an extended 'play' phase and to postpone final commitment until very late in the process. Examples of such tools include advanced computer modelling which allows for simulation and large-scale experiments, rapid prototyping which offers physical representations of form and substance, and simulation techniques which allow the workings of different options to be explored.[13]

We will explore this theme of building a business case more extensively in Chapter 8.

7.6 Building coalitions

Despite the presence of formal decision-making structures choices about which options to select are subjective in nature – leading to political and other behaviours.[14,15] Many of the problems in product and process innovation arise from the multifunctional nature of development and the lack of a shared perspective on the product being developed and/or the marketplace into which it will be introduced. A common problem is that 'X wasn't consulted, otherwise they would have told you that you can't do that. . .'. This places a premium on involving all groups at the earliest possible, i.e. the concept definition/product specification stage. Several structured approaches now exist for managing this, including quality function deployment (QFD) and functional mapping.[16]

The availability of simulation technology, especially computer-aided design, has helped facilitate this kind of early discussion and refinement of the concept. In process innovation early involvement of key users and the incorporation of their perspectives are strongly associated with improved overall performance and also with acceptability of the process in operation. This methodology has had a strong influence on, for example, the implementation of major integrated computer systems, which by definition cut across functional boundaries.[17,18]

Extending this idea of early involvement in concept development, an increasingly important routine is to bring suppliers of components and subsystems into the discussion. Their specialist expertise can often provide unexpected ways of saving costs and time in the subsequent development and production process. Increasingly, product development is being treated as a cooperative activity in which networks of players, each with a particular knowledge set, are coordinated towards a shared objective. Examples include automotive components, aerospace and electronic capital equipment, all of which make growing use of formal supplier involvement programmes.[19,20]

In process innovation this has also proved to be an important routine – both in order to secure acceptance (where early involvement of users is now seen as critical) and also in obtaining improved quality process design.[21] Interaction with outsiders also needs to take account of external regulatory frameworks, for example in product standards, environmental controls, safety legislation. Concept testing can be helped by close involvement with and participation in organizations which have responsibilities in these areas. A variety of tools for managing this process of securing involvement and input can be found with explanations and case descriptions on the website. And for products and services a central element in concept testing is the involvement and engagement of end users/customers themselves.

CASE STUDY 7.3

Concept testing a new supermarket approach

The UK supermarket giant Tesco has been engaged in considerable international expansion in recent years. A major challenge in 2006 was the decision to enter the highly competitive US market which it has done with its 'Fresh and Easy' chain of stores. However, it was concerned about the high failure rate of UK firms entering US retailing markets and so carried through extensive concept testing to ensure the format was appropriate. The strategy was one of 'probe and learn' with many modifications being made as a result of lessons learned from reactions to the concept test.

As part of the process the company made extensive use of ethnographic tools to try and understand how their target group of shoppers lived. As their corporate affairs director put it, *'spending time with people in their houses, looking in their cupboards and fridges and actually shopping with them is a great way to understand the market'* (*The Times*, 3 September 2006). The team of 'knowledge activities' responsible for this work developed the store concept and explored it via extensive trialling. To maintain secrecy senior executives from the company posed as film moguls and built a film set inside a West Coast warehouse. Members of the team posed as 'International

Research Resource Ltd', with its own website and other trappings, and then used this as a cover for building and stocking a complete mock-up of a potential Fresh and Easy store inside the warehouse. This gave them the opportunity to test a range of possible formats, layouts and product range; the team even prepared meals and fresh fruit dishes ready to 'sell' in their dummy store (*Business Week*, 27 February 2006). A more detailed analysis of this case is available on the website.

Source: Based on Michelle Lowe, Aspects of innovation in Tesco plc's market entry into the USA, AIM Working Paper, 2008.

7.7 Spreading the risk – building a portfolio

Even the smallest enterprise is likely to have a number of innovation activities running at any moment. It may concentrate most of its resources on its one major product/service offering or new process, but alongside this there will be a host of incremental improvements and minor change projects which also consume resources and require monitoring. For giant organizations like Procter & Gamble or 3M the range of products is somewhat wider – in 3M's case around 60 000. With pressures on increasing growth through innovation come challenges like 3M's '30% of sales to come from products introduced during the past three years' – implying a steady and fast-flowing stream of new product/service ideas running through, supported by other streams around process and position innovation. Even project-oriented organizations whose main task might be the construction of a new bridge or office block will have a range of subsidiary innovation projects running at the same time.

As we have seen, the innovation process has a funnel shape with convergence from a wide mouth of possibilities into a much smaller section which represents those projects to which resources will be committed. This poses the question of *which* projects and the subsidiary one of ensuring a balance between risk, reward, novelty, experience and many other elements of uncertainty. The challenge of building a portfolio is as much an issue in non-commercial organizations – for example, should a hospital commit to a new theatre, a new scanner, a new support organization around integrated patient care, or a new sterilization method? No organization can do everything so it must make choices and try to create a broad portfolio which helps with both the 'do what we do better' and the 'do different' agenda.

There are a variety of approaches which have developed to deal with the question of what is broadly termed 'portfolio management'. These range from simple judgements about risk and reward to complex quantitative tools based on probability theory. But the underlying purpose is the same – to provide a coherent basis on which to judge which projects should be undertaken, and to ensure a good balance across the portfolio of risk and potential reward. Failure to make such judgements can lead to a number of problem issues, as Table 7.1 indicates.

Portfolio methods try to deal with the issue of reviewing across a set of projects and look for a balance of economic and non-financial risk/reward factors. Chapter 8 discusses portfolio methods and tools in more detail. Some examples of portfolio management tools and matrices are given on the website.

TABLE 7.1 — Problems arising from poor portfolio management	
Without portfolio management there may be	**Impacts**
No limit to projects taken on	Resources spread too thinly
Reluctance to kill-off or 'de-select' projects	Resource starvation and impacts on time and cost – overruns
Lack of strategic focus in project mix	High failure rates, or success of unimportant projects and opportunity cost against more important projects
Weak or ambiguous selection criteria	Projects find their way into the mix because of politics or emotion or other factors – downstream failure rates high and resource diversion from other projects
Weak decision criteria	Too many 'average' projects selected, little impact downstream in market

Source: Based on Cooper, R. (1988) The new product process: a decision guide for management. *Journal of Marketing Management*, **3** (3), 238–55.

7.8 Decision making at the edge

When the innovation decision is about incremental – 'do what we do but better' – innovation there is relatively little difficulty. A business case with requisite information can be assembled, cost benefits can be argued and the 'fit' with the current portfolio demonstrated. But as the options move towards the more radical end so the degree of resource commitment and risk rises and decision making resembles more closely a matter of placing bets – and emotional and political influences become significant. At the limit the organization faces real difficulties in making choices about new trajectories – in moving 'outside the box' in which its prior experience and the dominant technological and market trajectories place it.[22–24] The website has an exercise designed to explore the challenges of discontinuous innovation.

Under such 'discontinuous' conditions – triggered, for example, by the emergence of a radical new technology or the emergence of a new market, or a shift in the regulatory framework – established incumbents often face a major challenge. Heuristics and internal rules for resource allocation are unhelpful and may actively militate against placing bets on the new options because they are far outside the firm's 'normal' framework. As Christensen argues, in his studies of disruption caused by emergence of new markets, the incumbent decision-making and underlying reward and reinforcement systems strongly favour the status quo, working with existing customers and suppliers. Such bounded decision making creates an opportunity for new entrants to colonize new market space – and then migrate towards incumbents' territory.[25] In similar fashion Henderson and Clark argue that shifting to new

Reframing the lighting business within Philips

Philips Lighting has been a dominant player in the industry for a long time but the business model has increasingly become one of managing a commodity. Recent changes include the emergence of solid-state lighting and the development of a new approach – a reframing of the business towards seing lighting as an 'atmosphere provider'. In the podcast and accompanying case on the website Dorothea Seebode and Gerard Harkin describe how the transition in the underlying mindset was managed when faced with radical innovation.

'architectures' – new configurations involving new knowledge sets and their arrangements – poses problems for established incumbents.[26]

Selection and reframing

A key part of this challenge lies in the difficulties organizations face with 'reframing' – viewing the world in different ways and changing the ways they make selection decisions as a result. Human beings cannot process all the rich and complex information coming at them and so they make use of a variety of simplifying frameworks – mental models – with which to make sense of the world. And the same is true for organizations – as collections of individuals they construct shared mental models through which the complex external world is experienced.[27] Of necessity such models are simplifications, for example, business models which provide lenses through which to make sense of the environment and guide strategic behaviour.

The problem with discontinuous innovation is that it presents challenges which do not fit the existing model and require a *reframing* – something that existing incumbents find hard to do. In a process akin to what psychologists call 'cognitive dissonance' in individuals, organizations often selectively perceive and interpret the new situation to match or fit their established world views. Since by definition discontinuous shifts usually begin as weak signals of major change, picked up on the edge of the radar screen, it is easy for the continuing interpretation of the signals in the old frame to persist for some time. By the time the disconnect between the two becomes apparent and the need for radical reframing is unavoidable it is often too late. As Dorothy Leonard-Barton puts it, core competencies become core rigidities.[28]

Igniting radical innovation

Cerulean is the market leader in electronic instrumentation for the tobacco industry and specialized tube-packing equipment. It's been around for about 40 years. It has a long history of incremental product improvement and the Quality Test Module (QTM) has been its core product for about 10 to 15 years. About four or five years ago, we got to the point where it was clear that that product was starting to run out of steam. We had been very good at incrementing that, improving it. It had several relaunches over the course of the 10 to 15 years, but we felt that we wanted to move beyond that. We wanted something new, something different. And we then set about how we were going to create a product that was different from what had gone before and that resulted in the innovation project.

The way we saw it was the incremental will run in the background. We're good at that. We do an awful lot of work with it. We had taken steps to improve our new product introduction process; we had stage-gate project management. We'd reviewed projects. We had a review process right at the beginning to look at new products. We reviewed them at each stage. If necessary, some of them get killed along the way. We track costs; we track product costs and we track project costs. So, it's fairly well managed. And that, if you like, was the underlying project management that we built up. What we wanted to do was to create something that sat on top of that, perhaps distinct, perhaps running separately, but perhaps slightly interwoven with it, that would allow us to do different projects, that would allow us to create projects that weren't an enhancement of what had gone before, but were something new, either coming from outside or coming from internal ideas. We had tried that once in the past, about seven or eight years ago, and the idea came from us going to an outside consultancy and paying a substantial amount of money for an idea that really should have come from within the company. All the consultancy did was talk to different people, play back those ideas in a form that would use a suitable product. I felt that we've got the raw materials; we've got the fuel for the product ideas within the business. We just need a mechanism to focus that and bring it to fruition as a project proposal, and the radical project, the radical innovation project, came from that. It wasn't intended to replace incremental; it was intended to, as I say, either sit alongside, on top or be interwoven with it. There was a feeling, I think initially, that we could plug them both into the same process and get a common output, but that quickly became apparent that that wasn't going to work.

(Interviewer: So, in a sense, you've got a problem with trying to create two different cultures: one that's there supporting incremental innovation; and a new one which, as you say, may sit alongside and may be a little separate, but which is about doing something rather different. Can you do it with the same people?)

Yes, you can . . . . . . those people have to be managed in a way that allows them to do things differently. One thing we didn't want to do was to lose our ability to do the incremental. We had continuous improvement, we had continuous development of our projects or products, and we wanted to retain that. But, at the same time, we wanted to be able to use that group of people to take ideas that had come from . . . ideas within the company, ideas from outside, and perhaps outside the industry, and say, right, here's a suitable product. And we didn't want to create something that sat outside. It would have been nice, but we're not big enough to have a skunk works. Also, we felt that if it was too remote, it became too detached. We're not in a position where we can do speculative development that might lead to something six or seven years down the road. We're a small business, we're relatively profitable, and we need to retain that profitability. And to retain that momentum, we needed this additional feature of two different products starting to flow through. We needed to revitalize the company and regain the reputation we had for being an innovation company.

Patrick McLaughlin, Managing Director, Cerulean

The full version of this video interview plus transcript is available on the website.

The case of Polaroid highlights the difficulty – an otherwise technologically successful company which had opened up the market for instant photography and had a strong reputation over 40 years suddenly found itself in Chapter 11 bankruptcy at the turn of the twenty-first century. According to Tripsas and Gavetti its difficulties in adapting to digital imaging *were mainly determined by the cognitive inertia of its corporate executives. As we have documented, managers directly involved with digital imaging developed a highly adaptive representation of the emerging competitive landscape. We speculate that the cognitive dissonance between senior management and digital imaging managers may have been exacerbated by the difference in signals that the two groups were receiving about the market.*[22] Bhide[29] and Christensen[25] support this view: that it is often the self-imposed barriers caused by inability to reframe which pose problems for established players. Both found that employees at incumbent companies often generated the ideas that went on to form the basis of discontinuous technologies. However, these were exploited and developed by competitors, or new organizations, and consequently adversely affected the incumbent.

The problem is not that such firms have weak or ineffective strategic resource allocation mechanisms for taking innovation decisions – but rather that these are too good. For as long as the decisions are taken within a framework – their 'box' – they are effective, but they break down when the challenge comes from outside that box. It is important to recognize that the justification for rejecting ideas that lie too far outside the framework is expressed in terms which are apparently 'rational' – that is, the reasons are clear and consistent with the decision rules and criteria associated with the framework. But they are examples of what the Nobel Prize-winning economist Herbert Simon called 'bounded rationality' – and underpinning them are a number of key psychological effects such as 'groupthink' and risky shift.[23]

Much of the difficulty in radical or discontinuous innovation selection arises from this framing problem. As Henderson and Clark point out innovation rarely involves dealing with a single technology or market but rather a bundle of knowledge, which is brought together into a configuration. Successful innovation management requires that we can get hold of and use knowledge about *components* but also about how those can be put together – what they termed the *architecture* of an innovation.[26] And the problem is that we are often unable to imagine alternative configurations, new and different architectures. In similar fashion Dosi uses the term 'paradigm' to describe the mental framework at a system level within which technological progress takes place,[30] whilst Abernathy and Utterback highlight the key role of the 'dominant design' in moving innovation from an experimental 'fluid' phase to a 'specific' and focused one within which firms follow similar pathways.[31] Markides[32] talks about 'strategic innovation': *'a fundamental re-conceptualization of what the business is all about that, in turn, leads to a dramatically different way of playing the game in an existing business.'* And Hamel[33] suggests the idea of business concept innovation that can be defined as *'the capacity to reconceive existing business models in ways that create new value for customers, rude surprises for competitors, and new wealth for competitors.'*

When there is a shift to a new mindset established players may have problems because of the required reorganization of their thinking. It is not simply adding new information but changing the structure of the frame through which they see and interpret that information. They need to 'think outside the box' within which their bounded exploration takes place – and this is difficult because it is highly structured and reinforced by organizational structures and processes.

So, for example, the famous 'not-invented-here' rejection is easier to understand if we see it as a problem of what makes sense within a specific context – the firm has little knowledge or experience in the proposed area, it is not its core business, it has no plans to enter that particular market, etc. Table 7.2 lists some examples of justifications which can be made to rationalize the rejection decision associated with radical innovation options.

> **TABLE 7.2** **Examples of justifications for non-adoption of radical ideas**

Argument	Underlying perceptions from within the established mental model
'It's not our business'	Recognition of an interesting new business idea but rejection because it lies far from the core competence of the firm
'It's not a business'	Evaluation suggests the business plan is flawed along some key dimension – often underestimating potential for market development and growth
'It's not big enough for us'	Emergent market size is too small to meet growth targets of large established firm
'Not invented here'	Recognition of interesting idea with potential but reject it – often by finding flaws or mismatch to current internal trajectories
'Invented here'	Recognition of interesting idea but rejection because internally generated version is perceived to be superior
'We're not cannibals'	Recognition of potential for impact on current markets and reluctance to adopt potential competing idea
'Nice idea but doesn't fit'	Recognition of interesting idea generated from within but whose application lies outside current business areas – often leads to inventions being shelved or put in a cupboard
'It ain't broke so why fix it'	No perceived relative advantage in adopting new idea
'Great minds think alike'	'Groupthink' at strategic decision-making level – new idea lies outside the collective frame of reference
'(Existing) customers won't/don't want it'	New idea offers little to interest or attract current customers – essentially a different value proposition
'We've never done it before'	Perception that risks involved are too high along market and technical dimensions
'We're doing OK as we are'	The success trap – lack of motivation or organizational slack to allow exploration outside of current lines
'Let's set up a pilot'	Recognition of potential in new idea but limited and insufficient commitment to exploring and developing it – lukewarm support

Arguably these are all ways of defending an established mental model – they may be 'correct' in terms of the criteria associated with the dominant framework but they may also be defensive. Importantly they can be cloaked in a shroud of 'rationality' – using numbers about market size to reject exploration of a new area, for example. They represent an 'immune system' response which rejects the strange in order to preserve the health of the current body unchanged. The website has a link to a famous innovation case, 'Gunfire at sea' which highlights the difficulties in introducing radically different ideas into an established institution.

It is important to understand the problem of reframing since it provides some clues as to where and how alternative routines might be developed to support decision making around selection under high uncertainty. Using 'rational' methods of the kind which work well for incremental innovation is likely to be ineffective because of the high uncertainty associated with this kind of innovation. Since there is a high degree of uncertainty it is difficult to assemble 'facts' to make a clear business case, whilst the inertia of the existing framework includes the capacity to make justifiable rejection arguments of the kind highlighted in Table 7.2. The problem is complicated by the potential for radical innovation options to conflict with mainstream projects (for example, risking 'cannibalization' of existing and currently profitable markets) and the need to acquire different resources to those normally available to the firm.

Instead some form of alternative approaches may be needed to handle the early-stage thinking and exploring of opportunities outside the 'normal' decision-making channels but bring them back into the mainstream when the uncertainty level has been lowered. Resolving these tensions may require development of parallel structures or even setting up of satellite ventures and organizations outside the normal firm boundary.

An alternative strategy is, of course, to adopt a 'wait and see' approach and allow the market to deal with early-stage uncertainty. By taking a 'fast-second' posture large and well-resourced firms are often capable of exploiting innovation opportunities more successfully than smaller early entrants.[34] Examples here might include Microsoft, which was not an early mover in fields like the Internet or GUI (graphical user interface), but which used its considerable resource base to play a successful 'fast-second' game. Similarly many of the major pharmaceutical firms are managing the high uncertainty in the bio-pharmaceutical world by watching and acquiring rather than direct involvement. Arguably such strategies depend on developing sophisticated early warning and scanning systems to search for such opportunities and monitor them, and also on some additional route into mainstream decision-making/resource allocation systems to allow for such 'managed reframing'.

7.9 Mapping the selection space

As we saw in Chapter 5 there is a balance to be struck between 'exploit' and 'explore' behaviour in the ways organizations search for innovation triggers. But there are also limits to what is 'acceptable' exploration – essentially organizations have 'comfort zones' beyond which they are reluctant or unable to search. In similar fashion their decision making, even around radical options, is often constrained – this gives rise to the anxiety often expressed about the need for 'out-of-the-box' thinking. Stage-gate and portfolio systems depend on using criteria which are 'bought into' by those bringing ideas – a perception that the resource allocation process is 'fair' and appropriate even if the decisions go the 'wrong' way. Under steady-state conditions these systems can and do work well and criteria are clearly established

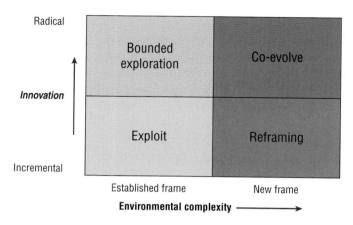

FIGURE 7.3: Outline map of innovation selection space

and perceived to be appropriate. But higher levels of uncertainty put pressure on the existing models – and one effect is that they reject ideas which don't fit – and over time build a 'self-censoring' aspect. As one interviewee in research on the way radical ideas were dealt with by his company's portfolio and stage-gate systems explained, '*around here we no longer have a funnel, we have a tube!*'

One way of looking at the innovation selection space is shown in Figure 7.3 – a model which we first saw in Chapter 5 when considering the innovation *search* space.[†] It is equally relevant to thinking about the challenges in innovation selection. The vertical axis refers to the familiar 'incremental/radical' dimension in innovation whilst the second relates to environmental complexity – the number of elements and their potential interactions. Rising complexity means that it becomes increasingly difficult to predict a particular state because of the increasing number of potential configurations of these elements. And it is here that problems of decision making become significant because of very high levels of uncertainty.

Zone 1 is essentially the 'exploit' domain in innovation literature. It presumes a stable and shared frame – 'business model'/architecture – within which adaptive and incremental development takes place. Selection routines – as we saw earlier in this chapter – are those associated with the 'steady-state' – portfolio methods, stage-gate reviews, clear resource allocation criteria, project management structures, etc. The structures involved in this selection activity are clearly defined with relevant actors, clear decision points, decision rules, criteria, etc. They correspond to widely accepted 'good practice' for product/service development and for process innovation.[3,9,35] As the sector matures so the tools and methods become ever more refined and subtle.

Zone 2 involves selection from exploration into new territory, pushing the frontiers of what is known and deploying different search techniques for doing so. But this is still taking place within the same basic cognitive frame – 'business model as usual'. Whilst the 'bets' may have longer odds the decision making is still carried out against an underlying strategic model and sense of core competencies. There may be debate and political behaviour at strategic level about choices between radical options

---

[†] The original idea of the matrix comes from Jean Boulton and is outlined in Boulton, J. and P. Allen (2004) Strategic management in a complex world. In *BAM Annual Conference*, St Andrews, Scotland.

but there is an underlying cognitive framework to define the arena in which this takes place and a sense of path dependency about the decisions taken. Often there is a sector-level trajectory, for example Moore's law shaping semiconductor, computer and related industry patterns. Although the activity is risky and exploratory it is still governed strongly by the frame for the sector – as Pavitt observed there are certain sectoral patterns which shape the behaviour of all the players in terms of their innovation strategies.[36]

The structures involved in such selection activity are, of necessity, focused at high level – these are 'big bets' – key strategic commitments rather than tactical investments. There are often tensions between the 'exploit' and the 'exploring' views and the boardroom battles between these two camps for resources are often tense. Since exploratory concepts carry high uncertainty the decision to proceed becomes more of an 'act of faith' than one which is matched by a clear, fact-based business case – and consequently emotional characteristics (such as passion and enthusiasm on the part of the proposer – 'champion' behaviour – or personal endorsement by a senior player ('sponsorship' behaviour) play a more significant role in persuading the decision makers.[37]

These first two zones represent familiar territory in discussion of exploit/explore in innovation selection. By contrast Zone 3 is associated with reframing. It involves searching and selecting from a space where alternative architectures are generated, exploring different permutations and combinations of elements in the environment. This process – essentially entrepreneurial – is risky and often results in failure but can also lead to emergence of new and powerful alternative business models (BMs). Significantly this often happens by working with elements in the environment not embraced by established BMs – but this poses problems for existing incumbents especially when the current BM is successful. Why change an apparently successful formula with relatively clear information about innovation options and well-established routines for managing the process? There is a strong reinforcing inertia about such systems for search and selection – the 'value networks' take on the character of closed systems which operate as virtuous circles and, for as long as they are perceived to create value through innovation, act as inhibitors to reframing.

The example of low-cost airlines here is relevant: it involved developing a new way of framing the transportation business based on rethinking many of the elements – turnaround times at airports, different plane designs, different Internet-based booking and pricing models, etc. – and also working with different new elements – essentially addressing markets like students and pensioners which had not been major elements in the 'traditional' BM. Other examples where a reframing of BM has taken place include hub and spoke logistics, digital imaging, digital music distribution, and mobile telephony/computing. The critical point here is that such innovation does not necessarily involve pushing the technological frontier but rather about working with new *architectures* – new ways for framing what is already there.

Selection under these conditions is difficult using existing routines which work well for zones 1 and 2. Whilst the innovations themselves may not be radical, they require consideration through a different lens and the kinds of information (and their perceived significance) that are involved may be unfamiliar or hard to obtain. For example, in moving into new under-served markets the challenge is that 'traditional' market research and analysis techniques may be inappropriate for markets which effectively do not yet exist. Many of the 'reasons' advanced for rejecting innovation proposals outlined in Table 7.2 can be mapped on to difficulties in managing selection in zone 3 territory – for example, '*it's not our business*' relates to the lack of perceived competence in analysis of new and unfamiliar variables. '*Not invented here*' relates to similar lack of perceived experience, competence or involvement in a technological field and the inability to analyse and take 'rational' decisions about it. '*It's not a business*' relates to apparent market size which in initial stages may appear small and unlikely to serve the growth needs of established incumbents. But such markets could grow – the challenge is seeing an alternative trajectory to the current dominant logic of the established business model.[38,39]

Here the challenge is seeing a new possible pattern and absorbing and integrating new elements into it. This is hard to do because it requires cognitive reframing – but also because it challenges the existing system – something Machiavelli was aware of many centuries ago. Powerful social forces towards conforming – groupthink, risky shift, etc. – come into play and reinforce a dominant line at senior levels.[23] This set of emotionally underpinned views is then rationalized with some of the statements in Table 7.2 – the 'immune system' we referred to earlier. Significantly where there are examples of radical changes in mindset and subsequent strategic direction these often come about as a result of crisis – which has the effect of shattering the mindset – or with the arrival from outside of a new CEO with a different world view.

Zone 4 is where new-to-the-world innovation takes place – and represents the 'edge of chaos' complex environment where such innovation emerges as a product of a process of co-evolution.[40,41] This is not the product of a predefined trajectory so much as the result of complex interactions between independent elements. Processes of amplification and feedback reinforce what begin as small shifts in direction – attractor basins – and gradually define a trajectory. This is the pattern we saw in Chapter 1 in the 'fluid' stage of the innovation life cycle before a dominant design emerges and sets the standard.[42] It is the state where all bets are potentially options – and high variety experimentation takes place. Selection strategies here are difficult since it is, by definition, impossible to predict what is going to be important or where the initial emergence will start and around which feedback and amplification will happen. Under such conditions the strategy breaks down into three core principles: be in there, be in there early and be in there influentially (i.e. in a position to be part of the feedback and amplification mechanisms).[43,44]

Examples here might be the emergence of product innovation categories for the first time, for example, the bicycle which emerged out of the nineteenth-century mix of possibilities created by iron-making technologies and social market demands for mass personal transportation.[45] The emergence of new techno-economic systems is essentially a process of *co-evolution* amongst a complex set of elements rather than a reframing of them. Change here corresponds to what Perez calls 'paradigm shift' and examples include the Industrial Revolution or the emergence of the Internet-based society.[46] The complex pattern of co-evolution which led to the emergence of the bicycle is explored in more detail in the case study on the website.

Once again this zone poses major challenges to an established set of selection routines – in this case they are equipped to deal with uncertainty but in the form of 'known unknowns' whereas zone 4 is essentially 'unknown unknowns' territory. Analytical tools and evidence-based decision making – for example reviewing business cases – are inappropriate for judging plays in a game where the rules are unclear and even the board on which it is played has yet to be designed! An example here might be the ways in which the Internet and the products/services which it will carry will emerge as a result of a complex set of interactions amongst users. Or the ways in which chronic diseases like diabetes will be managed in a future where the incidence is likely to rise, where the costs of treatment will increase faster than health budgets can cope and where many different stakeholders are involved – clinicians, drug companies, insurance companies, carers and patients themselves.

Table 7.3 summarizes the challenges posed across our selection space and highlights the need to experiment with new approaches for selection in zones 3 and 4.

TABLE 7.3	Selection challenges, tools and enabling structures

Zone	Selection challenges	Tools and methods	Enabling structures
1 'Business as usual' – innovation but under 'steady-state' conditions, little disturbance around core business model	Decisions taken on the basis of exploiting existing and understood knowledge and deploying in known fields. Incremental innovation aimed at refining and improving. Requires building strong ties with key players in existing value network and working with them	'Good practice' new product/service development Portfolio methods and clear decision criteria, stage-gate reviews along clear and established pathways	Formal and mainstream structures – established stage-gate process with defined review meetings High involvement across organization roles and functions in the decisionmaking
2 'Business model as usual' – bounded exploration within this frame	Exploration – pushing frontiers of technology and market via calculated risks – 'buying a look' at new options through strategic investments in further research. Involves risk taking and high uncertainty	Advanced tools for risk assessment, e.g. R&D options and futures. Multiple portfolio methods and 'fuzzy front end' toolkit – bubble charts, etc. Criteria used are a mix of financial and non-financial. Judgmental methods allow for some influence of passion and enthusiasm – the 'Dragon's Den' effect	May form part of existing stage-gate and review system with extra attention being devoted to higher risk projects at early stages. May also involve special meetings outside that frame – and decision making will be at strategic (board) level rather than operational
3 Alternative frame – taking in new/different elements in environment	Reframe – explore alternative options, introduce new elements. Challenge involves decision making under uncertainty but not simply a problem of lack of information and the need to take risky bets to learn more. Here there is also the issue of unfamiliar frames	May use variations of existing toolkit, e.g. portfolio methods, but extend the parameters, e.g. 'fuzzy front end', bubble charts, etc. Alternative futures and visioning tools Constructed crisis Prototyping – probe and learn	Unlikely to fit with established decision structures – stage gate and portfolio – since these are designed around established business model frame. Needs parallel or alternative evaluation structures – at least for early stage

(continued)

Zone	Selection challenges	Tools and methods	Enabling structures
TABLE 7.3 (Continued)			
	of reference and the difficulty of letting go of a dominant logic. Cognitive dissonance means that incumbents have trouble 'forgetting' enough to see the environment through 'new eyes'	Creativity techniques Use of internal and external entrepreneurs to decentralize development of early business case Alternative funding models and decentralized authority for early stage exploration	
4 Radical – new to the world – possibilities. New architecture around as yet unknown and established elements	Emergence – need to co-evolve with stakeholders Be in there Be in there early Be in there actively	Complexity theory – feedback and amplification, probe and learn, prototyping and use of boundary objects	Far from mainstream Satellite structures – skunk works or even outside the firm 'Licensed dreamers' Outside agents and facilitators

RESEARCH NOTE Tools to help with high uncertainty decision making[47]

Faced with the reframing and high uncertainty challenges of zones 3 and 4 how can organizations manage the selection process? Research within the 'Discontinuous Innovation Lab' – an experience-sharing network of 31 academic and 140 commercial organizations in 12 countries – suggests companies use a number of approaches, including the following.

Building alternative futures

One powerful approach lies in the area of 'futures studies', using tools such as forecasting, trend extrapolation and scenario building to create and explore alternative models of the future and the potential threats and opportunities which they contain.[48,49] Increasingly futures tools are being deployed in frameworks which are designed to open up new innovation space, for example, the 'GameChanger' programme has been widely used in organizations such as Shell and Whirlpool, whilst other companies like BMW, Novozymes and Nokia make extensive use of similar approaches.[33,50] They deploy a range of techniques including metaphors, storytelling and vision building and increasingly do so in a cross-sectoral fashion, recognizing that the future may involve blurring of traditional market or demographic boundaries. An important variant on this is the use of what is termed 'constructed crisis' – deliberately exploring radical and challenging futures to create a sense of unease – a 'burning platform' from which new directions forward can be developed.[51]

Prototyping as a way of building bridges in the selection process

When confronted by innovation trigger signals outside the 'normal' frame, organizations face the classic entrepreneur's challenge. It is possible to see something new but in order to take that forward, to make the idea a reality, the entrepreneur needs to mobilize resources and to do this he/she needs to convince people of the potential. The process involves building bridges in the minds of potential supporters between the current state of affairs and what might be. It is here that 'boundary objects' become important – things which can act as 'stepping stones' between the two. Prototyping offers a way of creating such stepping stones towards that new option – and importantly stepping stones which allow both building up of better understanding and also shaping the idea whilst it is still in its formative stages.[52]

There are many different ways of prototyping including physical models, simulation, etc. and these span both manufactured products and service concepts. The process can also involve the use of consultants who act in bridging fashion, helping reduce the risk by outsourcing the exploration to them. By employing consultants like IDEO or ?Whatif!, organizations can make a 'safe' experiment and then use their involvement with an external agency to develop and work with the emerging prototype.

Probe and learn

One way of dealing with the uncertainty problem is to use 'probe and learn' approaches – essentially making small steps into the fog and shining a torch (or swirling a fan) to illuminate enough of the pathway to see where it might lead next. Closely related to boundary objects, the idea here is to help move from outside the box to a new place outside the comfort zone by a series of planned experiments. These serve two functions: they provide new information about what does (and doesn't) work and so help build the case for selection along the 'rational' axis of the above diagram; they also represent ways of mapping 'unsafe' territory and reducing the emotional anxiety. In this sense they are investments in what Robert Cooper calls 'buying a look' – and they help assemble the beginnings of a case for further support and exploration.[2] Such investments in 'buying a look' may fail – progress on the pathway may end up confirming that this is not a good road to travel. But they may also help point in new and exciting directions – and in the process justify the investment.

Using alternative measurement and evaluation criteria

Within any selection system there is a need for criteria – and general acceptance of these as a good basis on which to take decisions. But this is difficult to do under conditions of high uncertainty – and so often the problem is resolved by adapting existing systems which may be only partially effective. For example, using conventional criteria but increasing the limits – e.g. the 'hurdle rate' for return on investment – in order to mitigate the risk associated with uncertainty or applying broad boundaries (maximum permissible losses) in which radical innovation can be nurtured.

Mobilizing networks of support

Much of the literature around radical innovation identifies the role of 'champions' of various kinds.[37] Importantly there are several kinds of champion roles, for example, 'power' promoters who can bring resources, backing, etc. and 'knowledge' promoters who have expertise and passion for a particular idea. These can be combined in the same individual – e.g. James Dyson – or in a team/tandem arrangement – e.g. Art Fry and Spence Silver at 3M. Rice *et al.* identify several types of champion: technical champions, project champions, senior management champions, business unit champions and in some cases a single individual champion who takes on multiple championing roles.[53]

Using alternative decision-making pathways

To help provide a pathway for developing radical ideas at least to the stage where they can stand up for themselves in the mainstream innovation funnel process, many organizations have experimented with parallel or alternative structures for radical innovation. They vary in shape and form but essentially have a 'fuzzy front end', which allows for building a potential portfolio of higher risk ideas and options, and some mechanisms for gradually building a business case, which can be subjected to increasingly critical criteria for resource allocation – essentially a parallel funnel structure. These systems may rejoin the mainstream funnel at a later stage or they may continue to operate in parallel – see Figure 7.2. And of course they may lead to very different options apart from progression as a mainstream project – spin off, license out, buy in, etc.

Deploying alternative funding structures

Just as the external financial markets recognize a place for 'venture capital' finance available for higher risk and potentially higher reward projects, so increasingly organizations are developing alternative and parallel funding arrangements which provide access to funding on different terms. These can take many forms, including special project teams, incubators, new venture divisions, corporate venture units and 'skunk works'.[54] Some have more formal status than others, some have more direct power or resource, whilst others are dependent on internal sponsors or patrons.

One key issue with such dual structures is the need to bring them back into the mainstream at some point. They can provide helpful vehicles for growing ideas to the point where they can be more fairly evaluated against mainstream criteria and portfolio selection systems, but they need to be seen as temporary rather than permanent mechanisms for doing so. Otherwise there is a risk of separation and at the limit a loss of leverage against the knowledge and other assets of the mainstream organization.

Using alternative/dedicated implementation structures

One strategy for dealing with the selection problem associated with radical ideas is to allow them to incubate elsewhere – off-line or at least away from the harsh environment of the normal resource allocation system. In essence this strategy bridges both the selection and implementation challenges and makes use of different mechanisms for incubation and early-stage development. These can take the form of *special external vehicles*, which operate outside the existing corporate structure – a good example is the famous 'skunk works' at Lockheed. Other variants include setting up *external ventures* where such incubation can take place, for example, Siemens makes use of 'satellite' SMEs in which it takes shares to act as incubator environments to take forward some of its more radical ideas. Others take stakes in start-ups to explore and develop ideas to the point where they might represent formal options for full acquisition – or spin out. Another approach is to use *third-party consultants* as a short-term environment in which more radical ideas can be developed and explored.

Mobilizing entrepreneurship

A number of organizations are trying to make explicit use of internal entrepreneurship – 'intrapreneurship' – to help with radical innovation. Creating the culture to enable this is not simple; it requires a commitment of resources but also a set of mechanisms to take bright ideas forward, including various internal development grants and an often complicated and fickle internal funding process. Many such schemes have a strong incentive scheme for those willing to take the lead in taking ideas into marketable products at their core. An additional incentive is often the opportunity to not only lead the development of the new idea but also get involved in the running of the new business.

Mechanisms for promoting entrepreneurship include provision of time or resources – 3M's 15% policy and more recent examples from Google underline the importance of this approach. Fostering a culture of 'bootlegging' can also help since it creates a difficult environment in which strong ideas can surface through the energy of entrepreneurs in spite of apparent rules and constraints.[55]

Summary and further reading

In this chapter we have looked at some of the challenges in making the selection decision – moving from considering all the possible trigger signals about what we could do in terms of innovation to committing resources to some particular projects. This quickly raises the issue of uncertainty and how we convert it to some kind of manageable risk – and build a portfolio of projects spreading this risk. Tools and techniques for doing so for incremental innovation are relatively straightforward (though there is never a guarantee of success), but as we increase the radical nature of the innovation so there is a need for different approaches. The problem is further compounded because of the simplifying assumptions we make when framing the complex world – and the risk is that in selecting projects which fit our frame we may miss important opportunities or challenges. For this reason we need techniques which help the organization look and make decisions 'outside the box' of its normal frame of reference.

The theme of innovation decision making, risk management and the use of the stage-gate concept is extensively covered in the work of Robert Cooper and colleagues (The invisible success factors in product innovation. *Journal of Product Innovation Management*, **16**(2), 1999; *Product Leadership*, Perseus Press, 2000). Tools for portfolio management and related approaches are discussed with good examples in

Goffin and Mitchell's book, *Innovation Management* (Pearson, 2005) and policy deployment approaches in Bessant (*High-Involvement Innovation*, John Wiley & Sons, Ltd, 2003) and Akao (*Hoshin Kanri: Policy Deployment for Successful TQM*, Productivity Press, 1991). Dodgson, Gann and Salter (*Think, Play, Do: Technology and Organization in the Emerging Innovation Process*, Oxford University Press, 2005) and Schrage (*Serious Play: How the World's Best Companies Simulate to Innovate*, Harvard Business School Press, 2000) explore the growing range of simulation and prototyping tools which can postpone the commitment decision point; whilst von Hippel and colleagues expand on the user involvement theme (E. Von Hippel, User toolkits for innovation. *Journal of Product Innovation Management*, **18**, 247–57, 2001; C. Herstatt and E. von Hippel, Developing new product concepts via the lead user method. *Journal of Product Innovation Management*, **9**(3), 213–21, 1992). Peter Koen's work (New concept development model: providing clarity and a common language to the 'fuzzy front end' of innovation. *Research Technology Management*, **44**(2), 46–55, 2001) provides useful insights on fuzzy front end tools and methods (a good source is the PDMA Handbook – P. Belliveau, A. Griffin, and S. Somermeyer, *The PDMA ToolBook for New Product Development*, John Wiley & Sons, Inc., 2002) and Julian Birkinshaw (Building ambidexterity into an organization. *Sloan Management Review*, **45** (4), 47–55, 2004) explores the challenges in developing 'ambidextrous' decision-making structures. A detailed review of the psychological issues and problems around reframing can be found in Hodgkinson and Sparrow (*The Competent Organization*, Open University Press, 2002), whilst the work of Karl Weick (*Making Sense of the Organization*, Blackwell, 2001) remains seminal in discussing the ways in which organizations try and make sense of complex worlds.

Useful websites include Innovation Tools (**www.innovationtools.com**), which provides case examples and links to a wide range of innovation support resources and the Product Development Management Association (**www.pdma.org**), which covers many of the decision tools used with practical examples of their application in the online '*Visions*' magazine. NESTA (**www.nesta.org.uk**) and AIM (**www.aimresearch.org**) provide reports and research papers around core innovation themes including many of the issues raised in this chapter.

Web links

Here are the full details of the resources available on the website flagged throughout the text:

Case studies:
ABC Electronics
Coloplast
Policy deployment cases
Tesco Fresh & Easy
Philips Lighting
Cerulean
Gunfire at sea
Evolution of the bicycle

Interactive exercises:
Managing discontinuous change

Tools:

Stage-gate

QFD

Portfolio management

Bubble charts

Matrix methods

Product mapping

Product profiling

Concept testing

Policy deployment

 Video podcast:

Cerulean

 Audio podcast:

Philips Lighting

References

1. **Keynes, J.M.** (2007) *The General Theory of Employment, Interest and Money*, Palgrave, Basingstoke.
2. **Cooper, R.** (2001) *Winning at New Products*, third edition, Kogan Page, London.
3. **Griffin, A., M. Rosenau, G. Castellion, and N. Anschuetz** (1996) *The PDMA Handbook of New Product Development*, John Wiley & Sons, Inc., New York.
4. **Cooper, R.** (1994) Third-generation new product processes. *Journal of Product Innovation Management*, **11** (1), 3–14.
5. **Floyd, C.** (1997) *Managing Technology for Corporate Success*, Gower, Aldershot.
6. **Bruce, M. and R. Cooper** (1997) *Marketing and Design Management*, International Thomson Business Press, London.
7. **Bessant, J. and D. Francis** (1997) Implementing the new product development process. *Technovation*, **17** (4), 189–97.
8. **Boer, H., A. Berger, R. Chapman, and F. Gertsen** (1999) *CI Changes: From Suggestion Box to the Learning Organisation*, Ashgate, Aldershot.
9. **Schroeder, A. and D. Robinson** (2004) *Ideas are Free: How the Idea Revolution is Liberating People and Transforming Organizations*, Berrett Koehler, New York.
10. **Bessant, J.** (2003) *High-Involvement Innovation*, John Wiley & Sons, Ltd, Chichester.
11. **Schroeder, D. and A. Robinson** (1991) America's most successful export to Japan – continuous improvement programs. *Sloan Management Review*, **32** (3), 67–81.
12. **Imai, M.** (1997) *Gemba Kaizen*, McGraw-Hill, New York.
13. **Dodgson, M., D. Gann, and A. Salter** (2005) *Think, Play, Do: Technology and Organization in the Emerging Innovation Process*, Oxford University Press, Oxford.
14. **Pettigrew, A.** (1974) *The Politics of Organizational Decision Making*, Tavistock, London.
15. **Van de Ven, A.** (1999) *The Innovation Journey*, Oxford University Press, Oxford.
16. **Wheelwright, S. and K. Clark** (1992) *Revolutionizing Product Development*, Free Press, New York.
17. **Mumford, E.** (1979) *Designing Human Systems*, Manchester Business School Press, Manchester.

18. **Bessant, J. and J. Buckingham** (1993) Organisational learning for effective use of CAPM. *British Journal of Management*, **4** (4), 219–34.

19. **Hines, P., P. Cousins, D. T. Jones, R. Lamming, and N. Rich** (1999) *Value Stream Management: The Development of Lean Supply Chains*, Financial Times Management, London.

20. **Rich, N. and P. Hines** (1997) Supply chain management and time-based competition: the role of the supplier association. *International Journal of Physical Distribution and Logistics Management*, **27** (3/4), 210–25.

21. **Legge, K.** *et al.*, eds (1991) *Case Studies in Information Technology, People and Organizations*, Oxford, Blackwell.

22. **Tripsas, M. and G. Gavetti** (2000) Capabilities, cognition and inertia: evidence from digital imaging. *Strategic Management Journal*, **21**, 1147–61.

23. **Hodgkinson, G. and P. Sparrow** (2002) *The Competent Organization*, Open University Press, Buckingham.

24. **White, A. and J. Bessant** (2006) Managerial responses to cognitive dissonance: causes of the mismanagement of discontinuous technological innovations. In T. Khalil, ed., *IAMOT 2004*, Elsevier, New York.

25. **Christensen, C.** (1997) *The Innovator's Dilemma*, Harvard Business School Press, Boston, MA.

26. **Henderson, R. and K. Clark** (1990) Architectural innovation: the reconfiguration of existing product technologies and the failure of established firms. *Administrative Science Quarterly*, **35**, 9–30.

27. **Weick, K.** (2002) Puzzles in organizational learning. *British Journal of Management*, **13**, S7–S16.

28. **Leonard-Barton, D.** (1995) *Wellsprings of Knowledge: Building and Sustaining the Sources of Innovation*, Harvard Business School Press, Boston, MA.

29. **Bhide, A.** (2000) *The Origin and Evolution of New Businesses*, Oxford University Press, Oxford.

30. **Dosi, G.** (1982) Technological paradigms and technological trajectories. *Research Policy*, **11**, 147–62.

31. **Utterback, J.** (1994) *Mastering the Dynamics of Innovation*, Harvard Business School Press, Boston, MA.

32. **Markides, C.** (1997) Strategic innovation. *Sloan Management Review*, **Spring**, 9–24.

33. **Hamel, G.** (2000) *Leading the Revolution*, Harvard Business School Press, Boston. MA.

34. **Markides, C. and P. Geroski** (2004). *Fast Second: How Smart Companies Bypass Radical Innovation to Enter and Dominate New Markets*, Jossey Bass, San Francisco.

35. **Roussel, P., K. Saad and T. Erickson** (1991) *Third Generation R&D: Matching R&D Projects with Corporate Strategy*, Harvard Business School Press, Boston. MA.

36. **Pavitt, K.** (1984) Sectoral patterns of technical change: towards a taxonomy and a theory. *Research Policy*, **13**, 343–73.

37. **Leifer, R., C. McDermott, G. O'Conner, L. Peters, M. Rice, and R. Veryzer** (2000) *Radical Innovation*, Harvard Business School Press, Boston. MA.

38. **Christensen, C. and R. Rosenbloom** (1995) Explaining the attacker's advantage: technological paradigms, organizational dynamics, and the value network. *Research Policy*, **24**, 233–57.

39. **Kim, W. and R. Mauborgne** (2005) *Blue Ocean Strategy: How to Create Uncontested Market Space and Make the Competition Irrelevant*, Harvard Business School Press, Boston. MA.

40. **McKelvey, B.** (2004) 'Simple rules' for improving corporate IQ: basic lessons from complexity science. In P. Andirani and G. Passiante, eds, *Complexity Theory and the Management of Networks*, Imperial College Press, London.

41. **Stacey, R.** (1993) *Strategic Management and Organizational Dynamics*, Pitman, London.

42. **Abernathy, W. and J. Utterback** (1975) A dynamic model of product and process innovation. *Omega*, **3** (6), 639–56.

43. **Allen, P.** (2001) A complex systems approach to learning, adaptive networks. *International Journal of Innovation Management*, **5**, 149–80.

44. **Boulton, J. and P. Allen** (2004) Strategic management in a complex world. In *BAM Annual Conference*, St Andrews, Scotland.

45. **Walsh, V., R. Roy, S. Potter, and M. Bruce** (1992) *Winning by Design: Technology, Product Design and International Competitiveness*, Basil Blackwell, Oxford.

46. **Perez, C.** (2002) *Technological Revolutions and Financial Capital*, Edward Elgar, Cheltenham.

47. **Bessant, J., B. Von Stamm, K. Moeslein, and A. Neyer** (2009) *Meeting the Selection Challenge*, Advanced Institute for Management Research, London.

48. **Wheelwright, S. and S. Makridakis** (1980) *Forecasting Methods for Management*, John Wiley & Sons, Inc., New York.

49. **Whiston, T.** (1979) *The Uses and Abuses of Forecasting*, Macmillan, London.

50. **Bessant, J. and B. Von Stamm** (2007) *Twelve Search Strategies Which Might Save Your Organization*, AIM Executive Briefing, London.

51. **Kim, L.** (1998) Crisis construction and organizational learning: capability building in catching-up at Hyundai Motor. *Organization Science*, **9** (4), 506–21.

52. **Schrage, M.** (2000) *Serious Play: How the World's Best Companies Simulate to Innovate*, Harvard Business School Press, Boston, MA.

53. **Rice, M., P. Mark, G. Colarelli Connor, and R. Pierantozzi** (2008) Implementing a Learning Plan to Counter Project Uncertainty, *MIT Sloan Management Review*, **49** (2), 54.

54. **Rich, B. and L. Janos** (1994) *Skunk Works*, Warner Books, London.

55. **Augsdorfer, P.** (1996) *Forbidden Fruit,* Avebury, Aldershot.

CHAPTER 8

Building the innovation case

The usual motive for developing a formal business plan is to secure support or funding for a project or venture. However, in practice business planning serves a much more important function, and can help to translate abstract or ambiguous goals into more explicit operational needs, and support subsequent decision making and identify trade-offs. A business plan can help to make more explicit the risks and opportunities, expose any unfounded optimism and self-delusion, and avoid subsequent arguments concerning responsibilities and rewards.

8.1 Developing the business plan

No standard business plan exists, but in many cases venture capitalists will provide a pro forma for their business plan. Typically a business plan should be relatively concise, say no more than 10–20 pages, begin with an executive summary, and include sections on the product, markets, technology, development, production, marketing, human resources, financial estimates with contingency plans, and the timetable and funding requirements. A typical formal business plan will include the following sections:[1]

1. Details of the product or service.
2. Assessment of the market opportunity.
3. Identification of target customers.
4. Barriers to entry and competitor analysis.
5. Experience, expertise and commitment of the management team.
6. Strategy for pricing, distribution and sales.
7. Identification and planning for key risks.
8. Cash-flow calculation, including breakeven points and sensitivity.
9. Financial and other resource requirements of the business.

Most business plans submitted to venture capitalists are strong on the technical considerations, often placing too much emphasis on the technology relative to other issues. As Roberts notes, 'entrepreneurs propose that they can do *it* better than anyone else, but may forget to demonstrate that anyone wants *it*'.[2] He identifies a number of common problems with business plans submitted to venture capitalists: marketing plan, management team, technology plan and financial plan. The management team will be assessed against their commitment, experience and expertise, normally in that order. Unfortunately, many potential entrepreneurs place too much emphasis on their expertise, but have insufficient experience in the team, and fail to demonstrate the passion and commitment to the venture (Table 8.1).

TABLE 8.1	Criteria used by venture capitalists to assess proposals		
Criteria	European (*n* = 195)	American (*n* = 100)	Asian (*n* = 53)
Entrepreneur able to evaluate and react to risk	3.6	3.3	3.5
Entrepreneur capable of sustained effort	3.6	3.6	3.7
Entrepreneur familiar with the market	3.5	3.6	3.6
Entrepreneur demonstrated leadership ability*	3.2	3.4	3.0
Entrepreneur has relevant track record*	3.0	3.2	2.9
Product prototype exists and functions*	3.0	2.4	2.9
Product demonstrated market acceptance*	2.9	2.5	2.8
Product proprietary or can be protected*	2.7	3.1	2.6
Product is 'high technology'*	1.5	2.3	1.4
Target market has high growth rate*	3.0	3.3	3.2
Venture will stimulate an existing market	2.4	2.4	2.5
Little threat of competition within 3 years	2.2	2.4	2.4
Venture will create a new market*	1.8	1.8	2.2
Financial return >10 times within 10 years*	2.9	3.4	2.9

(continued)

TABLE 8.1	(Continued)		
Criteria	European ($n = 195$)	American ($n = 100$)	Asian ($n = 53$)
Investment is easily made liquid* (e.g. made public or acquired)	2.7	3.2	2.7
Financial return >10 times within 5 years*	2.1	2.3	2.1

1 = irrelevant, 2 = desirable, 3 = important, 4 = essential. * Denotes significant at the 0.05 level.
Source: Adapted from Knight, R. (1992) Criteria used by venture capitalists. In T. Khalil and B. Bayraktar, eds, *Management of Technology III: The Key to Global Competitiveness* (pp. 574–83), Industrial Engineering & Management Press, Georgia.

There are common serious inadequacies in all four of these areas, but the worst are in marketing and finance. Less than half of the plans examined provide a detailed marketing strategy, and just half include any sales plan. Three-quarters of the plans fail to identify or analyse any potential competitors. As a result most business plans contain only basic financial forecasts, and just 10% conduct any sensitivity analysis on the forecasts. The lack of attention to marketing and competitor analysis is particularly problematic as research indicates that both factors are associated with subsequent success.

For example, in the early stages many new ventures rely too much on a few major customers for sales, and are therefore very vulnerable commercially. As an extreme example, around half of technology ventures rely on a single customer for more than half of their first-year sales. An overdependence on a small number of customers has three major drawbacks:

1. Vulnerability to changes in the strategy and health of the dominant customer.
2. A loss of negotiating power, which may reduce profit margins.
3. Little incentive to develop marketing and sales functions, which may limit future growth.

The case of Plaswood Recycling provides a good example of how to assess a new business concept.

Therefore it is essential to develop a better understanding of the market, and technological inputs to a business plan. The financial estimates flow from these critical inputs relatively easily, although risk and uncertainty still need to be assessed. This chapter focuses only on the most important, but often poorly executed, aspects of business planning for innovations. We first discuss approaches to forecasting markets and technologies, and then identify how a better understanding of the adoption and diffusion of innovations can help us to develop more successful business plans. Finally, we look at how to assess the risks and resources required to finalize a plan. We will return to the development of business plans in Chapter 10, in the specific context of new venture creation.

What is the 'fuzzy front end', why is it important, and how can it be managed?

Technically, new product development (NPD) projects often fail at the end of a development process. The foundations for failure, however, often seem to be established at the very beginning of the NPD process, often referred to as the 'fuzzy front end'. Broadly speaking, the fuzzy front end is defined as the period between when an opportunity for a new product is first considered, and when the product idea is judged ready to enter 'formal' development. Hence, the fuzzy front end starts with a firm having an idea for a new product, and ends with the firm deciding to launch a formal development project or, alternatively, decides not to launch such a project.

In comparison with the subsequent development phase, knowledge on the fuzzy front end is severely limited. Hence, relatively little is known about the key activities that constitute the fuzzy front end, how these activities can be managed, which actors participate, as well as the time needed to complete this phase. Many firms also seem to have great difficulties managing the fuzzy front end in practice. In a sense this is not surprising: the fuzzy front end is a crossroads of complex information processing, tacit knowledge, conflicting organizational pressures, and considerable uncertainty and equivocality. In addition, this phase is also often ill-defined and characterized by ad hoc decision making in many firms. It is therefore important to identify success factors which allow firms to increase their proficiency in managing the fuzzy front end. This is the purpose of this research note.

In order to increase knowledge on how the fuzzy front end can be better managed, we conducted a large-scale survey of the empirical literature on the fuzzy front end. In total, 39 research articles constitute the base of our review. Analysis of these articles identified 17 success factors for managing the fuzzy front end. The factors are not presented in order of importance, as the present state of knowledge makes such an ordering judgemental at best.

1. *The presence of idea visionaries or product champions.* Such persons can overcome stability and inertia and thus secure the progress of an emerging product concept.
2. *An adequate degree of formalization.* Formalization promotes stability and reduces uncertainty. The fuzzy front end process should be explicit, widely known among members of the organization, characterized by clear decision-making responsibilities, and contain specific performance measures.
3. *Idea refinement and adequate screening of ideas.* Firms need mechanisms to separate good ideas from the less good ones, but also to screen ideas by means of both business and feasibility analysis.
4. *Early customer involvement.* Customers can help to construct clear project objectives, reduce uncertainty and equivocality, and also facilitate the evaluation of a product concept.
5. *Internal cooperation among functions and departments.* A new product concept must be able to 'survive' criticism from different functional perspectives, but cooperation among functions and departments also creates legitimacy for a new concept and facilitates the subsequent development phase.

6. *Information processing other than cross-functional integration and early customer involvement.* Firms need to pay attention to product ideas of competitors, as well as legally mandated issues in their emerging product concepts.

7. *Senior management involvement.* A pre-development team needs support from senior management to succeed, but senior management can also align individual activities which cut across functional boundaries.

8. *Preliminary technology assessment.* Technology assessment means asking early whether the product can be developed, what technical solutions will be required, and at what cost. Firms need also to judge whether the product concept, once turned into a product, can be manufactured.

9. *Alignment between NPD and strategy.* New concepts must capitalize on the core competence of their firms, and synergy among projects is important.

10. *An early and well-defined product definition.* Product concepts are representations of the goals for the development process. A product definition includes a product concept, but in addition provides information about target markets, customer needs, competitors, technology, resources, etc. A well-defined product definition facilitates the subsequent development phase.

11. *Beneficial external cooperation with others than customers.* Many firms benefit from a 'value-chain perspective' during the fuzzy front end, e.g. through collaboration with suppliers. This factor is in line with the emerging literature on 'open innovation'.

12. *Learning from experience capabilities of the pre-project team.* Pre-project team members need to identify critical areas and forecast their influence on project performance, i.e. through learning from experience.

13. *Project priorities.* The pre-project team needs to be able to make trade-offs among the competing virtues of scope (product functionality), scheduling (timing) and resources (cost). In addition, the team also needs to use a priority criteria list, i.e. a rank ordering of key product features, should it be forced to disregard certain attributes due to e.g. cost concerns.

14. *Project management and the presence of a project manager.* A project manager can lobby for support and resources, and coordinate technical as well as design issues.

15. *A creative organizational culture.* Such a culture allows a firm to utilize the creativity and talents of employees, as well as maintaining a steady stream of ideas feeding into the fuzzy front end.

16. *A cross-functional executive review committee.* A cross-functional team for development is not enough – cross-functional competence is also needed when evaluating product definitions.

17. *Product portfolio planning.* The firm needs to assure sufficient resources to develop the planned projects, as well as 'balancing' its portfolio of new product ideas.

Although successful management of the fuzzy front end requires firms to excel in individual factors and activities, this is a necessary rather than sufficient condition. Firms must also be able to integrate or align different activities and factors, as reciprocal interdependencies exist among different success factors. This is often referred to as 'a holistic perspective', 'interdependencies among factors', or simply as 'fit'. To date, however, nobody seems to know exactly which factors should be integrated, and how this should be achieved. In addition, specific guidelines on how to measure performance in the fuzzy front end are also lacking. Hence, only fragments of a 'theory' for managing the fuzzy front end can be said to be in place.

To make things even more complicated, the fuzzy front end process seems to vary not only among firms, but also among projects within the same firm where activities, their sequencing, degree of overlap and relative time duration differ from project to project. Therefore, capabilities for managing the fuzzy front end are both highly valuable yet difficult to obtain. Developing firms therefore need first to obtain proficiency in individual success factors. Second, they need to integrate and arrange these factors into a coherent whole aligned to the circumstances of the firm. And finally, they need to master several trade-off situations which we refer to as 'balancing acts'.

As a first balancing act, firms need to ask if screening of ideas should be made gentle or harsh. On the one hand, firms need to get rid of bad ideas quickly, to save the costs associated with their further development. On the other hand, harsh screening may also kill good ideas too early. Ideas for new products often refine and gain momentum through informal discussion, a fact which forces firms to balance too gentle and too harsh screening. Another balancing act concerns formalization. The basic proposition is that formalization is good because it facilitates transparency, order and predictability. However, in striving to enforce effectiveness, formalization also risks inhibiting innovation and flexibility. Even if evidence is still scarce the relationship between formality and performance seems to obey an inverted U-shaped curve, where both too little and too much formality has a negative effect on performance. From this it follows that firms need to carefully consider the level of formalization they impose on the fuzzy front end.

A third balancing act concerns the trade-off between uncertainty and equivocality reduction. Market and technological uncertainty can often be reduced through environmental scanning and increased information processing in the development team, but more information often increases the level of equivocality. An equivocal situation is one where multiple meanings exist, and such a situation implies that a firm needs to construct, cohere or enact a reasonable interpretation to be able to move on, rather than to engage in information seeking and analysis. Therefore, firms need to balance their need to reduce uncertainty with the need to reduce equivocality, as trying to reduce one often implies increasing the other. Furthermore, firms need to balance the need for allowing for flexibility in the product definition, with the need to push it to closure. A key objective in the fuzzy front end is a clear, robust and unambiguous product definition as such a definition facilitates the subsequent development phase. However, product features often need to be changed during development as market needs change or problems with underlying technologies are experienced. Finally, a final balancing act concerns the trade-off between the competing virtues of innovation and resource efficiency. In essence, this concerns balancing competing value orientations, where innovation and creativity in the front end are enabled by organizational slack and an emphasis on people management, while resource efficiency is enabled by discipline and an emphasis on process management.

In addition, the fuzzy front end process needs to be adapted to the type of product under development. For physical products, different logics apply to assembled and non-assembled products. Emerging research shows that a third logic applies to the development of new service concepts. To conclude, managing the fuzzy front end is indeed no easy task, but can have an enormous positive impact on performance for those firms that succeed.

Source: Frishammar, J. and H. Florén (2008). Where new product development begins: success factors, contingencies and balancing acts in the fuzzy front end. Paper presented at the IAMOT conference in Dubai, 5–8 April. Reproduced by permission of Johan Frishammar (Luleå University of Technology, Sweden) and Henrik Florén (Halmstad University, Sweden).

8.2 Forecasting innovation

Forecasting the future has a pretty bad track record (see Box 8.1), but nevertheless has a central role in business planning for innovation. In most cases the outputs, that is the predictions made, are less valuable than the process of forecasting itself. If conducted in the right spirit, forecasting should provide a framework for gathering and sharing data, debating interpretations and making assumptions, challenges and risks more explicit.

The most appropriate choice of forecasting method will depend on:

- What we are trying to forecast.
- Rate of technological and market change.
- Availability and accuracy of information.
- The company's planning horizon.
- The resources available for forecasting.

BOX 8.1   **Limits of forecasting**

In 1986, Schnaars and Berenson published an assessment of the accuracy of forecasts of future growth markets since the 1960s, with the benefit of over 20 years of hindsight. The list of failures is as long as the list of successes. Below are some of the failures.

The 1960s were a time of great economic prosperity and technological advancement in the United States…One of the most extensive and widely publicized studies of future growth markets was TRW Inc. 'Probe of the Future'. The results . . . appeared in many business publications in the late 1960s…Not all…were released. Of the ones that were released, nearly all were wrong! Nuclear-powered underwater recreation centers, a 500 kilowatt nuclear power plant on the moon, 3D color TV, robot soldiers, automatic vehicle control on the interstate system, and plastic germproof houses were amongst some of the growth markets identified by this study.

In 1966, industry experts predicted, 'The shipping industry appears ready to enter the jet age.' By 1968, large cargo ships powered by gas turbine engines were expected to penetrate the commercial market. The benefits of this innovation were greater reliability, quicker engine starts and shorter docking times.

Even dentistry foresaw technological wonders…in 1968, the Director of the National Institute of Dental Research, a division of the US Public Health Service, predicted that 'in the next decade, both tooth decay and the most prevalent form of gum disease will come to a virtual end'. According to experts at this agency, by the late 1970s false teeth and dentures would be 'anachronisms' replaced by plastic teeth implant technology. A vaccine against tooth decay would also be widely available and there would be little need for dental drilling.

Source: Schnaars, S. and C. Berenson (1986) Growth market forecasting revisited: a look back at a look forward. *California Management Review*, **28**, 71–88.

In practice there will be a trade-off between the cost and robustness of a forecast. The more common methods of forecasting such as trend extrapolation and time series are of limited use for new products, because of the lack of past data. However, regression analysis can be used to identify the main factors driving demand for a given product, and therefore provide some estimate of future demand, given data on the underlying drivers.

For example, a regression might express the likely demand for the next generation of digital mobile phones in terms of rate of economic growth, price relative to competing systems, rate of new business formation, and so on. Data are collected for each of the chosen variables and coefficients for each derived from the curve that best describes the past data. Thus the reliability of the forecast depends a great deal on selecting the right variables in the first place. The advantage of regression is that, unlike simple extrapolation or time-series analysis, the forecast is based on cause and effect relations. Econometric models are simply bundles of regression equations, including their interrelationship. However, regression analysis is of little use where future values of an explanatory value are unknown, or where the relationship between the explanatory and forecast variables may change.

Leading indicators and analogues can improve the reliability of forecasts, and are useful guideposts to future trends in some sectors. In both cases there is a historical relationship between two trends. For example, new business start-ups might be a leading indicator of the demand for fax machines in six months' time. Similarly, business users of mobile telephones may be an analogue for subsequent patterns of domestic use.

Such 'normative' techniques are useful for estimating the future demand for existing products, or perhaps alternative technologies or novel niches, but are of limited utility in the case of more radical systems innovation. Exploratory forecasting, in contrast, attempts to explore the range of future possibilities. The most common methods are:

- customer or market surveys
- internal analysis, e.g. brainstorming
- Delphi or expert opinion
- scenario development.

Customer or market surveys

Most companies conduct customer surveys of some sort. In consumer markets this can be problematic simply because customers are unable to articulate their future needs. For example, Apple's iPod was not the result of extensive market research or customer demand, but largely because of the vision and commitment of Steve Jobs. In industrial markets, customers tend to be better equipped to communicate their future requirements, and consequently, business-to-business innovations often originate from customers. Companies can also consult their direct sales force, but these may not always be the best guide to future customer requirements. Information is often filtered in terms of existing products and services, and biased in terms of current sales performance rather than long-term development potential.

There is no 'one best way' to identify novel niches, but rather a range of alternatives. For example, where new products or services are very novel or complex, potential users may not be aware of, or able to articulate, their needs. In such cases traditional methods of market research are of little use, and there will be a greater burden on developers of radical new products and services to 'educate' potential users.

Our own research confirms that different managerial processes, structures and tools are appropriate for routine and novel development projects.[3] We discuss this in detail in Chapter 9, when we examine new product and service development. For example, in terms of frequency of use, the most common methods used for high novelty projects are segmentation, prototyping, market experimentation and industry experts; whereas for the less novel projects the most common methods are partnering customers, trend extrapolation and segmentation. The use of market experimentation and industry experts might be expected where market requirements or technologies are uncertain, but the common use of segmentation for such projects is harder to justify. However, in terms of usefulness, there are statistically significant differences in the ratings for segmentation, prototyping, industry experts, market surveys and latent needs analysis. Segmentation is more effective for routine development projects; and prototyping, industry experts, focus groups and latent needs analysis are all more effective for novel development projects.[4] Lead users are particularly effective for anticipating emerging market needs, as demonstrated by the case of 3M.

Internal analysis, e.g. brainstorming

Structured idea generation, or brainstorming, aims to solve specific problems or to identify new products or services. Typically, a small group of experts is gathered together and allowed to interact. A chairman records all suggestions without comment or criticism. The aim is to identify, but not evaluate, as many opportunities or solutions as possible. Finally, members of the group vote on the different suggestions. The best results are obtained when representatives from different functions are present, but this can be difficult to manage. Brainstorming does not produce a forecast as such, but can provide useful input to other types of forecasting.

We discussed a range of approaches to creative problem solving and idea generation in Chapter 3. Most of these are relevant here, and include ways of:[5]

- *Understanding the problem* – the active construction by the individual or group through analysing the task at hand (including outcomes, people, context and methodological options) to determine whether and when deliberate problem-structuring efforts are needed. This stage includes constructing opportunities, exploring data and framing problems.
- *Generating ideas* – to create options in answer to an open-ended problem. This includes generating and focusing phases. During the generating phase of this stage, the person or group produces many options (fluent thinking), a variety of possible options (flexible thinking), novel or unusual options (original thinking) or a number of detailed or refined options (elaborative thinking). The focusing phase provides an opportunity for examining, reviewing, clustering and selecting promising options.
- *Planning for action* – is appropriate when a person or group recognizes a number of interesting or promising options that may not necessarily be useful, valuable or valid. The aim is to make or develop effective choices, and to prepare for successful implementation and social acceptance.

External assessment, e.g. Delphi

The opinion of outside experts, or Delphi method, is useful where there is a great deal of uncertainty or for long time horizons.[6] Delphi is used where a consensus of expert opinion is required on the timing, probability and identification of future technological goals or consumer needs and the factors likely to affect their achievement. It is best used in making long-term forecasts and revealing how new technologies and other factors could trigger discontinuities in technological trajectories. The choice of experts and the identification of their level and area of expertise are important; the structuring of the questions is even more important. The relevant experts may include suppliers, dealers, customers, consultants and academics. Experts in non-technological fields can be included to ensure that trends in economic, social and environmental fields are not overlooked.

The Delphi method begins with a postal survey of expert opinion on what the future key issues will be, and the likelihood of the developments. The response is then analysed, and the same sample of experts resurveyed with a new, more focused questionnaire. This procedure is repeated until some convergence of opinion is observed, or conversely if no consensus is reached. The exercise usually consists of an iterative process of questionnaire and feedback among the respondents; this process finally yields a Delphi forecast of the range of experts' opinions on the probabilities of certain events occurring by a quoted time. The method seeks to nullify the disadvantage of face-to-face meetings at which there could be deference to authority or reputation, a reluctance to admit error, a desire to conform or differences in persuasive ability. All of these could lead to an inaccurate consensus of opinion. The quality of the forecast is highly dependent on the expertise and calibre of the experts; how the experts are selected and how many should be consulted are important questions to be answered. If international experts are used, the exercise can take a considerable length of time, or the number of iterations may have to be curtailed. Although seeking a consensus may be important, adequate attention should be paid to views that differ radically 'from the norm' as there may be important underlying reasons to justify such maverick views. With sufficient design, understanding and resources, most of the shortcomings of the Delphi technique can be overcome and it is a popular technique, particularly for national foresight programmes.

In Europe, governments and transnational agencies use Delphi studies to help formulate policy, usually under the guise of 'Foresight' exercises. In Japan, large companies and the government routinely survey expert opinion in order to reach some consensus in those areas with the greatest potential for long-term development. Used in this way, the Delphi method can to a large extent become a self-fulfilling prophecy.

Scenario development

Scenarios are internally consistent descriptions of alternative possible futures, based upon different assumptions and interpretations of the driving forces of change.[7] Inputs include quantitative data and analysis, and qualitative assumptions and assessments, such as societal, technological, economical, environmental and political drivers. Scenario development is not strictly-speaking prediction, as it assumes that the future is uncertain and that the path of current developments can range from the conventional to the revolutionary. It is particularly good at incorporating potential critical events which might result in divergent paths or branches being pursued.

Scenario development can be normative or explorative. The normative perspective defines a preferred vision of the future and outlines different pathways from the goal to the present. For example, this is commonly used in energy futures and sustainable futures scenarios. The explorative approach defines the drivers of change, and creates scenarios from these without explicit goals or agendas.

For scenarios to be effective they need to inclusive, plausible and compelling (as opposed to being exclusive, implausible or obvious), as well as being challenging to the assumptions of the stakeholders. They should make the assumptions and inputs used explicit, and form the basis of a process of discussion, debate, policy, strategy and ultimately action. The output is typically two or three contrasting scenarios, but the process of development and discussion of scenarios is much more valuable.

Scenario development may involve many different forecasting techniques, including computer-based simulation. Typically, it begins with the identification of the critical indicators, which might include use of brainstorming and Delphi techniques. Next, the reasons for the behaviour of these indicators are examined, perhaps using regression techniques. The future events which are likely to affect these indicators are identified. These are used to construct the best, worst and most-likely future scenarios. Finally, the company assesses the impact of each scenario on its business. The goal is to plan for the outcome with the greatest impact, or better still, retain sufficient flexibility to respond to several different scenarios. Scenario development is a key part of the long-term planning process in those sectors characterized by high capital investment, long lead times and significant environmental uncertainty, such as energy, aerospace and telecommunications.

RESEARCH NOTE   The pre-diffusion phase

The S-shaped diffusion curve is empirically observed for a broad range of new products such as the telephone, hybrid corn and the microwave oven. However, a critical, but under-researched issue in diffusion research is what happens *before* this well-known S-shaped diffusion curve. From a managerial perspective it is important to realize that diffusion requires that several conditions have to be met: for example, products have to be developed, produced, distributed and the necessary infrastructural arrangements have to be in place. It is seldom realized, however, that prior to any S-shaped diffusion curve, the market introduction of a new product is more typically followed by an erratic pattern of diffusion, referred to as the pre-diffusion phase. The lack of attention to this so-called pre-diffusion phase is one of the main limitations of mainstream research and practice.

1. The pre-diffusion phase for new products

We define the pre-diffusion phase to begin after the market introduction of the first new product and to end when the diffusion of this type of product takes off, i.e. when the regular S-shaped diffusion curve begins. After the introduction of the first product, instead of a smooth S-curve, in practice an erratic process of diffusion may occur. In this situation the market is unstable. In the field of telecommunications, for example, the diffusion of new communication products and services often starts with the periodic introduction, decline and re-introduction of product variants in multiple small-scale applications before mainstream applications and product designs appear and the diffusion takes off.

The table below shows estimates of the length of the pre-diffusion phase for a sample of products from different industries.

Length of the pre-diffusion phase of products from different industries

Product	Industry	Market introduction	Diffusion begins	Length of pre-diffusion phase (years)
Jet engine	Aerospace and defence	1941	1943	2
Radar		1934	1939	5
ABS	Automobile and parts	1959	1978	19
Airbag		1972	1988	16
Memory metal	Materials, compound and metals	1968	1972	4
Dyneema		1975	1990	15
Flash memory	IT and telecommunications hardware and software	1988	2001	13
Mobile telephony		1946	1983	37
Transistor	Electronic components & equipment	1949	1953	4
Television		1939	1946	7
Contraceptive pill	Medical equipment and medicines	1928	1962	34
MRI		1980	1983	3
Microwave oven	Personal goods and household equipment	1947	1955	8
Air conditioning		1902	1915	13

Average = 13

St dev = 11

Data in the table are derived from multiple sources and are based on original work from J.R. Ortt (2004, 2008).

From the table we can see that a significant pre-diffusion phase exists for most types of innovation. The average length of this phase for the sample of products is more than a decade. Moreover, the data shows that, even within industries, the variation in the length of the pre-diffusion phase is considerable.

2. Different perspectives on, and main causes of, the occurrence of the pre-diffusion phase

The pre-diffusion phase has been described from different scientific perspectives, each of which proposes alternative causes of this phase. Marx, for example, is an economist who more than 150 years ago described why it takes so long to implement new methods of production in companies and why these new methods at first diffuse remarkably slowly among companies in an industry. Marx focuses on the supply side of the market when describing the diffusion of these methods of production (so-called capital goods). From this perspective, the pre-diffusion phase is seen as a kind of trial-and-error process that is required to improve the production methods and to adapt these methods to the prevailing way of working in companies (and the other way around) before these methods become profitable.

About a century later, diffusion researchers took a different perspective and focused on the demand side of the market (Rogers, 2003). These researchers, mostly sociologists, tend to see the diffusion process as a communication process in a population or a segment of customers. The researchers have a bias towards the smooth S-shaped diffusion curve, but upon closer inspection their findings also indicate how demand-side factors may cause a pre-diffusion phase. Characteristics of subsequent groups of customers are often assessed in diffusion research. The very first group of customers, the innovators, are often deviant from the remainder of the potential customers and thereby might hamper the communication process that is required for diffusion.

Moore (2002) elaborates on this idea and concludes that a 'chasm' occurs between subsequent groups of customers. Moore focuses on the interaction of the demand and supply side of the market when he explains this chasm. The first types of customers, referred to as technology enthusiasts and visionaries, are customers willing to experiment with the product. Mainstream customers, however, hardly communicate with these sub-segments, so the diffusion does not proceed smoothly. Moreover, the mainstream customers want completely different product versions: they want reliable, foolproof and complete packages of products and services. Rather than testing these requirements themselves, they prefer to see how well-known companies or customers have already successfully implemented the product in their process of working. The technology enthusiasts or visionaries cannot fulfil this role and a chasm therefore occurs.

3. Main managerial consequences of the pre-diffusion phase

Each of these perspectives has its own way of explaining why this phase is managerially important. Marx' perspective implies that large-scale diffusion of new production methods is often preceded by considerable periods of experimentation. The costs incurred in this pre-diffusion phase can be considerable; the profits for the first company that in an economically viable way masters the application of these methods can be very large as well. Marx' perspective illustrates the importance of managing the innovation process before the implementation of new methods of production. Chasms in the diffusion process, noticed by Rogers and Moore, indicate that market introduction strategies of new products are crucially important as well. Segments of potential customers may be hard to distinguish and subsequent segments of customers may require completely different product variants and business models and thereby hamper the smooth diffusion process.

From a management perspective, the pre-diffusion phase is very risky. It is remarkable how many companies involved in the invention of new products lose out. About half of the pioneers that are first to introduce a *successful* product in the market, fail and vanish before their product diffuses on a large scale. One of the main reasons is that the pre-diffusion phase can last a very long time. In general, the pre-diffusion phase requires considerable investment yet does not generate the same amount of income. The existence of the pre-diffusion phase has profound managerial implications: it shows that introducing a new product usually is a matter of deep pockets and long breath.

From: J. Roland Ortt, and partly based on Ortt & Schoormans (2004) and Ortt & Delgoshaie (2008).

Marx, K. (1867) *Capital: A Critique of Political Economy*, Penguin edition, Middlesex, 1976; Moore, G.A. (2002) *Crossing the Chasm. Marketing and Selling Disruptive Products to Mainstream Customers*, HarperCollins, New York; Ortt, J.R. and N. Delgoshaie (2008) Why does it take so long before the diffusion of new high-tech products takes off? In B. Abu-Hijleh, M. Arif, T. Khalil and Y. Hosni, eds, *Proceedings of the 17th International Conference on Management of Technology* (6–10 April), Dubai; Ortt, J.R. and J.P.L. Schoormans (2004) The pattern of development and diffusion of breakthrough communication technologies. *European Journal of Innovation Management*, 7 (4), 292–302; Rogers, E.M. (2003) *Diffusion of Innovations*, fifth edition, Free Press, New York.

8.3 Estimating the adoption of innovations

A better understanding of why and how innovations are adopted (or not) can help us to develop more realistic plans. As the Research Note on the chasm between development and successful adoption shows, around half of all innovations never reach the intended markets. Conventional marketing approaches are fine for many products and services, but not for innovations. Marketing texts often refer to 'early adopters' and 'majority adopters', and even go so far as to apply numerical estimates of these, but these simple categories are based on the very early studies of the state-sponsored diffusion of hybrid-seed varieties in farming communities, and are far from universally applicable. To better plan for innovations we need a deeper understanding of what factors promote and constrain adoption, and how these influence the rate and level of diffusion within different markets and populations.

There are many barriers to the widespread adoption of innovations, including:

- *Economic* – personal costs versus social benefits, access to information, insufficient incentives.
- *Behavioural* – priorities, motivations, rationality, inertia, propensity for change or risk.
- *Organizational* – goals, routines, power and influence, culture and stakeholders.
- *Structural* – infrastructure, sunk costs, governance.

For these reasons, historically, large complex socio-technical systems tend to change only incrementally. However, more radical transformations can occur, but these often begin in strategic niches, with different goals, needs, practices and processes. As these niches demonstrate and develop the innovations, through social experimentation and learning, they may begin to influence or enter the mainstream. This may be through whole new market niches, or by forming hybrid markets between the niche and mainstream.

Rogers' definition of diffusion is used widely: '*the process by which an innovation is communicated through certain channels over time among members of a social system. It is a special type of communication, in that the messages are concerned with new ideas*'.[4] However, there are no generally accepted definitions of associated terms such as 'technology transfer', 'adoption', 'implementation' or 'utilization'. Diffusion usually involves the analysis of the spread of a product or idea in a given social system, whereas technology transfer is usually a point-to-point phenomenon. Technology transfer usually implies putting information to use, or more specifically moving ideas from the laboratory to the market. The distinction between adoption, implementation and utilization is less clear. Adoption is generally considered to be the decision to acquire something, whereas implementation and utilization imply some action and adaptation.

The literature on diffusion is vast and highly fragmented. However, a number of different approaches to diffusion research can be identified, each focusing on particular aspects of diffusion and adopting different methodologies. The main contributions have been from economics, marketing, sociology and anthropology. Economists have developed a number of econometric models on the diffusion of new products and processes in an effort to explain past behaviour and to predict future trends. Prediction is a common theme of the marketing literature. Marketing studies have adopted a wide range of different research instruments to examine buyer behaviour, but most recent research has focused on social and psychological factors. Development economics and rural sociology have both examined the adoption of agricultural innovations, using statistical analysis of secondary data and collection of primary data from surveys. Much of the anthropological research has been based on case studies of the diffusion of new ideas in tribes, villages or communities. Most recently, there has been a growing number of multidisciplinary studies which have examined the diffusion of educational, medical and other policy innovations.

Processes of diffusion

Research on diffusion attempts to identify what influences the rate and direction of adoption of an innovation. The diffusion of an innovation is typically described by an S-shaped (logistic) curve (Figure 8.1). Initially, the rate of adoption is low, and adoption is confined to so-called 'innovators'. Next to adopt are the 'early adopters', then the 'late majority', and finally the curve tails off as only the 'laggards' remain. Such taxonomies are fine with the benefit of hindsight, but provide little guidance for future patterns of adoption.[8]

Hundreds of marketing studies have attempted to fit the adoption of specific products to the S-curve, ranging from television sets to new drugs. In most cases mathematical techniques can provide a relatively good fit with historical data, but research has so far failed to identify robust generic models of adoption. In practice the precise pattern of adoption of an innovation will depend on the interaction of demand-side and supply-side factors:

- *Demand-side factors* – direct contact with or imitation of prior adopters, adopters with different perceptions of benefits and risk.
- *Supply-side factors* – relative advantage of an innovation, availability of information, barriers to adoption, feedback between developers and users.

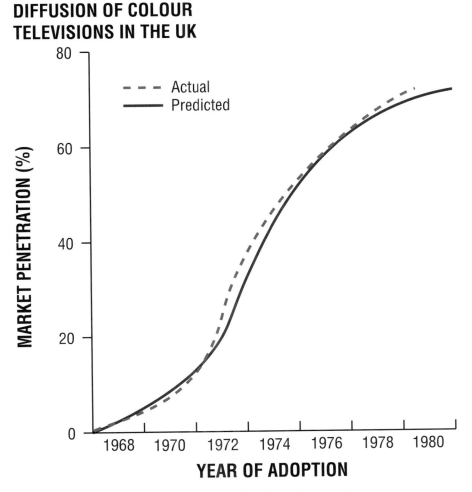

FIGURE 8.1: Typical diffusion S-curve for the adoption of an innovation

Source: Meade, N. and Islam, T. (2006), 'Modeling and forecasting the diffusion of innovation – a 25 year review', *International Journal of Forecasting*, 22 (3), 519–545.

The epidemic S-curve model is the earliest and is still the most commonly used. It assumes a homogeneous population of potential adopters, and that innovations spread by information transmitted by personal contact, observation and the geographical proximity of existing and potential adopters. This model suggests that the emphasis should be on communication, and the provision of clear technical and economic information. However, the epidemic model has been criticized because it assumes that all potential adopters are similar and have the same needs, which is unrealistic.

The Probit model takes a more sophisticated approach to the population of potential adopters. It assumes that potential adopters have different threshold values for costs or benefits, and will only adopt beyond some critical or threshold value. In this case differences in threshold values are used to explain different rates of adoption. This suggests that the more similar potential adopters are, the faster the diffusion.

However, adopters are assumed to be relatively homogeneous, apart from some difference in progressiveness or threshold values. Supply-side models do not consider the possibility that the rationality and the profitability of adopting a particular innovation might be different for different adopters. For example, local 'network externalities' such as the availability of trained skilled users, technical assistance and maintenance, or complementary technical or organizational innovations are likely to affect the cost of adoption and use, as distinct from the cost of purchase.

Also, it is unrealistic to assume that adopters will have perfect knowledge of the value of an innovation. Therefore Bayesian models of diffusion introduce lack of information as a constraint to diffusion. Potential adopters are allowed to hold different beliefs regarding the value of the innovation, which they may revise according to the results of trials to test the innovation. Because these trials are private, imitation cannot take place and other potential adopters cannot learn from the trials. This suggests better-informed potential adopters may not necessarily adopt an innovation earlier than the less well informed, which was an assumption of earlier models.[9]

Slightly more realistic assumptions, such as those of the Bass model, include two different groups of potential adopters: innovators, who are not subject to social emulation; and imitators, for whom the diffusion process takes the epidemic form. This produces a skewed S-curve because of the early adoption by innovators, and suggests that different marketing processes are needed for the innovators and subsequent imitators. The Bass model is highly influential in economics and marketing research, and the distinction between the two types of potential adopters is critical in understanding the different mechanisms involved in the two user segments.

Bandwagons may occur where an innovation is adopted because of pressure caused by the sheer number of those who have already adopted an innovation, rather than by individual assessments of the benefits of an innovation. In general, as soon as the number of adopters has reached a certain threshold level, the greater the level of ambiguity of the innovation's benefits, the greater the subsequent number of adopters. This process allows technically inefficient innovations to be widely adopted, or technically efficient innovations to be rejected. Examples include the QWERTY keyboard, originally designed to prevent professional typists from typing too fast and jamming typewriters; and the DOS operating system for personal computers, designed by and for computer enthusiasts.

Bandwagons occur due to a combination of competitive and institutional pressures.[10] Where competitors adopt an innovation, a firm may adopt because of the threat of lost competitiveness, rather than as a result of any rational evaluation of benefits. For example, many firms adopted flexible manufacturing systems (FMS) in the 1980s in response to increased competition, but most failed to achieve significant benefits. The main institutional pressure is the threat of lost legitimacy, for example, being considered by peers or customers as being less progressive or competent.[11]

The critical difference between bandwagons and other types of diffusion is that they require only limited information to flow from early to later adopters. Indeed, the more ambiguous the benefits of an innovation, the more significant bandwagons are on rates of adoption. Therefore the process of diffusion must be managed with as much care as the process of development. In short, better products do not necessarily result in more sales. Not everybody requires a better mousetrap.

Finally, there are more sociological and psychological models of adoption, which are based on interaction and feedback between the developers and potential adopters.[12] These perspectives consider how individual psychological characteristics such as attitude and perception affect adoption. Individual motivations, perceptions, likes and dislikes determine what information is reacted to and how it is processed. Potential adopters will be guided and prejudiced by experience, and will have 'cognitive maps' which filter information and guide behaviour. Social context will also influence individual behaviour. Social structures and meaning systems are locally constructed, and therefore highly context-specific. These can distort the way in which information is interpreted and acted upon. Therefore the perceived value of an innovation, and hence its subsequent adoption, is not some objective fact, but instead depends on individual psychology and social context. These factors are particularly important in the later stages of diffusion. For example, lifestyle aspirations, such as having more exercise and adopting a healthy diet have created the opportunity for many new products and services. Initially relying on local demand in North London and word-of-mouth recommendations, the fruit drink company Innocent Smoothies became a global phenomenon. It successfully anticipated a demand for convenient fruit consumption which the existing multinationals such as Unilever and Nestlé missed.

Initially, the needs of early adopters or innovators dominate, and therefore the characteristics of an innovation are most important. Innovations tend to evolve over time through improvements required by these early users, which may reduce the relative cost to later adopters. However, early adopters are almost by definition 'atypical', for example, they tend to have superior technical skills. As a result the preferences of early adopters can have a disproportionate impact on the subsequent development of an innovation, and result in the establishment of inferior technologies or abandonment of superior alternatives.

Factors influencing adoption

Numerous variables have been identified as affecting the diffusion and adoption of innovations, but these can be grouped into three clusters: characteristics of the innovation itself; characteristics of individual or organizational adopters; and the characteristics of the environment. Characteristics of an innovation found to influence adoption include relative advantage, compatibility, complexity, observability and trialability. Individual characteristics include age, education, social status and attitude to risk. Environmental and institutional characteristics include economic factors such as the market environment and sociological factors like communications networks. However, whilst there is a general agreement regarding the relevant variables, there is very little consensus on the relative importance of the different variables, and in some cases disagreements over the direction of relationships.

Characteristics of an innovation

A number of characteristics of an innovation have been found to affect diffusion and adoption:[4]

- relative advantage
- compatibility
- complexity
- trialability
- observability.

Relative advantage
Relative advantage is the degree to which an innovation is perceived as better than the product it supersedes, or competing products. Relative advantage is typically measured in narrow economic terms, for example cost or financial payback, but non-economic factors such as convenience, satisfaction and social prestige may be equally important. In theory, the greater the perceived advantage, the faster the rate of adoption.

It is useful to distinguish between the primary and secondary attributes of an innovation. Primary attributes, such as size and cost, are invariant and inherent to a specific innovation irrespective of the adopter. Secondary attributes, such as relative advantage and compatibility, may vary from adopter to adopter, being contingent upon the perceptions and context of adopters. In many cases, a so-called 'attribute gap' will exist. An attribute gap is the discrepancy between a potential user's perception of an attribute or characteristic of an item of knowledge and how the potential user would prefer to perceive that attribute. The greater the sum of all attribute gaps, the less likely a user is to adopt the knowledge. This suggests that preliminary testing of an innovation is desirable in order to determine whether significant attribute gaps exist. Not all attribute gaps require changes to the innovation itself – a distinction needs to be made between knowledge content and knowledge format. The idea of pre-testing information for the purposes of enhancing its value and acceptance is not widely practised.

Compatibility

Compatibility is the degree to which an innovation is perceived to be consistent with the existing values, experience and needs of potential adopters. There are two distinct aspects of compatibility: existing skills and practices; and values and norms. The extent to which the innovation fits the existing skills, equipment, procedures and performance criteria of the potential adopter is important, and relatively easy to assess.

However, compatibility with existing practices may be less important than the fit with existing values and norms.[13] Significant misalignments between an innovation and an adopting organization will require changes in the innovation or organization, or both. In the most successful cases of implementation, mutual adaptation of the innovation and organization occurs.[14] However, few studies distinguish between compatibility with value and norms, and compatibility with existing practices. The extent to which the innovation fits the existing skills, equipment, procedures and performance criteria of the potential adopter is critical. Few innovations initially fit the user environment into which they are introduced. Significant misalignments between the innovation and the adopting organization will require changes in the innovation or organization, or in the most successful cases of implementation, mutual adaptation of both. Initial compatibility with existing practices may be less important, as it may provide limited opportunity for mutual adaptation to occur.

Complexity

Complexity is the degree to which an innovation is perceived as being difficult to understand or use. In general, innovations which are simpler for potential users to understand will be adopted more rapidly than those which require the adopter to develop new skills and knowledge.

However, complexity can also influence the direction of diffusion. Evolutionary models of diffusion focus on the effect of 'network externalities', that is the interaction of consumption, pecuniary and technical factors which shape the diffusion process. For example, within a region the cost of adoption and use, as distinct from the cost of purchase, may be influenced by: the availability of information about the technology from other users, of trained skilled users, technical assistance and maintenance, and of complementary innovations, both technical and organizational.

Trialability

Trialability is the degree to which an innovation can be experimented with on a limited basis. An innovation that is trialable represents less uncertainty to potential adopters, and allows learning by doing. Innovations which can be trialled will generally be adopted more quickly than those which cannot. The exception is where the undesirable consequences of an innovation appear to outweigh the desirable characteristics. In general, adopters wish to benefit from the functional effects of an innovation, but avoid any dysfunctional effects. However, where it is difficult or impossible to separate the desirable from the undesirable consequences trialability may reduce the rate of adoption.

Developers of an innovation may have two different motives for involving potential users in the development process. First, to acquire knowledge from the users needed in the development process, to ensure usability and to add value. Second, to attain user 'buy-in', that is user acceptance of the innovation and commitment to its use. The second motive is independent of the first, because increasing user acceptance does not necessarily improve the quality of the innovation. Rather, involvement may increase users' tolerance of any inadequacies. In the case of point-to-point transfer, typically both motives are present.

However, in the case of diffusion it is not possible to involve all potential users, and therefore the primary motive is to improve usability rather than attain user buy-in. But even the representation of user needs must be indirect, using surrogates such as specially selected user groups. These groups can be problematic for a number of reasons. First, because they may possess atypically high levels of technical knowledge, and therefore are not representative. Second, where the group must represent diverse user needs, such as both experienced and novice users, the group may not work well together. Finally, when user representatives work closely with developers over a long period of time they may cease to represent users, and instead absorb the developer's viewpoint. Thus, there is no simple relationship between user involvement and user satisfaction. Typically, very low levels of user involvement are associated with user dissatisfaction, but extensive user involvement does not necessarily result in user satisfaction.

Observability

Observability is the degree to which the results of an innovation are visible to others. The easier it is for others to see the benefits of an innovation, the more likely it will be adopted. The simple epidemic model of diffusion assumes that innovations spread as potential adopters come into contact with existing users of an innovation.

Peers who have already adopted an innovation will have what communication researchers call 'safety credibility', because potential adopters seeking their advice will believe they know what it is really like to implement and utilize the innovation. Therefore early adopters are well positioned to disseminate 'vicarious learning' to their colleagues. Vicarious learning is simply learning from the experience of others, rather than direct personal experimental learning. However, the process of vicarious learning is neither inevitable nor efficient because, by definition, it is a decentralized activity. Centralized systems of dissemination tend to be designed and rewarded on the basis of being the source of technical information, rather than for facilitating learning among potential adopters.

Over time learning and selection processes foster both the evolution of the technologies to be adopted and the characteristics of actual and potential adopters. Thus an innovation may evolve over time through improvements made by early users, thereby reducing the relative cost to later adopters. In addition, where an innovation requires the development of complementary features, for example a specific infrastructure, late adopters will benefit. This suggests that instead of a single diffusion curve, a series of diffusion curves will exist for the different environments. However, there is a potential drawback to this model. The short-term preferences of early adopters will have a disproportionate impact on the subsequent development of the innovation, and may result in the establishment of inferior technologies and abandonment of superior alternatives. In such cases interventionist policies may be necessary to postpone the lock-in phenomenon.

From a policy perspective, high visibility is often critical. However, high visibility, at least initially, may be counter-productive. If users' expectations about an innovation are unrealistically high and adoption is immediate, subsequent disappointment is likely. Therefore in some circumstances it may make sense to delay dissemination or to slow the rate of adoption. However, in general researchers and disseminators are reluctant to withhold knowledge.

The choice between the different models of diffusion and factors that will most influence adoption will depend on the characteristics of the innovation and nature of potential adopters. The simple epidemic model appears to provide a good fit to the diffusion of new processes, techniques and procedures, whereas the Bass model appears to best fit the diffusion of consumer products. However, the mathematical structure of the epidemic and Bass models tends to overstate the importance of differences in adopter characteristics, and tends to underestimate the effect of macroeconomic and supply-side factors. In general, both these models of diffusion work best where the total potential market is known, that is for derivatives of existing products and services, rather than totally new innovations.

In the case of systemic or network innovations, a wider range of factors have to be managed to promote adoption and diffusion. In such cases a wider set of actors and institutions on the supply and demand side are relevant, in what has been called an adoption network.[15] On the supply side, other organizations may provide the infrastructure, support and complementary products and services which can promote or prevent adoption and diffusion. For example, in 2008 the two-year battle between the new high-definition DVD formats was decided not by price or any technical superiority, but rather because the Blu-ray consortium managed to recruit more film studios to its format than the competing HD-DVD format. As soon as the uncertainty over the future format was resolved, there was a step change increase in the rate of adoption.

On the demand side, the uncertainty of potential adopters, and communication with and between them needs to be managed. Whilst early adopters may emphasize technical performance and novelty above other factors, the mainstream mass market is more likely to be concerned with factors such as price, quality, convenience and support. This transition from the niche market and needs of early adopters, through to the requirements of more mass markets has been referred to as crossing the chasm by Moore.[16] Moore studied the successes and many more failures of Silicon Valley and other high-technology products, and argued that the critical success factors for early adopters and mass markets were fundamentally different, and most innovations failed to make this transition. Therefore the successful launch and diffusion of a systemic or network innovation demands attention to traditional marketing issues such as the timing and positioning of the product or service,[17] but also significant effort to demand-side factors such as communication and interactions between potential adopters.[18]

The continued improvement in health in the advanced economies over the past 50 years can be attributed in part to the supply of new diagnostic techniques, drugs and procedures, but also to changes in the demand side, such as increases in education, income and service infrastructure. However, the focus of innovation (and policy) in healthcare is too often on the development and commercialization of new pharmaceuticals, but this is only a part of the story. This is a clear case of systemic innovation, in which firm and public R&D are necessary, but not sufficient to promote improved health. The adoption network includes regulatory bodies, national health assessment and reference pricing schemes, regional health agencies, public and private insurers, as well as the more obvious hospitals, doctors, nurses and patients.[19] However, too often the management and policy for innovation in health is confined to regulation of prices and effects of intellectual property regimes.[20] There is a clear need for new methods of interaction, involvement and engagement in such cases.[21]

Diffusion research and practice has been criticized for an increasingly limited scope and methodology. Rogers identifies a number of shortcomings of research and practice:[4]

1. Diffusion has been seen as a *linear, unidirectional communication* activity in which the active source of research or information attempts to influence the attitudes and/or behaviours of essentially passive receivers. However, in most cases diffusion is an interactive process of adaptation and adoption.
2. Diffusion has been viewed as a *one-to-many communication* activity, but point-to-point transfer is also important. Both centralized and decentralized systems exist. Decentralized diffusion is a process of convergence as two or more individuals exchange information in order to move toward each other in the meanings they ascribe to certain events.
3. Diffusion research has been preoccupied with an *action-centred and issue-centred communication* activity, such as selling products, actions or policies. However, diffusion is also a social process, affected by social structure and position and interpersonal networks.
4. Diffusion research has used *adoption as the dependent variable* – the decision to use the innovation, rather than implementation itself – the consequences of the innovation. Most studies have used attitudinal change as the dependent variable, rather than change in overt behaviour.
5. Diffusion research has suffered from an implicit *pro-innovation bias*, which assumes that an innovation should be adopted by all members of a social system as rapidly as possible. Therefore the process of adaptation or rejection of an innovation has been overlooked, and there have been relatively few studies of how to prevent the diffusion 'bad' innovations.

RESEARCH NOTE **Why innovations fail to be adopted**

This research examined the factors which influence the adoption and diffusion of innovations drawing upon case studies of successful and less successful consumer electronics products, such as the Sony PlayStation and MiniDisc, Apple iPod and Newton, TomTom GO, TiVo and RIM Blackberry.

The study finds that a critical factor influencing successful diffusion is the careful management of acceptance by the early adopters, which in turn influences the adoption by the main market. Strategic issues such as positioning, timing and management of the adoption network are identified as being important. The adoption network is defined as a configuration of users, peers, competitors, and complementary products and services and infrastructure. However, the positioning, timing and adoption networks are different for the early and main market adopters, and failure to recognize these differences is a common cause of the failure of innovations to diffuse widely. Also, innovation contingencies such as the degree of radicalness and discontinuity affect how these factors interact and how these need to be managed to promote acceptance. The relevant assessment of the radicalness and discontinuity of an innovation is not based on the technological aspects, but rather the effects on user behaviour and consumption.

To promote use by early adopters, the research recommends that four enabling factors need to be managed: legitimize the innovation through reference customers and visible performance advantage; trigger word of mouth within specialist communities of practice; stimulate imitation to increase the user base and peer pressure; and collaborate with opinion leaders. Significantly, the study argues that the subsequent successful diffusion of an innovation into the mainstream market has very little to do with the merits of the product itself, and much more to do with the positive acceptance of early adopters and repositioning and targeting for the main market by influencing the relevant adoption network.

Source: Frattini, F. (2008) The commercialisation of innovation in high-tech markets, PhD thesis, Politecnico di Milano, Italy.

8.4 Assessing risk, recognizing uncertainty

Dealing with risk and uncertainty is central to the assessment of most innovative projects. Risk is usually considered to be possible to estimate, either qualitatively – high, medium, low – or ideally by probability estimates. Uncertainty is by definition unknowable, but nonetheless the fields and degree of uncertainty should be identified to help to select the most appropriate methods of assessment and plan for contingencies. Traditional approaches to assessing risk focus on the probability of foreseeable risks, rather than true uncertainty, or complete ignorance – what Donald Rumsfeld memorably called the 'unknown unknowns' (12 February 2002, US Department of Defense news briefing).

Research on new product development and R&D project management has identified a broad range of strategies for dealing with risk. Both individual characteristics and organizational climate influence perceptions of risk and propensities to avoid, accept or seek risks. Formal techniques such as failure mode and effects analysis (FMEA), potential problem analysis (PPA) and fault tree analysis (FTA) have a role, but the broader signals and support from the organizational climate is more important than the specific tools or methods used. For example, too many organizations emphasize project management in order to contain internal risks in the organization, but as a result fail to identify or exploit opportunities to take acceptable risks and to innovate.[22]

There are many approaches to risk assessment, but the most common issues to be managed include:

- Probabilistic estimates of technical and commercial success.
- Psychological (cognitive) and sociological perceptions of risk.
- Political and policy influences, such as the 'precautionary principle'.

Risk as probability

Research indicates that 30–45% of all projects fail to be completed, and over half of projects overrun their budgets or schedules by up to 200%. Figure 8.2 presents the results of a survey of R&D managers. Whilst most appear to be relatively confident when predicting technical issues such as the development time and costs, a much smaller proportion are confident when forecasting commercial aspects of the projects.

We examined how commonly different approaches to project assessment were used in practice. We surveyed 50 projects in 25 companies, and assessed how often different criteria were used, and how useful they were thought to be. Table 8.2 summarizes some of the results. Clearly probabilistic estimates of technical and commercial success are near universal, and considered to be of critical importance in all types of project assessment. These are usually combined with some form of financial assessment, and fit with the company strategy and capabilities.

Given the complexities involved, the outcomes of investments in innovation are uncertain, so that the forecasts (of costs, prices, sales volume, etc.) that underlie project and programme evaluations can

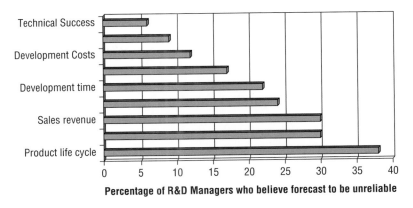

FIGURE 8.2: Uncertainty in project planning
Source: Derived from Freeman, C. and L. Soete (1997) *The Economics of Innovation*, MIT Press, Cambridge, MA.

Table 8.2	**Use and usefulness of criteria project screening and selection**			
	High novelty		**Low novelty**	
	Usage (%)	Usefulness	Usage (%)	Usefulness
Probability of technical success	100	4.37	100	4.32
Probability of commercial success	100	4.68	95	4.50
Market share*	100	3.63	84	4.00
Core competencies*	95	3.61	79	3.00
Degree of internal commitment	89	3.82	79	3.67
Market size	89	3.76	84	3.94
Competition	89	3.76	84	3.81
NPV/IRR	79	3.47	68	3.92
Payback period/break-even*	79	3.20	58	4.27

Usefulness score: 5 = critical; 0 = irrelevant. * denotes difference in usefulness rating is statistically significant at 5% level.
Source: Adapted from Tidd, J. & K. Bodley (2002) Effect of novelty on new product development processes and tools. *R&D Management,* 32 (2), 127–38. Based on 50 development projects.

be unreliable. According to Joseph Bower, management finds it easier, when appraising investment proposals, to make more accurate forecasts of reductions in production cost than of expansion in sales, whilst their ability to forecast the financial consequences of new product introductions is very limited indeed.[23] This last conclusion is confirmed by the study by Edwin Mansfield and his colleagues of project selection in large US firms.[24] By comparing project forecasts with outcomes, Mansfield showed that managers find it difficult to pick technological and commercial winners:

- Probability of *technical* success of projects $(Pt) = 0.80$
- Subsequent probability of *commercial* success $(Pc) = 0.20$
- Combined probability for all stages: $0.8 \times 0.2 = 0.16$

He also found that managers and technical managers cannot predict accurately the *development costs*, *time periods*, *markets* and *profits* of R&D projects. On average, costs were greatly *underestimated*, and time periods *overestimated* by 140–280% in incremental product improvements, and by 350–600% in major new products. Other studies have found that:

- About half business R&D expenditures are on *failed* R&D projects. The higher rate of success in *expenditures* than in *projects* reflects the weeding out of unsuccessful projects at their early stages and before large-scale commercial commitments are made to them.[25]
- R&D scientists and engineers are often deliberately overoptimistic in their estimates, in order to give the illusion of a high rate of return to accountants and managers.[26]

Trying to get involved in the right projects is worth an effort, both to avoid wasting time and resources in meaningless activities, and to improve the chances of success. Project appraisal and evaluation aims to:

1. Profile and gain an overall understanding of potential projects.
2. Prioritize a given set of projects, and where necessary reject projects.
3. Monitor projects, e.g. by following up the criteria chosen when the project was selected.
4. Where necessary, terminate a project.
5. Evaluate the results of completed projects.
6. Review successful and unsuccessful projects to gain insights and improve future project management, i.e. learning.

Project evaluation usually assumes that there is a choice of projects to pursue, but where there is no choice project evaluation is still important to help to assess the opportunity costs and what might be expected from pursuing a project. Different situations and contexts demand different approaches to project evaluation. We argued earlier that complexity and uncertainty are two of the most important dimensions for assessing projects. Different types of project will demand specific techniques, or at least different criteria for assessment. A common method used is the risk assessment matrix.

A large number of techniques have been developed over the years, and are still being developed and used today. Most of these can be described by means of some common elements which form the core of any project evaluation technique:

- *Inputs* into the assessment include likely costs and benefits in financial terms, probability of technical and market success, market attractiveness, and the strategic importance to the organization.

- *Weighting*: as certain data may be given more relevance than other (e.g. of market inputs compared with technical factors), in order to reflect the company's strategy or the company's particular views. The data is then processed to arrive at the outcomes.
- *Balancing* a range of projects, as the relative value of a project with respect to other projects is an important factor in situations of competition for limited resources. Portfolio management techniques are specifically devoted to deal with this factor.

Economic and cost-benefit approaches are usually based on a combination of expected utility or Bayesian assumptions. Expected utility theory can take into account probabilistic estimates and subjective preferences, and therefore deals well with risk aversion, but in practice utility curves are almost impossible to construct and individual preferences are different and highly subjective. Bayesian probability is excellent at incorporating the effects of new information, as we discussed earlier under the diffusion of innovations, but is very sensitive to the choice of relevant inputs and the weights attached to these.

As a result no technique should be allowed to determine outcomes, as these decisions are a management responsibility. Many techniques used today are totally or partially software based, which have some additional benefits in automating the process. In any case, the most important issue, for any method, is the managers' interpretation.

There is no single 'best' technique. The extent to which different techniques for project evaluation can be used will depend upon the nature of the project, the information availability, the company's culture and several other factors. This is clear from the variety of techniques that are theoretically available and the extent to which they have been used in practice. In any case, no matter which technique is selected by a company, it should be implemented, and probably adapted, according to the particular needs of that organization. Most of the techniques in practical use incorporate a mixture of financial assessment and human judgement.

Perceptions of risk

Probability estimates are only the starting point of risk assessment. Such relatively objective criteria are usually significantly moderated by psychological (cognitive) perceptions and bias, or overwhelmed altogether by sociological factors, such as peer pressure and cultural context. Studies suggest that different people (and animals) have different perceptions and tolerances for risk taking. For example, a study comparing the behaviours of chimpanzees and bonobo apes found that the chimps were more prepared to gamble and take risks.[27] At first sight this appears to support the personality explanation for risk taking, but actually the two types of ape share more than 99% of their DNA. A more likely explanation is the very different environments in which they have evolved: in the chimp environment food is scarce and uncertain, but in the bonobo habitats food is plentiful. We are not suggesting that entrepreneurs are chimp-like, or accountants are ape-like, but rather that experience and context have a profound influence on the assessment of, and appetite for, risk.

At the individual, cognitive level, risk assessment is characterized by overconfidence, loss aversion and bias.[28] Overconfidence in our ability to make accurate assessments is a common failing, and results in unrealistic assumptions and uncritical assessment. Loss aversion is well documented in psychology, and essentially means that we tend to prefer to avoid loss rather than to risk gain. Finally, cognitive bias is widespread and has profound implications for the identification and assessment of risk. Cognitive bias results in us seeking and overemphasizing evidence which supports our beliefs and reinforces our bias, but at the same time leads us to avoid and undervalue any information which contradicts our view.[29] Therefore we need to be aware of and challenge our own biases, and encourage others to debate and critique our data, methods and decisions.

Studies of research and development confirm that measures of cognitive ability are associated with project performance. In particular, differences in reflection, reasoning, interpretation and sense making influence the quality of problem formulation, evaluation and solution, and therefore ultimately the performance of research and development. A common weakness is the oversimplification of problems characterized by complexity or uncertainty, and the simplification of problem framing and evaluation of alternatives.[30] This includes adopting a single prior hypothesis, selective use of information that supports this, and devaluing alternatives, and illusion of control and predictability. Similarly, marketing managers are likely to share similar cognitive maps, and make the same assumptions concerning the relative importance of different factors contributing to new product success, such as the degree of customer orientation versus competitor orientation, and the implications of relationship between these factors, such as the degree of inter-functional coordination.[31] So the evidence indicates the importance of cognitive processes at the senior management, functional, group and individual levels of an organization. More generally, problems of limited cognition include:[32]

- *Reasoning by analogy*, which oversimplifies complex problems.
- *Adopting a single, prior hypothesis bias*, even where information and trials suggest this is wrong.
- *Limited problem set*, the repeated use of a narrow problem-solving strategy.
- *Single outcome calculation*, which focuses on a simple single goal and a course of action to achieve it, and denying value trade-offs.
- *Illusion of control and predictability*, based on an overconfidence in the chosen strategy, a partial understanding of the problem and limited appreciation of the uncertainty of the environment.
- *Devaluation of alternatives*, emphasizing negative aspects of alternatives.

At the group or social level, other factors also influence our perception and response to risk. How managers assess and manage risk is also a social and political process. It is influenced by prior experience of risk, perceptions of capability, status and authority, and the confidence and ability to communicate with relevant people at the appropriate times.[33] In the context of managing innovation, risk is less about personal propensity for risk taking or rational assessments of probability, and more about the interaction of experience, authority and context. In practice managers deal with risk in different ways in different situations. General strategies include delaying or delegating decisions, or sharing risk and responsibilities. Generally, when managers are performing well, and achieving their targets, they have less incentive to take risks. Conversely, when under pressure to perform, managers will often accept higher risks, unless these threaten survival.

Politics of risk

In most organizations risk has become a negative term, something which should be minimized or avoided, and implies hazard or failure. This view, particularly common in the policy domain, is enshrined in the 'precautionary principle' and the many regulatory regimes it has spawned, which, as the title suggests, wherever possible promotes the avoidance of risk taking.[34]

However, this interpretation perverts the nature of risk and opportunity, which are central to successful innovation, and promotes inaction and the status quo, rather than improvement or change.

The term 'risk' is derived from the Latin 'to dare', but has become associated with hazard or danger. We must also consider the 'risk' of success, or risks associated with *not* changing.[35] Berglund provides a good working definition of risk in the context of innovation, as *'the pursuit of perceived opportunities under conditions of uncertainty'*.[36]

In a corporate context, he identifies three aspects of risk which need to be managed:

- Compliance with formal project and process requirements, rather than innovation outcomes.
- Internal control and autonomy, and influence and use of external expertise.
- Flexibility of the business model and experimentation with alternative configurations and organization.

In any large organization, there will be formal process and project requirements. However, these may conflict with the goals of innovation. Risk taking requires a degree of tolerance of uncertainty and ambiguity in the workplace. In the high risk-taking climate, bold new initiatives can be taken even when the outcomes are unknown. People feel that they can 'take a gamble' on some of their ideas. People will often 'go out on a limb' and be first to put an idea forward. In a risk-avoiding climate there is a cautious, hesitant mentality. People try to be on the 'safe side'. They decide 'to sleep on the matter'. They set up committees and they cover themselves in many ways before making a decision. When risk taking is too low, employees offer few new ideas or few ideas that are well outside of what is considered safe or ordinary. In risk-avoiding organizations people complain about boring, low-energy jobs and are frustrated by a long, tedious process used to get ideas to action. These conditions can be caused by the organization not valuing new ideas or having an evaluation system that is bureaucratic, or people being punished for 'drawing outside the lines'. It can be remedied by developing a company plan that would speed 'ideas to action'. When risk taking is too high you will see that people are confused. There are many ideas floating around, but few are sanctioned. People are frustrated because nothing is getting done. There are many loners doing their own thing in the organization and no evidence of teamwork. These conditions can be caused by individuals not feeling they need a consensus or buy-in from others on their team in their department or organization. A remedy might include some team building and improving the reward system to encourage cooperation rather than individualism or competition.[5]

A recent study of organizational innovation and performance confirms the need for this delicate balance between risk and stability. Risk taking is associated with a higher relative novelty of innovation (how different it was to what the organization had done before), and absolute novelty (how different it was to what any organization had done before), and that both types of novelty are correlated with financial and customer benefits.[37] However, the same study concludes that *'incremental, safe, widespread innovations may be better for internal considerations, but novel, disruptive innovations may be better for market considerations . . . absolute novelty benefits customers and quality of life, relative innovation benefits employee relations (but) risk is detrimental to employee relations'*. In fact, many of the critical risks which need to be identified and managed are internal to organizations, rather than the more obviously anticipated external risks such as markets, competition and regulation.[38] For example, at 3M, 100 years of successful innovation was almost reversed following a change of CEO and an emphasis on six-sigma quality processes, rather than maintaining an innovative climate and products.

The inherent uncertainty in some projects limits the ability of managers to predict the outcomes and benefits of projects. In such cases changes to project plans and goals are commonplace, being

> ### TABLE 8.3   Management of conventional and risky projects
>
> | Conventional project management | Management of risky projects |
> | --- | --- |
> | Modest uncertainty | Major technical and market uncertainties |
> | Emphasis on detailed planning | Emphasis on opportunistic risk taking |
> | Negotiation and compromise | Autonomous behaviour |
> | Corporate interests and rules | Individualistic and ad hoc |
> | Homogeneous culture and experience | Heterogeneous backgrounds |

driven by external factors, such as technological breakthroughs or changes in markets, as well as internal factors, such as changes in organizational goals. Together the impact of changes to project plans and goals can overwhelm the benefits of formal project planning and management (Table 8.3).[22]

This is consistent with the real options approach to investing in risky projects, because investments are sequential and managers have some influence on the timing, resourcing and continuation or abandonment of projects at different stages. By investing relatively small amounts in a wide range of projects, a greater range of opportunities can be explored. Once uncertainty has been reduced, only the most promising projects should be allowed to continue. For a given level of investment this real option approach should increase the value of the project portfolio. However, because decisions and the options they create interact, a decision regarding one project can affect the option value of another project.[39,40] Nonetheless, the real options perspective remains a useful way of conceptualizing risk, particularly at the portfolio level. The goal is not to calculate or optimize, but rather to help to identify risks and payoffs, key uncertainties, decision points and future opportunities that might be created.[41] Combined with other methods, such as decision trees, a real options approach can be particularly effective where high volatility demands flexibility, placing a premium on the certainty of information and timing of decisions.

8.5 Anticipating the resources

Given their mathematical skills, one might have expected R&D managers to be enthusiastic users of quantitative methods for allocating resources to innovative activities. The evidence suggests otherwise: practising R&D managers have been sceptical for a long time (see Box 8.2). An exhaustive report by practising European managers on R&D project evaluation classifies and assesses more than 100 methods of evaluation and presents 21 case studies on their use.[42] However, it concludes that no method can guarantee success, that no single approach to pre-evaluation meets all circumstances, and that – whichever method is used – the most important outcome of a properly structured evaluation

---

BOX 8.2

A chief executive officer's completely perfect and absolutely quantitative method of measuring his R&D programme.

> I multiply your projects by words I can't pronounce,
> And weigh your published papers to the nearest half an ounce;
> I add a year-end bonus for research that's really pure,
> (And if it's also useful, your job will be secure).
>
> I integrate your patent-rate upon a monthly basis;
> Compute just what your place in the race to conquer space is;
> Your scientific stature I assay upon some scales
> Whose final calibration is the Company net-to-sales.
>
> And thus I create numbers where there were none before;
> I have lots of facts and figures – and formulae galore –
> And these quantitative studies make the whole thing crystal clear.
> Our research should cost exactly what we've budgeted this year.

Source: R. Landon, cited in Dr A. Bueche (vice-president for Research and Development of the US General Electric Company) in From laboratory to commercial application: some critical issues. Paper presented at the *17th International Meeting of the Institute of Management Sciences*, London, 2 July 1970.

---

is improved communication. These conclusions reflect three of the characteristics of corporate investments in innovative activities:

1. They are uncertain, so that success cannot be assured.
2. They involve different stages that have different outputs that require different methods of evaluation.
3. Many of the variables in an evaluation cannot be reduced to a reliable set of figures to be plugged into a formula, but depend on expert judgements: hence the importance of communication, especially between the corporate functions concerned with R&D and related innovative activities, on the one hand, and with the allocation of financial resources, on the other.

Financial assessment of projects

As we showed earlier, financial methods are still the most commonly used method of assessing innovative projects, but usually in combination with other, often more qualitative approaches. The financial methods range from simple calculation of payback period or return on investment, to more complex assessments of net present value (NPV) through discounted cash flow (DCF).

Project appraisal by means of DCF is based on the concept that money today is worth more than money in the future. This is not because of the effect of inflation, but reflects the difference in potential investment earnings, that is the opportunity cost of the capital invested.

The NPV of a project is calculated using:

$$NPV = \Sigma_0^T P_t / (1 + i)^t - C$$

where:

P_t = Forecast cash flow in time period t

T = Project life

i = Expected rate of return on securities equivalent in risk to project being evaluated

C = Cost of project at time $t = 0$

In practice, rather than use this formula, it is easy to create standard NPV templates in a spreadsheet package such as Excel.

How to evaluate learning?

However, the potential benefits of innovative activities are twofold. First, *extra profits* derived from increased sales and/or higher prices for superior products, and from lower costs and/or increased sales from superior production processes. Conventional project appraisal methods can be used to compare the value of these benefits against their cost. Second, *accumulated firm-specific knowledge* ('learning', 'intangible assets') that may be useful for the development of *future* innovations (e.g. new uses for solar batteries, carbon fibre, robots, word processing). This type of benefit is relatively more important in R&D projects that are more long term, fundamental and speculative.

Conventional techniques cannot be used to assess this second type of benefit, because it is an 'option' – in other words, it creates the *opportunity* for the firm to invest in a potentially profitable investment, but the realization of the benefits still depends on a decision to commit further resources. Conventional project appraisal techniques cannot evaluate options (see Box 8.3).

The inherent uncertainty in most R&D projects limits the ability of managers to predict the outcomes and benefits of projects. Research suggests that changes to R&D plans and goals are common,

---

BOX 8.3   **Why conventional financial evaluation methods do not work with investments in technology**

The following text was written by the Professor of Finance at the Sloan School of Management at MIT.

Suppose a firm invests in a negative *NPV* (net present value) project in order to establish a foothold in an attractive market. Thus a valuable second-stage investment is used to justify the immediate project. The second stage must depend on the first: if the firm could take the second project without having taken the first, then the future opportunity should have no impact on the immediate decision . . .

At first glance, this may appear to be just another forecasting problem. Why not estimate cash flows for both stages, and use discounted cash flow to calculate the *NPV* for the two stages taken together?

You would not get the right answer. The second stage is an option, and conventional discounted cash flow does not value options properly. The second stage is an option because the firm is not committed to undertaking it. It will go ahead if the first stage works and the market is still attractive. If the first stage fails, or if the market sours, the firm can stop after stage 1 and cut its losses. Investing in stage 1 purchases an intangible asset: a call option on stage 2. If the option's present value offsets the first stage's negative *NPV*, the first stage is justified . . .

DCF (discounted cash flow) is readily applied to 'cash cows' – relatively safe businesses held for the cash they generate… It also works for 'engineering investments', such as machine replacements, where the main benefit is reduced cost in a defined activity.

DCF is less helpful in valuing businesses with substantial growth opportunities or intangible assets. In other words, it is not the whole answer when options account for a large fraction of a business's value.

DCF is no help at all for pure research and development. The value of R&D is almost all option value. Intangible assets' value is usually options value.

Myers, S. (1984) Finance theory and financial strategy. *Interfaces*, 14, 126–37.

being driven by external factors, such as technological breakthroughs, as well as internal factors, such as changes in the project goals. Together the impact of changes to project plans and goals overwhelm the effects of the quality of formal project planning and management.[22] This reality is consistent with the real options approach to investing in R&D, because investments are sequential and managers have some influence on the timing, resourcing and continuation or abandonment of projects at different stages. By investing relatively small amounts in a wide range of projects, a greater range of technological opportunities can be explored. Once uncertainty has been reduced, only the most promising projects are allowed to continue. For a given level of R&D investment this real option approach should increase the value of the project portfolio. However, because options interact, a decision regarding one project can affect the option value of another project (unlike NPV calculations, which rarely include interaction effects). Therefore the creation of further options through R&D projects may not increase the overall option value of the R&D portfolio, and conversely the interaction of options arising from different projects can give rise to a nonlinear increase in the combined option value.[39]

However, in almost all cases it is impossible to calculate the value of R&D using real options, because unlike financial options it is difficult to predict technological breakthroughs, estimate future sales from products flowing from the R&D (or project payoff), or to identify and model project-specific risks, and the time-varying volatilities of the processes and eventual values.[40] Nonetheless, the real options perspective remains a useful way of conceptualizing R&D investment, particularly at the portfolio level. It can help to make more explicit and to identify future growth options created by R&D, even when these are not related to the (current) goals of the R&D. Combined with decision trees, a real options approach can help to identify risks and payoffs, key uncertainties, decision points and future branches (options).[41] It is particularly effective where high volatility demands flexibility, placing a premium on the certainty of information and timing of decisions (see Research Note).

In other words, the successful allocation of resources to innovation depends less on robustness of decision-making techniques than on the organizational processes in which they are embedded.

> ### RESEARCH NOTE   The value of uncertainty
>
> The real options approach has been used to evaluate R&D at both the project and firm levels. The idea is that investment in, or more strictly speaking spending on, R&D creates greater flexibility and a portfolio of options for future innovations, especially where the future is uncertain. Faced with uncertainty, managers can choose to commit additional resources to R&D to create an *option to grow*, or alternatively delay additional R&D to hold an *option to wait*.
>
> This study examined the different and combined effects of market and technological uncertainty on the financial valuation of firms' investments in R&D. They examined the behaviour and performance of 290 firms over 10 years, and found that the relationship between R&D and firm valuation depended on the source and degree of uncertainty. They identify a U-shaped relationship between market uncertainty and R&D capital: increasing market uncertainty initially reduces the value of any unit of investment in R&D until a point of inflection, beyond which it augments the value. The higher the rate of market growth, the lower the point of inflection. Conversely, the relationship between technological uncertainty and R&D capital is an inverted U-shape. This suggests that investors put a limit on the value of technology hedging: at low levels of technological uncertainty there is limited value in creating options, and at very high levels the cost of maintaining many alternatives is too high.
>
> Therefore it is important to identify the main sources of uncertainty, technology or market, in order to make better decisions about the potential value of investments in R&D options.
>
> *Source:* Oriani, R. and M. Sobrero (2008) Uncertainty and the market value of R&D within a real options logic. *Strategic Management Journal*, 29, 343–61.

According to Mitchell and Hamilton,[43] there are three (overlapping) categories of innovation that large firms must finance. Each category has different objectives and criteria for selection, the implications of which are set out in Table 8.4.

1. *Knowledge building* – This is the early-stage and relatively inexpensive research for nurturing and maintaining expertise in fields that could lead to future opportunities or threats. It is often treated as a necessary overhead expense, and sometimes viewed with suspicion (and even incomprehension) by senior management obsessed with short-term financial returns and exploiting existing markets, rather than creating new ones.

   With knowledge-building projects, the central question for the company is: 'What are the potential costs and risks of not mastering or entering the field?' Thus, no successful large firm in manufacture can neglect to explore the implications of development in IT, even if IT is not a potential core competence. And no successful firm in pharmaceuticals could avoid exploring recent developments in biotechnology. Decisions about such projects should be taken solely by technical staff on the basis of technical judgements, and especially those staff concerned with the longer term. Market analysis should not play any role. Outside financial linkages are likely to be with academic and other specialist groups, and to take the form of a grant.

2. *Strategic positioning* – These activities are in between knowledge building and business investment, and an important – and often neglected – link between them. They involve applied R&D and feasibility

TABLE 8.4	Resource allocation for different types of innovative project							
Objective	**Technical activity**	**Evaluation criteria (% of all R&D)**	**Decision takers**	**Market analysis**	**Nature of risk**	**Higher volatility**	**Longer time horizons**	**Nature of external alliances**
Knowledge building	Basic research, monitoring	Overhead cost allocation (2–10%)	R&D	None	Small = cost of R&D	Reflects wide potential	Increases search potential	Research grant
Strategic positioning	Focused applied research, exploratory development	'Options' evaluation (10–25%)	Chief executive R&D division	Broad	Small = cost of R&D	Reflects wide potential	Increases search potential	R&D contract, equity
Business Investment	Development and production engineering	'Net present value' analysis (70–99%)	Division	Specific	Large = total cost of launching	Uncertainty reduces net present value	Reduces present value	Joint venture Majority control

demonstration, in order to reduce technical uncertainties, and to build in-house competence, so that the company is capable of transforming technical competence into profitable investment. For this type of R&D, the appropriate question is: 'Is the project likely to create an option for a profitable investment at a later date?' Comparisons are sometimes made with financial stock options, where (for a relatively small sum) a firm can purchase the option to buy a stock at a specified price, before a specified date – in anticipation of increase in its value in future.

Decisions about this category of project should involve divisions, R&D directors and the chief executive, precisely because – as their description implies – these projects will help determine the strategic options open to the company at a later date. At this stage, market analysis should be broad (e.g. where could genetic engineering create new markets for vegetables in a food company?). A variety of evaluation methods may be used (e.g. the product–technology matrix), but they will be more judgemental than rigorously quantitative. Costs will be higher than those of knowledge building, but much lower than those of full-scale business investment. As with knowledge-building projects, both high volatility in predictions and expectations, and long time horizons, are not unwelcome signs of unacceptably high risk, but welcome signs are rich possibilities and sufficient time to explore them. Outside linkages require tighter management than those related to knowledge building, probably through a contract or equity participation.

3. *Business investment* – This is the development, production and marketing of new and better products, processes and services. It involves relatively large-scale expenditures, evaluated with conventional financial tools such as net present value. In such projects, the appropriate question is: 'What are the potential costs and benefits in continuing with the project?' Decisions should be taken at the level of the division bearing the costs and expecting the benefits. Success depends on meeting the precise requirements of specific groups of users, and therefore depends on careful and targeted marketing. Financial commitments are high, so that volatility in technological and market conditions is unwelcome, since it increases risk. Long time horizons are also financially unwelcome, since they increase the financial burden. Given the size and complexity of development and commercialization, external linkages need to be tightly controlled through majority ownership or a joint venture. Given the scale of resources involved, careful and close monitoring of progress against expectations is essential. For such projects most firms rely on financial methods to evaluate their project portfolio – around 77% of firms according to a recent survey. However, the same survey revealed that only 36% of the best performing firms rely on financial methods, compared to 39% which use strategic methods.[37] An explanation for the relatively poor performance of financial methods is that the sophistication of the models often far exceeds the quality of the data inputs, particularly at the early stages of a project's life.

Checklists are a commonly used example of a simple qualitative technique. A checklist is simply a list of factors which are considered important in making a decision in a specific case. These criteria include technical and commercial details, legal and financial factors, company targets and company strategy. Most useful criteria are essentially independent of the business field and the business strategy, but the precise criteria and their weights will differ in specific applications.

The requirements for the use of this technique are minimal, and the effort involved in using it is normally low. Another advantage of the technique is that it is very easily adaptable to the company's way of doing things. However, checklists can be a starting point for more sophisticated methods where the basic information can be used for better focus. One simple and useful example is a SWOT analysis, where projects are assessed for their strengths, weaknesses, opportunities and threats.

Therefore, this technique can be developed further and the analysis interaction and feedback can be easily managed using simple information technology. Ways to make the technique more sophisticated include:

- To include some quantitative factors among the whole list of factors.
- To assign different weights to different factors.
- To develop a systematic way of arriving at an overall opinion on the project, such as a score or index.

A simple checklist could be one made up of a range of factors which have been formed to affect the success of a project and which need to be considered at the outset. In the evaluation procedure a project is evaluated against each of these factors using a linear scale, usually 1 to 5 or 1 to 10. The factors can be weighted to indicate their relative importance to the organization.

The value in this technique lies in its simplicity, but by the appropriate choice of factors it is possible to ensure that the questions address, and are answered by, all functional areas. When used effectively this guarantees a useful discussion, an identification and clarification of areas of disagreement and a stronger commitment, by all involved, to the ultimate outcome. Table 8.5 shows an example of a checklist, developed by the Industrial Research Institute, which can be adapted to almost any type of project.

TABLE 8.5 List of potential factors for project evaluation

	Score (1-5)	Weight (%)	S × W
Corporate objectives			
Fits into the overall objectives and strategy			
Corporate image			
Marketing and distribution			
Size of potential market			
Capability to market product			
Market trend and growth			
Customer acceptance			
Relationship with existing markets			
Market share			
Market risk during development period			
Pricing trend, proprietary problem, etc.			

(continued)

	Score (1-5)	Weight (%)	S × W
TABLE 8.5 (Continued)			
Complete product line			
Quality improvement			
Timing of introduction of new product			
Expected product sales life			
Manufacturing			
Cost savings			
Capability of manufacturing product			
Facility and equipment requirements			
Availability of raw material			
Manufacturing safety			
Research and development			
Likelihood of technical success			
Cost			
Development time			
Capability of available skills			
Availability of R&D resources			
Availability of R&D facilities			
Patent status			
Compatibility with other projects			
Regulatory and legal factors			
Potential product liability			

(continued)

	Score (1-5)	Weight (%)	S × W
TABLE 8.5	**(Continued)**		
Regulatory clearance			
Financial			
Profitability			
Capital investment required			
Annual (or unit) cost			
Rate of return on investment			
Unit price			
Payout period			
Utilization of assets, cost reduction and cash-flow			

As with all techniques, there is a danger that project appraisal becomes a routine that a project has to suffer, rather than an aid to designing and selecting appropriate projects (Box 8.4). If this happens people may fail to apply the techniques with the rigour and honesty required, and can waste time and energy trying to 'cheat' the system. Care needs to be taken to communicate the reasons behind the methods and criteria used, and where necessary these should be adapted to different types of project and to changes in the environment.[44]

BOX 8.4   Limitations of conventional project and product assessment

Clayton Christensen and colleagues argue that three commonly used means of assessment discourage expenditure on innovation. Firstly, conventional means of assessing projects, such as discounted cash flow (DCF) and the treatment of fixed costs, favour the incremental exploitation of existing assets, rather than the more risky development of new capabilities. Secondly, methods such as the stage-gate process demand data on estimated markets, revenues and costs, which are much more difficult to generate for more radical innovations. Finally, senior managers and publically quoted firms are typically assessed by improvements in the earning per share (EPS), which encourages short-term investments and returns – most institutional investors hold shares for only 10 months in the USA, and the tenure of CEOs is shrinking.

Whilst they appreciate the benefits of such financial methods of assessment, they argue that such techniques should be adjusted to redress the balance for risk taking and expenditure on innovation. For example, when using DCF, comparative assessments should be made with the option of doing nothing, or not investing in an innovative project, rather than assuming a decision not to invest will result in no loss of competitiveness. Similarly, for the stage-gate process, they propose focusing less on the (unreliable) quantitative forecasts, and much more on challenging and testing the assumptions made in business planning. Finally, they believe that the use of short-term measures such as EPS is no longer appropriate because they provide perverse incentives. The original rationale for this type of approach was the principal–agent problem – to try to align the interests of the principals (owners/shareholders) and their agents (managers). However, the growth of collective institutional ownership of most public firms has created an agent–agent problem, and the interests of the agents need to be more aligned to promote innovation.

Source: Christensen, C.M, S.P. Kaufmann and W.C. Shih (2008) Innovation killers: how financial tools destroy your capacity to do new things. *Harvard Business Review*, January, 98–105.

Portfolio methods try to deal with the issue of reviewing across a set of projects and look for a balance of economic and non-financial risk/reward factors (Case study 8.1). A typical example is to construct some form of matrix measuring risk vs. reward, for example, on a 'costs of doing the project' vs. expected returns (Figure 8.3).

Rather than reviewing projects just on these two criteria it is possible to construct multiple charts to develop an overall picture, for example, comparing the relative familiarity of the market or technology – this would highlight the balance between projects that are in unexplored territory as opposed to those in familiar technical or market areas (and thus with a lower risk). Other possible axes include ease of entry vs. market attractiveness (size or growth rate), the competitive position of the organization in the project area vs. the attractiveness of the market or the expected time to reach the market vs. the attractiveness of the market. Some examples of portfolio management tools are given on the website.

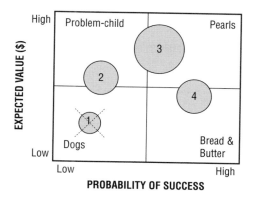

FIGURE 8.3: An example matrix-based portfolio

CASE STUDY 8.1

The Arthur D. Little matrix for technology decisions

A number of tools have been developed to help with strategic decision making around technology investments . Typical of these are those which make some classification of technologies in terms of their open availability and the ease with which they can be protected and deployed to strategic advantage. For example, the consultancy Arthur D. Little uses a matrix which groups technological knowledge into four key groups – base, key, emerging and pacing.

- Base technologies represent those on which product/service innovations are based and which are vital to the business. However they are also widely known about and deployed by competitors and offer little potential competitive advantage.
- Key technologies represent those which form the core of current products/services or processes and which have a high competitive impact – they are strategically important to the organization and may well be protectable through patent or other form.
- Pacing technologies are those which are at the leading edge of the current competitive game and may be under experimentation by competitors – they have high but as yet unfulfilled competitive potential.
- Emerging technologies are those which are at the technological frontier, still under development and whose impact is promising but not yet clear.

Making this distinction helps identify a strategy for acquisition based on the degree of potential impact plus the importance to the enterprise plus the protectability of the knowledge. For base technologies it may make sense to source outside whereas for key technologies an in-house or carefully selected strategic alliance may make more sense in order to preserve the potential competitive advantage. Emerging technologies may be best served by a watching strategy, perhaps through some pilot project links with universities or technological institutes.

Models of this can be refined, for example, by adding to the matrix information about different markets and their rate of growth or decline. A fast-growing new market may require extensive investment in the pacing technology in order to be able to build on the opportunities being created whereas a mature or declining market may be better served by a strategy which uses base technology to help preserve a position but at low cost.

For more detail on this approach see http://www.adlittle.com/.

A useful variant on this set of portfolio methods is the 'bubble chart' in which the different projects are plotted but represented by 'bubbles' – circles whose diameter varies with the size of the project (for example in terms of costs). This approach gives a quick visual overview of the balance of different-sized projects against risk and reward criteria. Case study 8.2 gives an example. However it is important to recognize that even advanced and powerful screening tools will only work if the corporate will is present to implement the recommended decisions, for example, Cooper and Kleinschmidt found that the majority of firms studied (885) performed poorly at this stage, and often failed to kill off weak concepts.[45]

Portfolio management of process innovation in Fruit of the Loom

The clothing manufacturer Fruit of the Loom reviewed its worldwide process innovation activities using a portfolio framework to help provide a clearer overview and develop focus. It used simple categories:

- 'Incremental' – essentially continuous improvement projects
- 'Radical' – using the same basic technology but with more advanced implementation
- 'Fundamental' – using different technology, for example, laser cutting instead of mechanical

Plotting on to a simple colour-coded bubble chart enabled a quick and easily communicable overview of their strategic innovation portfolio in this aspect of innovation.

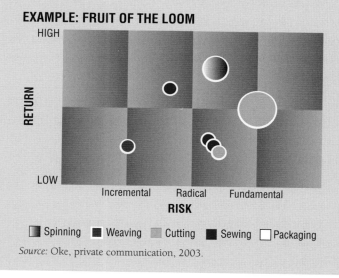

Source: Oke, private communication, 2003.

How practising managers cope

These two sets of difficulties – in evaluating the potential contributions of technological investments to firm-specific intangible assets, and in dealing with uncertainty – are reflected in how successful managers allocate resources to technological activities. In particular, they:

- Encourage *incrementalism* – step-by-step modification of objectives and resources, in the light of new evidence.
- Use *simple rules* models for allocating resources, so that the implications of changes can be easily understood.
- Make explicit from the outset criteria for *stopping* the project or programme.

- Use *sensitivity analysis* to explore if the outcome of the project is 'robust' (unchanging) to a range of different assumptions (e.g. 'What if the project costs twice as much, and takes twice as long, as the present estimates?').
- Seek the reduction of *key uncertainties* (technical and – if possible – market) before any irreversible commitment to full-scale – and costly – commercialization.
- Recognize that *different types* of innovation should be evaluated by *different criteria*.

VIEWS FROM THE FRONT LINE

Justifying value in R&D

A constant battle is being fought by R&D centres to obtain funds or prove that what they do receive is creating value for the company.

There are three distinct types of project defined by their anticipated duration before they contribute returns to a business:

1. Short term – incremental improvements to existing products.
2. Intermediate – substantial alterations or significant updates on well-founded products and markets.
3. Long term – speculative projects on something that may have a big future.

In our business of building power stations our products last for 40–50 years (with intermittent overhaul and servicing). Therefore, for us short term is 1–3 years, intermediate 3–7 years and long term can be over 20 years.

1. *Short term* – these are small continuous improvements or cost reduction projects. Each on its own is easy to cost but the return is difficult to quantify, e.g. improving a $10 wiper blade on a car is easy to define, but how many more cars do you sell as a result 1, 10, 100, 100 000 or zero? However, over time if these small changes are not made the car will become undesirable and thus less saleable compared to the competition.

   This is more difficult when the concept of fashion is introduced as this is more emotive than a relatively easy measurable such as an increase in performance.

   The motoring industry over time has become full of minor improvements that are now re-garded as essential – heaters, radio, electric windows and door mirrors, seats, air conditioning, satellite navigation, cruise control, iPod connections etc.
2. *Intermediate* – are the easiest to quantify and define as they are ringfenced projects for a known product in a relatively stable and an understood market.

   An example could be the moves from records, cassettes, CDs or video, DVD to Blu-ray HD. The demand from the market is fairly easy to quantify and one generation has more or less sub-stituted the previous one. The technology has been uncertain but understood. These types of projects can be compared and 'valued' via traditional evaluation tools such as NPV or option pricing.

In the power business such technologies would now encompass wind turbines and even nuclear power.

3. *Long term* – and sometimes very disruptive technologies and products. PCs and mobile communications are two such recent products.

The costs and time to market were long and adoption too was a drawn out affair. Costs of development were extremely hard to predict but the return was potentially enormous but equally hard to predict (see, for example, Microsoft and Vodafone).

Which companies could have run a NPV on these, how did Sony Walkman and iPod pass the financial hurdles, when both were new breakthroughs?

In the power business we are struggling with 'proving' the returns for Carbon Capture and Storage – with 10–20 year predictions for the development of the technology let alone commercialization versus the trillions of potential value – the race is on, but the NPV does not look realistic.

So where does that leave the R&D director? It's going to cost a lot, over an unknown duration (I don't know how we will invent the future) but it will be a massive market – trust me . . .

The best we can presently do is portfolio management – borrowed from the financial markets which basically translate as 'don't put all your eggs in one basket' – because we don't know what the future holds.

Source: Richard Dennis, Director R&D, Doosan-Babcock

Summary and further reading

The process of innovation is much more complex than technology responding to market signals. Effective business planning under conditions of uncertainty demands a thorough understanding and management of the dynamics of innovation, including conception, development, adoption and diffusion.

The adoption and diffusion of an innovation depend on the characteristics of the innovation, the nature of potential adopters and the process of communication. The relative advantage, compatibility, complexity, trialability and observability of an innovation all affect the rate of diffusion. The skills, psychology, social context and infrastructure of adopters also affect adoption. Epidemic models assume that innovations spread by communication between adopters, but bandwagons do not require this. Instead, early adopters influence the development of an innovation, but subsequent adopters may be more influenced by competitive and peer pressures. Forecasting the development and adoption of innovations is difficult, but participative methods such as Delphi and scenario planning are highly relevant to innovation and sustainability. In such cases the process of forecasting, including consultation and debate, is probably more important than the precise outcomes of the exercise.

More generally, the problems of forecasting the future development, adoption and diffusion of innovations are dealt with by many authors in the innovation field. Everett Roger's classic text the *Diffusion of Innovations*, first published in 1962, remains the best overview of this subject, the most recent and updated edition being published in 2003 (Simon and Schuster). More up-to-date accounts can be found in *Determinants of Innovative Behaviour*, edited by Cees van Beers, Alfred

Kleinknecht, Roland Ortt and Robert Verburg (Palgrave, 2008), and our own *Gaining Momentum: Managing the Diffusion of Innovations*, edited by Joe Tidd (Imperial College Press, 2009).

In *Democratizing Innovation* (MIT Press, 2005, and free online) Eric von Hippel builds on his earlier concept of 'lead users' in innovation, and argues that innovation is becoming more democratic, with users increasingly being capable of developing their own new products and services. He believes that such user innovation has a positive impact on social welfare as innovating users – both individuals and firms – often freely share their innovations with others, creating user-innovation communities and a rich intellectual commons. Examples provided range from surgical equipment to surfboards to software security. A broader review of user innovation is provided by the special issue of the *International Journal of Innovation Management*, **12** (3), 2008, edited by Steve Flowers.

Clayton Christensen's (with S.D. Anthony and E.A. Roth) *Seeing What's Next: Using the Theories of Innovation to Predict Industry Change* (Harvard Business School Press, 2005) is a useful up-to-date review of methods for forecasting radical and potentially disruptive innovations. A special issue of the journal *Long Range Planning*, **37** (2), 2004, is devoted to forecasting, and provides a good overview of current thinking. *Scenario Planning* by Gill Ringland (John Wiley & Sons, Ltd, 2nd edition, 2006) and *Scenario Planning: The Link Between Future and Strategy* by Mats Lindgren (Palgrave Macmillan, 2002) are both detailed and practical guides to conducting scenario planning, which is probably one of the most relevant methods for understanding innovation planning. For a comprehensive overview of international research and practice refer to *The Handbook of Technology Foresight*, edited by Luke Georghiou (Edward Elgar, 2008). A special issue of *R&D Management*, 2008, **38**(5), deals with managing risk and uncertainty in R&D.

Web links

Here are the full details of the resources available on the website flagged throughout the text:

Case studies:
Plaswood Recycling

Tools:
Risk assessment matrix

Video podcast:
3M: identifying emerging market needs

Audio podcast:
Richard Reed: Innocent Smoothies

References

1. **Kaplan, J.M. and A.C. Warren** (2007) *Patterns of Entrepreneurship*, John Wiley & Sons, Inc., New York.
2. **Roberts, E.B.** (1991) *Entrepreneurs in High Technology: Lessons from MIT and Beyond*, Oxford University Press, Oxford.

3. **Tidd, J. and K. Bodley** (2002) Effect of novelty on new product development processes and tools. *R&D Management*, **32** (2), 127–38.

4. **Rogers, E.M.** (2003) *Diffusion of Innovations*, Free Press, New York.

5. **Isaksen, S. and J. Tidd** (2006) *Meeting the Innovation Challenge: Leadership for Transformation and Growth*, John Wiley & Sons, Ltd, Chichester.

6. **Landeta, J.** (2006) Current validity of the Delphi method in social sciences. *Technological Forecasting and Social Change*, **73** (5), 467–82; **Fuller, T. and L. Warren** (2006) Entrepreneurship as foresight: a complex social network perspective on organisational foresight. *Futures*, **38** (8), 956–71; **Gupta, U.G. and R.E. Clarke** (1996) Theory and applications of the Delphi technique: a bibliography (1975–1994). *Technological Forecasting and Social Change*, **53** (2), 185–212.

7. **Ringland, G.** (2006) *Scenario Planning*, second edition, John Wiley & Sons, Ltd, Chichester; **Chermack, T.J.** (2005) Studying scenario planning: theory, research suggestions, and hypotheses. *Technological Forecasting and Social Change*, **72** (1), 59–73; **Burt, G. and K. van der Heijden** (2003) First steps: towards purposeful activities in scenario thinking and future studies. *Futures*, **35** (10), 1011–26.

8. **Geroski, P.A.** (2000) Models of technology diffusion. *Research Policy*, **29**, 603–25.

9. **Griffiths, T.L. and J.B. Tenebaum** (2006) Optimal predications in everyday cognition. *Psychological Science*, **45**, 56–63; **Lissom, F. and J.S. Metcalfe** (1994) Diffusion of innovation ancient and modern: a review of the main themes. In M. Dodgson and P.L. Rothwell, eds, *The Handbook of Industrial Innovation* (pp. 106–41), Edward Elgar, Cheltenham.

10. **Abrahamson, E. and L. Plosenkopf** (1993) Institutional and competitive band-wagons: using mathematical modelling as a tool to explore innovation diffusion. *Academy of Management Journal*, **18** (3), 487–517.

11. **Tidd, J.** (1991) *Flexible Manufacturing Technologies and International Competitiveness*, Pinter, London.

12. **Williams, F. and D.V. Gibson** (1990) *Technology Transfer: A Communications Perspective*, Sage, London.

13. **Leonard-Barton, D. and D.K. Sinha** (1993) Developer–user interaction and user satisfaction in internal technology transfer. *Academy of Management Journal*, **36** (5), 1125–39.

14. **Leonard-Barton, D.** (1990) Implementing new production technologies: exercises in corporate learning. In M.A. von Glinow and S.A. Mohmian, eds, *Managing Complexity in High Technology Organizations* (pp. 160–187), Cambridge University Press, Cambridge.

15. **Chakravorti, B.** (2003) *The Slow Pace of Fast Change: Bringing Innovation to Market in a Connected World*, Harvard Business School Press, Boston, MA; (2004) The new rules for bringing innovations to market. *Harvard Business Review*, **82** (3), 58–67; (2004) The role of adoption networks in the success of innovations. *Technology in Society*, **26**, 469–82.

16. **Moore, G.** (1991) *Crossing the Chasm: Marketing and Selling Technology Products to Mainstream Customers*, HarperBusiness, New York; (1998) *Inside the Tornado: Marketing Strategies from Silicon Valley's Cutting Edge*, John Wiley & Sons, Ltd, Chichester.

17. **Lee, Y. and G.C. O'Connor** (2003) New product launch strategy for network effects products. *Journal of the Academy of Marketing Science*, **31** (3), 241–55.

18. **Van den Bulte, C. and G.L. Lilien** (2001) Medical innovation revisited: social contagion versus marketing effort. *The American Journal of Sociology*, **106** (5), 1409–35; **Van den Bulte, C. and S. Stremersch** (2004) Social contagion and income heterogeneity in new product diffusion. *Marketing Science*, **23** (4), 530–44.

19. **Atun, R.A., I. Gurol-Urganci and D. Sheridan** (2007) Uptake and diffusion of pharmaceutical innovations in health systems. *International Journal of Innovation Management*, **11** (2), 299–322.

20. **Tidd, J.** (2006) Innovation management in the pharmaceutical industry: a case of restricted vision? *Innovation in Pharmaceutical Technology*, 16–19.

21. **Flowers, S.** (2008) Special issue on user-centered innovation. *International Journal of Innovation Management*, **12**(3).

22. **Dvir, D. and T. Lechler** (2004) Plans are nothing, changing plans is everything: the impact of changes on project success. *Research Policy*, **33**, 1–15.

23. **Bower, J.** (1986) *Managing the Resource Allocation Process*, Harvard Business School, Boston, MA.

24. **Mansfield, E., J. Raporport, J. Schnee, S. Wagner and M. Hamburger** (1972) *Research and Innovation in the Modern Corporation*, Macmillan, London.

25. **Booz Allen and Hamilton** (1982) *New Product Management in the 1980s*, New York.

26. **Freeman, C. and L. Soete** (1997) *The Economics of Industrial Innovation*, third edition, Pinter, London.

27. **Heilbronner, S.R., A.G. Rosati, J.R. Stevens, B. Hare, and M.D. Hauser** (2008) A fruit in the hand or two in the bush? Divergent risk preferences in chimpanzees and bonobos. *Biology Letters*, **4** (3) 246–9.

28. **Westland, J.C.** (2008) *Global Innovation Management: A Strategic Approach*, Palgrave Macmillan, Basingstoke.

29. **Gardner, D.** (2008) *Risk: The Science and Politics of Fear*, Virgin Books, London.

30. **Tenkasi, R.V.** (2000) The dynamics of cognitive oversimplification processes in R&D environments: an empirical assessment of some consequences. *International Journal of Technology Management*, **20**, 782–98.

31. **Tyler, B.B. and D.R. Gnyawali** (2002) Mapping managers' market orientations regarding new product success. *Journal of Product Innovation Management*, 259–76.

32. **Walsh, J.P.** (1995) Managerial and organizational cognition: notes from a field trip. *Organization Science*, **6** (1), 1–41.

33. **Genus, A. and A.M. Coles** (2006) Firm strategies for risk management in innovation. *International Journal of Innovation Management*, **10** (2), 113–26.

34. **Fischoff, B.** (1995) Risk perception and communication unplugged: twenty years of progress. *Risk Analysis*, **15** (2), 137–45; **Renn, O.** (1998) Three decades of risk research: accomplishments and new challenges. *Journal of Risk Research*, **1** (1), 49–72; **Stirling, A.** (1998) Risk at a turning point? *Journal of Risk Research*, **1** (2), 97–110.

35. **Sunstein, C.R.** (2005) *Laws of Fear: Beyond the Precautionary Principle*, Cambridge University Press, Cambridge; **Morris, J.** (2000) *Rethinking Risk and the Precautionary Principle*, Butterworth Heinemann, London.

36. **Berglund, H.** (2007) Risk conception and risk management in corporate innovation. *International Journal of Innovation Management*, **11** (4), 497–514.

37. **Totterdell, P., D. Leach, K. Birdi, C. Clegg, and T. Wall** (2002) An investigation of the contents and consequences of major organizational innovations. *International Journal of Innovation Management*, **6** (4), 343–68.

38. **Keizer, J.A., J.P. Vos and J.I.M. Halman** (2005) Risks in new product development: devising a reference tool. *R&D Management*, **35** (3), 297–306.

39. **McGrath, R.G. and Nerkar, A.** (2004) Real options reasoning and a new look at the R&D investment strategies of pharmaceutical firms. *Strategic Management Journal*, **25**, 1–21.

40. **Paxon, D.A.** (2001) Introduction to real R&D options. *R&D Management*, **31** (2), 109–13.

41. **Loch, C.H. and K. Bode-Greual** (2001) Evaluating growth options as sources of value for pharmaceutical research projects. *R&D Management*, **31** (2), 231–45.

42 **EIRMA** (1995) *Evaluation of R&D Projects*, European Industrial Research Management Association, Paris.

43. **Mitchell, G. and W. Hamilton** (1988) Managing R&D as a strategic option. *Research-Technology Management*, **31**, 15–22.

44. **Laslo, Z. and A.I. Goldberg** (2008) Resource allocation under uncertainty in a multi-project matrix environment: is organizational conflict inevitable? *International Journal of Project Management*, **26**(4); special issue of *R&D Management*, 2008, 38(5), on managing risk and uncertainly in R&D.

45. **Cooper, R. and E. Kleinschmidt** (1990) *New Products: The Key Factors in Success*, American Marketing Association, Chicago.

PART 5

Implement

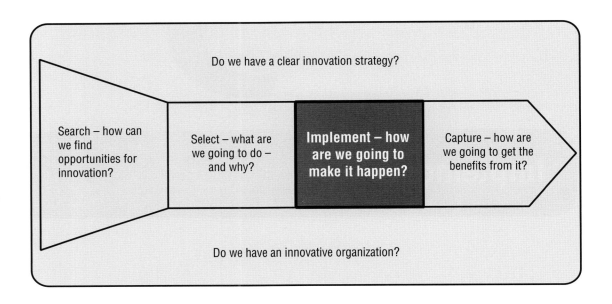

Here we look at the 'implementation' phase, where issues of how we move innovation ideas into reality become central. Chapter 9 looks at the ways in which innovation projects of various kinds are organized and managed and explores structures, tools and other support mechanisms to help facilitate this. Specifically, we focus on the development of new products and processes. Chapter 10 goes beyond product and service development, and examines the creation of new ventures: those arising from within the existing organization (corporate entrepreneurship); joint ventures between organizations; and those that involve setting up a new entrepreneurial venture outside the organization.

CHAPTER 9

Creating new products and services

In Chapter 5 we examined the range of sources and triggers for innovation, and identified some processes and tools to help to exploit these. In this chapter we focus on the more specific issue of developing new products and services. We begin by introducing the most common processes for development, the stage-gate and development funnel. We then review the generic factors that influence product and service success and failure. The central part of this chapter looks at how the market and technological context influence the process of development and commercialization, for example, how the development of radical is different from more common line extensions. Finally, we explore the similarities and differences between developing new products and services. In the most advanced service economies such as the USA and the UK, services create up to three-quarters of the wealth and 85% of employment, and yet most of what we know about managing innovation comes from research and experience in manufactured products.

9.1 Processes for new product development

We discussed the broader organizational factors to support innovation in Chapter 3, but here we explore the more specific needs of new product and service development. Successful product and service development requires much more than the application of a set of tools and techniques, and in addition requires an appropriate organization to support innovation and an explicit process to manage development. In this section we examine the critical role of organization, and the various options available in the case of new product and service development. The purpose of this section is not, however, to provide a more general overview of the theory and practice of organizational behaviour and development, and we assume that you are familiar with the basics of this field.

One of the key challenges facing the organization of new product and process development is that most organizations have not evolved or been designed to do this, but are structured for a different purpose, usually to serve some operational need. In most organizations new product or service development is a rather unusual and infrequent requirement, so the first decision is what sort of team to put together to do this.

Essentially the choice is between functional teams, cross-functional project teams or some form of matrix between the two. For example, the team might be within a single function or department such as research, marketing or design. Alternatively, a special cross-functional team might be established, including representatives from many (but not all) functional groups. In a matrix organization a dedicated team is not formed, but rather members remain in their functional or departmental groups, but are designated to a project group. Studies of new product development suggest four main types of team structure:

1. *Functional structure* – a traditional hierarchical structure where communication between functional areas is largely handled by function managers and according to standard and codified procedures.

2. *Lightweight product manager structure* – again a traditional hierarchical structure but where a project manager provides an overarching coordinating structure to the inter-functional work.
3. *Heavyweight product manager structure* – essentially a matrix structure led by a product (project) manager with extensive influence over the functional personnel involved but also in strategic directions of the contributing areas critical to the project. By its nature this structure carries considerable organizational authority.
4. *Project execution teams* – A full-time project team where functional staff leave their areas to work on the project, under project leader direction.

Project management structure is strongly correlated with product success, and of the available options the functional structures are the weakest. Associated with these different structures are different roles for team members and particularly for project managers. For example, the 'heavyweight project manager' has to play several different roles, which include extensive interpreting and communication between functions and players. Similarly, team members have multiple responsibilities. This implies the need for considerable efforts at team building and development, for example, to equip the team with the skills to explore problems, to resolve the inevitable conflicts that will emerge during the project, and to manage relationships inside and outside the project.

The process of new product or service development – moving from idea through to successful products, services or processes – is a gradual process of reducing uncertainty through a series of problem-solving stages, moving through the phases of scanning and selecting and into implementation – linking market- and technology-related streams along the way.

At the outset anything is possible, but increasing commitment of resources during the life of the project makes it increasingly difficult to change direction. Managing new product or service development is a fine balancing act, between the costs of continuing with projects which may not eventually succeed (and which represent opportunity costs in terms of other possibilities) – and the danger of closing down too soon and eliminating potentially fruitful options. With shorter life cycles and demand for greater product variety pressure is also placed upon the development process to work with a wider portfolio of new product opportunities and to manage the risks associated with progressing these through development to launch.

These decisions can be made on an ad hoc basis but experience and research suggests some form of structured development system, with clear decision points and agreed rules on which to base go/no-go decisions, is a more effective approach. Attention needs to be paid to reconfiguring internal mechanisms for integrating and optimizing the process such as concurrent engineering, cross-functional working, advanced tools, early involvement, etc. To deal with this attention has focused on systematic screening, monitoring and progression frameworks such as Cooper's 'stage-gate' approach, which we introduced in Chapter 7 (Figure 9.1).[1]

As Cooper suggests, successful product development needs to operate some form of structured, staging process. As projects move through the development process, there are a number of discrete stages, each with different decision criteria or 'gates' which they must pass. Many variations to this basic idea exist (e.g. 'fuzzy gates'), but the important point is to ensure that there is a structure in place which reviews both technical and marketing data at each stage. A common variation is the 'development funnel', which takes into account the reduction in uncertainty as the process progresses, and the influence of real resource constraints (Figure 9.2).[2]

There are numerous other models in the literature, incorporating various stages ranging from three to 13. Such models are essentially linear and unidirectional, beginning with concept development and

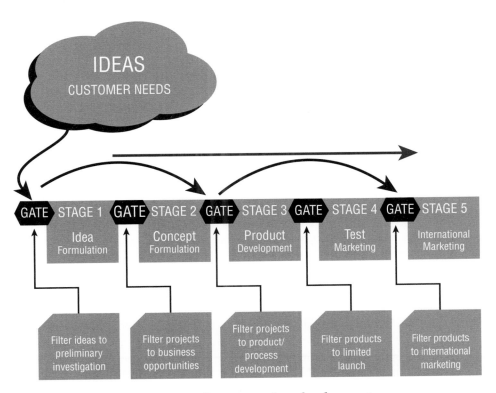

FIGURE 9.1: Stage-gate process for new product development

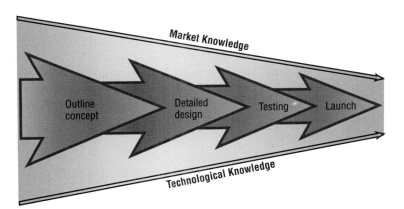

FIGURE 9.2: Development funnel model for new product development

ending with commercialization. Such models suggest a simple, linear process of development and elimination. However, in practice the development of new products and services is inherently a complex and iterative process, and this makes it difficult to model for practical purposes. For ease of discussion and analysis, we will adopt a simplified four-stage model which we believe is sufficient to discriminate between the various factors that must be managed at different stages:[3]

1. *Concept generation* – identifying the opportunities for new products and services.
2. *Project assessment and selection* – screening and choosing projects which satisfy certain criteria.
3. *Product development* – translating the selected concepts into a physical product (we'll discuss services later).
4. *Product commercialization* – testing, launching and marketing the new product.

Concept generation

Much of the marketing and product development literatures concentrate on monitoring market trends and customer needs to identify new product concepts. However, there is a well-established debate in the literature about the relative merits of 'market-pull' versus 'technology-push' strategies for new product development. A review of the relevant research suggests that the best strategy to adopt is dependent on the relative novelty of the new product. For incremental adaptations or product line extensions, 'market pull' is likely to be the preferred route, as customers are familiar with the product type and will be able to express preferences easily. However, there are many 'needs' that the customer may be unaware of, or unable

---

RESEARCH NOTE   Concept change in radical product development

Victor Seidel examined how concepts changed during the development of radical products using six case studies in consumer electronics, automotive and medical devices.

For a radical innovation, the initial product concept is more likely to be incomplete or vague, and the concept will evolve over time as more technical and market knowledge becomes available. In such cases formal, task-based development processes may be less effective. He observed that around half of all the final product concepts were developed *after* the initial definition stage. Therefore for more radical innovations the effort to develop clear concepts cannot be restricted to the early stages, and should continue throughout the project as new knowledge becomes available. For example, prototype testing may reveal new or alternative technical requirements, and user feedback may indicate unanticipated emerging market needs. However, the process of changing product concept is not iterative, as suggested by the literature. Rather than revise the entire concept in light of the new knowledge, the firms in this study focused on specific concept components, and chose to freeze some, substitute others, and in some cases maintain dual concepts in parallel. The strategy of allowing two concepts to coexist is very different to the prescription of stage-gate processes which aim to filter concepts in a stop/go fashion. For radical innovations, the dual concept allows development teams to continue to progress when faced with quite fundamental challenges, with the possibility of deferring decisions on specific concept components until uncertainty has been further reduced.

Source: Seidel, V.P. (2007) Concept shifting and the radical product development process. *Journal of Product Innovation Management*, **24**, 522–33.

to articulate, and in these cases the balance shifts to a 'technology-push' strategy. Nevertheless, in most cases customers do not buy a technology, they buy products for the benefits that they can receive from them; the 'technology push' must provide a solution for their needs. Thus some customer or market analysis is also important for more novel technology. We discussed the issue of concept development in detail in Chapter 8. This stage is sometimes referred to as the 'fuzzy front end' because it often lacks structure and order, but a number of tools are available to help systematically identify new product concepts, and these are described below. The research note on concept change for radical products illustrates this.

Project selection

This stage includes the screening and selection of product concepts prior to subsequent progress through to the development phase. Two costs of failing to select the 'best' project set are: the actual cost of resources spent on poor projects; and the opportunity costs of marginal projects which may have succeeded with additional resources.

There are two levels of filtering. The first is the aggregate product plan, in which the new product development portfolio is determined. The aggregate product plan attempts to integrate the various potential projects to ensure the collective set of development projects will meet the goals and objectives of the firm, and help to build the capabilities needed. The first step is to ensure resources are applied to the appropriate types and mix of projects. The second step is to develop a capacity plan to balance resource and demand. The final step is to analyse the effect of the proposed projects on capabilities, to ensure this is built up to meet future demands.

The second lower level filters are concerned with specific product concepts. The two most common processes at this level are the development funnel and the stage-gate system. The development funnel is a means to identify, screen, review and converge development projects as they move from idea to commercialization. It provides a framework in which to review alternatives based on a series of explicit criteria for decision making. Similarly, the stage-gate system provides a formal framework for filtering projects based on explicit criteria. The main difference is that where the development funnel assumes resource constraints, the stage-gate system does not. We discussed these in detail in Chapter 7.

Product development

This stage includes all the activities necessary to take the chosen concept and deliver a product for commercialization. It is at the working level, where the product is actually developed and produced, that the individual R&D staff, designers, engineers and marketing staff must work together to solve specific issues and to make decisions on the details (see research note for the critical role of cross-functional teams in product development). Whenever a problem appears, a gap between the current design and the requirement, the development team must take action to close it. The way in which this is achieved determines the speed and effectiveness of the problem-solving process. In many cases this problem-solving routine involves iterative design–test–build cycles, which make use of a number of tools.

Product commercialization and review

In many cases the process of new product development blurs into the process of commercialization. For example, customer co-development, test marketing and use of alpha, beta and gamma test sites yield data on customer requirements and any problems encountered in use, but also help to obtain customer buy-in and prime the market. It is not the purpose of this section to examine the relative efficacy of different marketing strategies, but rather to identify those factors which influence directly the process of new product development. We are primarily interested in what criteria firms use to evaluate the success of new products, and how these criteria might differ between low and high novelty projects. In the former case we would expect more formal and narrow financial or market measures, but in the latter case we find a broader range of criteria are used to reflect the potential for organizational learning and future new product options.

RESEARCH NOTE | **New product development using cross-function teams**

HighTech (a pseudonym) is a division of a global company which designs, builds and supports plant for semiconductor manufacture. Design work is strongly science based, ranging from experimental work in HighTech labs on two continents, through testing and troubleshooting on site and at suppliers, and commissioning and operational support in customers' plants worldwide. Over 24 weeks in real time we observed a process of decision making and development of the conceptual design for a new global release of an existing engineering product. Here we summarize the contribution of two types of critical cross-functional meeting.

Initial kick-off meeting

The kick-off meeting for the development programme was the most formal event that we observed. Participation in the meeting was very wide, and constituted the fullest showing of project stakeholders that we saw at any point during the 24-week period. Names of nominees for specified roles were entered into a computer-based project management system as part of the live business of the meeting. The programme manager presented images of the projected form of the new machine, and a bullet-pointed rationale for the design and launch of the machine. This kick-off event was formative in the sense that it allowed the programme manager subsequently to legitimately call on and deploy both financial resources and the intangible resources taken up by participation in cross-function meetings.

Review meetings

All development programmes currently active were required to present and discuss progress at formal review meetings held in a regular timeslot every two weeks. The central aim of reviews was to achieve planned and formally scheduled phase-exit events. Meetings were convened and led in a formal chairing style by the development process manager, a formal quality management role, occupied by a person who has no specific involvement with any actual development programme.

The flow of each meeting was organized in segments, each of which centred on the presentation and discussion of a 'dashboard' representation that was specific to this venue. It was a composite of three distinct representations: (i) a graphical timeline (showing critical specified events in the lifetime of a programme, on a week-by-week timeline, with the current week highlighted); (ii) a scorechart matrix showing status (good to bad, represented by standard 'smileys') against six specified dimensions of responsibility; and (iii) text bullet points to highlight

---

critical issues. These meetings were formally minuted, had a formal, pre-circulated agenda and pre-published the dashboards for each programme to be discussed. A series of such review meetings led eventually to sign-off for the programme, and a mandatory sign-off for beta release.

Overall, we observed that these strands of interaction were articulated through telling and elaborating (and challenging, amending, negotiating and confirming) 'stories' about the courses of action that participants were engaged in. There were stories about 'what this product will contribute to the business', 'how this product will be constituted, physically, financially and operationally to do this' and 'how we will organize this stream of events and outcomes to achieve a beta launch'. To emphasize the active nature of this, its utter strategic seriousness, and the highly focused and skilful attention that participants gave to this kind of activity, we might label it *story development* rather than story telling. In other words, story development appeared to be a central and intrinsic aspect – perhaps even the primary mode – of product development work.

Source: Hales & Tidd, 2009, 18(4).

Factors influencing product success or failure

There have been more than 200 studies that have investigated the factors affecting the success of new products. Most have adopted a 'matched-pair' methodology in which similar new products are examined, but one is much less successful than the other.[4] This allows us to discriminate between good and poor practice, and helps to control for other background factors. Table 9.1 summarizes some of the main research on the topic of product success and failure.

TABLE 9.1 **Some key studies of new product and service development**

Study name	Key focus	Further reference
Project SAPPHO	Success and failure factors in matched pairs of firms, mainly in chemicals and scientific instruments	5
Wealth from Knowledge	Case studies of successful firms – all were winners of the Queen's Award for Innovation	6
Post-innovation Performance	Looked at these cases 10 years later to see how they fared	7
Project Hindsight	Historical reviews of US government-funded work within the defence industry looking back over 20 years (from 1966) at key projects and success/failure factors	8

(continued)

TABLE 9.1 (Continued)		
Study name	**Key focus**	**Further reference**
TRACES	As Project Hindsight but with 50-year review and also exploring civilian projects. Main aims were to identify sources of successful innovation and management factors influencing success	9
Industry and Technical progress	Survey of UK firms to identify why some were apparently more innovative than others in the same sector, size range, etc. Derived a list of managerial factors which comprised 'technical progressiveness'	10
Minnesota Studies	Detailed case studies over an extended period of innovations. Derived a 'road map' of the innovation process and the factors influencing it at various stages	11
Project NEWPROD	Long-running survey of success and failure in product development and replications	12
Stanford Innovation Project	Case studies of (mainly) product innovations, emphasis on learning	13
Lilien and Yoon	Literature review of major studies of success and failure	14
Rothwell	25-year retrospective review of success and failure studies and models of innovation process	15
Mastering the Dynamics of Innovation	Five retrospective in-depth industry-level cases	16
Sources of Innovation	Case studies involving different levels and types of user involvement	17
Product Development Management Association	Handbook distilling key elements of good practice from a range of success and failure studies in product development	18
Ernst	Extensive literature review of success factors in product innovation	19

(continued)

TABLE 9.1	(Continued)	
Study name	**Key focus**	**Further reference**
Interprod	International study (17 countries) collecting data on the factors influencing new product success and failure	20
Christensen	Industry-level studies of disruptive innovation – includes disk drives, mechanical excavators, steel mini-mills	21
Eisenhardt and Brown	Detailed case studies of five semiconductor equipment firms	22
Revolutionizing Product Development	Case studies of product development	23
Winning by Design	Case studies of product design and innovation	24
Innovation Audits	Various frameworks synthesizing literature and reported key factors	25
Radical Innovation	Review of radical innovation practices in case study firms	26
Rejuvenating the Mature Business	Review of mature businesses in Europe and their use of innovation to secure competitive advantage	27
Innovation Wave	Case studies of manufacturing and service innovations based on experiences at the London Business School Innovation Exchange	28
Tidd and Bodley	Effects of product novelty on effectiveness of development tools, based on 50 development projects	3
SPOTS	Contribution and effectiveness of strategy, processes, organization, technology and systems for new service development in 108 firms	29

These studies have differed in emphasis and sometimes contradicted each other, but despite differences in samples and methodologies it is possible to identify some consensus of what the best criteria for success are:

- *Product advantage* – product superiority in the eyes of the customer, real differential advantage, high performance-to-cost ratio, delivering unique benefits to users – appears to be the primary factor separating winners and losers. Customer perception is the key.
- *Market knowledge* – the homework is vital: better predevelopment preparation including initial screening, preliminary market assessment, preliminary technical appraisal, detailed market studies and business/financial analysis. Customer and user needs assessment and understanding is critical. Competitive analysis is also an important part of the market analysis.
- *Clear product definition* – this includes defining target markets, clear concept definition and benefits to be delivered, clear positioning strategy, a list of product requirements, features and attributes or use of a priority criteria list agreed before development begins.
- *Risk assessment* – market-based, technological, manufacturing and design sources of risk to the development project must be assessed, and plans made to address them. Risk assessments must be built

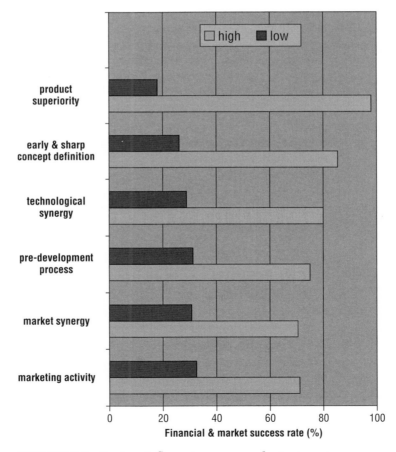

FIGURE 9.3: Factors influencing new product success
Source: Derived from Cooper, R. G. (2000) Doing it right: winning with new product *Ivey Business Journal*, **64** (6): 1–7.

into the business and feasibility studies so they are appropriately addressed with respect to the market and the firms' capabilities.

- *Project organization* – the use of cross-functional, multidisciplinary teams carrying responsibility for the project from beginning to end.
- *Project resources* – sufficient financial and material resources and human skills must be available; the firm must possess the management and technological skills to design and develop the new product.
- *Proficiency of execution* – quality of technological and production activities, and all pre-commercialization business analyses and test marketing; detailed market studies underpin new product success.
- *Top management support* – from concept through to launch. Management must be able to create an atmosphere of trust, coordination and control; key individuals or champions often play a critical role during the innovation process.

These factors have all been found to contribute to new product success, and should therefore form the basis of any formal process for new product development. Note from this list and the factors illustrated in Figures 9.3 and 9.4, that successful new product and service development requires the management of a blend of product or service characteristics, such as product focus, superiority and advantage, and organizational issues, such as project resources, execution and leadership. Managing only one of these key contributions is unlikely to result in consistent success.

The organizational issues appear to dominate in the case of more radical product or service offerings. This is probably because it is much more difficult in such cases to specify, in advance, the product or service characteristics in any detail, and instead managers have to rely more on getting the organization right and influencing the direction of development.

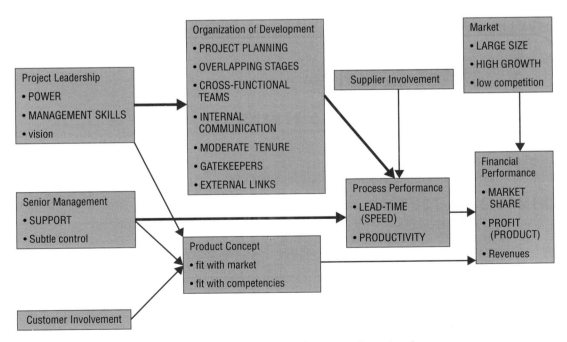

FIGURE 9.4: Key factors influencing the success of new product development
Source: Adapted from Brown, S.L. & Eisenhardt, K.M. (1995). Product development: Past research, present findings and future directions. Academy of Management Review, 20, 343–378.

RESEARCH NOTE | Factors influencing product success

Of the 200 or so systematic studies of new product development, many adopt the categories developed by Cooper in the famous NewProd research programme. For example, one study surveyed 126 development projects in 84 companies in China to try to better understand the effects ownership has on product success, and how factors influencing product success might be different in emerging and more mature economies.

The study found that the following factors were the most significant factors influencing success, ranked from the most to least important:

• Product advantage – e.g. unique features or higher quality.
• Market research proficiency – market segments, trends and competing products.
• Concept development and evaluation – development and screening.
• Market potential – large potential market and growth.
• Market information – customer needs and competitor intelligence.
• Technological synergy – adequate skills and resources.
• Marketing synergy – skills and resources.
• Market pre-testing – customer feedback, analysis and learning.
• Pre-development and planning – definition, cross-functional integration and clear timetable and milestones.
• Market launch – promotion, distribution and sales effort.
• Proficiency of technical activities – designing and testing.
• Strong financial and management support.

There are few surprises here, as these factors feature in most studies. However, the precise ranking and relative importance of different factors will vary with the type of product, technology and market.

Source: Jin, Z. and Z. Li (2007) Firm ownership and the determinants of success and failure in new product development. *International Journal of Innovation Management*, **11** (4), 539–64.

When we have asked managers to describe how radical products and services are developed, the answers include the mysterious and intuitive, and many highlight the importance of luck, accident and serendipity. Of course, there are examples of radical technologies or products that have begun life by chance, like the discovery of penicillin, but Pasteur's advice applies: 'luck favours the prepared mind'.

Gary Lynn and Richard Reilly have tried to identify in a systematic way the most common factors that contribute to successful product development, focusing on what they call 'blockbuster' products – more radical and successful than most new products. Over 10 years they studied more than 700 teams and nearly 50 detailed cases of some of the most successful products ever developed, and compared and contrasted these organizations with less successful counterparts. They identify five key practices that contribute to the successful development of 'blockbuster' products:[30]

• Commitment of senior management.
• Clear and stable vision.

- Improvisation.
- Information exchange.
- Collaboration under pressure.

All five practices operate as a system, and blockbuster development teams must adopt all five practices. Size of the organization did not seem to matter; neither did the type of product.

Commitment of senior management

Those teams that developed blockbusters had the full support and cooperation from senior management. These senior managers functioned as sponsors for the project and took on an active and intimate role. Senior managers would often provide more of a 'hit and run' kind of involvement for those teams that did not produce blockbusters.

Clear and stable vision

It is important for the development team to have a clear and stable vision to guide them, with specific and enduring parameters, something called 'project pillars'. These pillars are the key requirements, or 'must haves' for the new product. Mission awareness is a strong predictor of the success of R&D projects, the degree to which depends on the stage of the project. For example, in the planning and conceptual stage mission awareness explains around two-thirds of the subsequent project success. Leadership clarity is also associated with clear team objectives, high levels of participation, commitment to excellence and support for innovation. Leadership clarity, partly mediated by good team processes, is a good predictor of team innovation.

Improvisation

A clear and stable vision is necessary, but nobody is so brilliant that they can see the end product from the beginning. They may have a vision of what the end product may look like or what the experience of using it will be (or must be) like. It's more like having a dialogue with the product – in trying to get the end results you may ditch what you've done and try something else. You may just have to accept that you may come up with something you never thought you would produce and you might be better off for it. Teams that produce blockbuster products complete the traditional stages of product development, but they take a different approach to the process. Although this may appear to be undisciplined, the teams nearly always have to meet a hard and fast deadline, and are more likely to monitor their progress and costs than the less successful teams.

Information exchange

Effective communication and information exchange is another key practice. Many blockbuster outcomes require the use of cross-functional teams. Exchanging information openly and clearly on a cross-functional team can be challenging to say the least. Not only do specific functions have their own specialized language, they also often have conflicting interests. Team members call on each other through a variety of informal and personal ways like casual conversation, phone calls and meetings. In addition more formal knowledge exchange happens through a system for recording, storing, retrieving and reviewing information (see Chapter 11 for more on knowledge management). Both types of information exchange can be enabled for virtual team working, but all teams need some face-to-face time.

Collaboration under pressure

Blockbuster development teams are generally cross-functional, but must also often deal with outsiders to bring in a new perspective or expertise. Collaboration in the face of conflicting functions and other sources of internal and external pressure requires a number of facilitating factors. Teams that produced blockbuster products complete the traditional stages of product development, but take a different approach to the process. Rather than going through the gates step by step, waiting for a final decision to be made about going forward, they focus on getting an early prototype out quickly to learn how customers might respond. Once they learned how customers responded, they then continued to take out new prototypes for more continuous feedback. The teams need to be able to balance the insights they gained from the customers with the desired outcome. This constant balance allowed them to adjust and fine-tune their understanding of both the market need and the product concept. This fast, iterative process was critical to their success. 3M's application of the lead-user method is a good example of using external partners to develop radical products.

9.2 Influence of technology and markets on commercialization

So far we have described a generic process for new product development, and factors which we know affect success and failure. However, the type of innovation also influences the best way to develop and commercialize an innovation.

The innovation literature has long debated the relative merits of 'market pull' versus 'technology push' for explaining the success (or failure) of new products and services. The usual truce or compromise is to agree on a 'coupling model', whereby technological possibilities are coupled with market opportunities. However, this view is too simplistic. More than 40 years of research, case studies, surveys and econometric analysis are clear. In some cases clear market needs are unmet because of technological limitations, for example, the proverbial cure for cancer, but in other cases technological possibilities have no immediate or obvious commercial application, and anticipate or even create new markets. For example, lasers ('light amplification by the stimulated emission of radiation', if you ever wondered) were for many years simply a useful instrument in scientific experiments, initially used in various military applications, with mixed success, but later formed the basis of almost all optical recording and transmission of data, from broadband to DVD. In this section we try to provide an understanding of the influences the market and technological context has on new product and service development.

Marketing focuses on the needs of the customer, and therefore should begin with an analysis of customer requirements, and attempt to create value by providing products and services that satisfy those requirements. The conventional marketing mix is the set of variables that are to a large extent controllable by the company, normally referred to as the 'four Ps': product, price, place and promotion. All four factors allow some scope for innovation: product innovation results in new or improved products and services, and may change the basis of competition; product innovation allows some scope for premium pricing, and process innovation may result in price leadership; innovations in logistics may affect how a product or service is made available to customers, including distribution channels and nature of sales points; innovations in media provide new opportunities for promotion.

However, we need to distinguish between strategic marketing – that is whether or not to enter a new market – and tactical marketing, which is concerned mainly with the problem of differentiating existing products and services, and extensions to such products. There is a growing body of research that suggests that factors which contribute to new product success are not universal, but are contingent upon a range of technological and market characteristics. A study of 110 development projects found that complexity, novelty and whether the project was for hardware or software development affected the factors that contributed to success.[31] Our own research confirms that different managerial processes, structures and tools are appropriate for routine and novel development projects.[3] For example, in terms of frequency of use, the most common methods used for high novelty projects are segmentation, prototyping, market experimentation and industry experts, whereas for the less novel projects the most common methods are partnering customers, trend extrapolation and segmentation. The use of market experimentation and industry experts might be expected where market requirements or technologies are uncertain, but the common use of segmentation for such projects is harder to justify. However, in terms of usefulness, there are statistically significant differences in the ratings for segmentation, prototyping, industry experts, market surveys and latent needs analysis. Segmentation is the only method more effective for routine development projects, and prototyping, industry experts, focus groups and latent needs analysis are all more effective for novel development projects (Table 9.2). For example, IDEO, the global design and development consultancy, finds conventional market research methods insufficient and sometimes misleading for new products and services, and instead favours the use of direct observation and prototyping (see Case study 9.1).

TABLE 9.2 **The influence of product novelty on the effectiveness of tools used for product development**

	High novelty		Low novelty	
	Usage (%)	Usefulness	Usage (%)	Usefulness
Segmentation*	89	3.42	42	4.50
Prototyping*	79	4.33	63	4.08
Market experimentation	63	4.00	53	3.70
Industry experts*	63	3.83	37	3.71
Surveys/focus groups*	52	4.50	37	4.00
Trend extrapolation	47	4.00	47	3.44
Latent needs analysis*	47	3.89	32	3.67
User-practice observation	47	3.67	42	3.50
Partnering customers	37	4.43	58	3.67

(continued)

	High novelty		Low novelty	
	Usage (%)	Usefulness	Usage (%)	Usefulness
User-developers	32	4.33	37	3.57
Scenario development	21	3.75	26	2.80
Role-playing	5	4.00	11	1.00

TABLE 9.2   (Continued)

* Denotes difference in usefulness rating is statistically significant at 5% level ($n = 50$).
Source: Adapted from Tidd, J. and K. Bodley (2002) Effect of project novelty on the effectiveness of tools used to support new product development. *R&D Management*, **32**, 2, 127–138.

Clearly then, many of the standard marketing tools and techniques are of limited utility for the development and commercialization of novel or complex new products or services. A number of weaknesses can be identified:

- *Identifying and evaluating novel product characteristics.* Marketing tools such as conjoint analysis have been developed for variations of existing products or product extensions, and therefore are of little use for identifying and developing novel products or applications.
- *Identifying and evaluating new markets or businesses.* Marketing techniques such as segmentation are most applicable to relatively mature, well-understood products and markets, and are of limited use in emerging, ill-defined markets.
- *Promoting the purchase and use of novel products and services.* The traditional distinction between consumer and business marketing is based on the characteristics of the customers or users, but the characteristics of the innovation and the relationship between developers and users are more important in the case of novel and complex products and services.

CASE STUDY 9.1

Learning from users at IDEO

IDEO is one of the most successful design consultancies in the world, based in Palo Alto, California and London, UK, it helps large consumer and industrial companies worldwide to design and develop innovative new products and services. Behind its rather typical Californian wackiness lies a tried and tested process for successful design and development:

1. Understand the market, client and technology.
2. Observe users and potential users in real-life situations.

3. Visualize new concepts and the customers who might use them, using prototyping, models and simulations.
4. Evaluate and refine the prototypes in a series of quick iterations.
5. Implement the new concept for commercialization.

The first critical step is achieved through close observation of potential users in context. As Tom Kelly of IDEO argues, '*We're not big fans of focus groups. We don't much care for traditional market research either. We go to the source. Not the "experts" inside a (client) company, but the actual people who use the product or something similar to what we're hoping to create . . . we believe you have to go beyond putting yourself in your customers' shoes. Indeed we believe it's not even enough to ask people what they think about a product or idea . . . customers may lack the vocabulary or the palate to explain what's wrong, and especially what's missing.*'

The next step is to develop prototypes to help evaluate and refine the ideas captured from users. '*An iterative approach to problems is one of the foundations of our culture of prototyping . . . you can prototype just about anything – a new product or service, or a special promotion. What counts is moving the ball forward, achieving some part of your goal.*'

Source: Kelly, T. (2002) *The Art of Innovation: Lessons in Creativity from IDEO*, HarperCollinsBusiness, New York.

Therefore before applying the standard marketing techniques, we must have a clear idea of the maturity of the technologies and markets. Figure 9.5 presents a simple two-by-two matrix, with technological maturity as one dimension, and market maturity as the other. Each quadrant raises different issues and will demand different techniques for development and commercialization:

- *Differentiated*. Both the technologies and markets are mature, and most innovations consist of the improved use of existing technologies to meet a known customer need. Products and services are differentiated on the basis of packaging, pricing and support.
- *Architectural*. Existing technologies are applied or combined to create novel products or services, or new applications. Competition is based on serving specific market niches and on close relations with customers. Innovation typically originates or is in collaboration with potential users.
- *Technological*. Novel technologies are developed which satisfy known customer needs. Such products and services compete on the basis of performance, rather than price or quality. Innovation is mainly driven by developers.
- *Complex*. Both technologies and markets are novel, and co-evolve. In this case there is no clearly defined use of a new technology, but over time developers work with lead users to create new applications. The development of multimedia products and services is a recent example of such a co-evolution of technologies and markets.

Assessing the maturity of a market is particularly difficult, mainly due to the problem of defining the boundaries of a market. The real rate of growth of a market provides a good estimate of the stage in the product life cycle and, by inference, the maturity of the market. In general high rates of market growth are associated with high R&D costs, high marketing costs, rising investment in

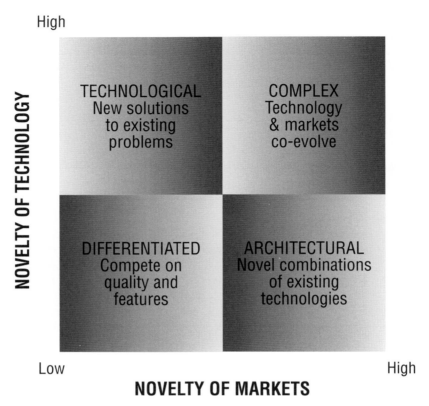

FIGURE 9.5: How technological and market maturity influence the commercialization process

capacity and high product margins (Figure 9.6). At the firm level there is a significant correlation between expenditure on R&D, number of new product launches and financial measures of performance such as value-added and market to book value.[32] Generally, profitability declines as a market matures as the scope for product and service differentiation reduces, and competition shifts towards price.

9.3 Differentiating products

In Chapter 4 we discussed generic corporate strategies based on price leadership or differentiation. Here we are concerned with the specific issue of how to differentiate a product from competing offerings where technologies and markets are relatively stable. It is in these circumstances that the standard tools and techniques of marketing are most useful. We assume the reader is familiar with the basics of marketing, so here we shall focus on product differentiation by quality and other attributes.

Differentiation measures the degree to which competitors differ from one another in a specific market. Markets in which there is little differentiation and no significant difference in the relative quality of competitors are characterized by low profitability, whereas differentiation on the basis of relative quality

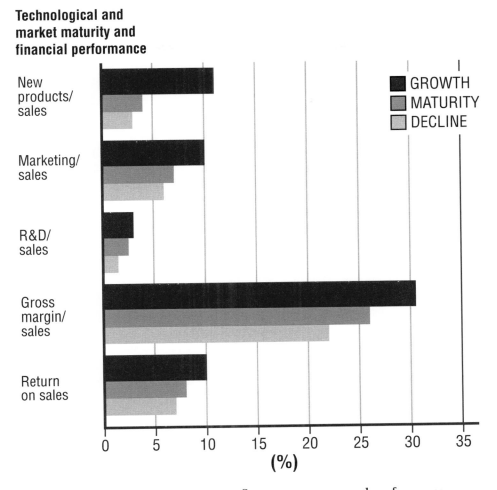

Technological and market maturity and financial performance

FIGURE 9.6: How market maturity influences resources and performance
Source: Derived from Buzzell, R.D. & Gale, B.T. (1987). *The PIMS Principle,* Free Press, New York.

or other product characteristics is a strong predictor of high profitability in any market conditions. Where a firm achieves a combination of high differentiation and high perceived relative quality, the return on investment is typically twice that of non-differentiated products. Analysis of the Strategic Planning Institute's database of more than 3000 business units helps us to identify the profit impact of market strategy (PIMS):[33]

- *High relative quality is associated with a high return on sales.* One reason for this is that businesses with higher relative quality are able to demand higher prices than their competitors. Moreover, higher quality may also help reduce costs by limiting waste and improving processes. As a result companies may benefit from both higher prices and lower costs than competitors, thereby increasing profit margins.
- *Good value is associated with increased market share.* Plotting relative quality against relative price provides a measure of relative value: high quality at a high price represents average value, but high quality at a low price represents good value. Products representing poor value tend to lose market share, but those offering good value gain market share.

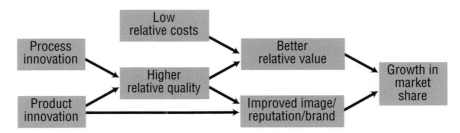

FIGURE 9.7: Relationship between innovation and performance in fast-moving consumer goods

Source: Adapted from Clayton, T. and Turner, G. (2006). *Brands, innovation and growth.* In Tidd, J. (ed) *From Knowledge Management to Strategic Competence: measuring technological, market and organizational innovation.* Imperial College Press, London.

- *Product differentiation is associated with profitability.* Differentiation is defined in terms of how competitors differ from each other within a particular product segment. It can be measured by asking customers to rank the individual attributes of competing products, and to weight the attributes. Customer weighting of attributes is likely to differ from that of the technical or marketing functions.

Analysis of the PIMS data reveals a more detailed picture of the relationships between innovation, value and market performance (Figure 9.7). Process innovation helps to improve relative quality and to reduce costs, thereby improving the relative value of the product. Product innovation also affects product quality, but has a greater effect on reputation and value. Together, innovation, relative value and reputation drive growth in market share. For example, there is an almost linear relationship between product innovation and market growth: businesses with low levels of product innovation – that is having less than 1% of products introduced in the last three years – experience an average real annual market growth of less than 1%; whereas businesses with high levels – that is having around 8% of products introduced in the past three years – experience real annual market growth of around 8%.[34] The compound effect of such differences in real growth can have a significant impact on relative market share over a relatively short period of time. However, in consumer markets maintaining high levels of new product introduction is necessary, but not sufficient. In addition, reputation, or brand image, must be established and maintained, as without it consumers are less likely to sample new product offerings whatever the value or innovativeness. Witness the rapid and consistent growth of Nokia in the mobile phone market (see Case study 9.2).

Quality function deployment (QFD) is a useful technique for translating customer requirements into development needs, and encourages communication between engineering, production and marketing. Unlike most other tools of quality management, QFD is used to identify opportunities for product improvement or differentiation, rather than to solve problems. Customer-required characteristics are translated or 'deployed' by means of a matrix into language which engineers can understand (Figure 9.8). The construction of a relationship matrix – also known as 'the house of quality' – requires a significant amount of technical and market research. Great emphasis must be made on gathering market and user data in order to identify potential design trade-offs, and to achieve the most appropriate balance between cost, quality and performance. The construction of a QFD matrix involves the following steps:[35]

1. Identify customer requirements, primary and secondary, and any major dislikes.
2. Rank requirements according to importance.
3. Translate requirements into measurable characteristics.

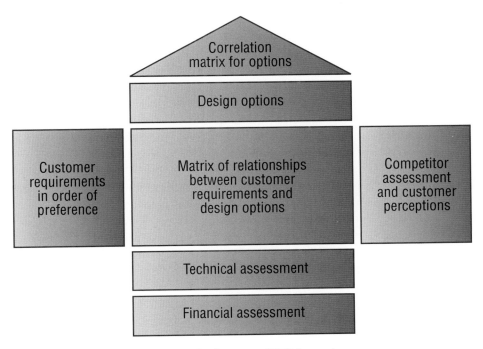

FIGURE 9.8: Quality function deployment (QFD) matrix

Nokia: differentiation by design and innovation

Founded in 1865, Nokia began as a forestry company, and for almost 100 years remained in the pulp and paper industry until a series of unrelated acquisitions in the 1960s. However, in the past decade Nokia has transformed itself from a sprawling conglomerate, with a wide range of mature, low-margin products, to the world's largest manufacturer of mobile phones, with around 30% of the global market for handsets, and 60% share of the industry profits.

When Jorma Ollila, a graduate of the LSE in London, became chief executive in 1992, his strategy consisted of four phrases: 'global', 'telecom-orientated', 'focus' and 'value-added'. In 1993 Ollila told the European Commission that he thought that Europe had lost the computer market to the USA, and consumer electronics to Japan, but could still dominate the emerging telecom market. By 1992 Nokia had become a conglomerate with businesses in aluminium, cables, paper, rubber, televisions, tyres, power generation and real estate. In the mid-1980s Nokia acquired its first interest in the electronics sector, the company Teleste, and in 1991 acquired the British telecoms company, Techophone. In 1985 the share of Nokia's revenues derived from telecoms was just 14%, but following disposal of almost all non-telecoms businesses, this share had grown to almost 90% by 1995. Its success has been based on a combination of technological innovation and product design. It has benefited from its early focus on digital and data traffic, rather than analogue and voice, and the co-development of the GSM (global system of mobile communications) standard, but also has developed a strong brand and carefully segmented the market by means of product design.

Annual growth has regularly been some 40%, and in 2000 the company sold more than 400 million phones. Despite the trend towards a high-volume mass market, Nokia achieves product margins of up to 25%, compared to its rivals' 1 to 3%. The company is now the fifth largest in Europe, employs 44 000 people in 11 countries, almost half of which are Finns. Due to the high rate of growth, half of Nokia's staff has worked for the company for less than three years, and the average age is only 32. It spends around 9% of revenues on R&D, and around a third of its staff work in design or R&D.

In 2004 Nokia began to lose sales and market share, faced with increased competition from Sony Ericsson and Samsung. One reason was a delay in developing newer clam-shell style handsets pioneered by competitors such as Samsung and Motorola; another reason is it being late in developing higher-end camera phones. As a result margins dropped to below 20%, and market share to below 30%, from a peak of 35% in 2003. Nokia successfully recovered its position and in 2007 sold 437 million phones, representing a global market share of 42%. Today around one in three of all handsets in use is a Nokia, reflecting its more recent success in emerging economies. Additional functions such as music and navigation have continued to stimulate demand in the more mature markets. For example, in 2007 Nokia sold 146 million MP3-enabled phones (compared to Apple's 52 million iPods that year), and acquired the mapping firm Navteq for $8.1 billion to develop navigational and location-sensitive services.

4. Establish the relationship between the customer requirements and technical product characteristics, and estimate the strength of the relationship.
5. Choose appropriate units of measurement and determine target values based on customer requirements and competitor benchmarks.

Symbols are used to show the relationship between customer requirements and technical specifications, and weights attached to illustrate the strength of the relationship. Horizontal rows with no relationship symbol indicate that the existing design is incomplete. Conversely, vertical columns with no relationship symbol indicate that an existing design feature is redundant as it is not valued by the customer. In addition, comparisons with competing products, or benchmarks, can be included. This is important because relative quality is more relevant than absolute quality: customer expectations are likely to be shaped by what else is available, rather than some ideal. The concept of QFD is straightforward, but the practice often reveals shortcomings in the product concept or development process.

In some cases potential users may have latent needs or requirements which they cannot articulate. In such cases three types of user need can be identified: 'must be's', 'one-dimensionals' and attractive features or 'delighters'.[36] Must be's are those features which must exist before a potential customer will consider a product or service. For example, in the case of an executive car it must be relatively large and expensive. One-dimensionals are the more quantifiable features which allow direct comparison between competing products. For example, in the case of an executive car, the acceleration and braking performance. Finally, the delighters, which are the most subtle means of differentiation. The inclusion of such features delights the target customers, even if they do not explicitly demand them. For example, delighters in the case of an executive car include ultrasonic parking aids, rain-sensitive windscreen wipers and photochromatic mirrors. Such features are rarely demanded by customers or identified by regular market research. However, indirect questioning can be used to help identify latent requirements.

QFD was originally developed in Japan, and is claimed to have helped Toyota to reduce its development time and costs by 40%. More recently many leading American firms have adopted QFD, including AT&T, Digital and Ford, but results have been mixed: only around a quarter of projects have resulted in any quantifiable benefit.[37] In contrast, there has been relatively little application of QFD by European firms. This is not the result of ignorance, but rather a recognition of the practical problems of implementing QFD.

Clearly, QFD requires the compilation of a lot of marketing and technical data, and more importantly the close cooperation of the development and marketing functions. Indeed, the process of constructing the relationship matrix provides a structured way of getting people from development and marketing to communicate, and therefore is as valuable as any more quantifiable outputs. However, where relations between the technical and marketing groups are a problem, which is too often the case, the use of QFD may be premature.

9.4 Building architectural products

Architectural products consist of novel combinations of existing technologies that serve new markets or applications. In such cases the critical issue is to identify or create new market segments.

Market share is associated with profitability: on average, market leaders earn three times the rate of return of businesses ranked fifth or less.[38] Therefore the goal is to segment a market into a sufficiently small and isolated segment which can be dominated and defended. This allows the product and distribution channels to be closely matched to the needs of a specific group of customers.

Market or buyer segmentation is simply the process of identifying groups of customers with sufficiently similar purchasing behaviour so that they can be targeted and treated in a similar way. This is important because different groups are likely to have different needs. By definition the needs of customers in the same segment will be highly homogeneous. In formal statistical terms the objective of segmentation is to maximize across-group variance and to minimize within-group variance.

In practice segmentation is conducted by analysing customers' buying behaviour and then using factor analysis to identify the most significant variables influencing behaviour – descriptive segmentation – and then using cluster analysis to create distinct segments which help identify unmet customer needs – prescriptive segmentation. The principle of segmentation applies to both consumer and business markets, but the process and basis of segmentation is different in each case.

Segmenting consumer markets

Much of the research on the buying behaviour of consumers is based on theories adapted from the social and behavioural sciences. Utilitarian theories assume that consumers are rational and make purchasing decisions by comparing product utility with their requirements. This model suggests a sequence of phases in the purchasing decision: problem recognition, information search, evaluation of alternatives and finally the purchase. However, such rational processes do not appear to have much influence on actual buying behaviour. For example, in the UK the Consumers' Association routinely tests a wide range of competing products, and makes buying recommendations based on largely objective criteria. If the majority of buyers were rational, and the Consumers' Association successfully identified all relevant criteria, these recommendations would become best-sellers, but this is not the case.

Behavioural approaches have greater explanatory power. These emphasize the effect of attitude, and argue that the buying decision follows a sequence of changing attitudes to a product – awareness, interest, desire and finally action. The goal of advertising is to stimulate this sequence of events. However, research suggests that attitude alone explains only 10% of decisions, and can rarely predict buyer behaviour.

In practice the balance between rational and behavioural influences will depend on the level of customer involvement. Clearly, the decision-making process for buying an aircraft or machine tool is different from the process of buying a toothpaste or shampoo. Many purchasing decisions involve little cost or risk, and therefore low involvement. In such cases consumers try to minimize the financial, mental and physical effort involved in purchasing. Advertising is most effective in such cases. In contrast, in high-involvement situations, in which there is a high cost or potential risk to customers, buyers are willing to search for information and make a more informed decision. Advertising is less effective in such circumstances, and is typically confined to presenting comparative information between rival products.

There are many bases of segmenting consumer markets, including by socioeconomic class, life-cycle groupings and by lifestyle or psychographic (psychological–demographic) factors. An example of psychographic segmentation is the Taylor–Nelson classification that consists of self-explorers, social registers, experimentalists, achievers, belongers, survivors and the aimless. Better-known examples include the *yuppy* (young upwardly mobile professional) and *dinky* (dual income, no kids), and the more recent *yappy* (young affluent parent), *sitcoms* (single income, two children, oppressive mortgage) and *skiers* (spending the kids' inheritance). There is often a strong association between a segment and particular products and services. For example, the yuppy of the 1980s was defined by a striped shirt and braces, personal organizer, brick-sized mobile phone and, of course, a BMW. Annual sales of Filofax were just £100 000 in the UK in 1980, but the deregulation of the City of London in 1986 created 50 000 new, highly paid jobs. As a result annual sales of Filofax reached a peak of £6 million in 1986, the year before the City crashed.

Such segmentation is commonly used for product development and marketing in fast-moving consumer goods such as foods or toiletries and consumer durables such as consumer electronics or cars (see Case study 9.3). It is of particular relevance in the case of product variation or extension, but can also be used to identify opportunities for new products, such as functional foods for the health-conscious, and emerging requirements such as new pharmaceuticals and healthcare services for the wealthy elderly.

CASE STUDY 9.3

The marketing of Persil Power

In 1994 the Anglo-Dutch firm Unilever launched its revolutionary new washing powder 'Persil Power' across Europe ('Omo Power' in some European markets). It was heralded as the first major technological breakthrough in detergents for 15 years. Development had taken 10 years and more than £100 million. The product contained a manganese catalyst, the so-called 'accelerator', which Unilever claimed washed whiter at lower temperatures. The properties of manganese were well known in the industry, but in the past no firm had been able to produce a catalyst which did not also damage clothes. Unilever believed that it had developed a suitable manganese catalyst, and protected its development with 35 patents. The company had test marketed the new product in some 60 000 households and more than 3 million washes, and was sufficiently confident to launch the product in April 1994. However, reports by Procter & Gamble, Unilever's main rival, and subsequent tests by the British Consumers' Association found that under certain conditions Persil Power significantly damaged clothes. After a fierce public relations battle Unilever was forced to withdraw the product, and wrote off some £300 million in development and marketing costs. What went wrong?

There were many reasons for this, but with the benefit of hindsight two stand out. First, the nature of the test marketing and segmentation. Unilever had conducted most of its tests in Dutch households. Typically, northern Europeans separate their whites from their coloured wash, and tend to read product instructions. In contrast, consumers in the South are more likely to wash whites and dyed fabrics together, and to wash everything on a hot wash irrespective of any instructions to the contrary. The manganese catalyst was fine at low temperatures for whites only, but reacted with certain dyes at higher temperatures. Second, the nature of the product positioning. Persil Power was launched as a broad-base detergent suitable for all fabrics, but in practice was only a niche product effective for whites at low temperatures. Unilever learned a great deal from this product launch, and has since radically reorganized its product development process to improve communication between the research, development and marketing functions. Now product development is concentrated in a small number of innovation centres, rather than being split between central R&D and the product divisions, and the whole company uses the formal new product development process based on the development funnel.

Segmenting business markets

Business customers tend to be better informed than consumers and, in theory at least, make more rational purchasing decisions. Business customers can be segmented on the basis of common buying factors or purchasing processes. The basis of segmentation should have clear operational implications, such as differences in preferences, pricing, distribution or sales strategy. For example, customers could be segmented on the basis of how experienced, sophisticated or price-sensitive they are. However, the process is complicated by the number of people involved in the buying process:

- The actual customer or buyer, who typically has the formal authority to choose a supplier and agree terms of purchase.
- The ultimate users of the product or service, who are normally, but not always, involved in the initiation and specification of the purchase.
- Gatekeepers, who control the flow of information to the buyers and users.
- Influencers, who may provide some technical support to the specification and comparison of products.

Therefore it is critical to identify all relevant parties in an organization, and determine the main influences on each. For example, technical personnel used to determine the specification may favour performance, whereas the actual buyer may stress value for money.

The most common basis of business segmentation is by the benefits customers derive from the product, process or service. Customers may buy the same product for very different reasons, and attach different weightings to different product features. For example, in the case of a new numerically controlled machine tool, one group of customers may place the greatest value on the reduction in unit costs it provides, whereas another group may place greater emphasis on potential improvements in precision or quality of the output (see Case study 9.4).

CASE STUDY 9.4

The marketing of Mondex

Mondex is a smart card which can be used to store cash credits – in other words, an electronic purse. The card incorporates a chip which allows cash-free transfers of monetary value from consumer to retailer, and from retailer to bank. NatWest bank first conceived of Mondex in 1990. The rationale for development of the system was the huge costs involved in handling small amounts of cash, estimated to be some £4.5 billion in the UK each year, and therefore the banks and retailers are the main potential beneficiaries. The benefits to consumers are less clear.

In 1991 NatWest created a venture to franchise the system worldwide, and in the UK entered alliances with Midland Bank and BT. Interviews with customer focus groups were conducted in the UK, USA, France, Germany and Japan to determine the likely demand for the service. The results of this initial market research suggested that up to 80% of potential customers would use Mondex, if available. Therefore internal technical trials went ahead in 1992, based on 6000 staff of NatWest. As a result, minor improvements were made, such as a key fob to read the balance remaining on a card, and a locking facility. Market trials began in Swindon in 1995, chosen for its demographic representativeness. Almost 70% of the town's retailers were recruited to the pilot, although several large multiple retailers declined to participate as they were planning their own cards. Some 14 000 customers of NatWest and Midland applied for a free card, but this represented just 25% of their combined customer base in the town. The main barrier to adoption appeared to be the lack of clear benefits to users, whereas the banks and retailers clearly benefited from reduced handling and security costs.

Nevertheless, in 1996 it was announced that Mondex would be offered to all students of Essex University, and cards were to include a broader range of functions including student identification and library access, as well as being accepted by all the banks, shops and bars on campus. University students are ideal consumers of such innovative services, and the campus environment represents a controllable environment in which to test the attractiveness of the service where universal acceptance is guaranteed. Five other universities were subsequently recruited to the three-year trial.

In 1996 Mondex was spun off from NatWest Bank, and is now owned by a consortium headed by Mastercard International. The main competing products are Visa Cash and Belgium's Proton technology. Only 2 million Mondex cards were in use in 2000, but many millions more are to be used by large credit card companies such as JCB of Japan which plans to replace 15 million credit, debit and loyalty cards over the next few years. In addition, Mondex technology, in particular its well-regarded operating system MultOS, has since successfully licensed its technology in more than 50 countries. In 2000 it was announced that Mondex technology was to be used in the Norwegian national lottery, and Mondex was part of a bid consortium for the UK national lottery. Thus the technology and associated business have evolved from a narrow focus on electronic cash, to the broader issue of smart card applications.

It is difficult in practice to identify distinct segments by benefit because these are not strongly related to more traditional and easily identifiable characteristics such as firm size or industry classification.[39] Therefore benefit segmentation is only practical where such preferences can be related to more easily observable and measurable customer characteristics. For example, in the case of the machine tool, analysis of production volumes, batch sizes, operating margins and value-added might help differentiate between those firms which value higher efficiency from those which seek improvements in quality.

This suggests a three-stage segmentation process for identifying new business markets:

1. First, a segmentation based on the functionality of the technology, mapping functions against potential applications.
2. Next, a behavioural segmentation to identify potential customers with similar buying behaviour, for example regarding price or service.
3. Finally, combine the functional and behavioural segmentations in a single matrix to help identify potential customers with relevant applications and buying behaviour.

In addition, analysis of competitors' products and customers may reveal segments not adequately served, or alternatively an opportunity to redefine the basis of segmentation. For example, existing customers may be segmented on the basis of size of company, rather than the needs of specific sectors or particular applications. However, in the final analysis segmentation only provides a guide to behaviour as each customer will have unique characteristics.

There is likely to be a continuum of customer requirements, ranging from existing needs, to emerging requirements and latent expectations, and these must be mapped on to existing and emerging technologies.[40] Whereas much of conventional market research is concerned with identifying the existing needs of customers and matching these to existing technological solutions, in this case the search has to be extended to include emerging and new customer requirements. There are three distinct phases of analysis:

1. Cross-functional teams including customers are used to generate new product concepts by means of brainstorming, morphology and other structured techniques.
2. These concepts are refined and evaluated, using techniques such as QFD.
3. Parallel prototype development and market research activities are conducted. Prototypes are used not as 'master models' for production, but as experiments for internal and external customers to evaluate.

Where potential customers are unable to define or evaluate product design features, in-depth interview clinics must be carried out with target focus groups or via antenna shops. In antenna shops, market researchers and engineers conduct interactive customer interviews, and use marketing research tools and techniques to identify and quantify perceptions about product attributes.

Product mapping can be used to expose the technological and market drivers of product development, and allows managers to explore the implications of product extensions. It helps to focus development efforts and limit the scope of projects by identifying target markets and technologies. This helps to generate more detailed functional maps for design, production and marketing. An initial product introduction, or 'core' product, can be extended in a number of ways:

- An enhanced product, which includes additional distinctive features designed for an identified market segment.
- An 'up-market' extension. This can be difficult because customers may associate the company with a lower quality segment. Also, sales and support staff may not be sufficiently trained or skilled for the new segments.
- A 'down-market' extension. This runs the risk of cannibalizing sales from the higher end, and may alienate existing customers and dealers.
- Custom products with additional features required by a specific customer or distribution channel.
- A hybrid product, produced by merging two core designs to produce a new product.

As we discussed in Chapter 2, in his detailed analysis of the disk drive industry, Clayton Christensen distinguishes between two types of architectural innovation.[41] The first, *sustaining* innovation, which continues to improve existing product functionality for existing customers and markets. The second, *disruptive* innovation, provides a different set of functions which are likely to appeal to a very different segment of the market. As a result, existing firms and their customers are likely to undervalue or ignore disruptive innovations, as these are likely to underperform existing technologies in terms of existing functions in established markets. This illustrates the danger of simplistic advice such as 'listening to customers', and the limitations of traditional management and marketing approaches. Therefore established firms tend to be blind to the potential of disruptive innovation, which is more likely to be exploited by new entrants. Segmentation of current markets and close relations with existing customers will tend to reinforce sustaining innovation, but will fail to identify or wrongly reject potential disruptive innovations. Instead firms must develop and maintain a detailed understanding of potential applications and changing users' needs.

A fundamental issue in architectural innovation is to identify the need to change the architecture itself, rather than just the components within an existing architecture. New product introduction is, up to a point, associated with higher sales and profitability, but very high rates of product introduction become counterproductive as increases in development costs exceed additional sales revenue. This was the case in the car industry, when Japanese manufacturers reduced the life cycle to just four years in the 1990s, but then had to extend it again. Alternatively, expectations of new product introductions can result in users skipping a generation of products in anticipation of the next generation. This has happened in both the PC and mobile phone markets, which has had knock-on effects in the chip industry. Put another way, there is often a trade-off between high rates of new product introduction and product life. The development of common product platforms and increased modularity is one way to try to tackle this trade-off in new product development. Incremental product innovation within an existing platform can either introduce benefits to *existing* customers, such as lower price or improved performance, or additionally attract *new* users and enter new market niches. A study of 56 firms and over 240 new products over a period of 22 years found that a critical issue in managing architectural innovation is the precise balance between the frequency of radical change of product platform, and incremental innovation within these platforms.[42] This suggests that a strategy of ever-faster new product development and

> **BOX 9.1** **Product strategies in services**
>
> Services differ from manufactured goods in many ways, but the two characteristics that most influence innovation management are their intangibility and the interaction between production and consumption. The intangibility of most services makes differentiation more difficult as it is harder to identify and control attributes. The near simultaneous production and consumption of many service offerings blurs the distinction between process (how) and product (what) innovation, and demands the integration of back- and front-end operations.
>
> For example, in our study of 108 service firms in the UK and USA, we found that a strategy of rapid, reiterative redevelopment (RRR) was associated with higher levels of new service development success and higher service quality. This approach to new service development combines many of the benefits of the polar extremes of radical and incremental innovation, but with lower costs and risks. This strategy is less disruptive to internal functional relationships than infrequent but more radical service innovations, and encourages knowledge reuse through the accumulation of numerous incremental innovations. For example, in 1995 the American Express Travel Service Group implemented a strategy of RRR. In the previous decade, the group had introduced only two new service products. In 1995 a vice-president of product development was created, cross-functional teams were established, a formal development process adopted, and computer tools, including prototyping and simulation, were deployed. Since then the group has developed and launched more than 80 new service offerings, and has become the market leader.
>
> *Source*: Tidd, J. and F. Hull (2006) Managing service innovation: the need for selectivity rather than 'best-practice'. *New Technology, Work and Employment*, **21** (2), 139–61; Tidd, J. and F. Hull (2003) *Service Innovation: Organizational Responses to Technological Opportunities and Market Imperatives*, Imperial College Press, London.

introduction is not sustainable, but rather the aim should be to achieve an optimum balance between platform change and new product based on existing platforms. This logic appears to apply to both manufactured products and services (see Box 9.1).

9.5 Commercializing technological products

Technological products are characterized by the application of new technologies in existing products or relatively mature markets. In this case the key issue is to identify existing applications where the technology has a cost or performance advantage.

The traditional literature on industrial marketing has a bias towards relatively low-technology products, and has failed largely to take into account the nature of high-technology products and their markets.

The first and most critical distinction to make is between a technology and a product.[43] Technologists are typically concerned with developing devices, whereas potential customers buy products, which marketing must create from the devices. Developing a product is much more costly and difficult than developing a device. Devices that do not function or are difficult to manufacture are relatively easy to identify and correct compared to an incomplete product offering. A product may fail or be difficult to sell due to poor logistics and branding, or difficult to use because insufficient attention has been paid to customer training or support. Therefore attempting to differentiate a product on the basis of its functionality or the performance of component devices can be expensive and futile.

For example, a personal computer (PC) is a product consisting of a large number of devices or sub-systems, including the basic hardware and accessories, operating system, application programs, languages, documentation, customer training, maintenance and support, advertising and brand development. Therefore a development in microprocessor technology, such as RISC (reduced instruction set computing) may improve product performance in certain circumstances, but may be undermined by more significant factors such as lack of support for developers of software and therefore a shortage of suitable application software.

Therefore in the case of high-technology products it is not sufficient to carry out a simple technical comparison of the performance of technological alternatives, and conventional market segmentation is unlikely to reveal opportunities for substituting a new technology in existing applications. It is necessary to identify why a potential customer might look for an alternative to the existing solution. It may be because of lower costs, superior performance, greater reliability, or simply fashion. In such cases there are two stages to identify potential applications and target customers: technical and behavioural.[44]

Statistical analysis of existing customers is unlikely to be of much use because of the level of detail required. Typically technical segmentation begins with a small group of potential users being interviewed to identify differences and similarities in their requirements. The aim is to identify a range of specific potential uses or applications. Next, a behavioural segmentation is carried out to find three or four groups of customers with similar situations and behaviour. Finally, the technical and behavioural segments are combined to define specific groups of target customer and markets that can then be evaluated commercially (Figure 9.9).

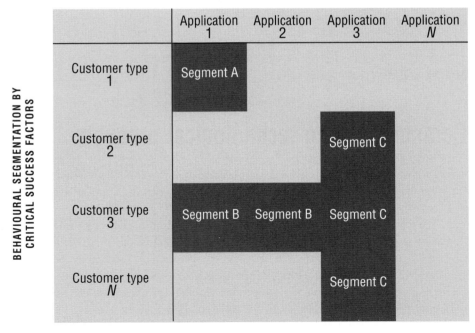

TECHNICAL SEGMENTATION BY APPLICATION

	Application 1	Application 2	Application 3	Application N
Customer type 1	Segment A			
Customer type 2			Segment C	
Customer type 3	Segment B	Segment B	Segment C	
Customer type N			Segment C	

BEHAVIOURAL SEGMENTATION BY CRITICAL SUCCESS FACTORS

FIGURE 9.9: Technical and behavioural segmentation for high-technology products and services

Several features are unique to the marketing of high-technology products, and affect buying behaviour:[45]

- Buyers' perceptions of differences in technology affect buying behaviour. In general, where buyers believe technologies to be similar, they are likely to search for longer than when they believe there to be significant differences between technologies.
- Buyers' perceptions of the rate of change of the technology affect buying behaviour. In general, where buyers believe the rate of technological change is high, they put a lot of effort in the search for alternatives, but search for a shorter time. In non-critical areas a buyer may postpone a purchase.
- Organizational buyers may have strong relationships with their suppliers, which increases switching costs. In general, the higher the supplier-related switching costs, the lower the search effort, but the higher the compatibility-related switching costs, the greater the search effort.

VIEWS FROM THE FRONT LINE

Managing risk in technology development

The precipitation and deposition of mineral scales in oil production systems can seriously restrict hydrocarbon flow and lead to marked reductions in well productivity. In addition, once deposited, these scales are often very difficult to remove, requiring costly well interventions and expensive mechanical removal methods. This is a particularly pernicious and costly problem for sulphate scales that arise when seawater, highly concentrated in sulphate ions, injected for secondary oil recovery, mixes with water already in the reservoir (so-called connate water) rich in divalent ions such as barium leading to the rapid formation of barium sulphate scales.

The nature of oil field scaling has led primarily to the development of two successful preventative approaches (for barium sulphate scale):

- altering the chemistry of the 'produced' water stream by the addition of chemical scale inhibitors to prevent the precipitation of scales; or
- removal of sulphate ions from the injection water using nanofiltration (a membrane-based process) thus eliminating the scale problem at source.

The former process requires treating production wells with scale inhibitors and slowly back producing the inhibitor (the so-called squeeze treatment). This results in oil production losses and, in deepwater and subsea fields, significant costs since well interventions can cost millions of dollars (these treatments may need to be performed several times per year per well and the well count can be a dozen or more). The sulphate removal process can eliminate the need for well interventions but entails considerable capital expenditure (both for investment in the nanofiltration membrane plant and for a larger offshore structure to house the treatment plant).

An innovative concept was developed that had the potential to remove the need for either scale inhibitor treatments or removal of sulphate ions from the injection water. This had the potential to save considerable sums worldwide for the company (and had very attractive net present value and rate of return metrics). The basic concept of the novel technology was to make microscopically small controlled-release particles of scale inhibitor which could be blended with the injection water in a water flood. The concept was that the particles would be transported with the injected water until they were close to a production well at which point they would slowly release scale inhibitor thus protecting the reservoir, the near-wellbore and the wellbore from scale formation. In principle, it goes further than any other currently available technology towards providing a totally intervention-less method of controlling downhole sulphate scale. Its only significant limitation is that it would not provide control in produced fluids prior to injection water breakthrough and additional control methods would still be required, e.g., for prevention of carbonate scaling. It possesses many advantages over the currently available conventional 'batch' (squeeze) methods of scale control.

The pre-screening studies suggested that the cost of the particle technology would make it economically competitive with squeeze treatments in deepwater sulphate removal. Being opex-based, it has the advantage of deferring costs to later in project life with minimal capital investment required. Clearly the economics are sensitive to the dose rate of the particles and the unit cost and thus viability would depend upon the type and cost of the solution finally adopted.

In developing the product, a staged process was adopted that allowed viability at each gate to be reviewed. The process used to develop the injector scale inhibitor technology followed the following format:

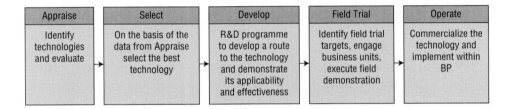

The adoption of this process led to the successful development of particulate scale inhibitors based upon cross-linking acid-based products with polyols to form a solid and processible product. The solid inhibitor was milled into particles small enough to be injected into an oil reservoir without blocking up the porous medium. Whilst the particles had no specific trigger to allow release of scale inhibitor, the rate of release (which occurred by hydrolysis) could be controlled allowing the majority of the scale inhibitor to be released close to the target production wells.

Unfortunately, having developed successful products in the laboratory, business unit engagement was poor and field trial opportunities were not forthcoming, leading to the technology eventually being abandoned

Why did the technology fail to achieve commerciality? Post analysis of the project suggested that the key reasons the technology failed to bridge the gap between laboratory and field demonstration were:

- It was not a complete scale management solution and thus not an attractive integrated solution (this is true of nano-filtration as well, however).
- The oil exploration and production business is conservative and risk averse and the particle technology is very novel.
- There was a perception that risk reduction was too complex (multiple field demonstrations would be required) and furthermore the technology could never be tested on a deepwater

development so the first adopter of the technology would be risking a multi-billion dollar investment on a technology unproven in their particular environment.
- Whilst less of an issue, the lack of field trial opportunities within BP was raised as a problem. This could have been solved by partnering with other companies that had more suitable field trial opportunities.
- Most of BP's production is offshore with large well spacings and any tests would have been on land with shorter well spacings leading to a risk that the response seen in trials would not happen if adopted in a new development.
- At the time we had a poor ability to simulate the process so it was difficult to predict with confidence the outcome of treatment.

Ian Collins, Technology Program Manager, BP Exploration & Production Technology Group

9.6 Implementing complex products

Complex products or systems are a special case in marketing because neither the technology nor markets are well defined or understood. Therefore technology and markets co-evolve over time, as developers and potential users interact. Note that technological complexity does not necessarily imply market complexity, or vice versa. For example, the development of a passenger aircraft is complex in a technological sense, but the market is well defined and potential customers are easy to identify. We are concerned here with cases where both technologies and markets are complex – for example, telecommunications, multimedia and pharmaceuticals.

The traditional distinction between consumer and industrial marketing in terms of the nature of users, rather than the products and services themselves, is therefore unhelpful. For example, a new industrial product or process may be relatively simple, whereas a new consumer product may be complex. The commercialization process for complex products has certain characteristics common to consumer and business markets:[46]

- Products are likely to consist of a large number of interacting components and subsystems, which complicates development and marketing.
- The technical knowledge of customers is likely to be greater, but there is a burden on developers to educate potential users. This requires close links between developers and users.
- Adoption is likely to involve a long-term commitment, and therefore the cost of failure to perform is likely to be high.
- The buying process is often lengthy, and adoption may lag years behind availability and receipt of the initial information.

The nature of complex products

Complex products typically consist of a number of components, or subsystems. Depending on how open the standards are for interfaces between the various components, products may be offered as bundled systems, or as subsystems or components. For bundled systems, customers evaluate purchases at the system level, rather than at the component level. For example, many pharmaceutical firms are now operating managed healthcare services, rather than simply developing and selling specific drugs. Similarly, robot manufacturers offer 'manufacturing solutions', rather than stand-alone robot manipulators. Bundled systems can offer customers enhanced performance by allowing a package of optimized components using proprietary interfaces of 'firmware', and in addition may provide the convenience of a single point of purchase and after-sales support. However, bundled systems may not appeal to customers with idiosyncratic needs, or knowledgeable customers able to configure their own systems.

The growth of systems integrators and 'turnkey' solutions suggests that there is additional value to be gained by developing and marketing systems rather than components: typically, the value added at the system level is greater than the sum of the value added by the components. There is, however, an important exception to this rule. In cases where a particular component or subsystem is significantly superior to competing offerings, unbundling is likely to result in a larger market.[47] The increased market is due to additional customers who would not be willing to purchase the bundled system, but would like to incorporate one of the components or subsystems into their own systems. For example, Intel and Microsoft have captured the dominant market shares of microprocessors and operating systems respectively, by selling components rather than by incorporating these into their own PCs.

Links between developers and users

The development and adoption process for complex products, processes and services is particularly difficult. The benefits to potential users may be difficult to identify and value, and because there are likely to be few direct substitutes available the market may not be able to provide any benchmarks. The choice of suppliers is likely to be limited, more an oligopolistic market than a truly competitive one. In the absence of direct competition, price is less important than other factors such as reputation, performance and service and support.

Innovation research has long emphasized the importance of 'understanding user needs' when developing new products,[48] but in the special case of complex products and services potential users may not be aware of, or may be unable to articulate, their needs. In such cases it is not sufficient simply to understand or even to satisfy existing customers, but rather it is necessary to lead existing customers and identify potential new customers. Conventional market research techniques are of little use, and there will be a greater burden on developers to 'educate' potential users. Hamel and Prahalad refer to this process as *expeditionary marketing*.[49] The main issue is how to learn as quickly as possible through experimentation with real products and customers, and thereby anticipate future requirements and pre-empt potential competitors.

The relationship between developers and users will change throughout the development and adoption process (Figure 9.10). Three distinct processes need to be managed, each demanding different linkages: development, adoption and interfacing. The process of diffusion and adoption was examined in Chapter 8. However, relatively little guidance is available for managing the interface between the developers and adopters of an innovation.

The interface process can be thought of as consisting of two flows: information flows and resource flows.[50] Developers and adopters will negotiate the inflows and outflows of both information and resources. Therefore developers should recognize that resources committed to development and resources committed to aiding adoption should not be viewed as independent or 'ring-fenced'. Both contribute to the successful commercialization of complex products, processes and services. Developers should also identify and manage the balance and direction of information and resource flows at different stages of the process of development and adoption. For example, at early stages managing information inflows

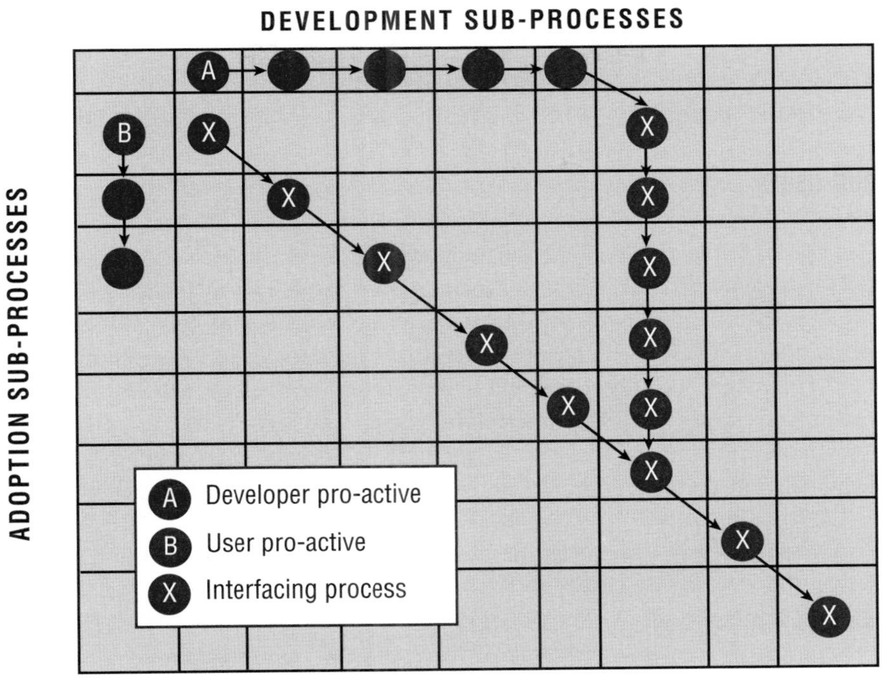

DEVELOPMENT SUB-PROCESSES

FIGURE 9.10: Developer–adopter relationship for complex products

may be most important, but at later stages managing outflows of information and resources may be critical. In addition, learning will require the management of knowledge flows, involving the exchange or secondment of appropriate staff.

Two dimensions help determine the most appropriate relationship between developers and users: the range of different applications for an innovation; and the number of potential users of each application:[51]

- *Few applications and few users.* In this case direct face-to-face negotiation regarding the technology design and use is possible.
- *Few applications, but many users.* This is the classic marketing case, which demands careful segmentation, but little interaction with users.
- *Many applications, but few users.* In this case there are multiple stakeholders amongst the user groups, with separate and possibly conflicting needs. This requires skills to avoid optimization of the technology for one group at the expense of others. The core functionality of the technology must be separated and protected, and custom interfaces developed for the different user groups.
- *Many applications and different users.* In this case developers must work with multiple archetypes of users and therefore aim for the most generic market possible, customized for no one group.

In general, where there are relatively few potential users, as is usually the case with complex products for business customers, customers are likely to demand that developers have the capability to solve their problems, and be able to transfer the solution to them. However, customer expectations vary by sector and nationality. For example, firms in the paper and pulp industry do not expect suppliers to have strong problem-solving capabilities, but do require solutions to be adapted to their specific needs. Conversely, firms in the speciality steel industry demand suppliers to possess strong problem-solving capabilities. Overall, German and Swedish customers expect suppliers to have problem-solving and adaptation capabilities, but British, French and Italian customers appear to be less demanding.[52]

Role of lead users

Lead users are critical to the development and adoption of complex products. As the title suggests, lead users demand new requirements ahead of the general market of other users, but are also positioned in the market to significantly benefit from the meeting of those requirements.[53] Where potential users have high levels of sophistication, for example in business-to-business markets such as scientific instruments, capital equipment and IT systems, lead users can help to co-develop innovations, and are therefore often early adopters of such innovations. The initial research by Von Hippel suggests lead users adopt an average of seven years before typical users, but the precise lead time will depend on a number of factors, including the technology life cycle. A recent empirical study identified a number of characteristics of lead users:[54]

- *Recognize requirements early* – are ahead of the market in identifying and planning for new requirements.
- *Expect high level of benefits* – due to their market position and complementary assets.
- *Develop their own innovations and applications* – have sufficient sophistication to identify and capabilities to contribute to development of the innovation.
- *Perceived to be pioneering and innovative* – by themselves and their peer group.

This has two important implications. First, those seeking to develop innovative complex products and services should identify potential lead users with such characteristics to contribute to the co-development and early adoption of the innovation. Second, that lead users, as early adopters, can provide insights to forecasting the diffusion of innovations. For example, a study of 55 development projects in telecommunications computer infrastructure found that the importance of customer inputs increased with technological newness and, moreover, the relationship shifted from customer surveys and focus groups to co-development because *'conventional marketing techniques proved to be of limited utility, were often ignored, and in hindsight were sometimes strikingly inaccurate'*.[55] Clayton Christensen and Michael Raynor make a similar point in their book *The Innovator's Solution*, and argue that conventional segmentation of markets by product attributes or user types cannot identify potentially disruptive innovations (Case study 9.5).

CASE STUDY 9.5

Identifying potentially disruptive innovations

In their book *The Innovator's Solution: Creating and Sustaining Successful Growth* (Harvard Business School Press, 2003), Clayton Christensen and Michael Raynor argue that segmentation of markets by product attributes or type of customer will fail to identify potentially disruptive innovations. Building on the seminal marketing work of Theodore Levitt, they recommend *circumstance*-based segmentation, which focuses on the 'job to be done' by an innovation, rather than product attributes or type of users. This perspective is likely to result in very different new products and services than traditional ways of segmenting markets. One of the insights this approach provides is the idea of innovations from *non-consumption*. So instead of comparing product attributes with competing products, identify target customers who are trying to get a job done, but due to circumstances – wealth, skill, location, etc. – do not have access to existing solutions. These potential customers are more likely to compare the disruptive innovation with the alternative of having nothing at all, rather than existing offerings. This can lead to the creation of whole new markets – for example, the low-cost airlines in the USA and UK, such as Southwest and Ryanair, or Intuit's QuickBooks. Similarly, in the MBA market, distance learning programmes were once considered inferior to conventional programmes, and instead leading business schools competed (and many still do) for funds for larger and ever-more expensive buildings in prestigious locations. However, improvements to technology, combined with other forms of learning to create 'blended' learning environments, have created whole new markets for MBA programmes, for those who are unable or unwilling to pursue more conventional programmes.

RESEARCH NOTE   Beyond lead users: the co-development of innovations

We are seeing a dramatic shift towards more open, democratized, forms of innovation that are driven by networks of individual users, not firms. Users are now visibly active within all stages of the innovation process and across many types of industrial output, and their influence is rippling out across many sectors. Users may now be actively engaged with firms in the co-development of products and services and the innovation agenda may no longer be entirely controlled by firms. This developing phenomenon has large implications for our understanding of the management of innovation.

The academic understanding of the role of the user as innovator tends to be fragmented, with different strands of literature focusing on particular aspects or perspectives. Within the innovation studies literature, the term 'user' generally takes a supplier-centric perspective and in this context the 'user' (e.g. lead user, final user, user innovation, learning by using) tends to be at the level of the firm. Users tend to be characterized as consumers whose needs must be understood, as 'tough customers' who make exacting demands, or as 'lead users', who may modify or develop existing products in response to their exacting and non-standard needs, potentially foreshadowing future demand. It is also understood that users may be drawn into firms' product development processes by developing and distributing supplier-designed 'toolkits'.

It has also been argued that the process of innovation is becoming democratized as improvements in ICT enable users to develop their own products and services. That users will often freely share their innovations with others, termed 'free revealing', has been widely documented and this forms a key element in the rapid dissemination of certain forms of user-led innovation. The potential for users, either as individuals or as groups, to become involved in the design and production of products has clearly been recognized for some time. However, these conceptions of user–supplier innovation all tend to depict a relationship in which suppliers are able, in some way or another, to harness the experience or ideas of users and apply them to their own product development efforts.

In contrast to the innovation studies literature, the Science and Technology Studies (STS) literature tends to adopt a more user-centric perspective, exploring how users actively shape technologies and are, in turn, shaped by them within the processes of innovation and diffusion. These processes are viewed as highly contested, with users, producers, policymakers and intermediary groups providing differing meanings and uses to technologies. The manner in which design and other activities attempt to define and constrain the ways in which a product can be used has been viewed as an attempt to configure the user. Within this literature, users are seen as having an active role in seeking to shape or reshape their relationship with technology, developing an agenda or 'anti-program' that conflicts with the designer, and going outside the scenario of use, or 'script', that is embodied in the product. Users' lack of compliance with designers and promoters of products and systems, far from being viewed as a deviant activity, is positioned as central to our understanding the processes of innovation and diffusion.

Drawing on both of these strands of literature it is clear that the boundary between producers and consumers has become less distinct and some users are able to develop and extend technologies or use them in entirely novel and unexpected ways. In this situation the boundary between consumer and producer, or between 'users' and 'doers' becomes harder to discern. Innovation becomes far more open, far more democratized, and far more complex. Users may be drawn into the linear model of innovation, but some forms of user activity may represent the emergence of a parallel system of innovation that does not share the same goals, drivers and boundaries of mainstream commercial activity. This has potentially significant implications for our understanding of innovation and key areas including industrial structures, business models, the operation of markets and intellectual property.

Source: S. Flowers and F. Henwood (2008) Special issue on user innovation. *International Journal of Innovation Management*, **12** (3).

Adoption of complex products

The buying process for complex products is likely to be lengthy due to the difficulty of evaluating risk and subsequent implementation. Perceived risk is a function of a buyer's level of uncertainty and the seriousness of the consequences of the decision to purchase. There are two types of risk: the performance risk, that is the extent to which the purchase meets expectations; and the psychological risk associated with how other people in the organization react to the decision. Low-risk decisions are likely to be made autonomously, and therefore it is easier to target decision makers and identify buying criteria. For complex products there is greater uncertainty, and the consequences of the purchase are more significant, and therefore some form of joint or group decision making is likely.

If there is general agreement concerning the buying criteria, a process of information gathering and deliberation can take place in order to identify and evaluate potential suppliers. However, if there is disagreement concerning the buying criteria, a process of persuasion and bargaining is likely to be necessary before any decision can be made.

In the case of organizational purchases, the expectations, perceptions, roles and ideas of risk of the main decision makers may vary. Therefore we should expect and identify the different buying criteria used by various decision makers in an organization. For example, a production engineer may favour the reliability or performance of a piece of equipment, whereas the finance manager is likely to focus on life-cycle costs and value for money (see Box 9.2). Three factors are likely to affect the purchase decision in an organization:[56]

1. *Political and legal environment.* This may affect the availability of, and information concerning, competing products. For example, government legislation might specify the tender process for the development and purchase of new equipment.
2. *Organizational structure and tasks.* Structure includes the degree of centralization of decision making and purchasing; tasks include the organizational purpose served by the purchase, the nature of demand derived from the purchaser's own business, and how routine the purchase is.
3. *Personal roles and responsibilities.* Different roles need to be identified and satisfied. Gatekeepers control the flow of information to the organization, influencers add information or change buying criteria, deciders choose the specific supplier or brand, and the buyers are responsible for the actual purchase. Therefore the ultimate users may not be the primary target.

BOX 9.2 **The EMI CAT scanner**

In 1972 the British firm EMI launched the first computer-assisted tomography (CAT) scanner for use in medical diagnosis. The CAT scanner converted conventional X-ray information into three-dimensional pictures which could be examined using a monitor. EMI had invented and patented all the key technologies of the CAT scanner. The initial slow scanning speed of early machines meant that they were only suitable for organs with minimal movement, such as the brain. In 1976 EMI introduced a faster machine which had a scan time of only 20 seconds, and therefore could be used for whole body scans. It was generally acknowledged that at that time the EMI CAT scanner provided a scanned image superior to that of competing machines, therefore allowing more detailed diagnosis.

Established suppliers of conventional X-ray equipment such as Siemens in Europe and General Electric in the USA responded by differentiating their CAT scanners from those offered by EMI. They competed with the technically superior machines of EMI by emphasizing the faster scan speed of their machines, which they claimed improved patient throughput times. EMI argued that there was a trade-off between scan time and image quality, and that in any case scan time was insignificant relative to the total consultation time required for a patient. However, in North American hospitals, which were the largest market for such machines, patient throughput was of critical importance. Worse still, early machines provided by EMI were highly complex and proved unreliable, and the company was unable to provide worldwide service and support until much later. Early users unfairly compared the reliability of the CAT scanners to more mature and less complex X-ray machines. As a result, the EMI scanner gained a reputation for being unreliable and slow. The machines supplied by its competitors were technically inferior in terms of scanning quality, but gained market share through clever marketing and better customer support. By 1977 the Medical Division of EMI was making a loss, and in 1979 the company was purchased by the Thorn Group.

EMI had invented the CAT scanner, but failed to identify the requirements of its key customers, and underestimated the technical and marketing response of established firms.

9.7 Service innovation

Employment trends in all the so-called advanced countries indicate a move away from manufacturing, construction, mining and agriculture, towards a range of services, including retail, finance, transportation, communication, entertainment, professional and public services. This trend is in part because manufacturing has become so efficient and highly automated, and therefore generates proportionately less employment; and partly because many services are characterized by high levels of customer contact and are reproduced locally, and are therefore often labour-intensive. In the most advanced service economies such as the USA and the UK, services create up to three-quarters of the wealth and 85% of employment, and yet we know relatively little about managing innovation in this sector. The critical role of services, in the broadest sense, has long been recognized, but service innovation is still not well understood.

Innovation in services is much more than the application of information technology (IT). In fact, the disappointing returns to IT investments in services have resulted in a widespread debate about its causes and potential solutions – the so-called 'productivity paradox' in services. Frequently service innovations, which make significant differences to the ways customers use and perceive the service delivered, will demand major investments in process innovation and technology by service providers, but also demand investment in skills and methods of working to change the business model, as well as major marketing changes. Estimates vary, but returns on investment on IT alone are around 15%, with a typical lag of two to three years, when productivity often falls, but when combined with changes in organization and management these returns increase to around 25%.[57]

In the service sector the impact of innovation on growth is generally positive and consistent, with the possible exception of financial services. The pattern across retail and wholesale distribution, transport and communication services, and the broad range of business services is particularly strong (Figure 9.11). More recent research has identified the 'hidden innovation' in the creative industries and media, for example, film and TV programme development, which is not captured by traditional policy or measures such as R&D or patents, as the case of the BBC shows.

Most research and management prescriptions have been based on the experience of manufacturing and high-technology sectors. Most simply assume that such practices are equally applicable to managing innovation in services, but some researchers argue that services are fundamentally different. There

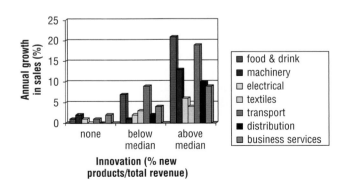

FIGURE 9.11: Innovation and growth in the service sector
Source: European Community Innovation Survey. Based on a survey of 2000 UK service businesses. © European Communities 2000. Reproduced with permission

is a clear need to distinguish what, if any, of what we know about managing innovation in manufacturing is applicable to services, what must be adapted, and what is distinct and different.

We will argue that generic good practices do exist, which apply to both the development of manufactured and service offerings, but that these must be adapted to different contexts, specifically the scale and complexity, degree of customization of the offerings, and the uncertainty of the technological and market environments. It is critical to match the configuration of management and organization of development to the specific technology and market environment. For example, service development in retail financial services is very similar to product development for consumer goods, as the case of Bank of Scotland illustrates.

The service sector includes a very wide range and a great diversity of different activities and businesses, ranging from individual consultants and shopkeepers, to huge multinational finance firms and critical non-profit public and third-sector organizations such as government, health and education. Therefore great care needs to be taken when making any generalization about the service sectors. We will introduce some ways of understanding and analysing the sector later, but it is possible to identify some fundamental differences between manufacturing and service operations:

- *Tangibility*. Goods tend to be tangible, whereas services are mostly intangible, even though you can usually see or feel the results.
- *Perceptions* of performance and quality are more important in services, in particular the difference between expectations and perceived performance. Customers are likely to regard a service as being good if it exceeds their expectations. Perceptions of service quality are affected by:
  - tangible aspects – appearance of facilities, equipment and staff
  - responsiveness – prompt service and willingness to help
  - competence – the ability to perform the service dependably
  - assurance – knowledge and courtesy of staff and ability to convey trust and confidence
  - empathy – provision of caring, individual attention.
- *Simultaneity*. The lag between production and consumption of goods and services is different. Most goods are produced well in advance of consumption, to allow for distribution, storage and sales. In contrast, many services are produced and almost immediately consumed. This creates problems of quality management and capacity planning. It is harder to identify or correct errors in services, and more difficult to match supply and demand.
- *Storage*. Services cannot usually be stored, for example a seat on an airline, although some, such as utilities, have some potential for storage. The inability to hold stocks of services can create problems matching supply and demand – capacity management. These can be dealt with in a number of ways. Pricing can be used to help smooth fluctuations in demand, for example by providing discounts at off-peak times. Where possible, additional capacity can be provided at peak times by employing part-time workers or outsourcing. In the worst cases, customers can simply be forced to wait for the services, by queuing.
- *Customer contact*. Most customers have low or no contact with the operations which produce goods. Many services demand high levels of contact between the operations and ultimate customer, although the level and timing of such contact varies. For example, medical treatment may require constant or frequent contact, but financial services only sporadic contact.
- *Location*. Because of the contact with customers and near simultaneous production and consumption of services, the location of service operations is often more important than for operations which produce goods. For example, restaurants, retail operations and entertainment services all favour

proximity to customers. Conversely, manufactured goods are often produced and consumed in very different locations. For these reasons the markets for manufactured goods also tend to be more competitive and global, whereas many personal and business services are local and less competitive. For example, only around 10% of services in the advanced economies are traded internationally.

These service characteristics should be taken into account when designing and managing the organization and processes for new service development, as some of the findings from research on new product development will have to be adapted or may not apply at all. Also, because of the diversity of service operations, we need also to tailor the organization and management to different types of service context.

In practice, most operations produce some combination of goods and services. It is possible to position any operation on a spectrum from 'pure' products or goods, through to 'pure' services. For example, a restaurant or retail operation both have real goods on offer, but in most cases the service provided is at least equally important. Conversely, most manufacturers now offer some after-sales service and support to customers.

However, the distinction between goods and services remains important because the differences in their characteristics demand a different approach to management and organization. It is perhaps better to think of any business or operation as offering a bundle of benefits, some of which will be tangible, some not, and from this decide the appropriate mix of products and services to be produced.

The service sector includes a wide range of very different operations, including low-skilled personal services such as cleaners, higher skilled personal services such as tradesmen, business services such as lawyers and bankers, and mass consumer services such as transportation, telecommunications and public administration. Critical dimensions which can be used to segment services include labour-intensity of the operations, that is the ratio of labour costs to equipment costs, and the degree of customization or interaction with customers.[58]

To identify common characteristics of service innovators, we have examined over 100 service businesses from the PIMS database, and separated out those which have the highest sustained new service content in their revenue (Table 9.3).

Not surprisingly, high innovators spend more on R&D, to change both what they deliver to customers, and how they deliver it. In addition they have often experienced technology change, and invested in fixed assets to do so. They usually take less than a year to bring new service concepts to market. Competition is also an important factor. The highest innovating firms are more than likely to have experienced entry into their markets by a significant new competitor. They are also much more likely to compete in open markets where international trade – both imports and exports – plays an important role.

The data also indicate that focus is an important discriminating factor between high and low service innovators. First, those businesses with the highest level of new service content tend to avoid overcomplicating their customer base. They are usually firms for which fewer key customer segments account for a higher proportion of their total revenue. This suggests that customer complexity can be a barrier to effective innovation in service businesses. This 'focus' service strategy is well demonstrated by the rise of 'no frills' air services in the USA and Europe since the mid-1990s, such as Southwest, Ryanair and easyJet. Second, it seems that focus in the procurement and service delivery process is also an aid to stronger innovation performance. High innovators tend to focus their purchases on fewer, larger suppliers, and are less vertically integrated – and therefore focused on fewer internal processes within the overall value chain.

Table 9.3	Characteristics of service 'high innovators'	
Business descriptor	**Low innovators**	**High innovators**
Innovation outcomes		
– % sales from services introduced < 3 years ago	<1%	17%
– % new services vs. competitors	>0%	5%
Customer base		
– Focus on key customers	Average	High
– Relative customer base	Similar to competitors	More focused than competitors
Value chain		
– Focus on key suppliers	Average	High/strategic
– Value-added/sales %	72%	60%
– Operating cost added/sales	36%	25%
– Vertical integration vs. competitors	Same or more	Same or less
Innovation input		
– 'What' R&D	0.1% sales	0.7% sales
– 'How' R&D	0.1% sales	0.5% sales
– Fixed assets/sales	growing at 10% p.a.	growing at >20% p.a.
– Overheads/sales %	8%	11%
Innovation context		
– Recent technology change	20%	40%
– Time to market	>1year	<1year
Competition		
– Competitor entry	10%	40%
– Imports/exports vs. market	2%	12%
Quality of offer		
– Relative quality vs. competitors	Declining	Improving
– Value for money	Just below competitors	Better than competitors
Output		
– Real sales	9%	15%

Source: Clayton (2003) in Tidd, J. and Hull, F.M., eds, *Service Innovation: Organizational Responses to Technological Opportunities and Market Imperatives*, Imperial College Press, London. Reproduced with permission

However, persuading customers to buy new services at a premium can be difficult. Most of our 'innovation winners' operate with a policy of parity pricing, of using their service advantage to go for growth, rather than to exploit it for maximum immediate profits. They grow real sales significantly faster, they grow share of their target markets faster than their direct competitors, and non-innovators generally, and in addition they increase their returns on capital employed and assets.

RESEARCH NOTE | Types of service organization for innovation

We studied over 100 service organizations through a series of surveys, interviews and workshops. The goal was to identify the relationships between service strategy, processes for service development, organization, technologies and performance. We found four distinct patterns or configurations which offered different advantages.

1. Client project orientated

Project leaders organize the involvement of everyone early on to reduce handovers, the essence of concurrent product development. Structured processes, such as QFD, are used to identify and influence customer requirements. Processes are mapped and continuously improved. The system is integrated by the voice of the customer and early involvement of the customer in need fulfillment. This configuration is strong on organization, but weaker on tools/technology, such as technological sophistication in either knowledge or IT. However, the art and craft of project management, which is somewhat analogous to batch production in goods industries, provides a strong yet flexible type of enabling control over the development and delivery of customer-focused services. It can achieve high levels of service delivery, and on time to market and cost reduction. These effects on performance are consistent with the inherent flexibility of project-based systems, and are effective in dynamic environments.

Many consultancies and technology-based firms fit this profile. For example, Arup is an international engineering consultancy firm that provides planning, designing, engineering and project management services. The business demands the simultaneous achievement of innovative solutions and significant time compression imposed by client and regulatory requirements. The organization has established a wide range of knowledge management initiatives to encourage sharing of know-how and experience across projects. These initiatives range from organizational processes and mechanisms, such as cross-functional communications meetings and skills networks, to technology-based approaches such as a project database and expert intranet. To date, the former have been more successful than the latter. This may be due to the difficulty of codifying tacit knowledge, which is difficult to store and retrieve electronically, and the unique environmental context of each project limiting the scope for the reuse of standardized knowledge and experience.

2. Mechanistic customization

This is organized by the involvement of external customers in product development and delivery process decisions. Standardization is a key factor in controlling the relationship, and electronic links are used to exchange data with customers and suppliers. Setting standards for projects and products is a key method of process control, and customers help set these standards in conformance with their requirements. The electronic interchange with customers provides the capability for routinely adapting them to market demand. In addition, this type also has a significant positive effect on product innovation and quality, and the locus in both cases is external – the customer.

For example, in British Gas Trading (BGT), standardized documentation and processes are used as an instrument of management control, and yet many different types of contract exist. Within BGT, there are formal procedures for assessing the financial performance of projects, and all projects over a certain threshold require the business owner to prepare a completion report within three months of completion. A project is complete when all physical work is completed, all costs relating to the work have been incurred, and all benefits have been delivered.

3. Hybrid knowledge sharing

In this type of organization people are cross-trained, co-rewarded and organized in groups, which reinforces their team identity. Electronic tools are distributed to all and enable team members to map processes, share best practices and communicate lessons learned online. Group systems are typically rather self-contained which may be one reason companies in this factor are more likely to value knowledge, reuse it and share it to achieve a balanced portfolio of performance advantages. It is strong in organization, tools and system integration, but lacks formal processes. Its use of tools compensates for a lack of processes, and these focus on knowledge management, e.g., distributed databases, templates for process mapping, etc. To the extent it represents a hybrid system, it can achieve different types of performance advantage simultaneously, but is not optimal for anything, and has only a weak association with product innovation and quality, time to market and service delivery. The hybrid knowledge-sharing configuration enables a relatively self-contained group of people to become experts in developing and delivering products as quasi-professionals. This type of organization thereby provides some of the advantages of codified knowledge with far less hierarchical control by bureaucratic forms, consistent with the view that most service innovations demand greater knowledge sharing than in conventional product development.

For example, Cable & Wireless Global Markets (CWGM), a division of the UK telecommunications operator Cable & Wireless, is a systems integrator and service provider which designs, integrates and operates telecommunications networks for multinational clients. CWGM was established to deal with the increasing number of non-standard and highly complex outsourcing projects. The common processes and standards developed by the parent company were found to be inappropriate for this type of business. In contrast to the formal business processes and matrix structure used for simpler management network services, CWGM has adopted a more flexible teaming approach, which includes a 'war room' to help build relationships and promote communication between team members and customers. In this way teams can more easily work closely with customers to develop innovative service packages of standardized products and customized applications to achieve the required service level agreements for outsourcing.

4. Integrated innovative

The integrated innovative organization is characterized by co-located, cross-functional teams in a flattened hierarchy. Communications are open regardless of rank, both face to face and via email. Its technical base utilizes expert systems and management information systems. Responsibility for work is shared and partnering is practised throughout the value chain. The organic design has many advantages for creativity and innovation. They have dense communications facilitated by cross-functional teams and physical co-location. Cross-functional teaming, whereby different specialists are assigned to work on the same project simultaneously, has been advocated and widely adopted in many companies as a strategy to improve their product development process. Collaboration among diverse functions typically provides better solutions to complex design problems. Physical co-location involves aggregating project team members in common space to enhance rich communications among group members. Accordingly, it ranks significantly higher than other configurations in innovation, but lowest in all other performance measures.

For example, in BBC Worldwide (BBCW) speed/timeliness is essential to the processes given its strategic nature. Processes are strongly time driven – indeed, diagrammatically they are captured in a timeline. A series of defined steps is defined, beginning with the initial receipt of programme treatment, to the final sign-off by a senior management committee. The process documentation at BBCW has in-built financial measures as well as benchmarks against the success of previous programmes. The quality of a bid is dependent on individuals and departments providing the required information on a timely basis, together with robust ROI analyses and sales projections. However, processes are able to evolve reactively to emergent business needs. For example, if a new means of exploiting programmes arises (video on demand, broadband video) these additional media can be included in the necessary documentation. In the case of an emergency item that requires urgent approval, informal contacts are exploited to minimize timescales, which is indicative of flexibility and the use of networking.

None of these different service organizations is optimal in every context, and instead different organizational configurations perform best in different cases or contingencies. The integrated innovative is the most innovative; the mechanistic customization is the most cost efficient; hybrid knowledge sharing is best for overall performance; and the client project orientated is best at service delivery.

Sources: Tidd, J. and F.M. Hull (2006) Managing service innovation: the need for selectivity rather than 'best-practice'. *New Technology, Work and Employment*, **21** (2), 139–61; Tidd, J. and F.M. Hull (2003) *Service Innovation: Organizational Responses to Technological Opportunities and Market Imperatives*, Imperial College Press, London. Reproduced with permission.

All four configurations in the Research Note have one or more significant effects on performance. Each appears to have evolved or acquired sufficient good practices to be viable at least in niche markets. The client project orientated reduces time to market and improves service delivery by focusing on customer requirements and project management; the mechanistic customization reduces costs by setting standards and through the involvement of suppliers and customers; the hybrid knowledge sharing provides a combination of innovation and efficiency by promoting team work and knowledge sharing; and the integrated innovative raises innovation and quality by means of cross-functional groups supported by groupware and other tools and technology, but this increased coordination raises the time and cost of service development.

Examination of the actual measures suggests that each of the four organizational configurations provide several common elements, including:

- organizational mode of bringing people together
- control mechanisms, either impersonal (standards, documentation, common software) or interpersonal (co-located teams)
- shared knowledge and/or technical information base
- external linkages, e.g., customers and/or partners/suppliers.

In terms of performance, innovation and quality appear to be improved by cross-functional teams and sharing information, raised by involvement with customers and suppliers, and by encouraging collaboration in teams. Service delivery is improved by customer focus and project management, and by knowledge sharing and collaboration in teams. Time to market is reduced by knowledge sharing and collaboration, and customer focus and project organization, but cross-functional teams can prolong the process. Costs are reduced by setting standards for projects and products, and by involvement of customers and suppliers, but can be increased by using cross-functional teams. Although individual practices can make a significant contribution to performance (Figure 9.12), it is clear that it is the coherent combination of practices and their interaction that creates superior performance in specific

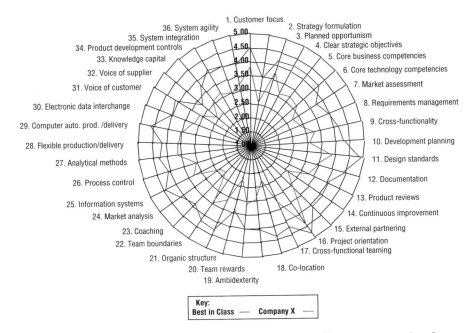

FIGURE 9.12: Factors influencing the effectiveness of new service development
Source: From Tidd, J. & Hull F.M (2006). Managing service innovation: the need for selectivity rather than 'best practice', *New Technology, Work and Employment*, **21** (2), 139–61; Tidd, J. and Hull, F.M. (2003) *Service Innovation: Organizational Responses to Technological Opportunities and Market Imperatives*, Imperial College Press, London.

contexts. These research findings can be used to help assess the effectiveness of existing strategies, processes, organization, tools, technology and systems (SPOTS), and to identify where and how to improve.

Summary and further reading

In this chapter we have examined how the maturity of technologies and markets affects the process of developing new products and services. Where both technologies and markets are relatively mature, the key issue is how to differentiate a product or service for competing offerings. In this case many of the standard marketing techniques can be applied, but other tools such as quality function deployment (QFD) are useful. Where existing technologies are applied to new markets, what we call architectural innovation, the key issue is the resegmentation of markets to identify potential new applications. Where new technologies are applied to existing markets, the key issue is to assess the advantage the technology may have over existing solutions in specific applications, and then identify target users based on behavioural characteristics. Finally, where both technologies and markets are complex, the key issue is the relationship between developers and potential users.

The classic texts on new product development are those by Robert Cooper, for example, *Winning at New Products: Accelerating the Process from Idea to Launch* (Perseus Books, 2001), or Cooper, R.G. (2000) Doing it right: winning with new products, *Ivey Business Journal*, **64** (6), 1–7, or anything by Kim Clark and Steven Wheelwright, such as Wheelwright, S.C. and Clark, K.B. (1997) Creating project plans to focus product development, *Harvard Business Review*, September–October, or their book *Revolutionizing Product Development* (Free Press, 1992). Paul Trott provides a good review of research in his text *Innovation Management and New Product Development* (FT Prentice Hall, fourth edition, 2008), but for a more concise review of the research see Panne, van der, G., Beers, C. van, Kleinknecht, A. (2003) Success and failure of innovation: a literature review, *International Journal of Innovation Management*, **7** (3), 309–38.

For more focused studies of new service development, see the recent article: Berry, L.L. *et al.* (2006) Creating new markets through service innovation, *MIT Sloan Management Review*, **47**, 2. More comprehensive overviews of service innovation are provided by Ian Miles in the Special Issue on innovation in services, *International Journal of Innovation Management*, December, 2000, or in the books by Tidd and Hull: *Service Innovation: Organizational Responses to Technological Opportunities and Market Imperatives* (Imperial College Press, London, 2003); and Normann: *Service Management – Strategy and Leadership in Service Business* (John Wiley & Sons, Ltd, third edition, 2001).

There are few texts which focus exclusively on how to apply conventional marketing tools and techniques to innovative new products and processes, but the best attempts to date are the chapter on 'Securing the future' in Gary Hamel and C. K. Prahalad's *Competing for the Future* (Harvard Business School Press, 1994) and the chapter on 'Learning from the market' in Dorothy Leonard-Barton's *Wellsprings of Knowledge* (Harvard Business School Press, 1995). Dawn Iacobucci has edited an excellent compilation of current theory and practice of business-to-business and other relationship-based marketing in *Networks in Marketing* (Sage, 1996), much of which is relevant to the development and marketing of complex products and services. It also provides a sound introduction to the more general subject of networks, which was discussed in Chapter 6. We discuss the special case of complex product systems in a special issue of *Research Policy*, **29**, 2000, and in *The Business of Systems Integration*, edited by Andrea Prencipe, Andy Davies and Mike Hobday (Oxford University Press, 2003).

A number of US texts cover the related but more narrow issue of marketing high-technology products, including William Davidow's *Marketing High Technology* (Free Press, 1986) and *Essentials of Marketing High Technology* by William Shanklin and John Ryans, Jr (Lexington, 1987). The former is written by a practising engineer/marketing manager, and therefore is strong on practical advice, and the latter is written by two academics, and provides a more coherent framework for analysis. Vijay Jolly's *Commercializing New Technologies* (Harvard Business School Press, 1998) provides a process model based on the experiences of leading firms such as 3M and Sony, which consists of five sub-processes or stages, but the framework is biased towards mass consumer markets. Geoffrey Moore has produced a series of useful guides based on the experience of technology-based firms, beginning with *Crossing the Chasm: Marketing and Selling Technology Products to Mainstream Customers* (Capstone, 1998).

Web links

Here are the full details of the resources available on the website flagged throughout the text:

Case studies:
  BBC
  Bank of Scotland

Interactive exercises:
  QFD Flash

Tools:
  SPOT analysis

Video podcast:
  3M: Breakthrough products and services

References

1. **Cooper, R.G.** (2000) Doing it right: winning with new products. *Ivey Business Journal*, **64** (6), 1–7.
2. **Wheelwright, S.C. and Clark, K.B.** (1997) Creating project plans to focus product development. *Harvard Business Review*, September–October.
3. **Tidd, J. and Bodley, K.** (2002) The effect of project novelty on the new product development process. *R&D Management*, **32** (2), 127–38.
4. **Rothwell, R.** (1977) The characteristics of successful innovators and technically progressive firms (with some comments on innovation research). *R&D Management*, **7** (3), 191–206; **Rothwell, R.** (1992) Successful industrial innovation: critical factors for 1990s. *R&D Management*, **22** (3), 221–39; **Balbontin, A., Yazdani, B., Cooper, R. and Souder, W.E.** (1999) New product development success factors in American and British firms. *International Journal of Technology Management*, **17** (3), 259–80; **Brown, S.L. and Eisenhardt, K.M.** (1995) Product development: past research, present findings, and future directions. *Academy of Management Review*, **20** (2), 343–78; **Mishra, S., Kim, D. and Lee, D.H.** (1996) Factors affecting new product success: cross-country comparisons.

Journal of Product Innovation Management, **13** (6), 530–50; **Ernst, H.** (2002) Success factors of new product development: a review of the empirical literature. *International Journal of Management Reviews*, **4** (1), 1–40.

5. **Rothwell, R.** (1977) The characteristics of successful innovators and technically progressive firms. *R&D Management*, **7** (3), 191–206.

6. **Langrish, J., M. Gibbons, W. Evans, and F. Jevons** (1972) *Wealth from Knowledge*, Macmillan, London.

7. **Georghiou, L., S. Metcalfe, M. Gibbons, T. Ray, and J. Evans** (1986) *Post-innovation Performance*, Macmillan, Basingstoke.

8. **Sherwin, C. and S. Isenson** (1967) Project Hindsight. *Science*, **156**, 571–7.

9. **Isenson, R.** (1968) *Technology in Retrospect and Critical Events in Science (Project TRACES)*, Illinois Institute of Technology/National Science Foundation.

10. **Carter, C. and B. Williams** (1957) *Industry and Technical Progress*, Oxford University Press, Oxford.

11. **Van de Ven, A., H. Angle and M. Poole** (1989) *Research on the Management of Innovation*, Harper & Row, New York.

12. **Cooper, R.** (2001) *Winning at New Products*, third edition. Kogan Page, London.

13. **Maidique, M. and B. Zirger** (1985) The new product learning cycle. *Research Policy*, **14** (6), 299–309.

14. **Lilien, G. and E. Yoon** (1989) Success and failure in innovation – a review of the literature. *IEEE Transactions on Engineering Management*, **36** (1), 3–10.

15. **Rothwell, R.** (1992) Successful industrial innovation: critical success factors for the 1990s. *R&D Management*, **22** (3), 221–39.

16. **Utterback, J.** (1994) *Mastering the Dynamics of Innovation*, Harvard Business School Press, Boston, MA.

17. **Von Hippel, E.** (1988) *The Sources of Innovation*, MIT Press, Cambridge, MA.

18. **Rosenau, M.A. Griffin, G.A. Castellio, and N.F. Anschuetz** (eds) (1996) *The PDMA Handbook of New Product Development*, John Wiley & Sons, Inc., New York.

19. **Ernst, H.** (2002) Success factors of new product development: a review of the empirical literature. *International Journal of Management Reviews*, **4** (1), 1–40.

20. **Souder, W. and S. Jenssen** (1999) Management practices influencing new product success and failure in the US and Scandinavia. *Journal of Product Innovation Management*, **16**, 183–204.

21. **Christensen, C.** (1997) *The Innovator's Dilemma*, Harvard Business School Press, Boston, MA.

22. **Eisenhardt, K. and S. Brown** (1997) The art of continuous change: linking complexity theory and time-paced evolution in relentlessly shifting organizations. *Administrative Science Quarterly*, **42** (1), 1–34.

23. **Wheelwright, S. and K. Clark** (1992) *Revolutionizing Product Development*, Free Press, New York.

24. **Walsh, V., R. Ray, M. Bruce, and S. Potter** (1992) *Winning by Design: Technology, Product Design and International Competitiveness*, Basil Blackwell, Oxford.

25. **Chiesa, V., P. Coughlan and C. Voss** (1996) Development of a technical innovation audit. *Journal of Product Innovation Management*, **13** (2), 105–136; **Design Council** (2002) *Living Innovation*, Design Council/Department of Trade and Industry, London; **Francis, D.** (2001) *Developing Innovative Capability*, University of Brighton, Brighton.

26. **Leifer, R., C.M. McDermott, G. Collarelli O'Connor, B.S. Peters, M. Rice, and R.W. Veryzer** (2000) *Radical Innovation*, Harvard Business School Press, Boston, MA.

27. **Baden-Fuller, C. and J. Stopford** (1995) *Rejuvenating the Mature Business*, Routledge, London.

28. **Von Stamm, B.** (2008) *Managing Innovation, Design and Creativity*; (2003) *The Innovation Wave*, John Wiley & Sons, Ltd, Chichester.

29. **Tidd, J. and Hull, F.M.** (2006) Managing service innovation: the need for selectivity rather than 'best-practice'. *New Technology, Work and Employment*, **21** (2), 139–61; (2003) *Service Innovation: Organizational Responses to Technological Opportunities and Market Imperatives*, Imperial College Press, London.

30. **Lynn, G.S. & Reilly, R.R.** (2002). *Blockbusters: The Five Keys to Developing Great New Products*, HarperBusiness, New York.

31. **Dvir, D., S. Lipovetskya, A. Shenharb and A. Tishler** (1998) In search of project classification: a non-universal approach to project success factors. *Research Policy*, **27**, 915–35.

32. **Tidd, J. and C. Driver** (2006) Technological and market competencies and financial performance. In J. Tidd, ed., *From Knowledge Management to Strategic Competence: Measuring Technological, Market and Organizational Innovation*, (pp. 94–125), Imperial College Press, London.

33. **Luchs, B.** (1990) Quality as a strategic weapon. *European Business Journal*, **2** (4), 34–47.

34. **Clayton, T. and G. Turner** (2006) Brands, innovation and growth. In J. Tidd, ed., *From Knowledge Management to Strategic Competence: Measuring Technological, Market and Organizational Innovation*, (pp. 77–93), Imperial College Press, London.

35. **Burn, G.** (1990) Quality function deployment. In B. Dale and J. Plunkett, eds, *Managing Quality*, (pp. 66–88), Philip Allan, London.

36. **Dimancescu, D. and K. Dwenger** (1995) *World-Class New Product Development*, American Management Association, New York.

37. **Griffin, A.** (1992) Evaluating QFD's use in US firms as a process for developing products. *Journal of Product Innovation Management*, **9**, 171–87.

38. **Buzzell, P. and B. Gale** (1987) *The PIMS Principle*, Free Press, New York.

39. **Moriarty, P. and D. Reibstein** (1986) Benefit segmentation in industrial markets. *Journal of Business Research*, **14**, 463–86.

40. **Lauglaug, A.** (1993) Technical-market research – get customers to collaborate in developing products. *Long Range Planning*, **26** (2), 78–82.

41. **Christensen, C.** (2000) *The Innovator's Dilemma*, HarperCollins, New York.

42. **Jones, N.** (2003) Competing after radical technological change: the significance of product line management strategy. *Strategic Management Journal*, **24**, 1265–87.

43. **Davidow, W.** (1986) *Marketing High Technology*, Free Press, New York.

44. **Millier, P.** (1989) *The Marketing of High-Tech Products: Methods of Analysis*, Editions d'Organisation, Paris (in French).

45. **Weiss, A. and J. Heide** (1993) The nature of organizational search in high technology markets. *Journal of Marketing Research*, **30**, 220–33.

46. **Hobday, M., H. Rush and J. Tidd** (2000) Complex product systems. *Research Policy*, **29**, 793–804.

47. **Wilson, L., A. Weiss and G. John** (1990) Unbundling of industrial systems. *Journal of Marketing*, **27**, 123–38.

48. **Cooper, R. and E. Kleinschmidt** (1993) Screening new products for potential winners. *Long Range Planning*, **26** (6), 74–81.

49. **Hamel, G. and C. Prahalad** (1994) *Competing for the Future*, Harvard Business School Press, Boston, MA.

50. **More, P.** (1986) Developer/adopter relationships in new industrial product situations. *Journal of Business Research*, **14**, 501–17.

51. **Leonard-Barton, D. and D. Sinha** (1993) Developer–user interaction and user satisfaction in internal technology transfer. *Academy of Management Journal*, **36** (5), 1125–39.

52. **Hakansson, H.** (1995) The Swedish approach to Europe. In D. Ford, ed., *Understanding Business Markets*, (pp. 232–61), The Dryden Press, London.

53. **Von Hippel, E.** (1986) Lead users: a source of novel product concepts. *Management Science*, **32** (7), 791–805; (1988) *The Sources of Innovation*, Oxford University Press, Oxford.

54. **Morrison, P., J. Roberts and D. Midgley** (2004) The nature of lead users and measurement of leading edge status. *Research Policy*, **33**, 351–62.

55. **Callahan, J. and E. Lasry** (2004) The importance of customer input in the development of very new products. *R&D Management*, **34** (2), 107–17.

56. **Webster, F. Jr** (1991) *Industrial Marketing Strategy*, third edition, John Wiley & Sons, Inc., New York.

57. **Crespi, G., Criscuolo, C., and Haskel, J.** (2006) Information technology, organisational change and productivity growth: evidence from UK firms. *The Future of Science, Technology and Innovation Policy: Linking Research and Practice*, SPRU 40th Anniversary Conference, Brighton, UK.

58. **Berry, L.L., V. Shankar, J. Turner Parish, S. Cadwaller, and D. Dotzel** (2006) Creating new markets through service innovation. *MIT Sloan Management Review*, **47**, 2; **Schmenner, R.W.** (1986) How can service businesses survive and prosper? *MIT Sloan Management Review*, **27** (3), 21–32.

CHAPTER 10

Exploiting new ventures

In Chapter 9 we examined the processes necessary to develop new products and services within the existing corporate environment, based on the strategy and capabilities identified in Chapter 4. In this chapter we explore how firms develop technologies, products and businesses outside their existing strategy and core competencies. We will discuss the role and management of a range of ventures in the creation and execution of new technologies, products and businesses, specifically:

- internal corporate ventures, or 'intrapreneurship'
- external joint ventures and alliances
- new ventures and spin-out firms

10.1 What is a venture?

Ventures, broadly defined, are a range of different ways of developing innovations, alternative to conventional internal processes for new product or service development. We discussed in Chapter 9 the many benefits of using structured approaches to new product and service development, such as stage-gate and development funnel processes. However, these approaches also have a major disadvantage, because decisions at the different gates are likely to favour those innovations close to existing strategy, markets and targets, and will filter out or reject those potential innovations further from the organization's comfort zone. For this reason other mechanisms of development and commercialization are necessary, ranging from internal corporate ventures, through joint ventures to spin-out businesses and new ventures.

Figure 10.1 suggests a range of venture types that can be used in different contexts. Corporate ventures are likely to be most appropriate where the organization needs to exploit some internal competencies and retain a high degree of control over the business. Joint ventures and alliances involve working with external partners, and therefore demand some release of control and autonomy, but in return introduce the additional competencies of the partners. Spin-out or new venture businesses are the extreme case, often necessary where there is little relatedness between the core competencies and new venture business. Note that these options are not mutually exclusive, for example, a spin-out business can become an alliance partner, or a corporate venture can spin-out. Also, all types of venture require a venture champion, a strong business case, and sufficient resources to be successful.

Profile of a venture champion

Research by Ed Roberts,[1] who studied 156 new technology-based firms (NTBFs), which were spin-offs from MIT in the USA (herein referred to as 'the US study'), and Ray Oakey,[2] who examined 131 NTBFs in the UK (herein referred to as 'the UK study'), provide a pretty consistent picture of the profile of a

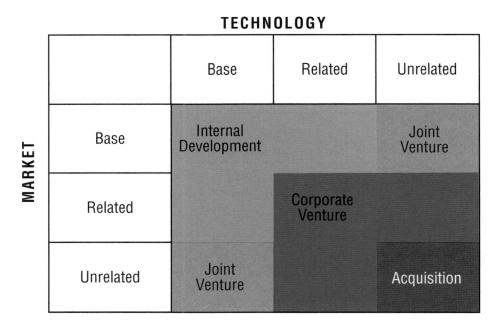

FIGURE 10.1: Role of venturing in the development and commercialization of innovations

Source: Adapted from Burgelman, R. (1984) Managing the internal corporate venturing process. *Sloan Management Review*, **25** (2), 33–48.

typical venture champion. Despite the obvious Anglo-Saxon bias of these two large studies, other research confirms the general relevance of these factors.

The creation of a venture is the interaction of individual skills and disposition and the technological and market characteristics. The US study emphasizes the role of personal characteristics, such as family background, goal orientation, personality and motivation; whereas the UK study stresses the role of technological and market factors. The decision to start an NTBF typically begins with a desire to gain independence and to escape the bureaucracy of a large organization, whether it be in the public or private sector. Thus the background, psychological profile and work and technical experience of a technical entrepreneur all contribute to the decision to form an NTBF (Figure 10.2).

Much of the American research on new ventures, and more general studies of entrepreneurs, tends to emphasize the background and characteristics of a typical entrepreneur. Factors found to affect the likelihood of establishing a venture include:

- family background
- religion
- formal education and early work experience
- psychological profile.

A number of studies confirm that both family background and religion affect an individual's propensity to establish a new venture. A significant majority of technical entrepreneurs have a self-employed or professional parent. Studies indicate that between 50% and 80% have at least one self-employed parent. For example, the US study found that four times as many technical entrepreneurs

FIGURE 10.2: Factors influencing the decision to establish a new venture

have a parent who is a professional, compared with other groups of scientists and engineers. The most common explanation for this observed bias is that the parent acts as a role model and may provide support for self-employment.

The effect of religious background is more controversial, but it is clear that certain religions are over-represented in the population of technical entrepreneurs. For example, in the USA and Europe, Jews are more likely to establish an NTBF; and Chinese are more likely to in Asia. Whether this observed bias is the result of specific cultural or religious norms, or the result of minority status, is the subject of much controversy but little research. The US study suggests that dominant cultural values are more important than minority status, but even this work indicates that the effect of family background is more significant than religion. In any case, and perhaps more importantly, there appears to be no significant relationship between family and religious background and the subsequent probability of success of an NTBF.

Education and training are major factors that distinguish the founders of NTBFs from other entrepreneurs. The median level of education of technical entrepreneurs in the US study was a master's degree and, with the important exception of biotechnology-based NTBFs, a doctorate was superfluous. Significantly, the levels of education of technical entrepreneurs do not differentiate them from other scientists and engineers. However, potential technical entrepreneurs tend to have higher levels of productivity than their technical work colleagues, measured in terms of papers published or patents granted:

6.35 versus 2.2 papers on average; and 1.6 versus 0.05 patents. This suggests that potential entrepreneurs may be more driven than their corporate counterparts.

In addition to a master's-level education, on average, a technical entrepreneur will have around 13 years of work experience before establishing an NTBF. In the case of the Route 128 technology cluster in Boston, the entrepreneurs' work experience is typically with a single incubator organization, whereas technical entrepreneurs in Silicon Valley tend to have gained their experience from a larger number of firms before establishing their own NTBF. This suggests that there is no ideal pattern of previous work experience. However, experience of development work appears to be more important than work in basic research. As a result of the formal education and experience required, a typical technical entrepreneur will be aged between 30 and 40 years when establishing their first NTBF. This is relatively late in life compared to other types of venture, and is due to a combination of ability and opportunity. On the one hand, it typically takes between 10 and 15 years for a potential entrepreneur to attain the necessary technical and business experience. On the other hand, many people begin to have greater financial and family responsibilities at this time. Thus there appears to be a window of opportunity to start an NTBF some time in the mid-thirties. Moreover, different fields of technology have different entry and growth potential. Therefore the choice of a potential entrepreneur will be constrained by the dynamics of the technology and markets. The capital requirements, product lead times and potential for growth are likely to vary significantly between sectors.

Much of the research on the psychology of entrepreneurs is based on the experience of small firms in the USA, so the generalizability of the findings must be questioned. However, in the specific case of technical entrepreneurs there appears to be some consensus regarding the necessary personal characteristics. The two critical requirements appear to be an internal locus of control and a high need for achievement. The former characteristic is common in scientists and engineers, but the need for high levels of achievement is less common. Entrepreneurs are typically motivated by a high need for achievement (so-called 'n-Ach'), rather than a general desire to succeed. This behaviour is associated with moderate risk taking, but not gambling or irrational risk taking. A person with a high n-Ach:

- likes situations where it is possible to take personal responsibility for finding solutions to problems
- has a tendency to set challenging but realistic personal goals and to take calculated risks
- needs concrete feedback on personal performance.

However, the US study of almost 130 technical entrepreneurs and almost 300 scientists and engineers found that not all entrepreneurs have high n-Ach, only some do. Technical entrepreneurs had only moderate n-Ach, but low need for affiliation (n-Aff). This suggests that the need for independence, rather than success, is the most significant motivator for technical entrepreneurs. Technical entrepreneurs also tend to have an internal locus of control. In other words, technical entrepreneurs believe that they have personal control over outcomes, whereas someone with an external locus of control believes that outcomes are the result of chance, powerful institutions or others. More sophisticated psychometric techniques such as the Myers–Briggs type indicators (MBTI) confirm the differences between technical entrepreneurs and other scientists and engineers.

Numerous surveys indicate that around three-quarters of technical entrepreneurs claim to have been frustrated in their previous job. This frustration appears to result from the interaction of the psychological predisposition of the potential entrepreneur and poor selection, training and development by the parent organization. Specific events may also trigger the desire or need to establish an NTBF, such as a major reorganization or downsizing of the parent organization. David Hall has developed an 'entrepreneurial code' based upon interviews with 40 successful entrepreneurs. He argues that entrepreneurship is more of an issue of personal rather than business process development, and emphasizes the personal characteristics needed and attention to networking.

Venture business plan

The primary reason for developing a formal business plan for a new venture is to attract external funding. However, it serves an important secondary function. A business plan can provide a formal agreement between founders regarding the basis and future development of the venture. A business plan can help reduce self-delusion on the part of the founders, and avoid subsequent arguments concerning responsibilities and rewards. It can help to translate abstract or ambiguous goals into more explicit operational needs, and support subsequent decision making and identify trade-offs. Of the factors *controllable* by entrepreneurs, business planning has the most significant positive effect on new venture performance. However, there are of course many *uncontrollable* factors, such as market opportunity, which have an even more significant influence on performance.[3] Pasteur's advice still applies, '. . . *chance favours only the prepared mind.*'

No standard business plan exists, but in many cases venture capitalists will provide a pro forma for the business plan. Typically a business plan should be relatively concise, say no more than 10 sides, begin with an executive summary, and include sections on the product, markets, technology, development, production, marketing, human resources, financial estimates with contingency plans, and the timetable and funding requirements. Most business plans submitted to venture capitalists are strong on the technical considerations, often placing too much emphasis on the technology relative to other issues. As Roberts notes, '*Entrepreneurs propose that they can do it better than anyone else, but may forget to demonstrate that anyone wants it.*'[1] He identifies a number of common problems with business plans submitted to venture capitalists: marketing plan, management team, technology plan and financial plan.

There were found to be serious inadequacies in all four of these areas, but the worst were in marketing and finance. Less than half of the plans examined provided a detailed marketing strategy, and just half included any sales plan. Three-quarters of the plans failed to identify or analyse any potential competitors. As a result most business plans contain only basic financial forecasts, and just 10% conducted any sensitivity analysis on the forecasts. The lack of attention to marketing and competitor analysis is particularly problematic as research indicates that both factors are associated with subsequent success. For new ventures it is critical to assess where and how value is to be created, and tools such as value analysis and value stream analysis can help to identify opportunities and weaknesses in business plans.

For example, the UK study found that in the early stages NTBFs rely too much on a few major customers for sales, and are therefore vulnerable. In the extreme case, half of NTBFs rely on a single customer for more than half of their first-year sales. This overdependence on a small number of customers has three major drawbacks:

1. Vulnerability to changes in the strategy and health of the dominant customer.
2. A loss of negotiating power, which may reduce profit margins.
3. Little incentive to develop marketing and sales functions, which may limit future growth.

Funding

New ventures are different from the relatively simple assessment of new products, as there is often no marketable product available before or shortly after formation. Therefore, initial funding of the venture cannot normally be based on cash flow derived from early sales. The precise cash-flow profile will be determined by a number of factors, including development time and cost, and the volume and profit margin of sales. Different development and sales strategies exist, but to some extent these factors are determined by the nature of the technology and markets (Figure 10.3(a)–(c)).

For example, biotechnology ventures typically require more start-up capital than electronics or software-based ventures, and have longer product development lead times. Therefore, from the perspective of a potential entrepreneur, the ideal strategy would be to conduct as much development work as possible within the incubator organization before starting the new venture. However, there are practical problems with this strategy, in particular ownership of the intellectual property on which the venture is to be based.

Given their strong desire for independence, most entrepreneurs seek to avoid external funding for their ventures. However, in practice this is not always possible, particularly in the latter growth stages. The initial funding required to form an NTBF includes the purchase of accommodation, equipment and other start-up costs, plus the day-to-day running costs such as salaries, heating, light and so on. Research in the USA and UK suggests that most NTBFs begin life as part-time ventures, and are funded by personal savings, loans from friends and relatives, and bank loans, in that order. Around half also receive some funding from government sources, but in contrast receive next to nothing from venture capitalists. Venture capital is typically only made available at later stages to fund growth on the basis of a proven development and sales record.

Research in the USA suggests that the initial capital needed to start an NTBF is relatively modest, typically less than $50 000 and in almost half of the cases less than $10 000 (1990 US dollars). However, the extent of the need for external funding will depend on the nature of the technology and the strategy of the NTBF. For example, an electronics or software-based venture will also demand high initial funding if a strategy of aggressive growth is to be achieved. The UK study shows that both the amount and source of initial funding for the formation of an NTBF vary considerably. For example, as software-based ventures typically require less start-up capital than either electronics or biotechnology ventures, it is more common for such firms to rely solely on personal funding. Biotechnology firms tend to have the highest R&D costs, and consequently most require some external funding. In contrast, software firms typically require little R&D investment, and are less likely to seek external funds. The UK study found that almost three-quarters of the software firms were funded by profits after three years, whereas only a third of the biotechnology firms had achieved this.

The initial funding to establish an NTBF is rarely a major problem. However, Peter Drucker suggests an NTBF requires financial restructuring every three years.[4] Other studies identify stages of development, each having different financial requirements:

1. Initial financing for launch.
2. Second-round financing for initial development and growth.
3. Third-round financing for consolidation and growth.
4. Maturity or exit.

In general, professional financial bodies are not interested in initial funding because of the high risk and low sums of money involved. It is simply not worth their time and effort to evaluate and monitor such ventures. However, as the sums involved are relatively small – typically of the order of tens of thousands of pounds – personal savings, remortgages and loans from friends and relatives are often sufficient. In contrast, third-round finance for consolidation is relatively easy to obtain, because by that time

(a) Research-based venture e.g. biotechnology

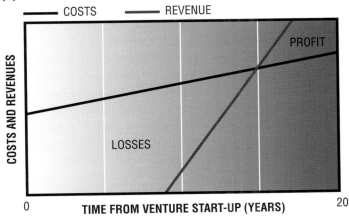

(b) Development-based venture e.g. electronics

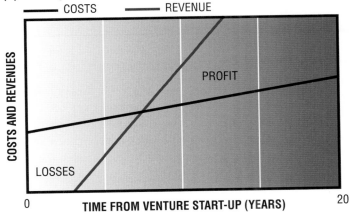

(c) Production-based venture e.g. software

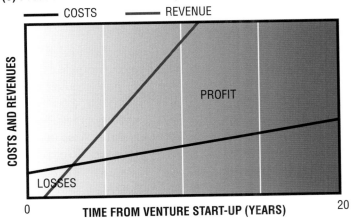

FIGURE 10.3: Cash-flow profiles for three types of technology-based venture: (a) Research-based, e.g. biotechnology; (b) Development-based, e.g. electronics; (c) Production-based, e.g. software

the venture has a proven track record on which to base the business plan, and the venture capitalist can see an exit route.

Venture capitalists are keen to provide funding for a venture with a proven track record and strong business plan, but in return will often require some equity or management involvement. Moreover, most venture capitalists are looking for a means to make capital gains after about five years. However, almost by definition technical entrepreneurs seek independence and control, and there is evidence that some will sacrifice growth to maintain control of their ventures. For the same reason, few entrepreneurs are prepared to 'go public' to fund further growth. Thus many entrepreneurs will choose to sell the business and found another NTBF. In fact, the typical technical entrepreneur establishes an average of three NTBFs. Therefore the biggest funding problem for an NTBF is likely to be for the second-round financing to fund development and growth. This can be a time-consuming and frustrating process to convince venture capitalists to provide finance. The formal proposal is critical at this stage. Professional investors will assess the attractiveness of the venture in terms of the strengths and personalities of the founders, the formal business plan and the commercial and technical merits of the product, probably in that order.

VIEWS FROM THE FRONT LINE

The role of venture capital in innovation

I was recently asked by a friend who works in the R&D group at a large corporation to summarize the role of venture capital in innovation. Trying to make it relevant to his own experience I explained that we simply provide the R&D budget for companies that would not ordinarily have one! I explained further that the companies we back are, on the whole, small self-contained R&D organizations generating intellectual property and ultimately new products that threaten the incumbents in any particular industry. Venture capitalists believe that to 'create value' a small firm should follow a strategy that means it will be needed by or become a threat to global corporations. That way, such corporations may be forced to bid against each other to acquire the small firm and obtain the new innovations (or remove the threat) thus providing the venture capitalist with a high value exit from its investment.

This goes to the very heart of the venture capital business model. Venture capitalists are professional fund managers who invest cash in early-stage high-risk ventures, in return for shares, with the aim of selling those shares at a later date through some form of exit event. The golden rule of investment 'buy low, sell high' is modified in the realm of venture capital to 'buy very low sell very high' to account for the extreme risk profile of the early-stage ventures they back.

The follow-up question to what venture capitalists do, is usually whether they provide value to early-stage ventures beyond pure financial investment. The question usually provokes a debate, sometimes heated, about the pros and cons of having venture capitalists involved in running a business. In my view the answer is simple – and is based around a philosophy within the venture capital industry to kill failure early. By allocating their capital only to companies that continue to demonstrate success, venture capitalists deprive underperforming ventures of cash and usually bring about its rapid demise. This is often not the case within the R&D groups of large corporations where underperforming or low potential projects can struggle on for years protected by managers' indecision and political sensitivity. Thus venture capitalists provide a rigorous and ongoing selection process for the innovation process holding the companies they back to strict targets and tight deadlines – there is no hiding place.

Thus, venture capital investment provides the cash to drive innovation forward within small companies at a faster rate than would ordinarily be possible and it provides a rigorous and ongoing monitoring process that responds by killing failure early. Ultimately, this is underpinned by the very simplest of selection criteria: will this investment make a significant financial return within 3–5 years' time? Answering that question clarifies even the most difficult of investment decisions.

Simon Barnes is managing partner of Tate & Lyle Ventures LP, an independent venture capital fund backed by Tate & Lyle, a global food ingredients manufacturer.

Corporate venture funding

A survey of corporate funding of NTBFs in the UK found that around 15% of large companies had made investments in external new ventures, mainly in their own sector.[5] As with internal corporate venturing, the funding of external ventures by large corporations is cyclical, reflecting the business environment. For example, surveys suggest that in 1998 the number of major corporations funding external ventures was around 110, but by 2000 this had grown to 350.[6] The typical investment (in 1997) was in excess of £500 000, and the investing companies preferred ventures requiring additional capital for expansion, rather than funds for start-up or early development. The most common problems encountered were agreement of the rate of return and details of corporate representation in the venture. The average period of investment was five to seven years, and corporate investors typically demanded a rate of return of 20 to 30%, which compares favourably with professional venture capitalists required returns of around 75%. Regarding professional venture capitalists, Figure 10.4 highlights two important issues. First, that the availability of venture capital varies worldwide, and that such disparities tend to be self-reinforcing as potential new ventures relocate to seek funding. The second point to note is the strong bias for finance for expansion, rather than start-ups, which is most significant in the UK. This creates a potential venture-funding gap, between the initial, usually self-financed stage, and the first involvement of professional venture capital. In the UK this gap is in the region of £200 000 to £750 000.[7]

Corporate investment in new ventures is increasingly popular in high-technology sectors, where large firms do not have access to all technologies in-house, and where emerging technologies remain unproven.[8] Investments in small biotechnology companies by pharmaceutical companies can be direct or indirect investment through specialist venture funds (see Case study 10.1). Direct investment is preferred where there is a high probability of technological success which is likely to impact the product pipeline in the near term. Indirect investments are concerned more with gaining windows on a range of early-stage technologies with the potential to impact the future direction of the product pipeline.[9] There has been a marked increase in the number of pharmaceutical companies investing through specialist venture funds, recent examples being Novartis (Novartis Ventures) and Bayer (Bayer Innovation). At the same time, pharmaceutical companies and their venture funds appear to be investing increasingly in independent seed capital funds focused on early-stage biotechnology, such as UK Medical

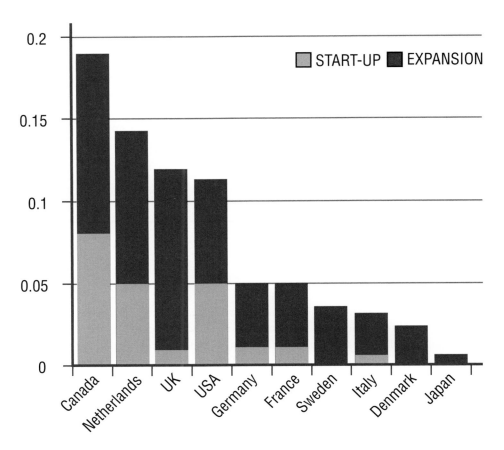

FIGURE 10.4: Venture capital as a percentage of GDP (1997)
Sources: CVCA (Canada), NCVA (USA), MITI (Japan), OECD (other nations)

Ventures (UK), New Medical Technologies (Switzerland) and Medical Technology Partners (USA). The precise objectives of such funds vary, but all share a common emphasis on strategic issues rather than purely financial. A principal investment criterion is 'no fit, no deal', the decision to invest being largely strategic, to 'scout for "out there" science'. The alternative mode of indirect venturing is participation in independent seed capital funds targeted at early-stage investments. This included direct investment in seed capital funds by the corporate parent, or indirect investment in seed funds via corporate venture funds. As may be expected, objectives include providing windows on early-stage technologies; for example, one company invested in a health informatics fund in order to place representatives on the board of the seed fund to gain competence in the field. Another reason for investing is to access 'deal flow' – that is the opportunity to participate directly in subsequent rounds of funding beyond the seed capital stage. This was aimed at investing more *directly* in later stage technologies should they appear to be strategically important for the corporate parent. Clearly then, the goals of pharmaceutical firms' investments in new ventures are fundamentally different from those of professional venture capital firms. A similar strategy applies in other sectors, such as information and communications technology (see Case study 10.2). The goals of corporate venture funds are largely

CASE STUDY 10.1

Johnson & Johnson Development Corporation

Johnson & Johnson Development Corporation (JJDC) is an independent venture capital firm within the Johnson & Johnson group of companies, and aims to identify and fund new technologies and businesses in the pharmaceutical and healthcare sector. JJDC was established in the USA 25 years ago and has since invested in more than 300 start-up businesses worldwide. In 1997 it created a dedicated European division, Johnson & Johnson Development Capital. Both companies exploit the scientific and market know-how of Johnson & Johnson and typically invest alongside professional venture capital firms in ventures in the start-up and early growth stages.

strategic, focusing on technology and potential new products, whereas the goals of venture capitalists are (rightly) purely financial.

Venture capital

Whilst there is general agreement about the main components of a good business plan, there are some significant differences in the relative weights attributed to each component. General venture capital firms typically only accept 5% of the technology ventures they are offered, and the specialist technology venture funds are even more selective, accepting around 3%. The main reasons for rejecting technology proposals compared to more general funding proposals are the lack of intellectual property, the skills of the management team, and size of the potential market. A survey of venture capitalists in North America, Europe and Asia found major similarities in the criteria used, but also identified several

CASE STUDY 10.2

Reuters' corporate venture funds

Reuters established its first fund for external ventures, Greenhouse 1, in 1995. It has since added a further two venture funds, which aim to invest in related businesses such as financial services, media and network infrastructure. By 2001 it had invested US$432 million in 83 companies, and these investments contributed almost 10% to its profits. However, financial return was not the primary objective of the funds. For example, it invested $1 million in Yahoo! in 1995, and consequently Yahoo! acquired part of its content from Reuters. This increased the visibility of Reuters in the growing Internet markets, particularly in the USA where it was not well known, and resulted in other portals following Yahoo!'s lead with content from Reuters. By 2001 Reuters' content was available on 900 web services, and had an estimated 40 million users per month.

Source: Loudon, A. (2001) *Webs of Innovation: The Networked Economy Demands New Ways to Innovate*, FT.com, Pearson Education, Harlow.

interesting differences in the weights attached to some criteria (see Chapter 8, Table 8.1). The criteria are similar to those discussed earlier, grouped into five categories:

1. The entrepreneur's personality
2. The entrepreneur's experience
3. Characteristics of the product
4. Characteristics of the market
5. Financial factors.

Overall, the study confirmed the importance of a bundle of personal, market and financial factors, which were consistently ranked as being most significant: a proven ability to lead others and sustain effort; familiarity with the market; and the potential for a high return within 10 years (see Case study 10.3). The personality and experience of the entrepreneurs were consistently ranked as being more important than either product or market characteristics, or even financial considerations. However, there were a number of significant differences between the preferences of venture capitalists from different

CASE STUDY 10.3

Andrew Rickman and Bookham Technology

Andrew Rickman founded Bookham Technology in 1988, aged 28. Rickman has a degree in mechanical engineering from Imperial College London, a PhD in integrated optics from Surrey University, an MBA and has worked as a venture capitalist. Unlike many technology entrepreneurs, he did not begin with the development of a novel technology and then seek a means to exploit it. Instead, he first identified a potential market need for optical switching technology for the then fledgling optical fibre networks, and then developed an appropriate technological solution. The market for optical components is growing fast as the use of Internet and other data-intensive traffic grows. Rickman aimed to develop an integrated optical circuit on a single chip to replace a number of discrete components such as lasers, lenses and mirrors. He chose to use silicon rather than more exotic materials to reduce development costs and exploit traditional chip production techniques. The main technological developments were made at Surrey University and the Rutherford Appleton Laboratory, where he had worked, and 27 patents were granted and a further 140 applied for. Once the technology had been proven, the company raised US$110 million over several rounds of funding from venture capitalist 3i, and leading electronics firms Intel and Cisco. The most difficult task was scale-up and production: '*Taking the technology out of the lab and into production is unbelievably tough in this area. It is infinitely more difficult than dreaming up the technology.*' Bookham Technology floated in London and on the NASDAQ in New York in April 2000 with a market capitalization of more than £5 billion, making Andrew Rickman, with 25% of the equity, a paper billionaire. Bookham is based in Oxford, and employs 400 staff. The company acquired the optical component businesses of Nortel and Marconi in 2002, and in 2003 the US optical companies Ignis Optics and New Focus, and the latter included chip production facilities in China. This puts Bookham in the top three in the global optoelectronics sector.

regions. Those from the USA placed greater emphasis on a high financial return and liquidity than their counterparts in Europe or Asia, but less emphasis on the existence of a prototype or proven market acceptance. Perhaps surprisingly, all venture capitalists are adverse to technological and market risks. Being described as a 'high-technology' venture was rated very low in importance by the US venture capitalists, and the European and Asian venture capitalists rated this characteristic as having a negative influence on funding. Similarly, having the potential to create an entirely new market was considered a drawback.

A study of venture capitalists in the UK compared attitudes to funding technology ventures over a 10-year period, and found that investment in technology-based firms as a percentage of total venture capital had increased from around 11% in 1990 to 25% by 2000 (by value).[10] Of the total venture capital investment in UK NTBFs of £1.6 billion in the year 2000, 30% was for early-stage funding (by value, or 47% by number of firms), 47% for expansion (by value, or 47% by number of firms), and the rest for management buy-outs (MBO). This increase was due to a combination of the growth of specialist technology venture capitalists, and greater interest by the more general venture capital firms. As venture capital firms have gained experience of this type of funding, and the opportunities for flotation have increased due to the new secondary financial markets in Europe such as the AIM, techMARK and Neuer Markt, their returns on investment have increased significantly. In the 1980s returns to UK early-stage technology investments were under 10%, compared to venture capital norms of twice that, but by 2000 the returns of technology ventures increased to almost 25%, which is higher than all other types of venture investment. However, this recent growth in venture capital funding of NTBFs needs to be put into perspective. Although the UK has the most advanced venture capital community in Europe, venture capital still only accounts for between 1 and 3% of the external finance raised by small firms.

An important issue is the influence of venture capitalists on the success of NTBFs. They can play two distinct roles. The first, to identify or select those NTBFs that have the best potential for success – that is, 'picking winners' or 'scouting'. The second role is to help develop the chosen ventures, by providing management expertise and access to resources other than financial – that is, a 'coaching' role. Distinguishing between the effects of these two roles is critical for both the management of and policy for NTBFs. For managers, it will influence the choice of venture capital firm; and for policy, the balance between funding and other forms of support. A study of almost 700 biotechnology firms over 10 years provides some insights to these different roles.[11] It found that when selecting start-ups to invest in, the most significant criteria used by venture capitalists were a broad, experienced top management team, a large number of recent patents and downstream industry alliances (but not upstream research alliances, which had a negative effect on selection). The strongest effect on the decision to fund was the first criterion, and the human capital in general. However, subsequent analysis of venture performance indicates that this factor has limited effect on performance, and that the few significant effects are split equally between improving and impeding the performance of a venture. The effects of technology and alliances on subsequent performance are much more significant and positive. In short, in the *selection* stage, venture capitalists place too much emphasis on human capital, specifically the top management team. In the development or coaching stages, venture capitalists do contribute to the success of the chosen ventures, and tend to introduce external professional management much earlier than in NTBFs not funded by venture capital. Taken together, this suggests that the coaching role of venture capitalists is probably as important, if not more so, than the funding role, although policy interventions to promote NTBFs often focus on the latter. For example, in the case of the Internet start-up ihavemoved.com, the founders raised the initial venture capital, partly due to their business experience and MBA training, but then faced the challenge of growing the new business.

10.2 Internal corporate venturing

The term corporate venturing or internal corporate venturing, is sometimes confusingly referred to as 'intrapreneurship', to distinguish it from venturing which takes the form of investments in external business. If managed effectively, a corporate venture has the resources of a large organization and the entrepreneurial benefits of a small one. A corporate venture differs from conventional R&D and product development activities in its objectives and organization. The former seeks to exploit existing technological and market competencies, whereas the primary function of a new venture is to learn new competencies. In practice the distinction may be less clear. For example, technical staff at 3M are expected to use 15% of their time exploring ideas outside their existing projects, and at the corporate level this translates into a target of at least 25% of sales from products less than five years old.

The Internet bubble of the late 1990s produced an ill-timed bandwagon for corporate venturing in large established companies in the information and communications technology sector as they attempted to capture some of the rapid growth of the dotcom start-up firms: in 1996 Nortel Networks created the Business Ventures Programme (see Case study 10.4), in 1997 Lucent established the Lucent New Ventures Group, in 2000 Ericsson formed Ericsson Business Innovation and British Telecom formed Brightstar.

CASE STUDY 10.4

Corporate venturing at Nortel Networks

Nortel Networks is a leader in a high-growth, high-technology sector, and around a quarter of all its staff are in R&D, but it recognizes that it is extremely difficult to initiate new businesses outside the existing divisions. Therefore in December 1996 it created the Business Ventures Programme (BVP) to help to overcome some of the structural shortcomings of the existing organization, and identify and nurture new business ventures outside the established lines of business: '*The basic deal we're offering employees is an extremely exciting one. What we're saying is "Come up with a good business proposal and we'll fund and support it. If we believe your business proposal is viable, we'll provide you with the wherewithal to realize your dreams."*' The BVP provides:

- guidance in developing a business proposal
- assistance in obtaining approval from the board
- an incubation environment for start-ups
- transition support for longer-term development.

The BVP selects the most promising venture proposals, which are then presented jointly by the BVP and employee(s) to the advisory board. The advisory board applies business and financial criteria in its decision whether to accept, reject or seek further development, and if accepted the most appropriate executive sponsor, structure and level of funding. The BVP then helps to incubate the new venture, including staff and resources, objectives and critical milestones. If successful, the BVP then assists the venture to migrate into an existing business division, if appropriate, or creates a new line or business or spin-off company:

'The programme is designed to be flexible. Among the factors determining whether or not to become a separate company are the availability of key resources within Nortel, and the suitability of Nortel's existing distribution channels . . . Nortel is not in this programme to retain 100% control of all ventures. The key motivators are to grow equity by maximizing return on investment, to pursue business opportunities that would otherwise be missed, and to increase employee satisfaction.'

In 1997 the BVP attracted 112 business proposals, and given the staff and financial resources available aimed to fund up to five new ventures. The main problems experienced have been the reaction of managers in established lines of business to proposals outside their own line of business:

'At the executive council level, which represents all lines of business, there is a lot of support . . . where it breaks down in terms of support is more in the political infrastructure, the middle to low management executive level where they feel threatened by it . . . the first stage of our marketing plan is just titled "overcoming internal barriers". That is the single biggest thing we've had to break through.'

Initially, there was also a problem capturing the experience of ventures that failed to be commercialized:

'Failures were typically swept under the rock, nobody really talked about them . . . that is changing now and the focus is on celebrating our failures as well as our successes, knowing that we have learned a lot more from failure than we do from success. Start-up venture experience is in high demand. Generally, it's the projects that fail, not the people.'

The most effective organization and management of a new venture will depend on two dimensions: the strategic importance of the venture for corporate development; and its proximity to the core technologies and business.[12] Typically, top management has risen through the ranks of the organization, and therefore will be familiar with the evaluation of proposals related to the existing lines of business. However, by definition, new venture proposals are likely to require assessment of new technologies and/or markets. The following checklist can be used to assess the strategic importance of a new venture:

- Would the venture maintain our capacity to compete in new areas?
- Would it help create new defensible niches?
- Would it help identify where not to go?
- To what extent could it put the firm at risk?
- How and when could the firm exit from the venture?

Assessment of the second dimension, the proximity to existing skills and capabilities, is more difficult. On the one hand, a new venture may be driven by newly developed skills and capabilities, but on the other a new venture may drive the development of new skills and capabilities. The former is consistent with an 'incremental' strategy in which diversification is a consequence of evolution, the latter with a 'rational' strategy which begins with the identification of new market opportunity. The relative merits and implications of these contrasting approaches were discussed in detail in Chapter 4.

Whatever the primary motive for establishing a new venture, the proposal should identify potential opportunities for positive synergies across existing technologies, products or markets. A checklist for assessing the proximity of the venture proposal to existing skills and capabilities would include:

- What are the key capabilities required for the venture?
- Where, how and when is the firm going to acquire the capabilities, and at what cost?
- How will these new capabilities affect current capabilities?
- Where else could they be exploited?
- Who else might be able to do this, perhaps better?

Assessment of a new venture along these two dimensions will help determine the organization and management of the venture. In particular, the strategic importance will determine the degree of administrative control required, and the proximity to existing skills and capabilities will determine the degree of operational integration that is desirable. In general, the greater the strategic importance, the stronger the administrative linkages between the corporation and venture. Similarly, the closer the skills and capabilities are to the core activities, the greater the degree of operational integration necessary for reasons of efficiency. Putting the two dimensions together creates a number of different options for the organization and management of a new venture (Figure 10.1). In this section we explore the design and management of internal corporate ventures, and in the next the role and management of joint ventures and alliances.

The management structures and processes necessary for routine operations are very different from those required to manage innovation. The pressures of corporate long-range strategic planning on the one hand, and the short-term financial control on the other, combine to produce a corporate environment that favours carefully planned and stable growth based on incremental developments of products and processes:

- Budgeting systems favour short-term returns on incremental improvements.
- Production favours efficiency rather than innovation.
- Sales and marketing are organized and rewarded on the basis of existing products and services.

Such an environment is unlikely to be conducive to radical innovation. An internal corporate venture attempts to exploit the resources of the large corporation, but provide an environment more conducive to radical innovation. A corporate venture is likely to be necessary when a firm attempts to enter a new market or to develop a new technology. The key factors that distinguish a potential new venture from the core business are risk, uncertainty, newness and significance. A corporate venture is a separate organization or system designed to be consistent with the needs of new, high-risk, but potentially high-growth businesses. The term 'intrapreneurship' is sometimes used to describe such entrepreneurial activity within a corporate setting. However, it is not sufficient to promote entrepreneurial behaviour within a large organization. Entrepreneurial behaviour is not an end in itself, but must be directed and translated into desired business outcomes. Entrepreneurial behaviour is not associated with superior organizational performance, unless it is combined with an appropriate strategy in a heterogeneous or uncertain environment.[13] This suggests the need for clear strategic objectives for corporate venturing and appropriate organizational structures and processes to achieve those objectives.

There are a wide range of motives for establishing corporate ventures:[14]

- Grow the business.
- Exploit underutilized resources.
- Introduce pressure on internal suppliers.
- Divest non-core activities.

- Satisfy managers' ambitions.
- Spread the risk and cost of product development.
- Combat cyclical demands of mainstream activities.
- Learn about the process of venturing
- Diversify the business.
- Develop new technological or market competencies.

We will discuss each of these motives in turn, and provide examples. The first three are primarily operational, the remainder primarily strategic.

To grow the business

The desire to achieve and maintain expected rates of growth is probably the most common reason for corporate venturing, particularly when the core businesses are maturing. Depending upon the time frame of the analysis, between only 5% and 13% of firms are able to maintain a rate of growth above the rate of growth in gross national product (GNP).[15] However, the pressure to achieve this for publicly listed firms is significant, as financial markets and investors expect the maintenance or improvement of rates of growth. The need to grow underlies many of the other motives for corporate venturing.

As a specific issue to be considered here, there is the drive to achieve growth in a corporate whose primary markets are maturing. For example, the UK water companies have a fully mature customer base, the regions are insular in their operation and the only opportunity for growth is beyond this mainstream and local activity, a situation which is driving their diversification into related and unrelated areas. Two water companies were recently reported to have shown interest in Tarmac's divestment of Econowaste, which could fetch up to £100 million. In seeking to grow the business, or make better commercial use of existing expertise, we have to challenge the more normal small scale of response to such initiatives.

Often the push is to analyse new or adapted products, processes or techniques, and relatively little emphasis is put on the expansion of the business considered as a whole. Severn Trent International, in contrast, has been set up with the task of taking Severn Trent Water's core business capabilities in water and waste water management to new international markets. This direct transfer of skills must surely imply less uncertainty than venturing into both new technologies and new markets.

To exploit underutilized resources in new ways

This includes both technological and human resources. Typically, a company has two choices where existing resources are underutilized – either to divest and outsource the process or to generate additional contribution from external clients. However, if the company wants to retain direct and in-house control of the technology or personnel it can form an internal venture team to offer the service to external clients. For example, at Cadbury Schweppes the Information Technology Division was extracted from Group Management Services forming a separate business unit to supply the internal needs of the group – Cadbury Limited and Coca-Cola and Schweppes Beverages Limited. With the mechanisms in place for operation as a trading entity ITnet Limited began to seek and develop external clients, who now provide a significant proportion of its revenues.

To introduce pressure on internal suppliers

This is a common motive, given the current fashion for outsourcing and market testing internal services. When a business activity is separated to introduce competitive pressure a choice has to be made – whether the business is to be subjected to the reality of commercial competition, or just to learn from it. If the corporate clients are able to go so far as to withdraw a contract, which is not conducive to learning, the business should be sold to allow it to compete for other work. For example, General Motors exposed its dominant supplier Delco to such competitive pressure by requiring it to earn a certain proportion of its sales from external sales.

To divest non-core activities

Much has been written of the benefits of strategic focus, 'getting back to basics', and creating the 'lean' organization–rationalization, which prompts the divestment of those activities that can be outsourced. However, this process can threaten the skill diversity required for an ever-changing competitive environment. New ventures can provide a mechanism to release peripheral business activities, but to retain some management control and financial interest.

To satisfy managers' ambitions

As a business activity passes through its life cycle it will require different management styles to bring out the maximum gain. This may mean that the management team responsible for a business area will need to change, whether between conception to growth, growth to maturity or maturity to decline phases. A paradoxical situation often arises because of the changing requirements of a business area: top managers in place who are ambitious and want to see growth, and managing businesses which are reaching the limits of that growth. To retain the commitment of such managers the corporation will have to create new opportunities for change or expansion. These managers are not only potential facilitators for venture opportunities, but also potential creators of venture opportunities. For example, Intel has long had a venture capital programme that invests in related external new ventures, but in 1998 it established the New Business Initiative to bootstrap new businesses developed by its staff: *'They saw that we were putting a lot of investment into external companies and said that we should be investing in our own ideas . . . our employees kept telling us they wanted to be more entrepreneurial.'* The initiative invests only in ventures unrelated to the core microprocessor business, and in 1999 attracted more than 400 proposals, 24 of which are being funded.

To spread the risk and cost of product development

Two situations are possible in this case: (i) where the technology or expertise needs to be developed further before it can be applied to the mainstream business or sold to current external markets; or (ii) where the volume sales on a product awaiting development must sell to a target greater than the existing customer groups to be financially justified. In both cases the challenge is to understand how to venture outside current served markets. Too often, when the existing customer base is not ready for a product, the research unit will just continue its development and refinement process. If intermediary markets were exploited these could contribute to the financial costs of development, and to the maturing of the final product.

To combat cyclical demands of mainstream activities

In response to the problem of cyclical demand Boeing set up two groups, Boeing Technology Services (BTS) and Boeing Associated Products (BAP), specifically with the function of keeping engineering and laboratory resources more fully employed when its own requirements waned between major development programmes. The remit for BTS was '*to sell off excess engineering laboratory capacity without a detrimental impact on schedules or commitments to major Boeing product-line activities*'; it has stuck carefully to this charter, and been careful to turn off such activity when the mainstream business requires the expertise. BAP was created to commercially exploit Boeing inventions that are usable beyond their application to products manufactured by Boeing. About 600 invention disclosures are submitted by employees each year, and these are reviewed in terms of their marketability and patentability. Licensing agreements are used to exploit these inventions; 259 agreements are currently active. Beyond the financial benefits to the company and to the employees of this programme it is seen to foster the innovation spirit within the organization.

To learn about the process of venturing

Venturing is a high-risk activity because of the level of uncertainty attached, and we cannot expect to understand the management process as we do for the mainstream business. If a learning exercise is to be undertaken, and a particular activity is to be chosen for this process, it is critical that goals and objectives are set, including a review schedule. This is important not just for the maximum benefit to be extracted but for the individuals who will pioneer that venture. For example, NEES Energy, a subsidiary of New England Electric Systems Inc., was set up to bring financial benefits, but was also expected to provide a laboratory to help the parent company learn about starting new ventures.[16]

Many companies develop hobby-size business activities to provide this 'learning by doing', but seldom is a time limit set on this learning stage, and as a consequence, no decision is formally made for the venture activities to be considered 'proper businesses'. The implications of this practice are to drain the enterprising managers of their enthusiasm and erode the value of potential opportunities.

To diversify the business

Whilst the discussion so far has implied that business development would be on a relatively small scale, this need not be the case. Corporate ventures are often formed in an effort to create new businesses in a corporate context, and therefore represent an attempt to grow via diversification. Therefore a decline in the popularity of internal ventures is associated with an emphasis on greater corporate focus and greater efficiency. For example, the identification and reengineering of *existing* business processes became fashionable in the mid-1990s, but as firms have begun to exhaust the benefits of this approach they are now exploring options for creating new businesses. Such diversification may be vertical, that is downstream or upstream of the current process in order to capture a greater proportion of the value added; or horizontal, that is by exploiting existing competencies across additional product markets.

For example, the fossil fuel energy sector has been facing for a long time the threat of extinction – coal, gas, oil reserves being exhausted. The cloud that has hung on the horizon is the collapse of its core business, and the temptation is therefore to diversify into other business areas that will ensure the continued existence of the firm. BP, Elf, Shell, Standard Oil and Total all set up corporate venturing initiatives when oil companies were first experiencing declining margins and an unfavourable world economic climate, with diversification as their main objective. These initiatives were primarily exercised through investment in external opportunities, although some have pursued venture opportunities based on internal technologies and expertise.

To develop new competencies

Growth and diversification are generally based on the exploitation of existing competencies in new products markets, but a corporate venture can also be used as an opportunity for learning new competencies. Organizational learning has four components:[17]

- knowledge acquisition
- information interpretation
- information distribution
- organizational memory.

An organization can acquire knowledge by experimentation, which is a central feature of formal R&D and market research activities. However, different functions and divisions within a firm will develop particular frames of reference and filters based on their experience and responsibilities, and these will affect how they interpret information. Greater organizational learning occurs when more varied interpretations are made, and a corporate venture can better perform this function as it is not confined to the needs of existing technologies or markets.

Similarly, a corporate venture can act as a broker or clearing house for the distribution of information within the firm. In practice, large organizations often do not know what they know. Many firms now have databases and groupware to help store, retrieve and share information, but such systems are often confined to 'hard' data. As a result, functional groups or business units with potentially synergistic information may not be aware of where such information could be applied. Organizational learning occurs when more of an organization's components obtain new knowledge and recognize it as of potential use.

Organizational memory is the process by which knowledge is stored for future use. Such information is stored either in the memories of members of an organization, or in the operating procedures and routines of the organization. The former suffers from all of the shortcomings of human memory, with the additional organizational problem of personnel loss or turnover. Organizational procedures and routines may provide a more robust memory, but difficulty in anticipating future needs means that much non-routine information is never stored in this way. Over time these routines create and are reinforced by artefacts such as organizational structures, procedures and policies which suit existing technologies and markets, but make it difficult to store and retrieve non-routine information. A corporate venture can act as a repository for such knowledge.

Organizational learning is more difficult in conventional product development activities because of the cost and time pressures. For example, there will be a trade-off between reducing the time and cost of development of a specific product, and documenting what has been learnt for future development projects.

In practice, the primary motives for establishing a corporate venture are strategic: to meet strategic goals and long-term growth in the face of maturity in existing markets (Table 10.1). However, personnel issues are also important. Sectorial and national differences exist. In the USA, new ventures are also used to stimulate and develop entrepreneurial management, and in Japan they help provide employment opportunities for managers and staff relocated from the core businesses (Table 10.2). Nonetheless,

| TABLE 10.1 | Objectives of corporate venturing in the UK |

Objective	Mean rank*
1. Long-term growth	4.58
2. Diversification	3.50
3. Promote entrepreneurial behaviour	2.68
4. Exploit in-house R&D	2.23
5. Short-term financial returns	2.08
6. Reduce/spread cost of R&D	1.81
7. Survival	1.76

(n = 90). * Scale: 1 = minimum, 5 = maximum importance.

Source: Gebbie, D. (1997) *Window on Technology: Corporate Venturing in Practice*, Withers, London.

| TABLE 10.2 | Comparison of motives for corporate venturing in the USA and Japan |

	US firms (n = 43)	Japanese firms (n = 149)
To meet strategic goals	76	73
Maturity of the base business	70	57
To provide challenges to managers*	46	15
To survive	35	28
To develop future managers*	30	17
To provide employment*	3	24

* Denotes statistically significant difference.

Source: Block, Z. and I. MacMillan (1993) *Corporate Venturing: Creating New Businesses Within the Firm*, NIA, Boston. Copyright ©1993 by the President and Fellows of Harvard College: all rights reserved. Reprinted by permission of Harvard Business School Press.

RESEARCH NOTE | Four approaches to corporate venturing

A study of corporate ventures at almost 30 large firms in the USA identified two critical dimensions which characterized four different approaches to venturing. The critical dimensions are the loci of ownership and funding: who and where in the company is responsible for venturing? For example, a central venture unit versus decentralized projects; and how are ventures funded and resourced? For example, central dedicated funding versus an ad hoc basis. These two dimensions create four distinct approaches, each with different management issues:

1. *Opportunistic* – no dedicated ownership or resources for venturing. This approach relies on a supportive organizational climate to encourage proposals which are developed and evaluated locally on a project-by-project basis. For example, Zimmer Medical Devices responded to a new hip replacement proposed by a trauma surgeon by creating the Zimmer Institute to train more than 6000 surgeons in the new minimally invasive procedure.
2. *Enabling* – no formal corporate ownership, but the provision of dedicated support, processes and resources. This approach works best where new ventures can be owned by existing divisions in the business. For example, Google provides time, funding and rewards for the development of ideas which extend the core business.
3. *Advocacy* – organizational ownership is clearly assigned, but little or no special funding is provided. This works when there are sufficient resources in the business, but insufficient specialist skills or support for venturing. For example, DuPont created the Market Driven Growth initiative which includes four-day business planning training and workshops and agreed access to and mentoring by senior staff.
4. *Producer* – includes both formal ownership and dedicated funding of ventures. This demands significant corporate resources and commitment to venturing, and therefore a critical mass of potential projects to justify this approach. Examples include IBM's Emerging Business Opportunities programme and Cargill's Emerging Business Accelerator initiative. In such cases the goal is to build new businesses, rather than just new products or services.

Source: Wolcott, R.C. and M.J. Lippitz (2007) The four models of corporate entrepreneurship. *MIT Sloan Management Review*, Fall, 74–82.

---

the primary objectives are strategic and long term, and therefore warrant significant management effort and investment.

Managing corporate ventures

A corporate venture is rarely the result of a spontaneous act or serendipity. Corporate venturing is a process that has to be managed. The management challenge is to create an environment that encourages and supports entrepreneurship, and to identify and support potential entrepreneurs. In essence, the venturing process is simple, and consists of identifying an opportunity for a new venture, evaluating that opportunity and subsequently providing adequate resources to support the new venture. There are six distinct stages, divided between definition and development.[18]

Definition stages

1. Establish an environment that encourages the generation of new ideas and the identification of new opportunities, and establish a process for managing entrepreneurial activity.
2. Select and evaluate opportunities for new ventures, and select managers to implement the venturing programme.
3. Develop a business plan for the new venture, decide the best location and organization of the venture and begin operations.

Development stages

4. Monitor the development of the venture and venturing process.
5. Champion the new venture as it grows and becomes institutionalized within the corporation.
6. Learn from experience in order to improve the overall venturing process.

Creating an environment which is conducive to entrepreneurial activity is the most important, but most difficult stage. Superficial approaches to creating an entrepreneurial culture can be counterproductive. Instead, venturing should be the responsibility of the entire corporation, and top management should demonstrate long-term commitment to venturing by making available sufficient resources and implementing the appropriate processes.

The conceptualization stage consists of the generation of new ideas and identification of opportunities that might form the basis of a new business venture. The interface between R&D and marketing is critical during the conceptualization stage, but the scope of new venture conceptualization is much broader than the conventional activities of the R&D or marketing functions, which understandably are constrained by the needs of existing businesses. At this stage three basic options exist:

1. Rely on R&D personnel to identify new business opportunities based on their technological developments, that is, essentially a 'technology-push' approach.
2. Rely on marketing managers to identify opportunities, and direct the R&D staff into the appropriate development work, essentially a 'market-pull' approach.
3. Encourage marketing and R&D personnel to work together to identify opportunities.

The technology-push approach has been described as being 'first-generation R&D', the 'market-pull' strategy as 'second generation' and the close coupling 'third generation', the implication being that firms should progress to close coupling.[19] The issue of strategic positioning was discussed in detail in Chapter 4. In theory, the third option is most desirable as it should encourage the coupling of technological possibilities and market opportunities at the concept stage, before substantial resources are committed to evaluation and development. However, in practice technology push appears to be the dominant strategy. This is because at the conceptualization stage highly specialized technical knowledge is required about what is feasible and what is not, and therefore what the characteristics of the final product are likely to be. Nevertheless, R&D personnel may become locked into a specific technical solution or address the needs of atypical users. Therefore management must ensure that R&D personnel are sufficiently flexible to modify or drop their proposals should technical issues or market requirements dictate.

Peter Drucker identifies a number of sources of ideas and opportunities, and argues that the search process should be systematic rather than relying on serendipity.[4] He suggests seven common sources of opportunities which should be monitored on a routine basis:

- demographic changes
- new knowledge
- incongruities (i.e. gaps between expectations and reality)
- changes in industry or market structure
- unexpected successes or failures
- process needs
- changes in perception.

Other sources of ideas include trade shows, exhibitions and trade journals. In the specific case of new business ventures there are four primary sources of ideas:

- the 'bright idea'
- customers requesting a new product or service
- internal analysis of a company's competencies and business processes
- scanning of external opportunities in related technologies, markets or services.

Contrary to popular perceptions, the 'bright idea' is the least common and most risky source of new business ventures, because the other sources are more directly stimulated by a market need, technological expertise or both together. These can be the initiative of someone at either operational or managerial level; the former may have difficulties finding an effective champion, whereas the latter may be too powerful, having the influence to force through an idea before it is exhaustively tested. A balance needs to be achieved between screening and championing the proposal. In contrast, a business venture based on a customer request has the highest chance of success as a potential market is to some extent predetermined. However, such ventures are typically based on an adaptation or extension of an existing product or service, and therefore less likely to spawn radical new businesses. These tend to be bottom-up initiatives, and the most difficult problem is to decide how the potential new business relates to the existing business or division. By far the two most promising corporate ventures are the result of systematic scanning of the internal and external environments, a process we advocate in Chapter 2.

Venture capital firms can help firms to monitor the external environment without distraction, and to take equity stakes in potential partners fairly anonymously. This practice is common in the pharmaceutical industry, where firms use a range of strategies to tap into the knowledge of biotechnology firms, including direct investment, licensing deals and indirect investment through professionally managed venture funds. Direct investments are favoured for technologies of high strategic importance, licensing for process and product developments, and indirect investments for windows on emerging technologies.[9]

Having identified the potential for a new venture, a product champion must convince higher management that the business opportunity is both technically feasible and commercially attractive and therefore justifies development and investment. Potential corporate entrepreneurs face significant political barriers:

- They must establish their legitimacy within the firm by convincing others of the importance and viability of the venture.
- They are likely to be short of resources, but will have to compete internally against established and powerful departments and managers.
- As advocates of change and innovation, they are likely to face at best organizational indifference, and at worst hostile attacks.

To overcome these barriers a potential venture manager must have political and social skills, in addition to a viable business plan. In addition, the product champion must be able to work effectively in a non-programmed and unpredictable environment. This contrasts with much of the R&D conducted in the operating divisions which is likely to be much more sequential and systematic. Therefore a product champion requires dedication, flexibility and luck to manage the transition from product concept to corporate venture, in addition to sound technical and market knowledge. The product champion is likely to require a complementary organizational champion, who is able to relate the potential venture to the strategy and structure of the corporation. A number of key roles must be filled when a new venture is established:[20]

- The technical innovator, who was responsible for the main technological development.
- The business innovator or venture manager, who is responsible for the overall progress of the venture.
- The product champion, who promotes the venture through the early critical stages.
- The executive champion or organizational champion, who acts as a protector and buffer between the corporation and venture.
- A high-level executive responsible for evaluating, monitoring and authorizing resources for the venture, but not the operation of specific ventures.

A new venture requires two types of skill: the technical knowledge necessary to develop the product, process or knowledge base; and the management expertise necessary to communicate and sell to the markets and parent organization (Table 10.3). The dilemma that has to be resolved in each case is whether to allow and develop technical experts to play a role in selling the product or managing the business, or to place managers above their heads to take the baton on.

To take project managers to venture manager status is often dangerous. Whilst these individuals understand the product fully, they may have difficulties in maximizing the cost/price differential, perhaps not always realizing the commercial value of the product and being less experienced in the negotiation process. It can be equally difficult to identify a manager who can communicate the product characteristics to customers with real needs, relay those needs to the product development team, and communicate and justify venture management needs to the corporate centre.

TABLE 10.3	**Systematic differences between technical and commercial orientations**	
	R&D personnel	**Marketing personnel**
Work environment		
Structure	Well defined	Ill defined
Methods	Scientific and codified	Ad hoc and intuitive
Data	Systematic and objective	Unsystematic and subjective
Pressures	Internal: How long will it take?	External: How long do we have?
Professional orientation		
Assumptions	Serendipity	Planning
Goals	New ideas: Can it be improved?	Big ideas: Does it work?
Performance criteria	Technical quality	Commercial value
Education and experience	Deep and focused	Broad

VIEWS FROM THE FRONT LINE

Identifying new opportunities at QinetiQ

Businesses tend to limit their strategic vision to the conventional boundaries of the existing industry. This they believe is an immutable given. When challenged to think 'out of the box' or to be more creative in their business models, because they do not explicitly acknowledge the boundaries in which they operate, they continue competing in traditional spaces.

Companies that do not permit themselves to be limited by current industry boundaries more often create new profitable spaces. In traditional strategy, pain points would be identified and solutions found. Here we use pain points to find the non-customer.

The boundary busting framework enables the process of exploration into unknown territory of the non-customer. By applying a set of six alternative 'lens' participants challenge the assumptions underpinning these traditional boundaries.

For each boundary type, we apply the 'Rule of Opposites', which is a set of specific critical questions performed to extract insight into potential new market spaces. Not all boundaries will yield new market opportunities, but may reveal insight which can be exploited across other boundaries.

Critical to identifying new market opportunities will be the ability to visualize and articulate the emergent previously ignored customer, to which a reconstructed value proposition can be offered.

The process undertaken includes:

1. Articulate the current bounds of the industry the product operates in across the dimensions of industry definition – strategic groups, chain of buyers, proposition, appeal, and time and trends.
2. For each existing customer, map out their buyer experience cycle to identify pain points.
3. Explicitly identify the core customer, then remove this customer from any further consideration.
4. Apply 'Rule of Opposites' to each boundary in turn to unearth whether new customer groups exist beyond the current boundary of the industry.
5. Once a new customer is articulated and brought to life undertake field work to find this person and prove the new opportunity.
6. Hypothesize a set of offerings that would meet this person's needs.
7. From the full range of new opportunities, distil down a set of propositions that minimally meet the needs of the largest catchment of non-customers.

Be aware that this process might initially feel strange, more like opening 'Pandora's box' than a structured analysis. The outcome of the market boundary analysis is a set of non-customer spaces. It is important to acknowledge that not all of the six dimensions of alternative marketplaces will yield results, typically two to four of the paths will present significant insight.

Carlos de Pommes, QinetiQ, www.qinetiq.com

Assessing the venture

The most appropriate filter to apply to a potential venture will depend on the motive for venturing. Roberts illustrates the point:

> The best time to detect if a CEO has a strategy or not is to observe the management team at work when trying to evaluate opportunities, especially those somewhat remote from the current business. On these occasions, we noticed that when faced with unfamiliar opportunities, management would put them through a hierarchy of different filters. The ultimate filter was always a fit between the products, customers and markets that the opportunity brought and one key element, or driving force, of the business. This is a clear signal that management had a sound filter for its decision.[21]

Without the type of strategic filters discussed in Chapter 4, managers are forced to rely on narrow financial methods of evaluation, such as potential sales growth, margins or net present value. The seduction of new opportunities is highest when a company is cash rich. For example, Daimler-Benz in the mid-1980s began an acquisition trail of many unrelated businesses, which have subsequently been disposed of. This distraction to top management may also be a contributing factor for Mercedes-Benz losing its hold in its home market to BMW. Similarly, Sony and Matsushita have made questionable decisions to diversify into US film production, respectively Columbia and MCA.

In assessing any venture it is essential to specify the purpose and criteria for success in the new market, business or technology. Ultimately the style of assessment adopted will depend on the size of the potential venture, the abilities of the people who currently understand the product and whether new partners or managers are expected to be introduced following assessment. See Case study 10.5 for a description of how Lucent Technologies approached this. A plan needs to be written by the managers involved in the venture, in part to test whether they understand the business as well as the technology. It is essential for in-house managers to be fully involved in the market research. The use of market research consultants should be limited to providing a first pass of potential markets. No one can know the product better, especially if it is new, and has niche applications, than the people who have worked on its development, and whose future careers may depend on it.

CASE STUDY 10.5

Lucent's new venture group

Lucent Technologies was created in 1996 from the break-up of the famous Bell Labs of AT&T. Lucent established the New Venture Group (NVG) in 1997 to explore how better to exploit its research talent by exploiting technologies which did not fit any of Lucent's current businesses, its mission was to '. . . *leverage Lucent technology to create new ventures that bring innovations to market more quickly . . . to create a more entrepreneurial environment that nurtures and rewards speed, teamwork, and prudent risk-taking.*' At the same time it took measures to protect the mainstream research and innovation processes within Lucent from the potential disruption NVG might cause. To achieve this balance, at the heart of the process are periodic meetings between NVG managers and Lucent researchers, where ideas are 'nominated' for assessment. These nominated ideas are first presented to the existing business groups within Lucent, and this creates pressure on the existing business groups to make decisions on promising technologies, as the vice president of the NVG notes: '*I think the biggest practical benefit of the (NVG) group was increasing the clockspeed of the system.*'

If the nominated idea is not supported or resourced by any of the businesses, the NVG can develop a business plan for the venture. The business plan would include an exit strategy for the venture, ranging from an acquisition by Lucent, external trade sale, IPO (initial public offering), or licence. The initial evaluation stage typically takes two to three months and costs US$50 000 to $100 000. Subsequent stages of internal funding reached $1 million per venture, and in later stages in many cases external venture capital firms are involved to conduct 'due diligence' assessments, contribute funds and management expertise. By 2001, 26 venture companies had been created by the NVG, and included 30 external venture capitalists who invested more than $160 million in these ventures. Interestingly, Lucent re-acquired at market prices three of the new ventures NVG had created, all based on technologies that existing Lucent businesses had earlier turned down. This demonstrates one of the benefits of corporate venturing – capturing false negatives – projects which were initially judged too weak to support, and that are rejected by the conventional development processes. However, following the fall in telecom and other technology equity prices, in 2002 Lucent sold its 80% interest in the remaining ventures to an external investor group for under $100 million.

Source: Chesbrough, H. (2003) *Open Innovation*, Harvard Business School Press, Boston, MA.

The purpose and nature of a business plan for a new venture differ from that for established businesses. The main purpose of the venture plan is to establish if and how to conduct the new business, and to attract key personnel and resources. The purpose of a plan for an existing business is to monitor and control performance. The technical and commercial aspects of a new venture plan will have much greater uncertainty than that for existing businesses. There are 10 essential elements of a new venture plan (Table 10.4).

Every new venture is an industrial experiment, and therefore the experiment must be designed to allow the assumptions and risks to be evaluated. The plan will be based on assumptions concerning the technology, product, market, economy, competition and the business environment, and given the inherent uncertainty of a new venture it is essential that testing of these assumptions is included in the go/no-go criteria. Similarly, contingency plans should be included to minimize any technological, market, management or financial risk.

The main criteria for assessing the business plan for a corporate venture are strategic fit and potential to enhance competitive position. But beyond such basic requirements, there appear to be significant differences between the criteria applied by American and Japanese firms (Table 10.5). American firms typically expect a high return on investment from corporate ventures, and therefore favour technologies that have high potential for premium pricing or rapid growth. Japanese firms appear to favour ventures that use related technologies to create new markets. This suggests that American firms view such ventures as opportunities to generate cash for the core business, whereas the Japanese see them as an opportunity to create new businesses. There is relatively little research on corporate venturing in Europe, but the experience of corporate ventures in the UK suggests a pattern closer to that of the American experience.

TABLE 10.4   Components of a typical business plan for a new venture

1. Description of the proposed business, including its objectives and characteristics

2. Strategic relationship between the new business and the parent firm

3. The target markets, including size, trends, reasons for purchase and specific target customers

4. Assessment of the present and anticipated competition

5. Human, physical and financial resources required

6. Financial projections, including assumptions and sensitivity analysis

7. Well-defined milestones and go/no-go conditions

8. Principal risks and how they will be managed

9. Definition of failure, and conditions under which the venture should be terminated

10. Description of the venture's management and compensation required

TABLE 10.5 Criteria for selecting corporate ventures		
	USA (*n* = 39)	**Japan (*n* = 126)**
Strategic fit	4.1	3.9
Competitive advantage	4.0	3.8
Potential return on investment*	3.9	3.6
Existence of market*	3.9	4.4
Potential sales	3.9	3.9
Risk/reward ratio	3.8	3.6
Presence of product champion	3.6	4.0
Synergy	3.5	3.7
Opportunity to create new market*	3.1	3.8
Closeness to present technology*	2.9	3.5
Patentability*	2.3	2.9

1 = unimportant, 5 = critical. * Denotes statistically significant difference.

Source: Block, Z. and I. MacMillan (1993) *Corporate Venturing: Creating New Businesses Within the Firm*, NIA, Boston. Copyright ©1993 by the President and Fellows of Harvard College: all rights reserved. Reprinted by permission of Harvard Business School Press.

Following development of an initial plan for the project, milestones for progress assessment should be developed and agreed upon by those responsible for monitoring and managing the venture. Milestones will ideally be a mix of financial and strategic. Strategic milestones can focus managers more on the long-term business direction rather than the (often) short-term financial criteria. Examples of milestones for the initial stages of a venture include: achievement of product development within time stated; product having been demonstrated on at least three sites; a sale to an important customer has been finalized.

Structures for corporate ventures

The choice of location and structure for a new venture will depend on a number of factors, the most fundamental being how close the activities are to the core business. How close a venture's focal activity is to the parent firm's technology, products and markets will determine the learning challenges the venture will face and the most appropriate linkages with the parent. In practice, there is likely to be some trade-off between the desire to optimize learning and the desire to optimize the use of existing resources. The venture will need to acquire resources, know-how and information from the corporate parent, get sufficient attention and commitment, but at the same time be protected politically and allowed optimal access to the target market. Consideration of these sometimes conflicting requirements will determine the best location and structure for the venture.

The classic study by Burgelman and Sayles of six internal ventures within a large American corporation demonstrated the managerial and administrative difficulties of establishing and managing internal ventures.[22] The study confirmed that no single organizational solution is optimal, and that different structures and processes are required in different circumstances. The choice of structure will depend on the level and urgency of the venturing activity, the nature and number of ventures to be established, and the corporate culture and experience. More fundamentally, it will depend on the balance between the desire to learn new competencies and the need to leverage existing competencies (Figure 10.5). For example, in e-business established firms are faced with the decision whether to develop separate businesses to exploit the opportunities, or to fully integrate e-business with the existing business. Neither strategy nor structure appears to be inherently superior, and depends on a

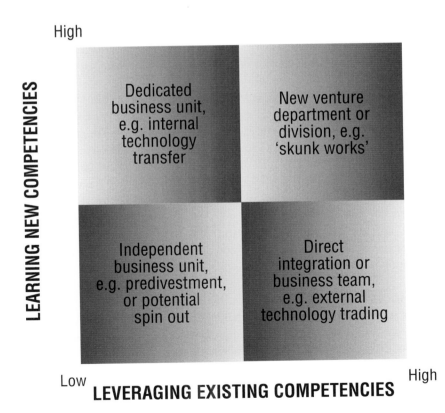

FIGURE 10.5: The most effective structure for a corporate venture depends on the balance between leverage or learning (exploit versus explore)
Source: Adapted from Tidd J. & Taurins, S. (1999). Learn or leverage? Strategic diversification and organisational learning through corporate ventures. *Creativity and Innovation Management*, **8** (2), 122–129.

consideration of the relatedness of the assets, operations, management and brand.[23] Design options for corporate ventures include:

- direct integration with existing business
- integrated business teams
- a dedicated staff function to support efforts company-wide
- a separate corporate venturing unit, department or division
- divestment and spin-off.

Each structure will demand different methods of monitoring and management – that is, procedures, reporting mechanisms and accountability. These choices are illustrated by more recent studies of venturing in the Europe and the USA,[24] and our own study of corporate venturing in the UK, as described in the following sections.[9,14]

Direct integration

Direct integration as an additional business activity is the preferred choice where radical changes in product or process design are likely to impact immediately on the mainstream operations and if the people involved in that activity are inextricably involved in day-to-day operations. For example, many engineering-based companies have introduced consultancy to their business portfolio, and in other technical organizations with large laboratory facilities these too have been sold out for analysis of samples, testing of materials, etc. In such cases it is not possible to outsource such activities because the same personnel and equipment are required for the core business.

The Natural History Museum (NHM) provides a good example. In the late 1980s funding was withdrawn from the museum, and it had to continue to reduce its costs and to find new sources of revenue. A principal aim was to generate sufficient revenue streams to avoid a massive reduction in research capacity, and avoid further redundancies. The NHM's resource of some 300 scientists has been developed in a commercial capacity from ground level up. The transformation from research to consultancy work required a complete cultural change. Heads of department ('keepers') were made responsible for identifying and developing the consultant activity, in addition to their normal operational management duties. Given the pervasive nature of the scientific expertise scattered throughout the organization and across departments, an integrated approach to venturing was the only realistic option. Science consultancy is now earning in excess of £1 million per annum, and the museum has been able to buy new capital equipment to support its work.

Integrated business teams

Integrated business teams are most appropriate where the expertise will have been nurtured within the mainstream operations, and may support or require support from those operations for development. Strategically, the product is sufficiently related to the mainstream business's key technologies or expertise that the centre wishes to retain some control. This control may either be to protect the knowledge that is intrinsic in the activity or to ensure a flow-back of future development knowledge. A business team of secondees is established to coordinate sourcing of both internal and external clients, and is usually treated as a separate accounting entity in order to ease any subsequent transition to a special business unit.

An example of this type of venture is provided by the development of an expert system within Welsh Water. The original system for planning water distribution and supply was devised by Cardiff University, but the system was co-developed and prototyped by a new business team working with the mainstream divisions. The new business team was created and supervised by the Enterprise Division, although it worked closely with the other divisions at operational level. The development team consisted of secondees released from other divisions for a duration of either six or 12 months. When the product was fully developed, it was passed to the Enterprise Division for sale externally.

New ventures department

A new ventures department is a group separate from normal line management that facilitates external trading. It is most suitable when projects are likely to emerge from the operational business on a fairly frequent basis and when the proposed activities may be beyond current markets or the type of product package sold is different. This is the most natural way for the trading of existing expertise to be developed when it lies fragmented through the organization, and each source is likely to attract a different type of customer. The group has responsibility for marketing, contracting and negotiation, but technical negotiation and supply of services take place at operational level.

Imperial College of Science, Technology and Medicine is a good example. Two separate venture organizations exist. One, IMPEL, is responsible for patenting and licensing technologies and products that emerge; and the other, ICON, for contracting out expertise or processes. ICON combs the organization for potential leading-edge process-oriented technologies, which are supported by available and willing expertise, and attempts to match these with market opportunities. ICON also offers solutions to problems identified by potential clients. For example, ICON manages external access to the ion implantation facility at Imperial, which has been applied to completely new market applications.

Techniques here are built very much on internal competencies that could not be separated from the mainstream activities and could not really be spun out or licensed to another company. In fact, development cycles are so short that it would not be realistic to spin off an individually developed technique. There are important additional benefits to the college through this activity: companies that initially commission a consultancy project may subsequently fund a related research project; also, researchers are brought far closer to the needs of technology users.

New venture division

A new venture division provides a safe haven where a number of projects emerge throughout the organization, and allows separate administrative supervision. Strategically, top management can retain a certain level of control until greater clarity on each venture's strategic importance is understood, but the efficiency of the mainstream business needs to be maintained without distraction, so some autonomy is required. Operational links are loose enough to allow information and know-how to be exchanged with the corporate environment. The origins of such a division vary:

- An effort to bring existing technologies and expertise throughout the company together for adaptation to new or existing markets.
- To combine research from different fields or locations, to accelerate the development of new products.
- To purchase or acquire expertise currently outside of the business for application to internal operations, or to assist new developments.
- To examine new market areas as potential targets for existing or adapted products within the current portfolio.

Where a critical mass of projects exist, a separate new venture division allows greater focus on the external environment, and the distance from the core corporation facilitates a global and cross-divisional view to be taken. Unfortunately, the division can often become a kind of dustbin for every new opportunity, and therefore it is critical to define the limits of its operation and its mission, in particular the criteria for termination or continued support of specific projects.

For example, British Gas established a division to exploit internal and external technologies. The division took technology that existed within the company and sought to exploit it commercially, in terms of a licence, new application or market. By means of a venture capital fund, the division also identified external technologies that might enhance the company's mainstream technical knowledge.

Similarly, Rolls-Royce Business Ventures Limited was set up to exploit Rolls-Royce technology in new product areas. Its mission was to lay the foundations for new businesses which had a good fit with the technology and skills available to the parent company, but which were outside the mandate of the parent company's existing business groups. The company, which was given its own site, has now been wound up. Nevertheless, it established two spin-off companies during its short life: Stresswave Technology Limited and Reflex Manufacturing Systems Limited.

Special business units

Special dedicated new business units are wholly owned by the corporation. High strategic relevance requires strong administrative control. Businesses like this tend to come about because the activity is felt to have enough potential to stand alone as a profit centre, and can thus be assessed and operated as a separate business entity. The requirement is that key people can be identified and extracted from their mainstream operational role.

For the business to succeed under the total ownership and control of a large corporate it must be capable of producing significant revenue streams in the medium term. On average, the critical mass appears to be around 12% of total corporate turnover – in a £200 million business perhaps £20 million – but in some cases the threshold for a separate unit is much higher. A potential new business must not only be judged on its relative size or profitability, but more importantly, by its ability to sustain its own development costs. For example, a profitable subsidiary may never achieve the status of a separate new business if it cannot support its own product development.

However, physically separating a business activity does not ensure autonomy. The greatest impediment to such a unit competing effectively in the market is a cosy corporate mentality. If managers of a new business are under the impression that the corporate parent will always assist, provide business and second its expertise and services at non-market rates, that business may never be able to survive commercial pressures. Conversely, if the parent plans to retain total ownership, the parent cannot realistically treat that unit independently. For example, a company which had been set up as a special business unit was undermined when the parent placed an order from an alternative supplier: the venture lost a large proportion of annual revenues, but more importantly that business lost all credibility in the eyes of its other customers. In 2000 Unilever established a ventures unit to invest in Internet businesses, established by staff or external start-ups in related businesses, and for the first time has allowed staff to take a financial interest in the new ventures.

Independent business units

Differing degrees of ownership will determine the administrative control over independent business units, ranging from subsidiary to minority interest. Control would only be exercised through a board presence if that were held. There are two reasons for establishing an independent business as opposed to divisionalizing an activity: to focus on the core business by removing the managerial and technical burden of activities unrelated to the mainstream business; or to facilitate learning from external sources in the case of enabling technologies or activities. This structure has benefits for both parent and venture:

- Defrayed risk for parent, greater freedom for venture.
- Less supervisory requirement for parent, less interference for venture.
- Reduced management distraction for parent, and greater focus for venture.
- Continued share of financial returns for parent, greater commitment from managers of the venture.
- Potential for flow-back or process improvements or product developments for parent, and learning for the venture.

Whilst the mainstream corporation may not be short of cash for investment, bringing in external funding provides additional advantages, such as sharing of risk with other parties and insulating the venture from changes in corporate investment policy. When release of ownership is allowed, venture capital firms can provide valuable help. For example, CharterRail, the spin-off from GICN, was provided with new sources of capital, and assisted in developing its management team and board by the venture capital investors.

By releasing some ownership, executives of the new business can hold a share of the business, indeed this could be a requirement of their involvement. Securing a capital commitment from these key people ensures their personal commitment to its commercial success. In spin-offs that have emerged from Thorn EMI, GICN and ICI, an equity interest has been maintained in each, but a board presence in only two. Part ownership does not have to involve the centre managerially, but when control is reduced the centre must perceive the holding more as an investment than a subsidiary.

The assignment of technical personnel is one of the most difficult problems when establishing an independent business unit. If the individuals necessary to coordinate future product development are unwilling to leave the relative security and comfort of a large corporate facility, which is understandable, the new business may be stopped in its tracks. It is critical to identify the most desirable individuals for such an operation, assessed in terms of their technical ability and personal characteristics. It is also important to assess the effect of these individuals leaving the mainstream development operations, as the capability of the parent's operations could be easily damaged.

Nurtured divestment

Nurtured divestment is appropriate where an activity is not critical to the mainstream business. The product or service has most likely evolved from the mainstream, and whilst supporting these operations it is not essential for strategic control. The design option provides a way for the corporate to release responsibility for a particular business area. External markets may be built up prior to separation, giving time to identify which employees should be retained by the corporate and providing a period of acclimatization for the venture. The parent may or may not retain some ownership.

Siemens has been very successful at developing nurtured divestments. It made a 25% stake in Tele-Processing Systems GmbH, which was started in 1986 by three Siemens employees who had developed products for the remote operation of computers, network auditing and in-house private branch exchanges (PBXs). Similarly, Siemens took a 15% stake in Micro Quartz GmbH, which was set up in 1988 in order to take a technology using fused silicon tubes for the chromatography of gas to the manufacturing level; and 38% of ECT GmbH, a spin-off based on electron beam manufacturing equipment developed in-house.

Instead of a formal equity stake, a parent may support a spin-off by contracted or seconded technologies or expertise. Manpower may be seconded and technologies or know-how may be sold through some legal agreement, whether licence or sale with a confidentiality tie. For example, British Gas developed a real-time expert system by forming a club of 35 companies. Once the prototype was ready, a new company was formed with the backing of a small number of the original club companies to market the product. British Gas provided the chairman and technical director of the new company. People who had been seconded from British Gas subsequently joined the new venture.

Complete spin-off

No ownership is retained by the parent corporation in the case of a complete spin-off. This is essentially a divest option, where the corporation wants to pass over total responsibility for activity, commercially and administratively. This may be due to strategic unrelatedness or strategic redundancy, as a consequence of changing corporate strategic focus. A complete spin-off allows the parent to realize the hidden value of the venture, and allows senior management of the parent to focus on their main business. We discuss these in greater detail in Section 10.4.

For example, in 1991 Quaker Oats Co. spun off Fisher Price Inc. Quaker management saw this as allowing the company to concentrate its efforts and resources on its core grocery business and to give shareholders more choice as to the industry segments in which they invest. The toy operations were separated from the divisional structure to become a separate entity. Once rationalized as a separate business, shares were issued to Quaker Oats shareholders on a one-for-five basis, and existing shareholders were effectively used as buyers, with no need for formal divestment. In this way Fisher Price stock was separately floated and market capitalizations were enhanced beyond the previous sum of the whole.

In addition to having the most appropriate structure for corporate venturing, Tushman and O'Reilly identify three other organizational aspects that have to be managed to achieve what they call the 'ambidextrous' organization – the coexistence of young, entrepreneurial, risky ventures with the more established, proven operations:[25]

- *Articulating a clear, emotionally engaging and consistent vision*. This helps to provide a strategic anchor for the diverse demands of the mainstream and venture businesses.
- *Building a senior team with diverse competencies*. The composition and demography of the senior team is critical. Homogeneity typically results in greater consensus, faster decision making and easier execution, but lowers levels of creativity and innovation; whereas heterogeneity can cause conflicts, but promotes more diverse perspectives. To achieve a balance, they suggest homogeneity by tenure/length of service, but diversity in backgrounds and perspectives. Alternatively, senior teams can be relatively homogeneous, but have more diverse middle management teams reporting to them.
- *Developing healthy team processes*. The need for creativity needs to be balanced with the need for execution, and team members must be able to resolve conflicts and to collaborate.

However, there is disagreement in the literature regarding the influences of the degree of integration of corporate ventures and the effects on their subsequent success. A study of almost 100 corporate ventures in Canada provided strong support for the need for high levels of integration between the corporate parent and the ventures. It found that the success of a venture was associated with a strong relationship with the corporate parent – specifically use of the parent firm's systems and resources – and conversely that the autonomy of ventures was associated with lower performance of the venture.[26] This appears to contradict the more general body of research which suggests that the managerial independence of ventures is associated with success. For example, a study of spin-offs from Xerox found that those ventures with high levels of funding and senior management from the parent were less successful than those funded more by professional venture capitalists and outside management.[27] One reason for this disagreement might be the period of assessment and measures of success: the Canadian study used the achievement of milestones as the measure of success, and the average age of the new ventures was less than five years; the Xerox study used two measures of success, average rates of growth and financial market value of the ventures, and assessed these over 20 years. In any case, this reflects the real difficulty of getting the right balance between autonomy and integration, as one study found:

> Internal entrepreneurs are faced with two choices: either go underground or spin-off a new venture, with or without the blessing of the parent company . . . it is therefore advisable to spin-off a company in agreement with the parent that contributes technology, personnel and possibly cash, in exchange for minority equity participation. The parent can hold one or more seats on the board of directors, provide advice, networking, and marketing support, share its R&D and pilot production facilities etc., *but must refrain from interfering with management* . . . continued cooperation with the parent also carries a price . . . with a seat on the board the parent is able to monitor and influence the evolution of the technology, and more importantly of the market.[28] (emphasis added)

This is critical as the Xerox study found that the eventual successful business models developed by the spin-offs evolved substantially from the initial plans at formation were very different to the business models of the parent company, and involved significant experimentation to explore the technologies and markets.

Learning through internal ventures

The success of corporate venturing varies enormously between firms, but on average around half of all new ventures survive to become operating divisions, which suggests that venturing may be a less risky strategy for diversification than acquisition or merger. Typically, a venture will achieve profitability within two to three years, and almost half are profitable within six years. However, the profitability of the overall corporate venturing process may be lower due to the effect of a few large failures. Four factors appear to characterize firms that are consistently successful at corporate venturing:

1. Distinguish between bad decisions and bad luck when assessing failed ventures.
2. Measure a venture's progress against agreed milestones, and if necessary redirect.
3. Terminate a venture when necessary, rather than make further investments.
4. View venturing as a learning process, and learn from failures as well as successes.

There are two main causes of failure of internal ventures: strategic reversal and the emergence trap. Strategic reversal occurs because of a conflict between the timescales of the new venture and the parent organization. An internal venture may be set up for a number of reasons: to support a strategy of diversification; because of a risk-taking top management; an excess of corporate cash; or a decline in the firm's main line of business. Whatever the reasons, the internal or external environment is unlikely to remain stable for the life of the new venture. A change of climate can result in the premature termination of a venture. Even normal business cycles may affect the fortunes of a new venture. For example, there appears to be a strong correlation between changes in corporate profits and the number of new ventures set up.[29]

The other, more subtle cause of venture failure is the emergence trap. As a venture expands it may lead to internal territorial infringements, and success leads to jealousy and may result in attempts to undermine the venture. Differences between the culture and style of managers in the parent firm and new venture are likely to amplify these problems (Table 10.6). In particular, new venture divisions are highly visible and represent a concentration of expenditure, and are therefore more vulnerable to changes in corporate performance or management sentiment.

In practice there is a trade-off between rapid growth and learning. A new venture will not have an indefinite period in which to prove itself, and in most cases corporate management will set high targets for growth and financial return in order to offset the risk and uncertainty inherent in a new venture. If successful, the venture will quickly achieve a track record and therefore attract further support from corporate management, resulting in a virtuous spiral of growth and investment. Conversely, if the venture fails to deliver early growth in sales or returns, it may be starved of further support, thus increasing the likelihood of subsequent failure, a vicious spiral of low investment and decline. There are a number of ways to help avoid these problems:[30]

- Make corporate and divisional managers aware of the long-term benefits of venture operations.
- Clearly specify the functions, procedures, boundaries and rewards of venture management.
- Establish a limited number of ventures with independent budgets.
- Establish and maintain multiple sources of sponsorship for ventures.

Therefore it is critical to define the purpose of a new venture, in order to apply the most appropriate financial and organizational structures. Firms may organize and manage new ventures in order to maximize exploitation of existing know-how, or to optimize learning, but not both. Therefore it is critical to define clearly scope and focal activity of a new venture, so that the appropriate linkages to other functions can be established. Where the primary motive is learning, systems and structures to support the new business must be established, rather than simply capitalizing on rapid sales growth. The precise structure and linkages with the parent firm will depend on the relatedness of product and process technologies and product markets (Table 10.7).

TABLE 10.6   Potential sources of conflict between corporate and venture managers

Corporate management	New venture management
Modest uncertainty	Major technical and market uncertainties
Emphasis on detailed planning	Emphasis on opportunistic risk taking
Negotiation and compromise	Autonomous behaviour
Corporate interests and rules	Individualistic and ad hoc
Homogeneous culture and experience	Heterogeneous backgrounds

TABLE 10.7 **Type of new venture and links with parent**

| | Relatedness of: | | | | |
Venture type	Product technology	Process technology	Product market	Focal activity of venture	Linkages with parent firm
Product development	Low	Low	High	Development and production	Marketing
Technological innovation	Low	High	High	R&D	Research, marketing and production
Market diversification	High	High	Low	Branding and marketing	Development and production
Technology commercialization	High	Low	Low	Marketing and production	Development
Blue-sky	Low	Low	Low	Development, production and marketing	Finance

---

The failure of the parent company to define and articulate the role of the venture is the proximate cause of most difficulties experienced with corporate ventures. Such conflicts can be minimized by ensuring that the primary motive for the venture is made explicit, and communicated to both corporate and venture management. In this way the most appropriate structure and management processes can be developed. Table 10.8 suggests the most appropriate links between the motives, structure and management of internal corporate ventures.

It is very difficult in practice to assess the success of corporate venturing. Simple financial assessments are usually based on some comparison of the investments made by the corporate parent and the subsequent revenue streams or market valuation of the ventures. Both of the latter are highly sensitive to the timing of the assessment. For example, at the height of the Internet bubble, financial market valuations suggested corporate venture returns of 70% or more, whereas a few years later these paper returns no longer existed. For example, a study of 35 spin-offs from Xerox over a period of 22 years reveals that the aggregate market value of these spin-offs exceeded those of the parent by a factor of two by 2001, and by a factor of five at the peak of the previous stock market bubble.[27] Assessment of the strategic benefits of corporate venturing is not much easier, but provided the time frames are sufficiently long these can be identified. An historical analysis of the development and commercialization of superconductor technologies at General Electric between 1960 and 1990 reveals how the technology began in internal research and development, but reached a point at which there was deemed to be insufficient market potential to justify any further internal investment. Two GE operating businesses were offered

| TABLE 10.8 | Motives, structure and management of corporate ventures | | |

Primary motive	Preferred structure	Key management task
Satisfy managers' ambition	Integrated business team	Motivation and reward
Spread cost and risk of development	Integrated business team	Resource allocation
Exploit economies of scope	Micro-venture department	Reintegration of venture
Learn about venturing	New venture division	Develop new skills
Diversify the business	Special business unit	Develop new assets
Divest non-core activities	Independent business unit	Management of intellectual property rights

Source: Adapted from Tidd, J. and S. Taurins (1999) Learn or leverage? Strategic diversification and organisational learning through corporate ventures. *Creativity and Innovation Management*, **8** (2), 122–9.

the technology, but declined to fund further development. Rather than abandon the technology altogether, in 1971 GE established a 40% owned venture called Intermagnetics General Corp. (IGC) to develop the technology further. GE became a major customer of IGC as demand for the technology grew in its Medical Systems business due to the growth of MRI (magnetic resonance imaging). However, by 1983 the need for the technology has become so central to GE business that GE had to redevelop its own core competencies in the field.[28]

10.3 Joint ventures and alliances

Almost all innovations demand some form of collaborative arrangement, for development or commercialization, but the failure rate of such alliances remains high. In Chapter 6 we reviewed the central role of innovation networks, but here we examine the more specific issue of bilateral alliances or joint ventures. We discuss the role of collaboration in the development of new technologies, products and businesses. Specifically, we address the following issues:

- Why do firms collaborate?
- What types of collaboration are most appropriate in different circumstances?
- How do technological and market factors affect the structure of an alliance?
- What organizational and managerial factors affect the success of an alliance?
- How can a firm best exploit alliances for learning new technological and market competencies?

www.managing-innovation.com

Why collaborate?

Firms collaborate for a number of reasons:

- To reduce the cost of technological development or market entry.
- To reduce the risk of development or market entry.
- To achieve scale economies in production.
- To reduce the time taken to develop and commercialize new products.
- To promote shared learning.

In any specific case, a firm is likely to have multiple motives for an alliance. However, for the sake of analysis it is useful to group the rationale for collaboration into technological, market and organizational motives (Figure 10.6). Technological reasons include the cost, time and complexity of development. In the current highly competitive business environment, the R&D function, like all other aspects of business, is forced to achieve greater financial efficiency, and to examine critically whether in-house development is the most efficient approach. In addition, there is an increasing recognition that one company's peripheral technologies are usually another's core activities, and that it often makes sense to source such technologies externally, rather than to incur the risks, costs and most importantly of all, timescale associated with in-house development. The rate of technological change, together with the increasingly complex nature of many technologies, means that few organizations can now afford to

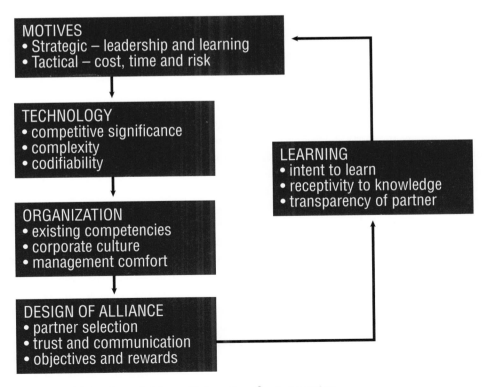

FIGURE 10.6: A model for collaboration for innovation

maintain in-house expertise in every potentially relevant technical area. Many products incorporate an increasing range of technologies as they evolve; for example, automobiles now include much computing hardware and software to monitor and control the engine, transmission, brakes and in some cases suspension. Therefore most R&D and product managers now recognize that no company, however large, can continue to survive as a technological island. For example, when developing the Jaguar XK Ford collaborated with Nippondenso in Japan to develop the engine management system and ZF in Germany to develop the transmission system and controls. In addition, there is a greater appreciation of the important role that external technology sources can play in providing a window on emerging or rapidly advancing areas of science. This is particularly true when developments arise from outside a company's traditional areas of business, or from overseas.

Two factors need to be taken into account when making the decision whether to 'make or buy' a technology: the transaction costs, and strategic implications.[31] Transaction cost analysis focuses on organizational efficiency, specifically where market transactions involve significant uncertainty. Risk can be estimated, and is defined in terms of a probability distribution, whereas uncertainty refers to an unknown outcome. Projects involving technological innovation will feature uncertainties associated with completion, performance and pre-emption by rivals. Projects involving market entry will feature uncertainties due to lack of geographical or product market knowledge. In such cases firms are often prepared to trade potentially high financial returns for a reduction in uncertainty.

However, sellers of technological or market know-how may engage in opportunistic behaviour. By opportunistic behaviour we mean high pricing or poor performance. Generally, the fewer potential sources of technology, the lower the bargaining power of the purchaser, and the higher the transaction costs. In addition, where the technology is complex it can be difficult to assess its performance. Therefore transaction costs are increased where a potential purchaser of technology has little knowledge of the technology. In this respect the acquisition of technology differs from subcontracting more routine tasks such as production or maintenance work, as it is difficult to specify contractually what must be delivered.[32]

As a result, the acquisition of technology tends to require a closer relationship between buyers and sellers than traditional market transactions, resulting in a range of possible acquisition strategies and mechanisms. The optimal technology acquisition strategy in any specific case will depend on the maturity of the technology, the firm's technological position relative to competitors and the strategic significance of the technology.[33] Some form of collaboration is normally necessary where the technology is novel, complex or scarce. Conversely, where the technology is mature, simple or widely available, market transactions such as subcontracting or licensing are more appropriate. However, the cumulative effect of outsourcing various technologies on the basis of comparative transaction costs may limit future technological options and reduce competitiveness in the long term.[34]

Therefore in practice, transaction costs are not the most significant factors affecting the decision to acquire external technology. Factors such as competitive advantage, market expansion and extending product portfolios are more important.[35] Adopting a more strategic perspective focuses attention on long-term organizational effectiveness, rather than short-term efficiency. The early normative strategy literature emphasized the need for technology development to support corporate and business strategies, and therefore technology acquisition decisions began with an evaluation of company strengths and weaknesses. The more recent resource-based approach emphasizes the process of resource accumulation or learning.[36] Competency development requires a firm to have an explicit policy or intent to use collaboration as an opportunity to learn rather than minimize costs. This suggests that the acquisition of external technology should be used to complement internal R&D, rather than being a substitute for it.

In fact, a strategy of technology acquisition is associated with diversification into increasingly complex technologies.[37]

Thus neither transaction costs nor strategic behaviour fully explain actual behaviour, and to some extent the approaches are complementary. For example, a survey of top executives found that the two most significant issues considered when evaluating technological collaboration were the strategic importance of the technology and the potential for decreasing development risk.[38] Thus both strategic and transaction cost factors appear to be significant. Strategic considerations suggest *which* technologies should be developed internally, and transaction costs influence *how* the remaining technologies should be acquired. Firms attempt to reduce transaction costs when purchasing external technology by favouring existing trading partners to other sources of technology.[39] In short, for successful technology acquisition the choice of partner may be as important as the search for the best technology. For both partners, the transaction costs will be lower when dealing with a firm with which they are familiar: they are likely to have some degree of mutual trust, shared technical and business information and existing personal social links.

There is also a growing realization that exposure to external sources of technology can bring about other important organizational benefits, such as providing an element of 'peer review' for the internal R&D function, reducing the 'not-invented-here' syndrome, and challenging in-house researchers with new ideas and different perspectives. In addition, many managers realize the tactical value of certain types of externally developed technology. Some of these are increasingly viewed as a means of gaining the goodwill of customers or governments, of providing a united front for the promotion of uniform industry-wide standards, and to influence future legislation.

The UMIST survey of more than 100 UK-based alliances confirms the relative importance of market-induced motives for collaboration (Table 10.9). Specifically, the most common reasons for collaboration for product development are in response to changing customer or market needs. However, these data provide only the motives for collaboration, not the outcomes. The same survey found that although many firms formed alliances in an effort to reduce the time, cost or risk of R&D, they did not necessarily realize these benefits from the relationship. In fact, the study concluded that around half of the respondents believed that collaboration made development more complicated and costly. However, it is important to relate benefits to the objectives of collaboration. For example, firms that entered into alliances specifically to reduce the cost or time of development often achieved this, whereas firms that formed alliances for other reasons were more likely to complain that the cost and time of development increased. The study also identified a number of potential risks associated with collaboration:

- leakage of information
- loss of control or ownership
- divergent aims and objectives, resulting in conflict.

Around a third of respondents claimed to have experienced such problems. The problem of leakage is greatest when collaborating with potential competitors, as it is difficult to isolate the joint venture from the rest of the business and therefore it is inevitable that partners will gain access to additional knowledge and skills. This additional information may take the form of market intelligence, or more tacit skills or knowledge. Consequently a firm may lose control of the venture, resulting in conflict between partners.

A study of the 'make or buy' decisions for sourcing technology in almost 200 firms concluded that product and process technology from external sources often provides immediate advantages, such as

TABLE 10.9	**Motives for collaboration**

	Mean score ($n = 106$)
In response to key customer needs	4.1
In response to a market need	4.1
In response to technology changes	3.8
To reduce risk of R&D	3.8
To broaden product range	3.7
To reduce R&D costs	3.7
To improve time to market	3.6
In response to competitors	3.5
In response to a management initiative	3.3
To be more innovative in product development	3.3

1 = low, 5 = high.

Source: Littler, D. A. (1993). *Risks and Rewards of Collaboration*, UMIST, Manchester.

lower cost or a shorter time to market, but in the longer term can make it harder for firms to differentiate their offerings, and difficult to achieve or maintain any positional advantage in the market.[40] Instead, successful strategies of cost leadership or differentiation (the two polar extremes of Porter's model, see Chapter 4) are associated with internal development of process and product technologies. However, in highly dynamic environments, characterized by market uncertainty and technological change, sourcing technology externally is a superior strategy to relying entirely on internal capabilities.

For example, high levels of collaboration in the information and communications technology and biotechnology industries, but lower levels in more mature sectors. In the more high-technology sectors, organizations generally seek *complementary* resources – for example, the many relationships between biotechnology firms (for basic research), and pharmaceutical firms (for clinical trials, production and marketing and distribution channels). In the pharmaceutical sector the number of *exploration* alliances with biotechnology firms is predictive of the number of products in development, which in turn is predictive of the number of *exploitation* alliances for sales and distribution.[41] In more mature sectors, more often partners pool *similar* resources to share costs or risk, or to achieve critical mass or economies of scale. There are also differences in the choice of partner. Firms in higher technology sectors tend to favour *horizontal* relationships with their peers and competitors, whereas those in more mature sectors more commonly have *vertical* relations with suppliers and customers.[42] At the firm level, R&D intensity is still associated with the propensity to collaborate, but firms developing products 'new to the market' are much more likely to collaborate than those developing products only 'new to the firm'.[43] This is because the more novel innovations demand more inputs or novelty of inputs, and are associated with greater market uncertainty.

Forms of collaboration

Joint ventures, whether formal or informal, typically take the form of an agreement between two or more firms to co-develop a new technology or product. Whereas research consortia tend to focus on more basic research issues, strategic alliances involve near-market development projects. However, unlike more formal joint ventures, a strategic alliance typically has a specific end goal and timetable, and does not normally take the form of a separate company. There are two basic types of formal joint venture: a new company formed by two or more separate organizations, which typically allocate ownership based on shares of stock controlled; or a more simple contractual basis for collaboration. The critical distinction between the two types of joint venture is that an equity arrangement requires the formation of a separate legal entity. In such cases management is delegated to the joint venture, which is not the case for other forms of collaboration. Doz and Hamel identify a range of motives for strategic alliances and suggest strategies to exploit each:[44]

- To build critical mass through co-option.
- To reach new markets by leveraging co-specialized resources.
- To gain new competencies through organizational learning.

In a co-option alliance, critical mass is achieved through temporary alliances with competitors, customers or companies with complementary technology, products or services. Through co-option a company seeks to group together other relatively weak companies to challenge a dominant competitor. Co-option is common where scale or network size is important, such as mobile telephony and airlines. For example, Airbus (see Case study 10.6) was originally created in response to the dominance of Boeing, and Symbian and Linux in response to Microsoft's dominance. Greater international reach is a common related motive for co-option alliances. Fujitsu initially used its alliance with ICL to develop a market presence in Europe, as did Honda with Rover. However, co-option alliances may be inherently unstable and transitory. Once the market position has been achieved, one partner may seek to take control through acquisition, as in the case of Fujitsu and ICL, or to go unilateral, as in the case of Honda and Rover.[45]

In a co-option alliance, partners are normally drawn from the same industry, whereas in co-specialization partners are usually from different sectors. In a co-specialized alliance, partners bring together unique competencies to create the opportunity to enter new markets, develop new products or build new businesses. Such co-specialization is common in systems or complex products and services. However, there is a risk associated with co-specialization. Partners are required to commit to partners' technology and standards. Where technologies are emerging and uncertain and standards are yet to be established, there is a high risk that a partner's technology may become redundant. This has a number of implications for co-specialization alliances. First, that at the early stages of an emerging market where the dominant technologies are still uncertain, flexible forms of collaboration such as alliances are preferable, and at later stages when market needs are clearer and the relevant technological configuration better defined, more formal joint ventures become appropriate.[46] Second, to restrict the use of alliances to instances where the technology is tacit, expensive and time consuming to develop. If the technology is not tacit, a licence is likely to be cheaper and less risky, and if the technology is not expensive or time consuming to develop, in-house development is preferable.[47]

CASE STUDY 10.6

Airbus Industrie

Airbus Industrie was formed in France in 1969 as a joint venture between the German firm MBB (now DASA) and French firm Aérospatiale, to be joined by CASA of Spain in 1970 and British Aerospace (now BAe Systems) in 1979. Airbus is not a company, but a Groupment d'Intérêt Economique (GIE), which is a French legal entity that is not required to publish its own accounts. Instead, all costs and any profits or losses are absorbed by the member companies. The partners make components in proportion to their share of Airbus Industrie: Aérospatiale and DASA each have 37.9%, BAe 20% and CASA 4.2%.

At that time the international market for civil aircraft was dominated by the US firm Boeing, which in 1984 accounted for 40% of the airframe market in the non-communist world. The growing cost and commercial risk of airframe development had resulted in consolidation of the industry and a number of joint ventures. In addition, product life cycles had shortened due to more rapid improvements in engine technology. The partners identified an unfilled market niche for a high-capacity/short-medium-range passenger aircraft, as more than 70% of the traffic was then on routes of less than 4600 km. Thus the Airbus A300 was conceived in 1969. The A300 was essentially the result of the French and German partners, the former insisting on final assembly in France, and the latter gaining access to French technology. The first A300 flew in 1974, followed by a series of successful derivatives such as the A310 and the A320. The British partner played a leading role in the subsequent projects, bringing both capital and technological expertise to the venture. Airbus has since proved to be highly innovative with the introduction of fly-by-wire technology, and common platforms and control systems for all its aircraft to reduce the cost of crew training and aircraft maintenance. In 2000 the group announced plans to develop a double-decker 'super' jumbo, the A380, with seats for 555 passengers and costing an estimated US$12 billion to develop. Airbus estimates a global market of 1163 very large passenger aircraft and an additional 372 freighters, but needs to sell only 250 A380s to achieve breakeven. This would challenge Boeing in the only market it continues to dominate. (However, Boeing predicts a market of just 320 very large aircraft, as it assumes a future dominance of point-to-point air travel by smaller aircraft, whereas Airbus assume a growth in the hub-and-spoke model, which demands large aircraft for travel between hubs.)

In 1998 Airbus outsold Boeing for the first time in history. In 1999 Daimler-Chrysler (DASA), Aérospatiale and CASA merged to form the European Aeronautic Defence and Space Company (EADS), making BAe Systems, formerly British Aerospace, the only non-EADS member of Airbus. The group plans to move from the unwieldy GIE structure to become a company. This would allow streamlining of its manufacturing operations, which are currently geographically dispersed across the UK, France, Germany and Spain, and more importantly help create financial transparency to help identify and implement cost savings. Also, some customers have reported poor service and support as Airbus has to refer such work to the relevant member company.

Airbus demonstrates the complexity of joint ventures. The primary motive was to share the high cost and commercial risk of development. On the one hand, the French and German participation was underwritten by their respective governments. This fact has not escaped the attention of Boeing and the US government, which provides subsidies indirectly via defence contracts. On the other hand, all partners had to some extent captive markets in the form of national airlines, although almost three-quarters of all Airbus sales were ultimately outside the member countries. Finally, there were also technology motives for the joint venture. For example, BAe specializes in development of the wings, Aérospatiale the avionics, DASA the fuselages and CASA the tails. However, as suggested above, there are now strong financial, manufacturing and marketing reasons for combining the operations within a single company.

In 2006 Airbus sold 434 planes against Boeing's 398. The first commercial service of the A380 began in 2007 with Singapore Airlines, followed by Emirates. Airbus planned to deliver 25 A380 Airbus aircraft in 2009, below early forecasts.

There has been a spectacular growth in strategic alliances, and at the same time more formal joint ventures have declined as a means of collaboration. In the mid-1980s less than 1000 new alliances were announced each year, but by the year 2000 this had grown to almost 10000 per year (based on data from Thomson Financial). There are a number of reasons for the increase in alliances overall, and more specifically the switch from formal joint ventures to more transitory alliances:[48]

- *Speed: transitory alliances versus careful planning.* Under turbulent environmental conditions, speed of response, learning and lead time are more critical than careful planning, selection and development of partnerships.
- *Partner fit: network versus dyadic fit.* Due to the need for speed, partners are often selected from existing members of a network, or alternatively reputation in the broader market.
- *Partner type: complementarity versus familiarity.* Transitory alliances increasingly occur across traditional sectors, markets and technologies, rather than from within. Microsoft and LEGO to develop an Internet-based computer game, Deutsche Bank and Nokia to create mobile financial services.
- *Commitment: aligned objectives versus trust.* The transitory nature of relationships make the development of commitment and trust more difficult, and alliances rely more on aligned objectives and mutual goals.
- *Focus: few, specific tasks versus multiple roles.* To reduce the complexity of managing the relationships, the scope of the interaction is more narrowly defined, and focused more on the task than the relationship.

Patterns of collaboration

Research on collaborative activity has been plagued by differences in definition and methodology. Essentially there have been two approaches to studying collaboration. The approach favoured by economists and strategists is based on aggregate data and examines patterns within and across different sectors. This type of research provides useful insights into how technological and market characteristics affect the level, type and success of collaborative activities. The other type of research is based on structured case studies of specific alliances, usually within a specific sector, but sometimes across national boundaries, and provides richer insights into the problems and management of collaboration.

Industry structure and technological and market characteristics result in different opportunities for joint ventures across sectors, but other factors determine the strategy of specific firms within a given sector. At the industry level, high levels of R&D intensity are associated with high levels of technologically oriented joint ventures, probably as a result of increasing technological rivalry. This suggests that technologically oriented joint ventures are perceived to be a viable strategy in industries characterized by high barriers to entry, rapid market growth and large expenditures on R&D. However, within a specific sector, joint venture activity is not associated with differences in capital expenditure or R&D intensity. A study of joint ventures in the USA found that technologically oriented alliances tend to increase with the size of firm, capital expenditure and R&D intensity.[49] Similarly, the number of marketing and distribution-oriented joint ventures increases with firm size and capital expenditure, but is not affected by R&D intensity. At the level of the firm, different factors are more important. For example, there are significant differences in the motives of small and large firms. In general, large firms use joint ventures to acquire technology, whilst smaller firms place greater emphasis on the acquisition of market knowledge and financial support.

Joint venture activity is high in the chemical, mechanical and electrical machinery sectors, as firms seek to acquire external technological know-how in order to reduce the inherent technological uncertainty in those sectors. In contrast, joint ventures are much less common in consumer goods industries, where market position is the result of product differentiation, distribution and support. If obtaining complementary assets or resources are a primary motive for collaboration, we would expect alliances to be concentrated in those sectors in which mutual ignorance of the partner's technology or markets is likely to be high.[50] Similarly, joint ventures would occur more frequently between partners who are in industries relatively unrelated to one another, and that such alliances are likely to be short-lived as firms learn from each other. Surveys of alliances in so-called high-technology sectors such as software and automation appear to confirm that access to technology is the most common motive. Market access appears to be a more common motive for collaboration in the computer, microelectronics, consumer electronics and telecommunications sectors.

However, these data need to be treated with some caution as in many cases partners exchange market access for technology access or vice versa. For example, Japanese firms rarely sell technology, but are often prepared to exchange technology for access to markets. Conversely, European firms commonly trade market access for technology.[51] In this way firms limit the potential for paying high price premiums for market or technologies because of their lack of knowledge.

A breakdown of alliances by region provides some further explanation. Patterns within and between triad regions are very different. Alliances between US firms appear to be common in all fields. Alliances between European firms are concentrated in software development and telecommunications, but there is relatively little collaborative activity within the European automation, microelectronics and computing industries. Alliances between Japanese firms appear to be much less common than expected. This may reflect the weakness of the database, but is more likely to reflect the rationale for strategic alliances. The most common reason for international alliances is market access, whereas the most common reason for intra-regional alliances is technology acquisition.

The patterns of collaboration between the different triad regions provide some support for this argument. The data do not provide any indication of the direction of technology transfer, but knowledge of national strengths and weaknesses allows some analysis. Alliances between American and European firms are significant in all fields. Alliances between American and Japanese firms are only significant in computers and microelectronics, presumably the former dominated by the US partners, and the latter by the Japanese. There appears to be relatively little collaboration between Japanese and European companies, perhaps reflecting the weakness of the European electronics industry.

Given the problems of management and organization, potential for opportunistic behaviour and the limited success of alliances it might be expected that the popularity of alliances might decline as firms gain experience of such problems. However, according to the Cooperative Agreements and Technology Indicators (CATI) database the number of technology alliances increased from fewer than 300 in 1990 to more than 500 by 2000. It is possible to identify a number of significant trends in recent years (Figure 10.7).

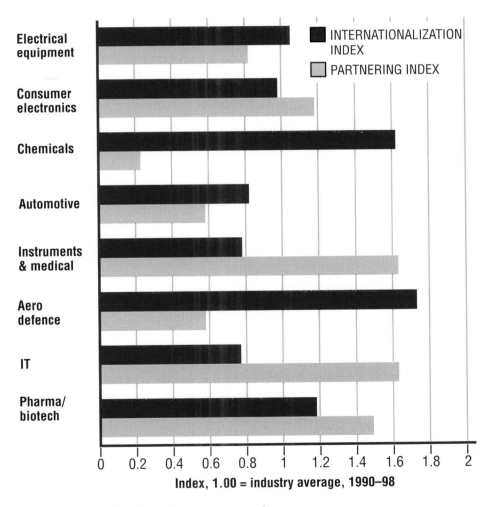

FIGURE 10.7: Collaboration by sector and region
Source: Derived from Hagedoorn, J. (2002) Inter-firm R&D partnerships. *Research Policy,* **31**, 477–92

Overall, the number of alliances has increased over time, and networks of collaboration appear to have become more stable, being based around a number of nodal firms in different sectors. These networks are not necessarily closed, but rather represent the dynamic partnering behaviour of large, leading firms in each of the sectors. The nodal firms are relatively stable, but their partners change over time. Contrary to the claims of globalization, the number of domestic alliances has increased faster than international ones. As a result, international partnerships fell from around 80% of all new agreements in 1976, to below 50% by 2000. This trend is particularly strong in the USA. Distinct sectoral patterns exist. In the more high-technology sectors such as pharmaceuticals, biotechnology and information and communications technologies, most of the collaborative activity is confined within each of the triad regions: Europe, Japan and North America, the exceptions being aerospace and defence. In contrast, most of the activity in the chemical and automotive sectors is across the triad regions. This suggests that the primary motive for collaborating with domestic firms is access to technology, but market access is more important in the case of cross-border alliances. This concentration of high-technology collaboration within regions appears to be more problematic for some regions than others. For example, a study of European electronics firms found that intra-European R&D agreements had no effect on firm patenting, even when sponsored by the EU. However, R&D collaboration with extra-European firms had a positive effect, which in this case means with US partners.[52]

The most recent data from the MERIT-CATI database indicate that flexible forms of collaboration such as strategic alliances have become more popular than the more formal arrangements such as joint ventures. In 1970 more than 90% of the relationships were formal equity joint ventures, but this had fallen to 50% by the mid-1980s and is currently only 10%, the balance being contractual joint ventures and more transitory alliances of some type. This trend has been most marked in high-technology sectors where firms seek to retain the flexibility to switch technology. Together, the pharmaceutical (including biotechnology) and information and communications technology sectors account for almost 80% of the growth in technology collaboration since the mid-1980s. The other most common sectors are aerospace and instrumentation and medical equipment, but collaboration in the aerospace and defence industries has declined. Collaboration in 'mid-technology' sectors such as chemicals, automotive and electronics has shown little or no increase over the same period.

Effects of technology and organization

One study of how 23 UK and 15 Japanese firms acquired technology externally identified the conditions under which each particular method is most common.[53] It is possible to identify two dimensions which affect companies' attitudes towards technology acquisition: the characteristics of the technology and the organization's 'inheritance'. Together, the eight factors in Table 10.10 determine the knowledge acquisition strategy of a firm. The relevant characteristics of the technology include:

- competitive significance of the technology
- complexity of the technology
- codifiability, or how easily the technology is encoded
- credibility potential, or political profile of the technology.

TABLE 10.10	Technological and organizational factors which influence acquisition mechanisms	
Organizational and technological factors	**Acquisition mechanism (most favoured/alternative)**	**Rationale for decision**
I. Characteristics of the organization		
Corporate strategy:		
Leadership	In-house R&D/equity acquisition	Differentiation, first-mover, proprietary technology
Follower	Licence/customers and suppliers/contract	Low-cost imitation
Fit with competencies:		
Strong	In-house R&D	Options to leverage competencies
Weak	Contract/licence/consortia	Access to external technology
Company culture:		
External focus	Various	Cost-effectiveness of source
Internal focus	In-house/joint venture	Learning experience
Comfort with new technology:		
High	In-house corporate/university	High risk and potential high reward
Low	Licence/customers and suppliers/consortia	Lowest risk option
II. Characteristics of the technology		
Base	Licence/contract/customers/suppliers	Cost-effective/secure source
Key	In-house R&D/joint venture	Maximize competitive advantage

(continued)

TABLE 10.10 **(Continued)**		
Organizational and technological factors	**Acquisition mechanism (most favoured/alternative)**	**Rationale for decision**
Pacing	In-house corporate/university	Future position/learning
Emerging	University/in-house corporate	Watching brief
Complexity:		
High	Consortia/universities/suppliers	Specialization of know-how
Low	In-house R&D/contract/suppliers	Division of labour
Codifiability:		
High	Licence/contract/university	Cost-effectiveness of source
Low	In-house R&D/joint venture	Learning/tacit know-how
Credibility potential:		
High	Consortia/customer/government	High profile source
Low	University/contract/licence	Cost-effectiveness of source

Source: Adapted from Tidd, J. and M. Trewhella (1997) Organizational and technological antecedents for knowledge acquisition. *R&D Management,* **27** (4), 359–75.

An organization's inheritance encompasses those characteristics which, at least in the short run, are fixed and therefore represent constraints within which the R&D function develops its strategies for acquiring technology. These include:

- corporate strategy, for example, a leadership versus follower position
- capabilities and existing technical know-how
- culture of the firm, including receptivity to external knowledge
- 'comfort' of management with a given technical area.

Competitive significance

Without doubt, the competitive significance of the technology is the single most important factor influencing companies' decisions about how best to acquire a given technology.

Strategies for acquiring pacing technologies – i.e. those with the potential to become tomorrow's key technologies – vary. For example, some organizations, such as AEA Technology, seek to develop and maintain at least some in-house expertise in many pacing technologies, so they will not be 'wrong-footed' if conditions change or unexpected advances occur. In the past, this policy enabled the company to recognize the importance of finite element analysis to its modelling core competence, and to acquire the necessary aspects of this technology before its competitors. Other firms, such as Kodak, also recognize the need to monitor developments in a number of pacing technologies, but see universities or joint ventures as the most efficient means of achieving this. The company sponsors a large amount of research in leading universities throughout the world, and has also set up a number of joint venture programmes with firms in complementary industries. Guinness, for example, identified genetic engineering as a pacing technology and seconded a member of staff to work at a leading university for three years. The outcome of this initiative was a new biological product, protected by a confidentiality agreement with the university.

Extensions to existing in-house research typically involve using universities to conduct either fundamental research, aimed at gaining a better understanding of an underlying area of science, or more speculative extensions to existing in-house programmes which cannot be justified internally because of their high risk, or because of limited in-house resources. For example, Zeneca has made extensive use of universities to undertake fundamental studies into the molecular biology of plants and the cloning of genes. Although not key technologies, access to state-of-the-art knowledge in these areas is vital to support a number of the organization's core agricultural activities.

University-funded research can also be used as windows on emerging or rapidly advancing fields of science and technology. Companies view access to such information as being critical in making good decisions about if or when to internalize a new technology. For example, Azko launched a series of university-funded research programmes in the USA during the late 1980s. During its first three years, these programmes yielded 40 patent applications.

Most companies look to acquire base technologies externally or, in the case of noncompetitive technologies, by cooperative efforts. Companies recognize that their base technologies are often the core competencies of other firms. In such cases, the policy is to acquire specific pieces of base technology from these firms, who can almost always provide better technology, at less cost, than could have been obtained from in-house sources. Materials testing, routine analysis and computing services are common examples of technical services now acquired externally.

Complexity of the technology

The increasingly interdisciplinary nature of many of today's technologies and products means that, in many technical fields, it is not practical for any firm to maintain all necessary skills in-house. This increased complexity is leading many organizations to conclude that, in order to stay at the forefront of their key technologies, they must somehow leverage their in-house competencies with those available externally. For example, the need to acquire external technologies appears to increase as the number of component technologies increases. In extreme cases of complexity, networks of specialist developers may emerge which serve companies that specialize in systems integration and customization for end users.

Alliances between large pharmaceutical firms and smaller biotechnology firms have received a great deal of management and academic attention over the past few years. On the one hand, pharmaceutical firms have sought to extend their technological capabilities through alliances with and the acquisition of specialist biotechnology firms. Each of the leading drug firms will at any time have about 200 collaborative projects, around half of which are for drug discovery. On the other hand, small biotechnology firms have sought relationships with pharmaceutical firms to seek funding, development, marketing and distribution. In general, pharmaceutical and biotechnology firms each use alliances to acquire complementary assets, and such alliances are found to contribute significantly to new product development and firm performance.[54] For the pharmaceutical firms, there is a strong positive correlation between the number of alliances and markets sales. For the biotechnology firms the benefits of such relationships are less clear. Two trajectories coexist. The first is based on increasing specification of biological hypotheses. The second is based on platform technologies related to the generation and screening of compounds and molecules, such as combinational chemistry, genomic libraries, bioinformatics and proteomics. The former type of biotechnology firm remains dependent upon the complementary assets of the pharmaceutical firms, whereas the latter type appears to have the capacity to benefit from a broader range of network relationships.[55] A biotechnology firm's *exploration* alliances with pharmaceutical firms is a significant predictor of products in development (along with technological diversity), and in turn products in development are a predictor of *exploitation* alliances with pharmaceutical firms, and these exploitation alliances predict a firm's products in the market.[56]

However, different forms of alliance yield different benefits. Research contracts and licences with biotechnology firms are associated with an increase in biotechnology-based *patents* by pharmaceutical firms, whereas the acquisition of biotechnology firms is associated with an increase in biotechnology-related *products* from pharmaceutical firms. This increase in biotechnology-related products includes only those products developed subsequent to the acquisition, and does not include those products directly acquired with the biotechnology firms. Interestingly, minority equity interests in biotechnology firms and joint ventures between pharmaceutical and biotechnology firms are associated with a reduction in biotechnology-related patents and products.[41] This may be due to the very high organizational costs of joint ventures, or the fact that joint ventures tend to tackle more complex and risky projects than simpler licensing or research contracts.

Codifiability of the technology

The more that knowledge about a particular technology can be codified, i.e. described in terms of formulae, blueprints and rules, the easier it is to transfer, and the more speedily and extensively such technologies can be diffused. Knowledge that cannot easily be codified – often termed 'tacit' – is, by contrast, much more difficult to acquire, since it can only be transferred effectively by experience and face-to-face interactions. All else being equal, it appears preferable to develop tacit technologies in-house. In the absence of strong intellectual property rights (IPR) or patent protection, tacit technologies provide a more durable source of competitive advantage than those which can easily be codified.

For example, the design skills of many Italian firms have allowed them to remain internationally competitive despite significant weaknesses in other dimensions. The difficulty of maintaining a competitive advantage when technology is easily codifiable is highlighted by Guinness, which developed a small, plastic, gas-filled device that gives canned beer the same creamy head as keg beer. This 'widget' initially provided the company with a source of competitive advantage and extra sales, but the innovation was soon copied widely throughout the industry, to the extent that widgets are now almost a requirement for any premium canned beer.

Credibility potential

The credibility given to the company by a technology, or by the source of the technology, is a significant factor influencing the way companies decide to acquire a technology. Particular value is placed on gaining credibility or goodwill from governments, customers, market analysts, and even from the company's own top management, academic institutions and potential recruits. For example, Celltech's collaboration with a large US chemical firm appears to have enhanced the former's market credibility. Not only did the collaboration demonstrate the organization's ability to manage a multimillion-dollar R&D project, but the numerous patents and academic publications that arose from it were also felt to have improved the company's scientific standing. Similarly, in Japan the mobile telecommunications services provider DoCoMo worked closely with the national telephone services provider NTT, although it had the depth and range of technologies required to develop telephony equipment and products. The rationale for the relationship was to influence future standards and to increase the credibility of its consumer telephone products in a market in which it was increasingly difficult to differentiate by means of product or service (see Case study 10.7).

Corporate strategy

One of the most important factors affecting the balance between in-house generated, and externally acquired, technology is the degree to which company strategy dictates that it should pursue a policy of technological differentiation or leadership (see Chapter 4). For example, Kodak distinguishes between two types of technical core competencies: strategic, i.e. those activities in which the company must be a

CASE STUDY 10.7

Symbian

Symbian is a joint venture between Psion, Nokia, Ericsson and Motorola which develops and licenses operating systems for wireless devices such as cell phones and electronic organizers. The members, all electronics manufacturers, formed Symbian in 1998 to help reduce the chance that they would become low-margin 'box-shifters', as has happened in the PC market. The venture was initially established to exploit the EPOC operating system developed by Psion, which owned 40% of Symbian. The two Scandinavian telecommunications companies each invested £57.5 million for their original 30% share. Motorola joined Symbian later. It is estimated that in 2000 Psion invested £8 million in Symbian, but accounted for as much as 90% of Psion's stock market value. Psion sold its share in 2003, and Nokia now owns 48%, and Panasonic, Siemens and Sony Ericsson have all increased their shares of Symbian. Licensees are required to pay a royalty per device. By 2003 Symbian's operating systems were in 6.7 million handsets globally, and in 2004 were in 41% of all personal organizers and smart phones sold, compared to 23% with PalmSource, and just 5% with Microsoft software. Symbian, based in London, planned to increase its staff from 900 to 1200 in 2004. However, two doubts remain. First, the competitive response of companies such as Microsoft, with Windows CE; and second, strategic conflicts between members of Symbian, in particular the growing dominance of Nokia.

world leader because they represent such an important source of competitive advantage; and enabling, i.e. skills required for success, but which do not have to be controlled internally. Although all strategic activities are retained in-house, the company is prepared to access enabling technologies externally, if the overall technology is sufficiently complex.

Some companies adopt a policy of intervention in the technology supply market, until the market becomes sufficiently competitive to ensure reliable sources of technology continue to be available at reasonable prices. For example, the extent to which BP is prepared to rely on external sources of technology depends, amongst other things, on the nature of the supply market. When only a few suppliers exist, BP will develop key items of technology itself, and pass these on to its suppliers in order to ensure their availability. However, once sufficient suppliers have entered the market to make it competitive, its policy is to conduct no further in-house development in that area. Indeed, one of the declared aims of BP's in-house R&D activities is to 'force the pace' at which the industry as a whole innovates.

Firm competencies

An organization's internal technical capabilities are another factor influencing the way in which it decides to acquire a given technology. Where these are weak, a firm normally has little choice but to acquire from outside, at least in the short run, whereas strong in-house capabilities often favour the internal development of related technologies, because of the greater degree of control afforded by this route. In such cases, the main driving force behind the acquisition strategy is speed to market. For example, speed to market is a critical success factor for many firms in consumer markets. Such firms select the technology acquisition method that provides the fastest means of commercialization. When the required expertise is available in-house, this route is normally preferred because it allows greater control of the development process, and is therefore usually quicker. However, where suitable in-house capabilities are lacking, external sourcing is almost always faster than building the required skills internally. Gillette, for example, found that one of its new products required laser spot-welding competencies that the company lacked and, given the limited market window, was forced to go outside to acquire this technology.

Company culture

Every company has its own culture, that is 'the way we do things around here'. We will discuss culture in more detail in the next chapter, but here we are concerned with the underlying values and beliefs that play an important role in technology acquisition policies. A culture of 'we are the best' is likely to contribute towards a rather myopic view of external technology developments, and limit the potential for learning from external partners. Some organizations, however, consistently reinforce the philosophy that important technical developments can occur almost anywhere in the world. Consequently, staff in these companies are encouraged to identify external developments, and to internalize potentially important technologies before the competition. However, in practice few firms have formal 'technology scouting' personnel or functions.

For example, GSK emphasizes that companies need to guard against becoming captives of their own in-house expertise, since this limits the scope of its activities to what can be achieved through internal resources. With this in mind, the company has expanded its research effort by placing many of its more specialized R&D activities overseas. This, it is claimed, allows its research to benefit from different cultural and scientific approaches, and from being brought into intimate contact with the many different markets it serves. Local perspectives are particularly important for product development, but international networks can also be used to acquire access to basic research.

Kodak's philosophy is that world-class organizations must access technology wherever it resides, and that a culture of 'not invented here' is a prescription for second-class citizenship in the global marketplace. Like Glaxo, this firm also has a number of foreign research laboratories. For example, Japan is now the centre of this organization's worldwide efforts in molecular beam epitaxy, a method of growing crystals for making gallium arsenide chips. A key role for overseas laboratories is to monitor technology developments in host countries. Local champions from around the world are closely networked so that technical advances made in one geographical location are rapidly disseminated around the organization as a whole. Such is this company's determination to maintain a 'window' on potential sources of technology that it has set up joint ventures with many large and small companies worldwide, including links with Matsushita, Canon, Nikon, Minolta, Fuji and Apple.

Management comfort

The degree of comfort that management has with a given technology manifests itself at the level of the individual R&D manager or management team, rather than at the level of the organization as a whole. Management comfort is multifaceted. One aspect is related to a management team's familiarity with the technology. Another reflects the degree of confidence that the team can succeed in a new technical area, perhaps because of a research group's track record of success in related fields. Attitude to risk is also a factor.[57]

All else being equal, the more comfortable a company's managers feel with a given technology, the more likely that technology is to be developed in-house. For example, AEA Technology's core technologies of plant life extension, environmental sciences, modelling and land remediation treatment all derive from its nuclear industry background. Top management's comfort with these technologies has led them to encourage staff to build on these skills, and to use these as a springboard for diversification into new scientific areas.

Managing alliances for learning

So far we have discussed collaboration as a means of accessing market or technological know-how, or acquiring assets. However, alliances can also be used as an opportunity to learn new market and technological competencies, in other words to internalize a partner's know-how. Seen in this light, the success of an alliance becomes difficult to measure.

Collaboration is an inherently risky activity, and less than half achieve their goals. A study of almost 900 joint ventures found that only 45% were mutually agreed to have been successful by all partners.[58] Other studies confirm that the success rate is less than 50%.[59]

It is difficult to assess the success of a collaborative venture, and in particular termination of a partnership does not necessarily indicate failure if the objectives have been met. For example, around half of all alliances are terminated within seven years, but in some cases this is because the partners have subsequently merged. It is common for a collaborative arrangement to evolve over time, and objectives may change. For example, a licensing agreement may evolve into a joint venture. Finally, an apparent failure may result in knowledge or experience that may be of future benefit. An alliance is likely to have a number of different objectives – some explicit, others implicit – and outcomes may be planned or unplanned. Therefore any measure of success must be multidimensional and dynamic in order to capture the different objectives as they evolve over time. Reasons for failure include strategic divergence, procedural problems and cultural mismatch. Table 10.11 presents the most common reasons for the failure of alliances, based on a meta-analysis of the 16 studies. The studies reviewed differ in their samples and methodologies, but 11 factors appear in a quarter of the studies, which provides some level of confidence.

Firms have different expectations of alliances and these affect their evaluation of success. Those firms which view product development collaboration as discrete events with specific aims and objectives are more likely to evaluate the success of the relationship in terms of the project cost and time and ultimate product performance. However, a small proportion of firms view collaboration as an opportunity to learn new skills and knowledge and to develop longer term relationships. In such cases measures of success need to be broader. If learning is a major goal, it is necessary for partners to have complementary skills and capabilities, but an even balance of strength is also important. The more equal the partners, the more likely an alliance will be successful. Both partners must be strong financially and in the technological, product or market contribution they make to the venture. A study of 49 international alliances by management consultants

---

TABLE 10.11   **Common reasons for the failure of alliances (review of 16 studies)**

Reason for failure	% studies reporting factor ($n = 16$)
Strategic/goal divergence	50
Partner problems	38
Strong–weak relation	38
Cultural mismatch	25
Insufficient trust	25
Operational/geographical overlap	25
Personnel clashes	25
Lack of commitment	25
Unrealistic expectations/time	25
Asymmetric incentives	13

Source: Derived from Duysters, G., G. Kok and M. Vaandrager (1999) Crafting successful strategic technology partnerships. *R&D Management*, **29** (4), 343–51.

McKinsey found that two-thirds of the alliances between equally matched partners were successful, but where there was a significant imbalance of power almost 60% of alliances failed.[60] Consequently in the case of a formal joint venture equal ownership is the most successful structure, 50–50 ownership being twice as likely to succeed as other ownership structures. This appears to be because such a structure demands continuous consultation and communication between partners, which helps anticipate and resolve potential conflicts, and problems of strategic divergence. Our own study of Anglo–Japanese joint ventures identified three sources of strategic conflict between parent firms: product strategy; market strategy; and pricing policy. These were primarily the result of coupling complementary resources with divergent strategies, what we refer to as the 'trap of complementarity'. In essence, parents with complementary resources almost inevitably have different long-term strategic objectives. Too many joint ventures are established to bridge gaps in short-term resources, rather than for long-term strategic fit.[61]

This suggests that firms must learn to design alliances with other firms, rather than pursue ad hoc relationships. By design we do not mean the legal and financial details of the agreement, but rather the need to select a partner which can contribute what is needed, and needs what is offered, of which there is sufficient prior knowledge or experience to encourage trust and communication, to allow areas of potential conflict such as overlapping products or markets to be designed out. Partners must specify mutual expectations of respective contributions and benefits. They should agree on a business plan, including contingencies for possible dissolution, but allow sufficient flexibility for the goals and structure of the alliance to evolve. It is important that partners communicate on a routine basis, so that any problems are shared. Without such explicit design, collaboration may make product development more costly, complex and difficult to control (Table 10.12). Thus whilst the *failure* of an alliance is most likely to be the result of strategic divergence, the *success* of an alliance depends to a large extent on what can be described as operational and people-related factors, rather than strategic factors such as technological, market or product fit (Table 10.13).

TABLE 10.12 The effects of collaboration on product development		
	Agree/ strongly agree	Disagree/ strongly disagree
Makes product development more costly	51	22
Complicates product development	41	35
Makes development more difficult to control	41	38
Makes development more responsive to supplier needs	36	26
Allows development to adapt better to uncertainty	27	43
Accelerates product development	25	58
Makes development more responsive to customer needs	22	50

(continued)

	Agree/ strongly agree	Disagree/ strongly disagree
Allows development to respond better to market opportunities	15	63
Enhances competitive benefits arising through development	12	65
Facilitates the incorporation of new technology in development	7	70

TABLE 10.12 **(Continued)**

Source: Adapted from Bruce, M., F. Leverick and D. Littler (1995) Complexities of collaborative product development. *Technovation*, **15** (9), 535–52, with kind permission from Elsevier Science Ltd, The Boulevard, Langford Lane, Kidlington OX5 1GB, UK.

TABLE 10.13   **Factors influencing success of collaboration**

Factor	Respondents freely mentioning factor ($n = 106$)
Establishing ground rules	67
Clearly defined objectives agreed by all parties	41
Clearly defined responsibilities agreed by all parties	19
Realistic aims	10
Defined project milestones	11
People factors	54
Collaboration champion	22
Commitment at all levels	11
Top management commitment	10

(continued)

TABLE 10.13	(Continued)	

Factor	Respondents freely mentioning factor ($n = 106$)
Personal relationships	10
Staffing levels	3
Process factors	45
Frequent communication	20
Mutual trust/openness/honesty	17
Regular progress reviews	13
Deliver as promised	9
Flexibility	3
Ensuring equality	42
Mutual benefit	22
Equality in power/dependency	11
Equality of contribution	9
Choice of partner	39
Culture/mode of operation	13
Mutual understanding	12
Complementary strengths	12
Past collaboration experience	2

Source: Adapted from Bruce, M., F. Leverick and D. Littler (1995) A management framework for collaborative product development. In M. Bruce and W.G. Biemans, eds, *Product Development: Meeting the Challenge of the Design–Marketing Interface*, John Wiley & Sons, Chichester, p. 171.

The most important operational factors are agreement on clearly stated aims and responsibilities, and the most important people factors are high levels of commitment, communication and trust. A survey of 135 German firms gives us a better idea of the relative importance of these different factors.[62] The study found that firms take people-related, economic and technological factors into consideration, but that these three groups of variables are largely independent of each other. Factor analysis confirms that the people-related factors are more significant than either the economic or technological considerations, specifically creation of trust, informal networking and learning. However, managers often put greater effort into the 'harder' technical and operational issues, than the 'softer' but more important people issues, and focus more on 'deal making' to form alliances, than the processes necessary to sustain them. One study of alliances between high-technology firms found that more than half of the problems in the first year of an alliance relate to the relationship, rather than the strategic or operational factors. The most common problems were poor communication – quality and frequency – and conflicts due to differences in national or corporate cultures.[63] The study identified three strategies for minimizing these cultural mismatches. First, for one partner to adopt the culture of the other (unlikely outside an acquisition). Second, to limit the degree of cultural contact necessary through the operational design of the project. Finally, to appoint cultural translators or liaisons to help identify, interpret and communicate different cultural norms.

Other factors which contribute to the success of an alliance include:[64]

- The alliance is perceived as important by all partners.
- A collaboration 'champion' exists.
- A substantial degree of trust between partners exists.
- Clear project planning and defined task milestones are established.
- Frequent communication between partners, in particular between marketing and technical staff.
- The collaborating parties contribute as expected.
- Benefits are perceived to be equally distributed.

Mutual trust is clearly a significant factor, when faced with the potential opportunistic behaviour of the partners; for example, failure to perform or the leakage of information. Trust may exist at the personal and organizational levels, and researchers have attempted to distinguish different levels, qualities and sources of trust.[65] For example, the following bases of trust in alliances have been identified:

- Contractual – honouring the accepted or legal rules of exchange, but can also indicate the absence of other forms of trust.
- Goodwill – mutual expectations of commitment beyond contractual requirements.
- Institutional – trust based on formal structures.
- Network – because of personal, family or ethnic/religious ties.
- Competence – trust based on reputation for skills and know-how.
- Commitment – mutual self-interest, committed to the same goals.

These types of trust are not necessarily mutually exclusive, although over-reliance on contractual and institutional forms may indicate the absence of the types of trust. Goodwill is normally a second-order effect based on network, competence or commitment. In the case of innovation, problems may occur where trust is based on the network, rather than competence or commitment, as discussed earlier. Clearly, high levels of interpersonal trust are necessary to facilitate communication and learning in collaboration, but inter-organizational trust is a more subtle issue. Organizational trust may be defined in terms of organizational routines, norms and values which are able to survive changes in individual personnel. In this way organizational learning can take place, including new ways of doing things (operational or lower-level learning) and doing new things through diversification (strategic or higher-level learning). Organizational trust requires a longer time horizon to ensure that reciprocity can occur, as for any particular collaborative project one partner is likely to benefit disproportionately. In this way organizational trust may mitigate against opportunistic behaviour. However, in practice this may be difficult where partners have different motives for an alliance or differential rates of learning.

In Chapter 4 we examined the nature of core competencies. Conceiving of the firm as a bundle of competencies, rather than technology or products, suggests that the primary purpose of collaboration is the acquisition of new skills or competencies, rather than the acquisition of technology or products. Therefore a crucial distinction must be made between acquiring the skills of a partner, and simply gaining access to such skills. The latter is the focus of contracting, licensing and the like, whereas the internalization of a partner's skills demands closer and longer contact, such as formal joint ventures or strategic alliances. An example would be Kodak which for many years dominated the photographic market exploiting its competencies based on 'wet chemistry in the dark'. However, the advent of digital photography threatened many (but not all) these competencies, and through a combination of corporate ventures, alliances and acquisitions Kodak successfully managed the transition from chemistry to digital photography, unlike its rival Polaroid (see Case study 10.8).

CASE STUDY 10.8

Kodak develops digital competencies through alliances and acquisitions

Faced with developments in digital imaging technology, Kodak redefined its business as 'pictures, not technology', stressing that the market competencies were still relevant to the digital photographic markets, but it lacked the relevant technological competencies. The board hired George Fisher from Motorola to be the new CEO. Fisher pursued a two-tier strategy for new business development. For the medical imaging business, Kodak acquired a number of specialist digital technology firms, including Imation Corporation, which had developed a hybrid dry laser imaging technology. It combined these new competencies with its existing market knowledge, as it accounted for around 30% of the global medical imaging market at that time.

For the consumer imaging market, Kodak established a new Digital and Applied Imaging division, but this suffered from the parent company's organizational routines, which had evolved to monitor relatively stable mass markets and slow-moving technology, and were therefore inappropriate for digital imaging at that time. As a result of organizational problems, the division was made organizationally independent in 1997, and in 1998 formed a joint venture with Intel to develop the 'Picture CD' project. Similarly, initial attempts to develop digital cameras in the existing Consumer Imaging division resulted in cameras that failed to meet the technological and market demands. These developments were also later moved to the new Digital and Applied Imaging division, which had routines more suited to the needs of emerging technologies and markets. A series of successful products followed, and by 2004 Kodak had 20% of the global market share in digital cameras.

Source: Derived from Jeffrey T. Macher and Barak D. Richman (2004) Organizational responses to discontinuous innovation. *International Journal of Innovation Management*, **8** (1), 87–114.

It is possible to identify three factors that affect learning through alliances: intent, transparency and receptivity (Table 10.14). Intent refers to a firm's propensity to view collaboration as an opportunity to learn new skills, rather than to gain access to a partner's assets. Thus where there is intent, learning takes place by design rather than by default, which is much more significant than mere leakage of information. Transparency refers to the openness or 'knowability' of each partner, and therefore the potential for learning. Receptivity, or absorptiveness, refers to a partner's capacity to learn. Clearly, there is much a firm can do to maximize its own intent and receptivity, and minimize its transparency. Intent to learn will influence the choice of partner and form of collaboration. Transparency will depend on the penetrability of the social context, attitudes towards outsiders, i.e. clannishness, and the extent to which the skills are discrete and encodable. Explicit knowledge, such as designs and patents, are more easily

TABLE 10.14 **Determinants of learning through alliances**	
Factors which promote learning	
A. Intent to learn	
1. Competitive posture	Cooperate now, compete later
2. Strategic significance	High, to build competencies, rather than to fix a problem
3. Resource position	Scarcity
4. Relative power balance	Balance creates instability, rather than harmony
B. Transparency or potential for learning	
5. Social context	Language and cultural barriers
6. Attitude towards outsiders	Exclusivity, but absence of 'not invented here'
7. Nature of skills	Tacit and systemic, rather than explicit
C. Receptivity or absorptive capacity	
8. Confidence in abilities	Realistic, not too high or too low
9. Skills gap	Small, not too substantial
10. Institutionalization of learning	High, transfer of individual learning to organization

Source: Adapted from Hamel, G. (1991) Learning in international alliances. *Strategic Management Journal,* **12**, 91.

encoded than tacit knowledge. This suggests that a harmonious alliance may not necessarily represent a win–win situation. On the contrary, where two partners attempt to extract value from their alliance in the same form, whether in terms of short-term economic benefits or longer-term skills acquisition, managers are likely to frequently engage in arguments over value sharing. Where partners have different goals, for example one partner seeks short-term benefits whereas the other seeks the acquisition of new skills, the relationship tends to be more harmonious, at least until one partner is no longer dependent on the other. For example, where a firm works with a university or commercial research organization, the goals of the alliance are likely to be very different, and therefore the factors influencing a successful outcome may differ (Table 10.15).

Therefore the preferred structure for an alliance will depend on the nature of the knowledge to be acquired, whereas the outcome will be determined largely by a partner's ability to learn, which is a function of skills and culture. Tactical alliances are most appropriate to obtain migratory or explicit knowledge, but more strategic relationships are necessary to acquire embedded or tacit knowledge.[66] Alliances for explicit knowledge focus on trades in designs, technologies or products, but by the very nature of such knowledge this provides only temporary advantages because of its ease of codification and movement. Alliances for embedded knowledge present a more subtle management challenge. This involves the transfer of skills and capabilities, rather than discrete packages of know-how. This requires personnel to have direct, intimate and extensive exposure to the staff, equipment, systems and culture of the partnering organization. However, the absorptive capacity of an organization is not a constant,

TABLE 10.15 Factors influencing the success of relationships between firms and contract research organizations

Significant factor	For firm	For research organization
Previous links	Significant	Significant
Commitment	Significant	Significant
Partner's reputation	Not significant	Significant
Definition of objectives	Significant	Not significant
Communication	Not significant	Significant
Conflict	Significant	Not significant
Organizational design	Not significant	Not significant
Geographical proximity	Not significant	Not significant

Source: Derived from Mora-Valentin, E.M., A. Montoro-Sanchez and L.A. Guerras-Martin (2004) Determining factors in the success of R&D cooperative agreements between firms and research organizations. *Research Policy*, **33**, 17–40.

and depends on the fit with the partner's knowledge base, organizational structures and processes, such as the degree of management formalization and centralization of decision making and research.[67] Studies suggest that knowledge creation in an alliance is more likely to occur where there is a clear intent and specific goals exist, but conversely individual autonomy within a joint project is associated with a reduction in knowledge creation. One of the most significant factors influencing knowledge creation and learning in an alliance is the use of formal environmental scanning, and this effect increases with the complexity of projects.[68] There appear to be two reasons for the importance of scanning in such alliances. First, the need to identify relevant knowledge in the environment, and second, to ensure that the developments continue to be relevant to the changing environment.

The conversion of tacit to explicit knowledge is a critical mechanism underlying the link between individual and organizational learning.[69] Through a process of dialogue, discussion, experience sharing and observation, individual knowledge is amplified at the group and organizational levels. This creates an expanding community of interaction, or 'knowledge network', which crosses intra- and inter-organizational levels and boundaries. These knowledge networks are a means to accumulate knowledge from outside the organization, share it widely within the organization, and store it for future use. Therefore the interaction of groups with different cultures, whether within or beyond the boundaries of the organization, is a potential source of learning and innovation.

Organizational structure and culture will determine absorptive capacity in inter-organizational learning. Culture is a difficult concept to grasp and measure, but it helps to distinguish between national, organizational, functional and group cultures.[70] Differences in national culture have received a great deal of attention in studies of cross-border alliances and acquisitions, and the consensus is that national differences do exist and that these affect both the intent and ability to learn. In general, British and American firms focus more on the legal and financial aspects of alliances, but rarely have either the intent or ability to learn through alliances. In contrast, French, German and Japanese firms are more likely to exploit opportunities for learning.[71] The issue of national stereotypes aside, there may be structural reasons for these differences in the propensity to learn.

For example, Japanese firms have good historical reasons for exploiting alliances as opportunities for learning. Initially, Western firms typically entered Japan through alliances in which they provided technology in return for access to Japanese sales and distribution channels. This exchange of technology for market access appeared to offer value to both sides. However, while the Western partner often remained dependent on the Japanese partner for distribution and sales, the Japanese partner typically built up its technological skills and became less reliant on the Western partner. As a result, in the 1980s European and American partners began to lose technological leadership in many fields, and were forced to trade distribution and sales channels at home for access to the Japanese market. Therefore collaboration has shifted from relatively simple and well-defined licensing agreements or joint ventures, to more complex and informal relationships, which are much more difficult to manage.

Most recently, firms from the USA and Europe have begun to use alliances for operational learning. Operational learning provides close exposure to what competitors are doing in Japan and how they are doing it. For example, to learn how Japanese partners manage their production facilities, supplier base or product development process. This is not possible from a distance, and requires close alliances with potential competitors. However, fewer firms in the West have exploited fully the potential of alliances for strategic learning, that is the acquisition of new technological and market competencies.

In contrast, many American and British firms find it difficult to learn through alliances. This appears to be because firms focus on financial control and short-term financial benefits, rather than the longer-term potential for learning. For example, firms will attempt to minimize the number and quality of people they contribute to a Japanese joint venture, and the time committed. As a result, little learning takes place and little or no corporate memory is built up.

At the lower level of analysis, different functional groups and project teams may have different cultures. For example, the differences between technical and marketing cultures are well documented, and are a major barrier to communication within an organization.[72] When such groups are required to communicate across organizations the potential for problems is even greater. There is some evidence that employees attempt to trade information based on the perceived economic interests of their firms, but that these perceptions differ. A study of 39 managers involved in alliances in the steel industry identified three clusters of behaviour regarding information trading: value-oriented, competition-oriented and complex decision makers.[73] Value-oriented employees base their behaviour on the importance of the information to their own firm, independent of its potential value to the partner. Competition-oriented employees base their behaviour solely on the value of the information to competitors. The complex decision makers include both considerations, and also the potential for trading information. Some firms develop reputations for being very secretive, while others are seen as more open. No doubt this contrasting approach to knowledge sharing will interest enthusiasts of game theory, but the empirical evidence suggests that firms that share their knowledge with their peers and competitors – for example, through conferences and journals – have a higher innovative performance than those that do not share, controlling for the level of R&D spending and number of patents.[74] The reasons for this apparent reward for generosity include the need to motivate and recruit researchers, and a strategy to be perceived as a technology leader to influence technological trajectories and attract alliance partners.

10.4 Spin-outs and new ventures

Much of what we know about spin-out ventures and NTBFs is based on the experience of firms in the USA, in particular, the growth of biotechnology, semiconductor and software firms. Many of these originated from a parent or 'incubator' organization, typically either an academic institution or large well-established firm. Examples of university incubators include Stanford which spawned much of Silicon Valley, the Massachusetts Institute of Technology (MIT) which spawned Route 128 in Boston, and Imperial and Cambridge in the UK. MIT in particular has become the archetype academic incubator, and in addition to the creation of Route 128, its alumni have established some 200 NTBFs in northern California, and account for more than a fifth of employment in Silicon Valley.[1] The so-called MIT model has been adopted worldwide, so far with limited success. For example, in 1999 Cambridge University in the UK formed a UK government-sponsored joint venture with MIT to help develop spin-offs in the UK. However, to put such initiatives into perspective, Hermann Hauser, a venture capitalist, notes 'Stanford alumni have produced companies worth a trillion dollars. MIT half a trillion dollars. If Cambridge is getting to $20 billion we will be lucky.' One reason is the differences in scale. Mike Lynch, founder of the software company Autonomy, observes, 'Silicon Valley is 60 miles long and in the last few months there will have been 70 to 80 money raisings in the $50 million to $200 million range. In Cambridge we might think of one, perhaps.'

Examples of large incubator firms include the Xerox PARC (see Case study 10.9) and Bell Laboratories in the USA which spawned Fairchild Semiconductor, which in turn led to numerous spin-offs including Intel, Advanced Memory Systems, Teledyne and Advanced Micro-Devices.

Similarly, Engineering Research Associates (ERA) led to more than 40 new firms, including Cray, Control Data Systems, Sperry and Univac (see Case study 10.10). In many cases, incubator firms provide the technical entrepreneurs, and the associated academic institutions provide the additional qualified staff.

CASE STUDY 10.9

Spin-off companies from Xerox's PARC Labs

Xerox established its Palo Alto Research Center (PARC) in California in 1970. PARC was responsible for a large number of technological innovation in the semiconductor lasers, laser printing, Ethernet networking technology and web indexing and searching technologies, but it is generally acknowledged that many of its most significant innovations were the result of individuals who left the company and firms which spun-off from PARC, rather than developed via Xerox itself. For example, many of the user–interface developments at Apple originated at Xerox, as did the basis of Microsoft's Word package. By 1998 Xerox PARC had spun-out 24 firms, including 10 which went public such as 3Com, Adobe, Documentum and SynOptics. By 2001 the value of the spin-off companies was more than twice that of Xerox itself.

A debate continues to the reasons for this, most attributing the failure to retain the technologies in-house to corporate ignorance and internal politics. However, most of the technologies did not simply 'leak out', but instead were granted permission by Xerox, which often provided non-exclusive licences and an equity stake in the spin-off firms. This suggests that Xerox's research and business managers saw little potential for exploiting these technologies in its own businesses. One of the reasons for the failure to commercialize these technologies in-house was that Xerox had been highly successful with its integrated product-focused strategy, which made it more difficult to recognize and exploit potential new *businesses*.

Source: Chesbrough, H. (2003) *Open Innovation: The New Imperative for Creating and Profiting from Technology*, Harvard Business School Press, Boston, MA.

CASE STUDY 10.10

Mike Lynch and Autonomy

Mike Lynch founded the software company Autonomy in 1994, a spin-off from his first start-up Neurodynamics. Lynch, a grammar-school graduate, studied information science at Cambridge where he carried out PhD research on probability theory. He rejected a conventional research career as he had found his summer job at GEC Marconi a '*boring, tedious place*'. In 1991, aged 25, he approached the banks to raise money for his first venture, Neurodynamics, but '*met a nice chap who laughed a lot and admitted that he was only used to lending money to people to open newsagents*'. He subsequently raised the initial £2000 from a friend of a friend. Neurodynamics developed pattern recognition software which it sold to specialist niche users such as the UK police force for matching fingerprints and identifying disparities in witness statements, and banks to identify signatures on cheques.

Autonomy was spun off in 1994 to exploit applications of the technology in Internet, intranet and media sectors, and received the financial backing of venture capitalists Apax, Durlacher and ENIC. Autonomy was floated on the EASDAQ in July 1998, on the NASDAQ in 1999, and in February 2000 was worth US$5bn, making Lynch the first British software billionaire. Autonomy creates software which manages unstructured information, which accounts for 80% of all data. The software applies Bayesian probabilistic techniques to identify patterns of data or text, and compared to crude keyword searches can better take into account context and relationships. The software is patented in the USA, but not in Europe as patent law does not allow patent protection of software. The business generates revenues through selling software for cataloguing and searching information direct to clients such as the BBC, Barclays, BT, Eli Lilly, General Motors, Merrill Lynch, News Corporation, Nationwide, Procter & Gamble and Reuters. In addition, it has more than 50 licence agreements with leading software companies to use its technology, including Oracle, Sun and Sybase. A typical licence will include a lump sum of US$100 000 plus a royalty on sales of 10–30%. By means of such licence deals Autonomy aims to become an integral part of a range of software and the standard for intelligent recognition and searching. In the financial year ending March 2000 the company reported its first profit of US$440 000 on a turnover of $11.7 million. The company employs 120 staff, split between Cambridge in the UK and Silicon Valley, and spends 17% of its revenues on R&D. In 2004 sales were around $60 million, with an average licence costing $360 000, and high gross margins of 95%. New customers include AOL, BT, CitiBank, Deutsche Bank, Ford, the 2004 Greek Olympics, and the defence agencies in the USA, Spain, Sweden and Singapore. Repeat customers accounted for 30% of sales.

NTBF spin-offs tend to cluster around their respective incubator organizations, forming regional networks of expertise. The firms tend to remain close to their parents for a number of technical and personal reasons. Most NTBFs retain contacts with their parent organizations to gain financial and technical support, and are often reluctant to disrupt their social and family lives whilst establishing a new venture. Perhaps surprisingly, the mortality rate of NTBFs is lower than that of most types of new firm, around 20 to 30% in 10 years compared to more than 80% for other types of new business.[29] One explanation for the higher survival rate of NTBFs is that the barriers to entry are higher than for many other businesses, in terms of expertise and capital. Therefore those NTBFs which are able to overcome such barriers are more likely to survive. The concentration of start-ups in a region can create positive feedback, through demonstration effects and by increasing the demand for, and experience of, supporting institutions, such as venture capitalists, legal services and contract research and production, thereby improving the environment and probability of success of subsequent start-ups. Failures are an inherent part of such a system, and provided a steady stream of new venture proposals exists and venture capitalists maintain diverse investment portfolios and are ruthless with failed ventures, the system continues to learn from both good and bad investments.

However, the unique circumstances of the US environment in the 1970s and 1980s question the generalizability of the lessons of Silicon Valley and Route 128. Specifically, the role of the defence industry investment, liberal tax regimes and sources of venture capital were unique. In addition, it is important to distinguish the evolutionary growth of such regional clusters of NTBFs, from more recent attempts to establish science parks based around universities. For example, success of science parks in Europe and Asia in the 1990s, and other attempts to emulate the early US experiences, has been limited.[75] This is partly because NTBFs are often very unwilling to share their knowledge with other firms or organizations, including universities. A survey comparing high-technology firms located on and off university science parks concluded that there were no statistically significant differences between their technological inputs, such as expenditure on R&D, and outputs, such as new products and patents.[76]

RESEARCH NOTE   Factors influencing venture success

A study of 11 259 new technology ventures in the USA over a period of five years found that 36% survived after four years, and 22% after five years. To try to explain the success and failure of these ventures, the researchers reviewed 31 other key studies of technology ventures, and found only eight factors that were consistently found to influence success:

1. *Value chain management* – cooperation with suppliers, distribution, agents and customers.
2. *Market scope* – variety of customers and market segments, and geographic reach.
3. *Firm age* – number of years in existence.
4. *Size of founding team* – likely to bring additional and more diverse expertise to the ventures, and better decision making.
5. *Financial resources* – venture assets and access to funding.
6. *Founders' marketing experience* – but not technical experience, or prior experience of start-ups (see below).
7. *Founders' industry experience* – in related markets or sectors.
8. *Existence of patent rights* – in product or process technology, but R&D investment was not found to be significant.

The first three factors were by far the most significant predictors of success. However, clearly there is also some interaction between these effects, for example, the founders' marketing and industry experience is likely to influence the attention to market scope and the value chain, and patent rights make raising finance easier, and vice versa.

In addition, they found that some commonly cited factors had no effect, including founders' experience of R&D or prior start-ups. The importance of other factors depended on the precise context of the venture, for example, for independent start-ups R&D alliances and product innovation both had a negative effect on performance, but for ventures of mixed origins R&D alliances and product innovation both had a positive effect on performance.

Source: Song, M., K. Podoynitsyna, H. van der Bij and J.I.M. Halman (2008) Success factors in new ventures: a meta-analysis. *Journal of Product Innovation Management,* **25,** 7–27.

University incubators

The creation and sharing of intellectual property is a core role of a university, but managing it for commercial gain is a different challenge. Most universities with significant commercial research contracts understand how to license, and the roles of all parties – the academics, the university and the commercial organization – are relatively clear. In particular, the academic will normally continue with the research whilst possibly having a consultancy arrangement with the commercial company. However, forming an independent company is a different matter. Here both the university and the scientist must agree that spin-out is the most viable option for technology commercialization and must negotiate a spin-out deal. This may include questions of, for example, equity split, royalties, academic and university investment in the new venture, academic secondment, identification and transfer of intellectual property and use of university resources in the start-up phase. In short, it is complicated. As Chris Evans, founder of Chiroscience (see Case study 10.11) and Merlin Ventures notes: 'Academics and universities . . . have no management, no muscle, no vision, no business plan and that is 90% of the task of exploiting science and taking it to the market place. There is a tendency for universities to think, "we invented the thing so we are already 50% there". The fact is they are 50% to nowhere' (Times Higher, 27 March 1998). A characteristically provocative statement, but it does highlight the gulf between research and successful commercialization. Many universities have accepted and followed the fashion for the commercial exploitation of technology, but typically put too much emphasis on the importance of the technology and ownership of the intellectual property, and 'fail to recognize the importance and sophistication of the business knowledge and expertise of management and other parties who contribute to the non-technical aspects of technology shaping and development . . . the linear model gives no insight into the interplay of technology push and market pull'.[77]

CASE STUDY 10.11

Chris Evans and Chiroscience

Chiroscience plc is one of the nearly 20 biotechnology firms founded by the microbiologist/entrepreneur Chris Evans. Evans, PhD, and since OBE, formed his first new venture, Enzymatix Ltd, in 1987, aged 30. His business plan was rejected by venture capitalists, so he was forced to sell his house for £40 000 to raise the initial finance. Subsequent finance of £1 million was provided by the commodities group Berisford International, but following financial problems in the property market, the company was divided into Celsis plc, which makes contamination testing equipment, and Chiroscience, which exploits chiral technology, the basis of which is that most molecules have mirror images that have different properties, essentially a right-hand sense and a left-hand sense. Isolating the more effective mirror image in an existing drug formulation can improve its efficacy, or reduce unwanted side effects.

Chiroscience was formed in 1990, other directors being recruited from large established pharmaceutical firms such as Glaxo, SmithKline Beecham and Zeneca. The company was floated on the London Stock Exchange in 1994. This was only possible because in 1992 the Stock Exchange relaxed its requirements for market entry, and no longer required three consecutive years' profits before listing. The biotechnology company applies chiral technology to the purification of existing drugs and design of new drugs. Chiroscience has three potential applications of chiral technology: first, and most immediately, the improvement of existing drugs by isolating the most effective sense of molecules; second, the development of alternative processes for the production of existing drugs as they come off patent; and finally, the design of new drugs by means of single isomer technology.

Chiroscience was the first British biotechnology firm to be granted approval for sale of a new product, Dexketoprofen, in 1995. This is a non-steroidal anti-inflammatory drug, based on a right-handed version of the older drug ketoprofen. The drug is marketed by the Italian firm Menarini. Chiroscience has been involved in a number of collaborative development and marketing deals. In 1995 it formed an alliance with the Swedish pharmaceutical group Pharmacia, to develop and market its local anaesthetic, Levobupivacaine. It also forged a more general strategic alliance with Medeva, the pharmaceutical group which performs no primary research, but specializes in taking products to market.

Biotechnology stocks are more volatile than most other investments, and it is difficult to use conventional techniques to assess their current value or future potential. Expenditure on R&D in the initial years typically results in significant losses, and sales may be negligible for up to 10 years. Therefore there are no price–earnings ratios or future revenues to discount. For example, in its first two years after flotation Chiroscience reported cumulative losses of £3.7 million, due largely to research spending of £12.4 million. Nevertheless, Chiroscience has outperformed the financial markets, and most other biotechnology stock. The company was floated in 1994 at 150p, and quickly fell to below 100p. However, by December 1995 shares had reached 364p. As a result, Chris Evans's personal fortune was estimated to have reached £50 million by 1995.

In January 1999 Chiroscience merged with Celltech to form Celltech Chiroscience, which subsequently acquired Medeva to become the Celltech Group. The new company has some 400 research staff, an R&D budget of £51 million and adds much-needed sales and marketing competencies with a sales force of 550. Celltech Group is three times the size of Chiroscience, and reached a market capitalization of £3 billion in 2000. It is one of the few British biotechnology companies to gain regulatory approval for its products in the USA, and the first to achieve profitability. Sir Chris Evans (he was knighted in 2001) now runs the biotechnology venture capital firm Merlin Biosciences.

Since the mid-1980s the role of universities in the commercialization of technology has increased significantly. For example, the number of patents granted to US universities doubled between 1984 and 1989, and doubled again between 1989 and 1997. In 1979 the number of patents granted to US universities was only 264, compared to 2436 in 1997. There are a number of explanations for this significant increase in patent activity. Changes in government funding and intellectual property law played a role, but detailed analysis indicates that the most significant reason was technological opportunity. For example, changes in funding and law in the 1980s clearly encouraged many more universities to establish licensing and technology transfer departments, but the impact of these has been relatively small. For example, there is strong evidence that the scientific and commercial quality of patents has fallen since the mid-1980s as a result of these policy changes, and that the distribution of activity has a very long tail. Measured in terms of the number of patents held or exploited, or by income from patent and software licenses, commercialization of technology is highly concentrated in a small number of elite universities, which were highly active prior to changes to funding policy and law: the top 20 US universities account for 70% of the patent activity.[78] Moreover, at each of these elite universities a very small number of key patents account for most of the licensing income, the five most successful patents typically account for 70 to 90% of total income.[79] This suggests that a (rare) combination of research excellence and critical mass is required to succeed in the commercialization of technology. Nonetheless, technological opportunity has reduced some of the barriers to commercialization. Specifically, the growing importance of developments in the biosciences and software present new opportunities for universities to benefit from the commercialization of technology.

University spin-outs are an alternative to exploitation of technology through licensing, and involve the creation of an entirely new venture based upon intellectual property developed within the university. Estimates vary, but between 3 and 12% of all technologies commercialized by universities are via new ventures. As with licensing, the propensity and success of these ventures varies significantly. For example, MIT and Stanford University each create around 25 new start-ups each year, whereas Columbia and Duke Universities rarely generate any start-up companies. Studies in the USA suggest that the financial returns to universities are much higher from spin-out companies than from the more common licensing approach. One study estimated that the average income from a university license was $63 832, whereas the average return from a university spin-out was more than 10 times this – $692 121. When the extreme cases were excluded from the sample, the return from spin-outs was still $139 722, more than twice that for a licence.[80] Apart from these financial arguments, there are other reasons why forming a spin-out company may be preferable to licensing technology to an established company:

- No existing company is ready or able to take on the project on a licensing basis.
- The invention consists of a portfolio of products or is an 'enabling technology' capable of application in a number of fields.
- The inventors have a strong preference for forming a company and are prepared to invest their time, effort and money in a start-up.

As such they involve the 'academic entrepreneur' more fully in the detail of creating and managing a market entry strategy than is the case for other forms of commercialization. They also require major career decisions for the participants. Consequently, they highlight most clearly the dilemmas faced as the scientist tries to manage the interface between academe and industry. The extent to which an individual is motivated to attempt the launch of a venture depends upon three related factors – antecedent influences, the incubator organization and environmental factors:

- *Antecedent influences,* often called the 'characteristics' of the entrepreneur, include genetic factors, family influences, educational choices; and previous career experiences all contribute to the entrepreneur's decision to start a venture.
- Individual incubator experiences immediately prior to start-up include the nature of the physical location, the type of skills and knowledge acquired, contact with possible fellow founders, the type of new venture or small business experience gained.
- Environmental factors include economic conditions, availability of venture capital, entrepreneurial role models, availability of support services.

There are relatively few data on the characteristics of the academic entrepreneur, partly due to the low numbers involved, but also because the traditional context within which they have operated, particularly as they apply to IPR and equity sharing, has meant that many have been unwilling to be researched. It is also probable that this is compounded by inadequate university data capture systems. Nevertheless, it is clear that in the USA, scientists and engineers working in universities have long become disposed towards the commercialization of research. A study of American universities in 1990 observed: '*Over the last eight years we have seen increasing legitimizing of university–industry research interactions.*'[81] A study of 237 scientists working in three large national laboratories in the USA found clear differences between the levels of education in inventors in national laboratories and those in a study of technical entrepreneurs from MIT.[82] The study found significant differences between entrepreneurs and non-entrepreneurs in terms of situational variables such as the level of involvement in business activities outside the laboratory or the receipt of royalties from past inventions. A study of scientists in four research institutes in the UK identified a relationship between attitudes to industry, number of industry links and commercial activity.[83] This begs the question: what is the direction of causation? Do entrepreneurial researchers seek more links outside the organization, or do more links encourage entrepreneurial behaviour?

Entrepreneurs, academic or otherwise, require a supportive environment. Surveys indicate that two-thirds of university scientists and engineers now support the need to commercialize their research, and half the need for start-up assistance.[84] There are two levels of analysis of the university environment: the formal institutional rules, policies and structures; and the 'local norms' within the individual department. There are a number of institutional variables which might influence academic entrepreneurship:

1. Formal policy and support for entrepreneurial activity from management.
2. Perceived seriousness of constraints to entrepreneurship, e.g. IPR issues.
3. Incidence of successful commercialization, which demonstrates feasibility and provides role models.

Formal policies to encourage and support entrepreneurship can have both intended and unintended consequences. For example, a university policy of taking an equity stake in new start-ups in return for paying initial patenting and licensing expenses seems to result in a higher number of start-ups, whereas granting generous royalties to academic entrepreneurs appears to encourage licensing activity, but tends to suppress significantly the number of start-up companies.[85] Similarly, encouraging commercially oriented, or industry-funded research, appears to have no effect on the number of start-ups, whereas a university's intellectual eminence has a very strong positive effect. A reason for the former effect is that typically such research restricts the ownership of formal intellectual property, and narrows the choice of route to market. There are two reasons for this: more prestigious universities typically attract better researchers and higher funding; and other commercial investors use the prestige or reputation of the institution as a signal or indicator of quality. In addition, some very common university policies appear to have little or no positive effect on the number of subsequent successes of start-ups, including university incubators and local venture capital funding. Moreover, badly targeted and poorly monitored financial support may encourage 'entrepreneurial academics', rather than academic entrepreneurs – scientists in the public sector who are not really committed to creating start-ups, but rather are seeking alternative support for their own research agendas.[86] This can result in start-ups with little or no growth prospects, remaining in incubators for many years.

A survey of 778 life scientists working in 40 US universities concluded that developing formal policies may send a signal, but the effect on individual behaviour depends very much on whether these policies are reinforced by behavioural expectations.[87] They found that individual characteristics and local norms appear to be equally effective predictors of entrepreneurial activity, but only provided 'weak and unsystematic predictions of the forms of entrepreneurship'. Where successful, this can create a virtuous circle, the demonstration effect of a successful spin-out encouraging others to try. This leads to clusters of spin-outs in space and time, resulting in entrepreneurial departments or universities, rather than isolated entrepreneurial academics. Local norms or culture at the departmental level will influence the effectiveness of formal policies by providing a strong mediating effect between the institutional context and individual perceptions. Local norms evolve through self-selection during recruitment, resulting in staff with similar personal values and behaviour, and reinforced by peer pressure or behavioural social-ization resulting in a convergence of personal values and behaviour. However, there is a real potential conflict between the pursuit of knowledge and its commercial exploitation, and a real danger of lower-ing research standards exists. Therefore it is essential to have explicit guidelines for the conduct of business in a university environment:[88]

1. Specific guidelines on the use of university facilities, staff and students and intellectual property rights.
2. Specific guidelines for, and periodic reviews of, the dual employment of scientist entrepreneurs, including permanent part-time positions.
3. Mechanisms to resolve issues of financial ownership and the allocation of research contracts between the university and the venture.

A recent study of nine university spin-off companies in the UK identified a number of common stages of development, each demanding different capabilities, resources and support:[89]

- *Research phase* – all of the academic entrepreneurs were at the forefront of their respective fields, were focused on their research, respected by their academic communities and had high levels of publica-tion. This contributes to the generation of know-how and the likelihood of generating more formal intellectual property.
- *Opportunity framing phase* – the development of an understanding of how best to create commercial value from the science. In most cases the opportunities are defined imprecisely, targeted ambiguously and prove impracticable. In particular, there is a need to define the complementary resources necessary for commer-cialization, including human, financial, physical and technological resources. Therefore the framing process is usually iterative and slow, taking many months or even years.
- *Pre-organization phase* – decisions made at this early stage often have a significant impact upon the entire future success of the venture, since they direct the path of development and constrain future options. At this stage access to networks of expertise and prior entrepreneurial experience are critical.
- *Re-orientation phase* – once the venture has gained sufficient resource and credibility to start-up, the venture must 'repackage' its technology and acquire new information and resources to create some-thing of value to some target customer group.
- *Sustainable returns phase* – with an emphasis on business capabilities, winning orders, selling products or services, and making a return. This demands professional management, greater financial resources and a broader range of capabilities.

At each of these stages there are different significant challenges to overcome in order to make a success-ful transition to the next stage, what the researchers call 'critical junctures':

- *Opportunity recognition* – at the interface of the research and opportunity framing phases. This requires the ability to connect a specific technology or know-how to a commercial application, and is based on a rather rare combination of skill, experience, aptitude, insight and circumstances. A key issue here is the ability to synthesize scientific knowledge and market insights, which increases with the entrepreneur's social capital – linkages, partnerships and other network interactions.

• *Entrepreneurial commitment* – acts and sustained persistence that bind the venture champion to the emerging business venture. This often demands difficult personal decisions to be made – for example, whether or not to remain an academic – as well as evidence of direct financial investments to the venture.

• *Venture credibility* – is critical for the entrepreneur to gain the resources necessary to acquire the finance and other resources for the business to function. Credibility is a function of the venture team, key customers and other social capital and relationships. This requires close relationships with sponsors, financial and other, to build and maintain awareness and credibility. Lack of business experience, and failure to recognize their own limitations are a key problem here. One solution is to hire the services of a 'surrogate entrepreneur'. As one experienced entrepreneur notes, '*The not so smart or really insecure academics want their hands over everything. These prima donnas make a complete mess of things, get nowhere with their companies and end up disappointed professionally and financially.*'

In the UK, the Lambert Review of Business–University Collaboration reported in December 2003. It reviewed the commercialization of intellectual property by universities in the UK, and also made international comparisons of policy and performance. The UK has a similar pattern of concentration of activity as the USA: in 2002 80% of UK universities made no patent applications, whereas 5% filed 20 or more patents; similarly, 60% of universities issued no new licences, but 5% issued more than 30. However, in the UK there has been a bias towards spin-outs rather than licensing, which the Lambert Report criticizes. It argues that spin-outs are often too complex and unsustainable, and of low quality – a third in the UK are fully funded by the parent university and attract no external private funding. In 2002 universities in the UK created over 150 new spin-out firms, compared to almost 500 by universities in the USA; the respective figures for new licenses that year were 648 and 4058. As a proportion of R&D expenditure, this suggests that British universities place greater emphasis on spin-outs than their North American counterparts, and less on licensing. Lambert argues that universities in the UK may place too high a price on their intellectual property, and that contracts often lack clarity of ownership. Both of these problems discourage businesses from licensing intellectual property from universities, and may encourage universities to commercialize their technologies through wholly owned spin-outs.

Growth and performance of innovative small firms

There has been a great deal of economic and management research on small firms, but much of this has been concerned with the contribution all types of small firms make to economic, employment or regional development. Relatively little is known about innovation in small firms, or the more narrow issue of the performance of NTBFs. As demonstrated by the preceding discussion, almost all research on innovative small firms has been confined to a small number of high-technology sectors, principally microelectronics and more recently biotechnology. A notable exception is the survey of 2000 SMEs conducted by the Small Business Research Centre in the UK. The survey found that 60% of the sample claimed to have introduced a major new product or service innovation in the previous five years.[90] Whilst this finding demonstrates that the management of innovation is relevant to the majority of small firms, it does not tell us much about the significance of such innovations, in terms of research and investment, or subsequent market or financial performance.

Research over the past decade or so suggests that the innovative activities of SMEs exhibit broadly similar characteristics across sectors.[91] They:

- are more likely to involve product innovation than process innovation
- are focused on products for niche markets, rather than mass markets
- will be more common amongst producers of final products, rather than producers of components
- will frequently involve some form of external linkage
- tend to be associated with growth in output and employment, but not necessarily profit.

The limitations of a focus on product innovation for niche or intermediate markets were discussed earlier, in particular problems associated with product planning and marketing, and relationships with lead customers and linkages with external sources of innovation. Where an SME has a close relationship with a small number of customers, it may have little incentive or scope for further innovation, and therefore will pay relatively little attention to formal product development or marketing. Therefore SMEs in such dependent relationships are likely to have limited potential for future growth, and may remain permanent infants or subsequently be acquired by competitors or customers.[92] Moreover, an analysis of the growth in the number of NTBFs suggests that the trend has as much to do with negative factors, such as the downsizing of larger firms, as it does with more positive factors such as start-ups.[93]

Innovative SMEs are likely to have diverse and extensive linkages with a variety of external sources of innovation, and in general there is a positive association between the level of external scientific, technical and professional inputs and the performance of an SME.[94] The sources of innovation and precise types of relationship vary by sector, but links with contract research organizations, suppliers, customers and universities are consistently rated as being highly significant, and constitute the 'social capital' of the firm. However, such relationships are not without cost, and the management and exploitation of these linkages can be difficult for an SME, and overwhelm the limited technical and managerial resources of SMEs.[95] As a result, in some cases the cost of collaboration may outweigh the benefits[96] and in the specific case of collaboration between SMEs and universities there is an inherent mismatch between the short-term, near-market focus of most SMEs and the long-term, basic research interests of universities.[97] There are a number of structured ways to identify the need for external expertise and partners.

In terms of innovation, the performance of SMEs is easily exaggerated. Early studies based on innovation counts consistently indicated that when adjusted for size, smaller firms created more new products than the larger counterparts. However, methodological shortcomings appear to undermine this clear message. When the divisions and subsidiaries of larger organizations are removed from such samples,[98] and the innovations weighted according to their technological merit and commercial value, the relationship between firm size and innovation is reversed: larger firms create proportionally more significant innovations than SMEs.[99] The amount of expenditure by SMEs on design and engineering has a positive effect on the share of exports in sales,[100] but formal R&D by SMEs appears to be only weakly associated with profitability,[101] and is not correlated with growth.[102] Similarly, the high growth rates associated with NTBFs are not explained by R&D effort,[103] and investment in technology does not appear to discriminate between the success and failure of NTBFs. Instead, other factors have been found to have a more significant effect on profitability and growth, in particular the contributions of technically qualified owner managers and their scientific and engineering staff, and attention to product planning and marketing.[104]

A large study of start-ups in Germany found that the founder's level of management experience was a significant predictor of the growth of a venture. However, innovation, broadly defined, was found to be statistically three times more important to growth than founder attributes or any other of the factors measured.[105] Another study, of Korean technology start-ups, also found that innovativeness, defined as a propensity to engage in new idea generation, experimentation and R&D, was associated with performance. So was proactiveness, defined as the firm's approach to market opportunities through active market research and the introduction of new products and services.[106] The same study also found that what it referred to as sponsorship-based linkages had a positive effect on performance. This included links with venture capital firms, which reinforces the developmental role these can play, as discussed earlier.

The size and location of NTBFs also has an effect on performance. Geographic closeness increases the likelihood of informal linkages and encourages the mobility of skilled labour across firms. However, the probability of a start-up benefiting from such local knowledge exchanges appears to decrease as the venture grows.[107] This growing inability to exploit informal linkages is a function of organizational size, not the age of the venture, and suggests that as NTBFs grow and become more complex, they begin to suffer many of the barriers to innovation discussed in Chapter 3, and therefore the explicit processes and tools to help overcome these (see Chapter 12) become more relevant. This interpretation is reinforced by other, cross-sectional research. Larger SMEs are associated with a greater spatial reach of innovation-related linkages, and with the introduction of more novel product or process innovations for international markets. In contrast, smaller SMEs are more embedded in local networks, and are more likely to be engaged in incremental innovations for the domestic market.[108] It is always difficult to untangle cause and effect relationships from such associations, but it is plausible that as the more innovative start-ups begin to outgrow the resources of their local networks, they actively replace and extend their networks, which both creates the opportunity and demand for higher levels of innovation. Conversely, the less innovative start-ups fail to move beyond their local networks, and therefore are less likely to have either the opportunity or need for more radical innovation.

In short, most of what we have discussed in this book about managing innovation is relevant to innovative small firms, but more research needs to be done on how to manage the particular problems they face. As we have argued throughout the book, we believe that there is a generic core innovation process which represents management 'best practice', but that this has to be modified in different organizational, technological and market contingencies. In the specific case of small innovative firms, best practice would include scanning the external environment and developing appropriate relationships with sources of innovation and lead users. However, different contingencies will demand different innovation strategies. For example, a study of 116 software start-ups identified five factors that affected success: level of R&D expenditure; how radical new products were; the intensity of product upgrades; use of external technology; and management of intellectual property.[109] In contrast, a study of 94 biotechnology start-ups found that three factors were associated with success: location within a significant concentration of similar firms; quality of scientific staff (measured by citations); and the commercial experience of the founder.[110] The number of alliances had no significant effect on success, and the number of scientific staff in the top management team had a negative association, suggesting that the scientists are best kept in the laboratory. Other studies of biotechnology start-ups confirm this pattern, and suggest that maintaining close links with universities reduces the level of R&D expenditure needed, increases the number of patents produced, and moderately increases the number of new products under development. However, as with more general alliances, the *number* of university links has no effect on the success or performance of biotechnology start-ups, but the *quality* of such relationships does.[111]

Such sector-specific studies confirm that the environment in which small firms operate significantly influences both the opportunity for innovation, in a technological and market sense, and the most appropriate strategy and processes for innovation. For example, an NTBF may have a choice of whether to use its intellectual assets by translating its technology into product and services for the market, or alternatively it may exploit these assets through a larger, more established firm, through licensing, sale of IPR or by collaboration. More specifically the NTBF needs to consider two environmental factors:[112]

- *Excludability* – to what extent the NTBF can prevent or limit competition from incumbents who develop similar technology?
- *Complementary assets* – to what extent do the complementary assets – production, distribution, reputation, support, etc. – contribute to the value proposition of the technology?

Combining these two dimensions creates four strategy options:

- *Attacker's advantage* – where the incumbent's complementary assets contribute little or no value, and the start-up cannot preclude development by the incumbent (e.g. where formal intellectual property is irrelevant, or enforcement poor), NTBFs will have an opportunity to disrupt established positions, but technology leadership is likely to be temporary as other NTBFs and incumbents respond, resulting in fragmented niche markets in the longer term. This pattern is common in computer components businesses.
- *Ideas factory* – in contrast, where incumbents control the necessary complementary assets, but the NTBF can preclude effective development of the technology by incumbents, cooperation is essential. The NTBF is likely to focus on technological leadership and research, with strong partnerships downstream for commercialization. This pattern tends to reinforce the dominance of incumbents, with the NTBFs failing to develop or control the necessary complementary assets. This pattern is common in biotechnology.
- *Reputation-based* – where incumbents control the complementary assets, but the NTBF cannot prevent competing technology development by the incumbents, NTBFs face a serious problem of disclosure and other contracting hazards from incumbents. In such cases the NTBF will need to seek established partners with caution, and attempt to identify partners with a reputation for fairness in such transactions. Cisco and Intel have both developed such a reputation, and are frequently approached by NTBFs seeking to exploit their technology. This pattern is common in capital-intensive sectors such as aerospace and automobiles. However, these sectors have a lower 'equilibrium', as established firms have a reputation for expropriation, therefore discouraging start-ups.
- *Greenfield* – where incumbents' assets are unimportant, and the NTBF can preclude effective imitation, there is the potential for the NTBF to dominate an emerging business. Competition or cooperation with incumbents are both viable strategies, depending upon how controllable the technology is – for example, through establishing standards or platforms, and where value is created in the value chain.

A high proportion of new ventures fail to grow and prosper. Estimates vary by type of business and national context, but typically 40% of new businesses fail in their first year, and 60% within the first two. In other words, around 40% survive the first two years. Common reasons for failure include:

- Poor financial control.
- Lack of managerial ability or experience.
- No strategy for transition, growth or exit.

There are many ways that a new venture can grow and create additional value:

- Organic growth through additional sales and diversification.
- Acquisition of or merger with another company.
- Sale of the business to another company, or private equity firm.
- An initial public offering (IPO) on a stock exchange.

For example, The UK Sunday Times Profit Track estimates that of the 500 fastest growing private firms in the UK, over five years around 100 have merged with or been acquired by other companies or private equity firms, but only 10 or so have been floated (Table 10.16). Some of the best-performing have been based upon ICT, others on service innovation. A separate survey of technology-based start-ups reveals a dominance of web-based businesses, which demonstrates how much has changed since the Internet bubble burst (Case study 10.12).

TABLE 10.16   **Some of the fastest growing private firms in the UK**

Name	Date founded	Business	Profit, 2005, £ million	Annual growth, %
Betfair	1999	Online bookmaker	23.2	146
Invotec	2001	Circuit boards	3.4	88
Azzurri	2000	Telecoms services	8.0	77
UNiCOM	1998	Telecoms services	3.3	86
Regard	1994	Care homes	4.0	76
Spearhead	2000	Farm produce	5.2	74
Baxter	2000	Contract caterer	4.1	66
Ingenious Media	1998	Media adviser	35.7	56
INEOS	1998	Chemicals	191	56
ESRI	1993	Software	5.2	79

Source: Sunday Times Profit Track, April 2006.

CASE STUDY 10.12

Technology-based high-growth ventures

Since 2001 the Oxford-based research company Fast Track has complied a report for the newspaper the *Sunday Times* on the top 100 technology-based new ventures in the UK, sponsored by consultants PriceWaterhouseCoopers and Microsoft.

Following the collapse of the dotcom boom and bust, the annual survey provides an excellent barometer of the more robust and consistent technology-based new ventures, which, without reaching the headlines, continue to be created, grown and prosper.

Of the 100 firms studied, 48 have been funded by venture capital or private equity funds. As might be expected, many of the most successful new ventures are based on software or telecommunications technologies, so-called information communication telecommunications (ICT) technologies, but the commercial applications are increasingly dynamic and diverse, including gaming, gambling, music, film, fashion and education. Although most of these firms are only five or six years old, annual sales average £5 million, with annual growth of 60%. Examples include:

- Gamesys, a gaming website operator created in 2001, now with 50 staff and sales of £9.4 million.
- The Search Works, an advertising consultant for search engines, founded in 1999, now employing more than 50 staff, with sales of $18.6 million.
- Redtray, an e-learning software developer, formed in 2002, now has 30 staff and sales of £4.5 million.
- Ocado, the delivery business for online orders to supermarket Waitrose, created in 2000, and now employing almost 1000 staff, with 3 million deliveries each week, and turnover of $143 million.
- Wiggle, an online retailer of sports goods, founded in 1998, now with 50 staff and sales of £9.2 million.
- Betfair, an online bookmaker and betting website, established in 1999, with turnover of £107 million and employing more than 400 staff.

Source: Sunday Times Tech Track 100, 24 September, 2006, www.fasttrack.co.uk, www.pwc.com.

Summary and further reading

A venture represents an opportunity to grow new businesses based on new technologies, products or markets, where conventional processes for new product or service development are inappropriate. In this chapter we have explored the rationale, characteristics and management of venturing to exploit innovation, ranging from internal corporate ventures, through joint ventures and alliances, to new ventures and spin-out businesses.

Like any new business, a venture requires a clear business plan, strong champion and sufficient resources. Any venture champion must identify the opportunity for a new venture, raise the finance and manage the development and growth of the business. The individuals involved in internal and external new ventures are likely to have similar backgrounds, levels of education and personalities; they tend to be highly motivated and demand a high level of autonomy. However, unlike external entrepreneurs, the corporate entrepreneur requires a high degree of political and social skill. This is because the corporate entrepreneur has the advantage of the financial, technical and marketing resources of the parent firm, but must deal with internal politics and bureaucracy.

Joint ventures are a more special case. Essentially, firms collaborate to reduce the cost, time or risk of access to unfamiliar technologies or markets. The precise form of collaboration will be determined by the motives and preferences of the partners, but their choice will be constrained by the nature of the technologies and markets, specifically the degree of complexity and tacitness. The success of an alliance depends on a number of factors, but organizational issues dominate, such as the degree of mutual trust and level of communication. The transaction costs approach better explains the relationship between the reason for collaboration, and the preferred form and structure of an alliance. The strategic learning approach better explains the relationship between the management and organization of an alliance and the subsequent outcomes.

There are thousands of books and journal articles on the more general subject of entrepreneurship, but relatively little has been produced on the more specific subject of new technology-based entrepreneurism. Ed. Roberts's *Entrepreneurs in High Technology: Lessons from MIT and Beyond* (Oxford University Press, 1991) is an excellent study of the MIT experience, although perhaps places too much emphasis on the characteristics of individual entrepreneurs. For a broader analysis of technology ventures in the USA see Martin Kenny (ed.), *Understanding Silicon Valley: Anatomy of an Entrepreneurial Region* (Stanford University Press, 2000). For a more recent analysis of technological entrepreneurs, see *Inventing Entrepreneurs: Technology Innovators and their Entrepreneurial Journey*, by Gerry George and Adam Bock (Prentice Hall, 2008). Ray Oakey's *High-Technology New Firms* (Paul Chapman, 1995) is a similar study of NTBFs in the UK, but places greater emphasis on how different technologies constrain the opportunities for establishing NTBFs, and affect their management and success. For a review of recent research on the broader issue of innovative small firms see 'Small firms, R&D, technology and innovation: a literature review' by Kurt Hoffman *et al.*, published in *Technovation*, **18** (1), 39–55, 1998. A special issue of the *Strategic Management Journal* (volume 22, July 2001) examined entrepreneurial strategies, and includes a number of papers on technology-based firms, and a special issue of the journal *Research Policy* (volume 32, 2003) features papers on technology spin-offs and start-ups. A special issue of the *Journal of Product Innovation Management* examined technology commercialization and entrepreneurship (volume 25, 2008), and a special issue of *Industrial and Corporate Change* focused on university spin-outs (**16** (4), 2007). Francois Therin provides a summary of selected recent research in *Handbook of Research on Techno-Entrepreneurship* (Edward Elgar, 2007).

On the subject of internal corporate venturing Burgelman and Sayles's *Inside Corporate Innovation* (Macmillan, London, 1986) remains the best combination of theory and case studies, but the more recent book by Block and MacMillan, *Corporate Venturing: Creating New Businesses Within the Firm* (Harvard Business School Press, 1995), provides a better review of research on internal corporate ventures. More recent books which include some interesting examples of venturing in the information and telecommunications sectors are *Webs of Innovation* by Alexander Loudon (FT.com, 2001), which despite its title has several chapters related to venturing, and Henry Chesbrough's *Open Innovation* (Harvard Business School Press, 2003), which includes case studies of the usual suspects such as IBM, Xerox,

Intel and Lucent. The book *Inventuring* by W. Buckland, A. Hatche and J. Birkinshaw (McGraw-Hill, 2003) is also a good review of corporate venture initiatives, including those at GE, Intel and Lucent, which suggest a range of successful venture models and common reasons for failure. The text *Corporate Entrepreneurship* by Paul Burns provides a useful framework and case examples (Palgrave Macmillan, 2008).

For joint ventures and collaboration, a good review of the theoretical issues is provided by a compilation of papers edited by Rod Coombs *et al.*, *Technological Collaboration* (Edward Elgar, 1996). The book includes a discussion of both economic and sociological analyses of collaboration, including transaction costs and evolutionary theories, as well as a review of the more recent network approaches. The literature on innovation networks is large and growing, but the following provide a good introduction: O. Jones, S. Conway and F. Steward, *Social Interaction and Organizational Change: Aston Perspectives on Innovation Networks* (Imperial College Press, London, 2001); *International Journal of Innovation Management*, Special Issue on Networks, **2** (2) (1998); R. Gulati, 'Alliances and networks', *Strategic Management Journal*, **19**, 293–317 (1998); and F. Belussi and F. Arcangeli, 'A typology of networks', *Research Policy*, **27**, 415–28 (1998). The open innovation movement includes a lot of relevant work on collaboration and networks, and Henry Chesbrough, Wim Vanhaverbeke and Joel West have edited a good overview of the main research themes in *Open Innovation: Researching a New Paradigm* (Oxford University Press, 2008).

For a less academic treatment of alliances, Bleeke and Ernst provide a practical guide, albeit a little dated, for managers of collaborative projects in *Collaborating to Compete* (John Wiley & Sons, Inc., 1993), written by two management consultants at McKinsey & Co., and based on a survey of international alliances and acquisitions. In *Alliance Advantage* (Harvard Business School Press, 1998) Yves Doz and Gary Hamel develop a framework to help understand and better manage alliances, drawing on their earlier work on learning through alliances. On the more specific subject of customer–supplier alliances, Jordan Lewis provides a practical guide based on studies of a number of American and British present and past exemplars such as Motorola and Marks & Spencer in *The Connected Corporation* (Free Press, 1995). The example of Marks & Spencer is now even more interesting as it demonstrates some of the limitations of supplier partnerships. More academic and rigorous treatments of customer–supplier alliances are provided by Richard Lamming in *Beyond Partnership* (Prentice-Hall, 1993) and Toshihiro Nishiguchi in *Strategic Industrial Sourcing: The Japanese Advantage* (Oxford University Press, 1994), both based mainly on the experience of the automobile industry.

Web links

Here are the full details of the resources available on the website flagged throughout the text:

 Case studies:
ihavemoved.com

 Interactive exercises:
Partner search
Acquiring technological knowledge

Tools:
Value analysis
Value stream analysis

Video podcast:

Audio podcast:
David Hall on entrepreneurship

References

1. **Roberts, E.** (1991) *Entrepreneurs in High Technology: Lessons from MIT and Beyond*, Oxford University Press, Oxford.
2. **Oakey, R.** (1995) *High-Technology New Firms*, Paul Chapman, London.
3. **Delmar, F. and S. Shane** (2003) Does business planning facilitate the development of new ventures? *Strategic Management Journal*, **24**, 1165–85.
4. **Drucker, P.** (1985) *Innovation and Entrepreneurship*, Harper & Row, New York.
5. **Withers** (1997) *Window on Technology: Corporate Venturing in Practice*, Withers, London.
6. **Loudon, A.** (2001) *Webs of Innovation: The Networked Economy Demands New Ways to Innovate*, FT.com, Pearson Education, Harlow.
7. **Harding, R.** (2000) Venture capital and regional development: towards a venture capital system. *Venture Capital*, **2** (4), 287–311.
8. **Binding, K., C. McCubbin and L. Doyle** (1998) *Technology Transfer in the UK Life Sciences*, Arthur Andersen, London.
9. **Tidd, J. and S. Barnes** (1999) Spin-in or spin-out? Corporate venturing in life sciences. *International Journal of Entrepreneurship and Innovation*, **1** (2), 109–16.
10. **Lockett, A., G. Murray and M. Wright** (2002) Do UK venture capitalists still have a bias against investment in new technology firms? *Research Policy*, **31**, 1009–30.
11. **Baum, J. and B. Silverman** (2004) Picking winners or building them? Alliance, intellectual and human capital as selection criteria in venture financing and performance of biotechnology start-ups. *Journal of Business Venturing*, **19**, 411–36.
12. **Burgelman, R.** (1984) Managing the internal corporate venturing process. *Sloan Management Review*, **25** (2), 33–48.
13. **Dess, G., G. Lumpkin and J. Covin** (1997) Entrepreneurial strategy making and firm performance. *Strategic Management Journal*, **18** (9), 677–95.
14. **Tidd, J. and S. Taurins** (1999) Learn or leverage? Strategic diversification and organisational learning through corporate ventures. *Creativity and Innovation Management*, **8** (2), 122–9.
15. **Christensen, C. and M. Raynor** (2003) *The Innovator's Solution: Creating and Sustaining Successful Growth*, Harvard Business School Press, Boston, MA.
16. **Kanter, R.** (1985) Supporting innovation and venture development in established companies. *Journal of Business Venturing*, **1**, 47–60.
17. **Huber, G.** (1996) Organizational learning: the contributing processes and the literatures. In M. Cohen and L. Sproull, eds, *Organizational Learning* (pp. 124–62), Sage, London.

18. **Block, Z. and I. MacMillan** (1993) *Corporate Venturing: Creating New Businesses Within the Firm.* Harvard Business School Press, Boston, MA.

19. **Roussel, P., K. Saad and T. Erickson** (1991) *Third-Generation R&D: Managing the Link to Corporate Strategy*, Harvard Business School Press, Boston, MA.

20. **Maidique, M.** (1980) Entrepreneurs, champions and technological innovation. *Sloan Management Review*, **21** (2), 59–76.

21. **Roberts, M.** (1992) The do's and don'ts of strategic alliances. *Journal of Business Strategy*, March/April, 50–3.

22. **Burgelman, K. and L. Sayles** (1986) *Inside Corporate Innovation*, Macmillan, London.

23. **Gulati, R. and J. Garino** (2000) Get the right mix of bricks and clicks. *Harvard Business Review*, May–June, 107–66.

24. **Wolcott, R.C. and M.J. Lippitz** (2007) The four models of corporate entrepreneurship. *MIT Sloan Management Review*, **49** (1), 74–82; **Buckland, W., A. Hatche and J. Birkinshaw** (2003) *Inventuring*, McGraw-Hill, New York.

25. **Tushman, M. and C. O'Reilly** (2002) *Winning through Innovation: A Practical Guide to Leading Organizational Change and Renewal*, Harvard Business School Press, Boston, MA.

26. **Thornhill, S. and Amit, R.** (2000) A dynamic perspective of external fit in corporate venturing. *Journal of Business Venturing*, **16**, 25–50.

27. **Chesbrough, H.** (2002) The governance and performance of Xerox's technology spin-off companies. *Research Policy*, **32**, 403–21.

28. **Abetti, P.** (2002) From science to technology to products and profits: superconductivity at General Electric and Intermagnetics General (1960–1990). *Journal of Business Venturing*, **17**, 83–98.

29. **Martin, M.** (1994) *Managing Innovation and Entrepreneurship in Technology*, John Wiley & Sons, Inc., New York.

30. **Fast, N.** (1979) A visit to a new venture graveyard. *Research Management*, **22** (2), 18–22.

31. **McGee, J. and M. Dowling** (1994) Using R&D cooperative arrangements to leverage managerial experience. *Journal of Business Venturing*, **9**, 33–48.

32. **Hauschildt, J.** (1992) External acquisition of knowledge for innovations – a research agenda. *R&D Management*, **22** (2), 105–10; **Brusconi, S., A. Prencipe and K. Pavitt** (2002) Knowledge specialization and the boundaries of the firm. *Administrative Science Quarterly*, **46** (4), 597–621.

33. **Welch, J. and P. Nayak** (1992) Strategic sourcing: a progressive approach to the make or buy decision. *Academy of Management Executive*, **6** (1), 23–31.

34. **Bettis, R., S. Bradley and G. Hamel** (1992) Outsourcing and industrial decline. *Academy of Management Executive*, **6** (1), 7–21.

35. **Atuaheme-Gima, K. and P. Patterson** (1993) Managerial perceptions of technology licensing as an alternative to internal R&D in new product development: an empirical investigation. *R&D Management*, **23** (4), 327–36; **Chiesa, V., R. Manzini, and E. Pizzurno** (2004) The externalization of R&D activities and the growing market of product development services. *R&D Management*, **34**, 65–75.

36. **Robins, J. and M. Wiersema** (1995) A resource-based approach to the multibusiness firm. *Strategic Management Journal*, **16** (4), 277–300.

37. **Granstrand, O., E. Bohlin, C. Oskarsson, and N. Sjoberg** (1992) External technology acquisition in large multi-technology corporation. *R&D Management*, **22** (2), 111–33.

38. **Tyler, B. and H. Steensma** (1995) Evaluating technological collaborative opportunities: a cognitive modeling perspective. *Strategic Management Journal*, **16**, 43–70.

39. **Bidault, F. and T. Cummings** (1994) Innovating through alliances: expectations and limitations. *R&D Management,* **24** (2), 33–45.

40. **Scott Swan, K. and B. Allred** (2003) A product and process model of the technology-sourcing decision. *Journal of Product Innovation Management,* **20**, 485–96.

41. **Rothaermel, F. and D. Deeds** (2004) Exploration and exploitation alliances in biotechnology: a system of new product development. *Strategic Management Journal,* **25**, 201–21.

42. **Miotti, L. and F. Sachwald** (2003) Cooperative R&D; why and with whom? An integrated framework of analysis. *Research Policy,* **32**, 1481–99.

43. **Teher, B.** (2002) Who cooperates for innovation, and why? An empirical analysis. *Research Policy,* **31**, 947–67.

44. **Doz, Y. and G. Hamel** (1998) *Alliance Advantage: The Art of Creating Value through Partnering,* Harvard Business School Press, Boston, MA.

45. **Carr, C.** (1999) Globalisation, strategic alliances, acquisitions and technology transfer: lessons from ICL/Fujitsu and Rover/Honda and BMW. *R&D Management,* **29** (4), 405–21.

46. **Mauri, A. and G. McMillan** (1999) The influence of technology on strategic alliances. *International Journal of Innovation Management,* **3** (4), 367–78.

47. **Jay Lambe, C. and R. Spekman** (1997) Alliances, external technology acquisition, and discontinuous technological change. *Journal of Product Innovation Management,* **14**, 102–16.

48. **Duysters, G. and A. de Man** (2003) Transitionary alliances: an instrument for surviving turbulent industries? *R&D Management,* **33**, 49–58.

49. **Berg, S., J. Duncan and P. Friedman** (1982) *Joint Venture Strategies and Corporate Innovation,* Gunn & Ham, Cambridge, MA.

50. **Balakrishnan, S. and M. Koza** (1995) An information theory of joint ventures. In L. Gomez-Mejia and M. Lawless, eds, *Advances in Global High Technology Management: Strategic Alliances in High Technology* (Vol. 5, Part B, pp. 59–72), JAI Press, Greenwich, CN.

51. **Krubasik, E. and H. Lautenschlager** (1993) Forming successful strategic alliances in high-tech businesses. In J. Bleeke and D. Ernst, eds, *Collaborating to Compete* (pp. 55–65), John Wiley & Sons, Inc., New York.

52. **Duysters, G., G. Kok and M. Vaandrager** (1999) Crafting successful strategic technology partnerships. *R&D Management,* **29** (4), 343–51; Hagedoorn, J. (2002) Inter-firm R&D partnerships: an overview of major trends and patterns since 1960. *Research Policy,* **31**, 477–92; **Giarrantana, M. and S. Torrisi** (2002) Competence accumulation and collaborative ventures: evidence from the largest European electronics firms and implications for EU technological policies. In S. Lundan, ed., *Network Knowledge in International Business* (pp. 196–215), Edward Elgar, Cheltenham.

53. **Tidd, J. and M. Trewhella** (1997) Organizational and technological antecedents for knowledge acquisition and learning. *R&D Management,* **27** (4), 359–75.

54. **Rothaermel, F.** (2001) Complementary assets, strategic alliances, and the incumbent's advantage: an empirical study of industry and firms effects in the biopharmaceutical industry. *Research Policy,* **30**, 1235–51.

55. **Orsenigo, L., F. Pammolli and M. Riccaboni** (2001) Technological change and network dynamics: lessons from the pharmaceutical industry. *Research Policy,* **30**, 485–508.

56. **Nicholls-Nixon, C. and C. Woo** (2003) Technology sourcing and the output of established firms in a regime of encompassing technological change. *Strategic Management Journal,* **24**, 651–66.

57. **Harrigan, K.** (1986) *Managing for Joint Venture Success*, Lexington Books, Lexington, MA.

58. **Dacin, M., M. Hitt and E. Levitas** (1997) Selecting partners for successful international alliances. *Journal of World Business*, **32** (1), 321–45.

59. **Spekmen, R., A. Lynn, T.C. MacAvoy, and I. Forbes** (1996) Creating strategic alliances which endure. *Long Range Planning*, **29** (3), 122–47.

60. **Bleeke, J. and D. Ernst** (1993) *Collaborating to Compete*, John Wiley & Sons, Inc., New York.

61. **Tidd, J. and Y. Izumimoto** (2001) Knowledge exchange and learning through international joint ventures: an Anglo-Japanese experience. *Technovation*, **21** (2).

62. **Brockhoff, K. and T. Teichert** (1995) Cooperative R&D partners' measures of success. *International Journal of Technology Management*, **10** (1), 111–23.

63. **Bruce, M., F. Leverick and D. Littler** (1995) Complexities of collaborative product development. *Technovation*, **15** (9), 535–52.

64. **Kelly, M., J. Schaan and H. Joncas** (2002) Managing alliance relationships: key challenges in the early stages of collaboration. *R&D Management*, **32** (1), 11–22.

65. **Hoecht, A. and P. Trott** (1999) Trust, risk and control in the management of collaborative technology development. *International Journal of Innovation Management*, **3** (3), 257–70.

66. **Lane, P. and M. Lubatkin** (1998) Relative absorptive capacity and inter-organizational learning. *Strategic Management Journal*, **19**, 461–77.

67. **Nonaka, I. and H. Takeuchi** (1995) *The Knowledge-Creating Company*, Oxford University Press, Oxford.

68. **Johnson, W.** (2002) Assessing organizational knowledge creation theory in collaborative R&D projects. *International Journal of Innovation Management*, **6** (4), 387–418.

69. **Levinson, N. and M. Asahi** (1995) Cross-national alliances and interorganizational learning. *Organizational Dynamics*, Autumn, 50–63.

70. **Hamel, G.** (1991) Competition for competence and inter-partner learning within international strategic alliances. *Strategic Management Journal*, **12**, 83–103.

71. **Jones, K. and W. Shill** (1993) Japan: allying for advantage. In J. Bleeke and D. Ernst, eds, *Collaborating to Compete* (pp. 115–44), John Wiley & Sons, Inc., New York; **Sasaki, T.** (1993) What the Japanese have learned from strategic alliances. *Long Range Planning*, **26** (6), 41–53.

72. **Biemans, W.** (1995) Internal and external networks in product development. In M. Bruce and W. Biemans, eds, *Product Development: Meeting the Challenge of the Design–Marketing Interface*, (pp. 137–59), John Wiley & Sons, Ltd, Chichester.

73. **Schrader, S.** (1995) Informal alliances: information trading between firms. In L. Gomez-Mejia and M. Lawless, eds, *Advances in Global High-Technology Management: Strategic Alliances in High Technology*, (Vol. 5, Part B, pp. 31–55), JAI Press, Greenwich, CN.

74. **Spencer, J.** (2003) Firms' knowledge-sharing strategies in the global innovation system: evidence from the flat panel display industry. *Strategic Management Journal*, **24**, 217–33.

75. **Massey, D., D. Wield and P. Quintas** (1991) *High-Tech Fantasies: Science Parks in Society, Science and Space,* Routledge, London.

76. **Oakey, R.** (2007) Clustering and the R&D management of high-technology small firms: in theory and in practice. *R&D Management*, **37** (3), 237–48; **Westhead, P.** (1997) R&D 'inputs' and 'outputs' of technology-based firms located on and off science parks. *R&D Management*, **27** (1), 45–61.

77. **Bower, J.** (2003) Business model fashion and the academic spin out firm. *R&D Management*, **33** (2), 97–106.

78. **Henderson, R., A. Jaffe and M. Trajtenberg** (1998) Universities as a source of commercial technology: a detailed analysis of university patenting 1965–1988. *Review of Economics and Statistics*, 119–127.

79. **Mowery, D., R. Nelson, B. Sampat and A. Ziedonis** (2001) The growth of patenting and licensing by US universities: an assessment of the effects of the Bayh–Dole Act of 1980. *Research Policy*, **30**, 99–119.

80. **Bray, M. and J. Lee** (2000) University revenues from technology transfer: licensing fees versus equity positions. *Journal of Business Venturing*, **15**, 385–92.

81. **Peters, L. and H. Etzkowitz** (1990) University–industry connections and academic values. *Technology in Society*, **12**, 427–40.

82. **Kassicieh, S., R. Radosevich and J. Umbarger** (1996) A comparative study of entrepreneurship incidence among inventors in national laboratories. *Entrepreneurship Theory and Practice*, Spring, 33–49.

83. **Butler, S. and S. Birley** (1999) Scientists and their attitudes to industry links. *International Journal of Innovation Management*, **2** (1), 79–106.

84. **Lee, Y.** (1996) Technology transfer and the research university: a search for the boundaries of university–industry collaboration. *Research Policy*, **25**, 843–63.

85. **Di Gregorio, D. and S. Shane** (2003) Why do some universities generate more start-ups than others? *Research Policy*, **32**, 209–27.

86. **Meyer, M.** (2004) Academic entrepreneurs or entrepreneurial academics? Research-based ventures and public support mechanisms. *R&D Management*, **33** (2), 107–15.

87. **Seashore, L., D. Blumenthal, M.E. Gluck, and M.A. Stoto** (1989) Entrepreneurs in academe: an exploration of behaviors among life scientists. *Administrative Science Quarterly*, **34**, 110–31.

88. **Samson, K. and M. Gurdon** (1993) University scientists as entrepreneurs: a special case of technology transfer and high-tech venturing. *Technovation*, **13** (2), 63–71.

89. **Vohora, A., M. Wright and A. Lockett** (2004) Critical junctures in the development of university high-tech spinout companies. *Research Policy*, **33**, 147–75.

90. **Small Business Research Centre** (1992) *The State of British Enterprise: Growth, Innovation and Competitiveness in Small and Medium Sized Firms*, SBRC, Cambridge.

91. **Hoffman, K., M. Parejo, J. Bessant, and L. Perren** (1998) Small firms, R&D, technology and innovation in the UK: a literature review. *Technovation*, **18** (1), 39–55.

92. **Calori, R.** (1990) Effective strategies in emerging industries. In R. Loveridge and M. Pitt, eds, *The Strategic Management of Technological Innovation*, (pp. 21–38), John Wiley & Sons, Ltd, Chichester; **Walsh, V., J. Niosi and P. Mustar** (1995) Small firms formation in biotechnology: a comparison of France, Britain and Canada. *Technovation*, **15** (5), 303–28; **Westhead, P., D. Storey and M. Cowling** (1995) An exploratory analysis of the factors associated with survival of independent high technology firms in Great Britain. In F. Chittenden, M. Robertson and I. Marshall, eds, *Small Firms: Partnership for Growth in Small Firms*, (pp. 63–99), Paul Chapman, London.

93. **Tether, B. and D. Storey** (1998) Smaller firms and Europe's high technology sectors: a framework for analysis and some statistical evidence. *Research Policy*, **26**, 947–71.

94. **MacPherson, A.** (1997) The contribution of external service inputs to the product development efforts of small manufacturing firms. *R&D Management*, **27** (2), 127–43.

95. **Rothwell, R. and M. Dodgson** (1993) SMEs: their role in industrial and economic change. *International Journal of Technology Management*, Special Issue, 8–22.

96. **Moote, B.** (1993) *Financial Constraints to the Growth and Development of Small High Technology Firms*. Small Business Research Centre, University of Cambridge; **Oakey, R.** (1993) Predatory networking: the role of small firms in the development of the British biotechnology industry. *International Small Business Journal*, **11** (3), 3–22.

97. **Storey, D.** (1992) United Kingdom: case study. In *Small and Medium Sized Enterprises, Technology and Competitiveness*, OECD, Paris; **Tang, N.** *et al.* (1995) Technological alliances between HEIs and SMEs: examining the current evidence. In D. Bennett and F. Steward, eds, *Proceedings of the European Conference on the Management of Technology: Technological Innovation and Global Challenges*, Aston University, Birmingham.

98. **Tether, B.** (1998) Small and large firms: sources of unequal innovations? *Research Policy*, **27**, 725–45.

99. **Tether, B., J. Smith and A. Thwaites** (1997) Smaller enterprises and innovations in the UK: the SPRU Innovations Database revisited. *Research Policy*, **26**, 19–32.

100. **Strerlacchini, A.** (1999) Do innovative activities matter to small firms in non-R&D-intensive industries? *Research Policy*, **28**, 819–32.

101. **Hall, G.** (1991) Factors associated with relative performance amongst small firms in the British instrumentation sector. Working Paper No. 213, Manchester Business School.

102. **Oakey, R., R. Rothwell and S. Cooper** (1988) *The Management of Innovation in High Technology Small Firms*, Pinter, London.

103. **Keeble, D.** (1993) *Regional Influences and Policy in New Technology-based Firms: Creation and Growth*, Small Business Research Centre, University of Cambridge.

104. **Dickson, K., A. Coles and H. Smith** (1995) Scientific curiosity as business: an analysis of the scientific entrepreneur. Paper presented at the 18th National Small Firms Policy and Research Conference, Manchester; **Lee, J.** (1993) Small firms' innovation in two technological settings. *Research Policy*, **24**, 391–401.

105. **Bruderl, J. and P. Preisendorfer** (2000) Fast-growing businesses. *International Journal of Sociology*, **30**, 45–70.

106. **Lee, C., K. Lee and J. Pennings** (2001) Internal capabilities, external networks, and performance: a study of technology-based ventures. *Strategic Management Journal*, **22**, 615–40.

107. **Almeida, P., G. Dokko and L. Rosenkopf** (2003) Startup size and the mechanisms of external learning: increasing opportunity and decreasing ability? *Research Policy*, **32**, 301–15.

108. **Freel, M.** (2003) Sectoral patterns of small firm innovation, networking and proximity. *Research Policy*, **32**, 751–70.

109. **Zahra, S. and W. Bogner** (2000) Technology strategy and software new ventures performance. *Journal of Business Venturing*, **15** (2), 135–73.

110. **Deeds, D., D. DeCarolis and J. Coombs** (2000) Dynamic capabilities and new product development in high technology ventures: an empirical analysis of new biotechnology firms. *Journal of Business Venturing*, **15** (3), 211–29.

111. **George, G., S. Zahra and D. Robley Wood** (2002) The effects of business-university alliances on innovative output and financial performance: a study of publicly traded biotechnology companies. *Journal of Business Venturing*, **17**, 577–609.

112. **Gans, J. and S. Stern** (2003) The product and the market for 'ideas': commercialization strategies for technology entrepreneurs. *Research Policy*, **32**, 333–50.

PART 6

Capture

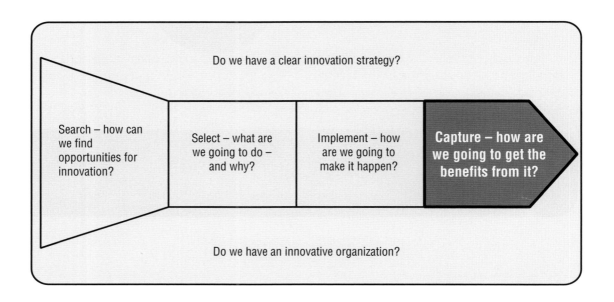

In this final phase we examine how can we ensure that we capture value from our efforts at innovation. Chapter 11 looks at the relationships between different types of knowledge, innovation and performance. Managing intellectual property has become an increasingly significant issue in a world where knowledge production approaches $1 billion per year worldwide, and where the ability to generate knowledge may be less significant than the ability to trade and use it effectively. However, innovation can also create significant social benefits, including core contributions to global development and sustainability.

Finally Chapter 12 looks at how we can assess the ways in which we organize and manage innovation and use these to drive a learning process to enable us to do it better next time. The concern here is not just to build a strong innovation management capability but to recognize that – faced with the moving target which innovation represents in terms of technologies, markets, competitors, regulators and so on – the challenge is to create a learning and adaptive approach which constantly upgrades this capability. In other words we are concerned to build 'dynamic capability'.

CHAPTER 11

Capturing the benefits of innovation

In this chapter we examine how organizations, private and public, can better capture the benefits of innovation, and minimize the drawbacks of change. We begin with a discussion of the classic, but rather narrow, view of economists who identify some of the ways in which firms appropriate the benefits of innovation, in particular through returns on product and process innovation. In the second section we identify the relationships between different types of innovation and various forms of financial and market performance. Next we broaden the scope to include the competitive advantages of exploiting knowledge, both tacit and more formal types, including intellectual property. Finally, we review the more fundamental contributions innovation can make to economic and social change, focusing on the potential for economic development, improvement in social services and greater sustainability.

11.1 Creating value through innovation

One of the central problems of managing innovation is how to create and capture value. For example, in Chapter 1 we discussed the recent transitions in the music industry, and changes in how music is produced, distributed, consumed and paid for (or not in many cases). Video content is facing a similar challenge to the dominant business model, and the producers, distributors and users are experimenting with a range of new ways of generating an income to pay for the production and distribution of video content (Case study 11.1).

CASE STUDY 11.1

Profiting from digital media

The business model for capturing the value from video is simple but conservative: own and enforce the copyright, global cinema release, followed by DVD rental and sale, and lastly TV and other broadcast. The DVD stage is critical, as it generates income of $23.4 billion in the USA, compared to $9.6 billion from cinema release. Note that when DVD was introduced in 1997, three of the major studios initially refused to publish on it, as they feared losing revenue from the existing proven VHS tape format.

However, annual DVD sales have begun to stabilize at around nine billion units worldwide, and in some markets have begun to decline. Therefore the industry has begun to promote the successor to DVD, the high-definition DVD. After a format war, Blu-ray became the new standard for high-definition disks early in 2008. Initial sales of the new format have been slow, not helped by uncertainty of the format war, with nine million Blu-ray disks shipped in 2007, compared to nine billion conventional DVDs – just 0.1% of the market (in addition some 40 million Blu-ray PS3 games were sold – since its launch in 2006 the Sony PlayStation 3 has sold some 11 million games consoles which also play Blu-ray disks). Surveys in the USA and Europe suggest that 80% of consumers are happy with the picture and sound quality of DVD and standard definition broadcast. Therefore formats such as Blu-ray and high-definition satellite and cable broadcasts are aimed at the 20% 'early adopters' who value (i.e. are prepared to pay a premium for) higher definition pictures and sound, primarily for films and sports coverage.

However, for the majority who favour cost and convenience over quality, the Internet is the current preferred medium, legal or otherwise. Illegal sites lead the way, such as ZML which offers 1700 movies for (illegal) download, whereas to date the legal services like MovieFlix and FilmOn tend to be restricted to independent or amateur content. Hollywood has been slow to adapt its business model, and still relies on cinema releases, followed by DVD rental and sales, and finally broadcast. Legal download and streaming offer the potential for lower cost (and prices), as this removes much of the cost of creating, distributing and selling physical media, as well as greater convenience for consumers in terms of choice and flexibility. However, DVD sales depend on the major chain stores for distribution, for example, in the USA Wal-Mart accounts for around 40% of sales, and this represents a powerful resistance to change. As a result, in 2008 legal online film distribution was only around $58 million in the USA, less than 5% of total film sales. Television broadcasters have been faster to adopt such services, such as the BBC i-Player in the UK, mainly because their current business model is based on subscription or advertising, without the film studios' legacy of reliance on physical media and retail distributors. In the USA, Apple iTunes and TV and the Microsoft Xbox have begun to dominate the emerging market for download video rental, but copyright issues have restricted the legal sale of video by download.

As a result of the growing importance of Internet sales of video material, in 2007 the Writers' Guild of America went on strike for better payment terms for electronic distribution and sales. The Hollywood studios' offer was for the payments for Internet sales to be based on the precedent set by DVD – 1.2% of gross receipts – whereas the writers wanted something closer to book or film publishing – 2.5% of gross. The final settlement, reached in February 2008, was a compromise, with a royalty on download rentals of 1.2% of gross, and 0.36–0.70% of gross on download sales, and up to 2% where video streaming is part-funded by advertising. A partial victory for the authors, but this compares with 20% of gross receipts claimed by some leading actors of blockbusters. Clearly there is work to be done on the final business model for the creation, sales and distribution of digital video. Greater clarity of the regime for managing intellectual property is a start, and faster broadband will soon make higher quality download practical for the mass markets, so all that remains is a little innovation in the business model.

Sources: *The Economist*, 23 February 2008, Volume 386, Issue 8568; *ALCS News*, Spring 2008.

At the level of the firm, there is only a weak relationship between innovation and performance. For example, according to the conventional wisdom of strategic management, firms must decide between two broad innovation strategies:

1. *Innovation 'leadership'* – where firms aim at being first to market, based on technological leadership. This requires a strong corporate commitment to creativity and risk taking, with close

linkages both to major sources of relevant new knowledge, and to the needs and responses of customers.

2. *Innovation 'followership'* – where firms aim at being late to market, based on imitating (learning) from the experience of technological leaders. This requires a strong commitment to competitor analysis and intelligence, to reverse engineering (i.e. testing, evaluating and taking to pieces competitors' products, in order to understand how they work, how they are made and why they appeal to customers), and to cost cutting and learning in manufacturing.

However, in practice the distinction between 'innovator' and 'follower' is much less clear. For example, market pioneers often continue to have high expenditures on R&D, but this is most likely to be aimed at minor, incremental innovations. A pattern emerges where pioneer firms do not maintain their historical strategy of innovation leadership, but instead focus on leveraging their competencies in minor incremental innovations. Conversely, late entrant firms appear to pursue one of two very different strategies. The first is based on competencies other than R&D and new product development, for example, superior distribution or greater promotion or support. The second, more interesting strategy, is to focus on major new product development projects in an effort to compete with the pioneer firm.

It is not necessarily a great advantage to be a technological leader in the early stages of the development of radically new products, when the product performance characteristics, and features valued by users, are not always clear, either to the producers or to the users themselves. Especially for consumer products, valued features emerge only gradually through a process of dynamic competition, which involves a considerable amount of trial, error and learning by both producers and users. New features valued by users in one product can easily be recognized by competitors and incorporated in subsequent products. This is why market leadership in the early stages of the development of personal computers was so volatile, and why pioneers are often displaced by new entrants. In such circumstances, product development must be closely coupled with the ability to monitor competitors' products and to learn from customers. In fact, pioneers in radical consumer innovations rarely succeed in establishing long-term market positions. Success goes to so-called 'early entrants' with the vision, patience and flexibility to establish a mass consumer market. For example, studies of the PIMS (Profit Impact of Market Strategy) database indicate that (surviving) product pioneers tend to have higher quality and a broader product line than followers, whereas followers tend to compete on price, despite having a cost disadvantage. A pioneer strategy appears more successful in markets where the purchasing frequency is high, or distribution important (e.g. fast-moving consumer goods), but confer no advantage where there are frequent product changes or high advertising expenditure (e.g. consumer durables).

Therefore technological leadership in firms does not necessarily translate itself into economic benefits. The capacity of the firm to appropriate the benefits of its investment in technology depends on: its ability to translate its technological advantage into commercially viable products or processes, for example, through complementary assets or capabilities in marketing and distribution; and its capacity to defend its advantage against imitators, for example, through secrecy, standards or intellectual property. Some of the factors that enable a firm to benefit commercially from its own technological lead can be strongly shaped by its management: for example, the provision of complementary assets to exploit the lead. Other factors can be influenced only slightly by the firm's management, and depend much more on the general nature of the technology, the product market and the regime of intellectual property rights: for example, the strength of patent protection.

The early work on this was by economists who argued that under perfect market conditions there would be no incentive for individual entrepreneurs or firms to innovate, as ease of imitation would make it difficult to achieve returns from the risky investment in innovation.[1] Subsequently, the focus was on what conditions were optimal to encourage risk taking and innovation, but prevent monopoly positions emerging. For example, as we discussed in Chapter 4, David Teece argues that three groups of factors influence the ability of a firm to capture value from innovation: the *appropriability regime*, which includes the strength of formal intellectual property rights, nature of the knowledge (tacit versus codified), secrecy, ease of imitation and lead times; *complementary assets*, such as brand, position, distribution, support and services; and the *dominant design*.[2]

However, simplistic arguments in favour of ever-stronger intellectual property rights (IPR), in particular patents and copyright, fail to understand the evidence of their limited effectiveness, both in terms of encouraging innovation, and in creating and capturing value from innovation. For example, in the USA the number of patents granted to firms during the 1990s more than doubled, and the cases of legal enforcement of IPR more than tripled, resulting in legal expenditures equivalent to 25% of the R&D of the firms involved, but without any associated step-change in the levels of innovation or profitability.[3]

There are a number of other empirical reasons to believe that IPR have only a minor role to play in the creation and capture of value from innovation. Firstly, the propensity to use, and more importantly to enforce, IPR varies by sector significantly. In some industries (and countries) the IPR regime is strong, such as pharmaceuticals, in others much weaker, such as information and communications technologies (ICTs). However, these differences in the strength of IPR are not reflected in the rates of innovation or profitability across these sectors.[4] In each case, other aspects, such as sales and distribution, service and support, are much more important explanatory factors. Secondly, the high variation in innovation and performance within the same sectors and within similar IPR regimes indicates that other, firm-level factors are also at work. For example, in services, differences in the external linkages with suppliers, consultants, customers and other partners are associated with differences in innovation and growth.[5]

In fact an over-reliance on using IPR for protection can limit the benefits derived from innovation. Firms need to balance the desire to protect their knowledge with the need to share aspects of this knowledge to promote innovation. This is particularly necessary for systemic innovations, which may demand externalities and complementary products and services to be successful, or where potential network externalities exist. Network externalities arise when increases in the number of users result in reduced costs but greater benefits, like many Internet products and services (Case study 11.2). A degree of IPR is associated with network externalities. In such cases IPR may indicate that there is knowledge in a codified form, which makes it easier to transfer or share within a network, and the security offered by the IPR can encourage collaboration and licensing.[6] By influencing the shape or architecture of an emerging innovation in this way, a firm can capture a small proportion of a potentially very large pie, rather than focusing on the protection of a much smaller pie. Where imitation is likely, investment in complementary assets can result in higher returns in the longer term.[7] In fact, the research indicates that use of IPR has a *negative* effect on a strategy of long-term value creation, and that lead time, secrecy and the tacitness of knowledge are more strongly associated with creating value.[8]

In summary, theoretical arguments and empirical research suggest that from both a policy and management perspective, only a limited level of IPR is desirable to encourage risk taking and innovation, and that a broader repertoire of strategies is necessary to create and capture the economic and social benefits of innovation. Economic and social value are created by innovation in many different ways, and tools such as value analysis and value stream analysis can help to identify alternative ways to create and capture value.

The disruptive business model of Skype

Skype successfully combined two emerging technologies to create a new service and business model for telecommunications. The two technologies were Voice over Internet Protocol (VoIP) and peer-to-peer (P2P) file sharing. The first allowed the transfer of voice over the Internet, rather than conventional telecommunications networks, and the other exploited the distributed computing power of users' computers to avoid the need for a dedicated centralized server or infrastructure.

Skype was created in 2003 by the Swedish serial entrepreneur Niklas Zennström. Zennström was previously (in)famous for his pioneering web company Kazaa, which provided a P2P service, mainly used for the (illegal) exchange of MP3 music files. He sold Kazaa to the USA company Sharman Networks to concentrate on the development of Skype. He teamed up with the Dane Janus Friis and together they built Skype. Unlike other VoIP firms like Vonage, which charges a subscription for use and is based on proprietary hardware, Skype was available for free download and use for free voice communication between computers. Additional premium pay services were subsequently added, such as Skype-Out to connect to conventional telephones, and Skype-In, to receive conventional calls. The service was made available in 15 different languages which covered 165 countries, and partnerships were made with Plantronics to provide headsets, and Siemens and Motorola for handsets. Happy users quickly recruited family and friends to the service which grew rapidly.

Given the provision of free software and free calls between computers, the business model had to be innovative. There were several ways in which revenues were generated. The premium services like Skype-In and Skype-Out proved to be very popular with small- and medium-sized firms for business and conference calls, and the licensing of the software to specialist providers and the hardware partnership deals were also lucrative. Later, the large user base also attracted web advertising.

By 2005 there were 70 million users registered, but despite this rapid growth the core model of providing a free service meant that revenues were a rather more modest US$7 million, equivalent to only 10 cents per user. In 2008 Skype had around 310 million registered users, 12 million of which were online at any time. Its revenues were estimated to be US$126 million, equivalent to 40 cents per user. This does represent an improvement in financial performance, especially as costs remain low, but the business model remains unproven, except for the founders of Skype. They sold the company to eBay Inc. in October 2005 for US$2.6 billion, with further performance-based bonuses of $1.5 billion by 2009. For eBay, the plan is to use Skype to increase trading turnover by introducing voice bargaining and pay-per-call advertising, and exploit its previous acquisition PayPal to provide improved billing for Skype customers.

Source: Derived from Rao, B., B. Angelov and O. Nov (2006) Fusion of disruptive technologies: lessons from the Skype case. *European Management Journal*, **24** (2 & 3), 174–88.

11.2 Innovation and firm performance

There are several difficulties in constructing a model of the effects of innovation on the financial perform-ance of the firm.[9] First, at the firm level, the relationship between inputs and outputs is much weaker than at the industry level. The weakness in the relationship may be caused simply by the random unpredictabil-ity of innovation. If this were the case, then firms spending more on inputs could be said to be more inno-vative in a probabilistic sense even if they did not actually innovate strongly. However, if firms differ in their technological opportunities, it may not make sense for one firm to innovate more than another – it would mean a misallocation of resources. Even if spillover was believed to be particularly strong so that innova-tion was likely to be suboptimal in general, it would not be clear, without looking at the specifics of a firm, whether it was over- or underinvesting in R&D. Any comparison must, therefore, be across homogeneous firms and this may be difficult to arrange. Secondly, the reporting behaviour of firms may change in respect of any variable that is monitored to be used in an index of innovation. This reflects the so-called 'Goodhart' law phenomenon whereby monetary indicators devised by the government become subverted as behav-iour changes in response to measurement. Thirdly, an objective of the indicators may be to influence financial markets and lending behaviour. However, these markets at present give a lot of attention to the management and efficiency of technological inputs, which are assessed almost entirely by track record. It is not clear that any index of innovation activity is likely to supplant this. Furthermore, financial markets will concern themselves only with the gain appropriable by the firm itself.

In order to determine whether inputs (or outputs) measure anything of relevance, it is necessary to look for correlations between indicators, such as R&D expenditure, productivity growth, profitability or the stock-market value of the firm. For example, there is quite a strong relationship between R&D and the number of patents at the cross-sectional level, across firms and industries. However, at the level of the firm, the relationship is much weaker over time. More promising are econometric studies of the relationship between patents and financial performance. An example is the use of patent numbers as a proxy for 'intangible' capital in stock-market value of firm regressions.

Econometric techniques can be used to assess the impact of innovation inputs, specifically the expenditure on R&D, and on some measure of performance, typically productivity or patents. Research shows that *product* R&D is significantly less productive than *process* R&D.[10] Other studies using the SPRU significant innovations database found that the impact of the *use* of innovations was around four times that of their *generation*.[11,12] The same study found that the productivity increases took 10–15 years to be fully effected. Using R&D as a proxy for *inputs* to the innovation process, and patents as an indi-cator of *outputs*, at the national level, patents and R&D are correlated and, also, to some extent at the sectoral level, but as Pavitt notes, the extent of unexplained variation is high at the level of cross-company analysis.[13] Part of the difficulty in obtaining stable relationships between patents and R&D lies in the fact that firms have different propensities to patent their discoveries. This partly reflects the ease of protecting the gains from innovation in other ways, such as secrecy and first-mover advantages. Furthermore, the effectiveness of patents varies across industries, for example, being strong in pharma-ceuticals but weak in consumer electronics.[14]

R&D statistics also display industry-specific bias with some sectors classifying their development work as design or production.[15] The fact that weaker relationships between outputs and inputs are observed at the firm level, rather than at the industry level, suggests that there is a lot of variability in the productivity of technological inputs, and that there may be some point in studying the particular conditions under which the inputs are used most effectively.

The most likely explanatory factors are *scale, technological opportunity* and *management*.[16] The evidence on scale is mixed. There are two linked hypotheses – that the size of the R&D effort counts, and that the size of the firm makes R&D more effective, say, because of economies of scope between projects.[17] Studies suggest that the scale of R&D effort is important only in chemicals and pharmaceuticals.[18] Firm size is a more difficult issue to study because the interpretation of R&D and patents differ between class sizes of firms. One study compared over 600 manufacturing firms between 1972 and 1982 in the UK, matched to the SPRU database of significant technical innovations.[19] It suggests that large firms tend to innovate more because they have a higher incentive to do so: a doubling of market share from the mean of 2.5% will increase the probability of innovation in the next period by 0.6%. This result is qualified by noting that less competitive firms (higher concentration and lower import ratios) innovate less. Technological opportunity at the industry level has been examined in the context of relative appropriability.[14] Technological opportunity also exists at the firm level via the spillover effects from other firms.[12] Such spillovers are not automatic, and demand explicit attention to technology transfer and search for external sources of innovation, as advocated by us throughout this book. The classic study of the managerial efficiency of R&D inputs is the SPRU project SAPPHO, best summarized in Freeman, which found that commitment to the project by senior management and good communications are crucial to success.[20]

A major problem with measuring inputs and outputs is: how do we take account of the 'spillover' of innovation benefits or information to other firms or industries? For example, if we are looking at a particular sector's industrial output or productivity in relation to its R&D spending, how do we take account of spillover from other sectors or non-industry R&D?[21] The question really relates to the appropriate level of investigation – is it the company or industry or entire economy? Freeman discusses the question of spillover, arguing that the appropriate connection to make is not so much company R&D and productivity as industry R&D and productivity.[20] For example, the whole electronics industry benefited from Bell's work on semiconductors, and only a small part was recovered by Bell in the form of

---

RESEARCH NOTE   Exploiting (nearly) new technologies

A study of the relationships between the age of patents and financial performance appears to provide some additional support for a 'fast-follower' strategy, rather than a 'first-mover' approach. It found that the median age of the patents of a firm is correlated with its stock-market value, but not in a linear way. For firms utilizing very recent patents or older patents, the relationship is negative, resulting in below-average performance over time, whereas firms using patents close to the median age outperform the average over time.

The study examined 288 firms over 20 years, and 204 000 patents. When patents are filed they must list the other patents which they cite, by patent number and year of filing. This data allows the median age of the patent to be calculated – the median difference between the patent application date and the dates of the prior patents cited. This provides an indication of the age of the technological inputs used, but needs to be compared to the average within different technology patents classes, as the technology life cycle varies significantly between the 400 patent classes, from months to decades. This comparison reveals a variation in the median ages of technologies used by different firms operating in the same technical fields, indicating different technology strategies. Finally, this data is compared with the financial performance, in this case share performance, of the firms over time. The results show that firms at the technological frontier, defined as one or more standard deviations ahead of their industry, or for those using mature technologies, that is 1.3 or more standard deviations behind the industry average, the stock returns underperform. However, the stock-market returns outperform for firms exploiting median-age technologies.

One interpretation of this observed relationship is that the firms with the very new patents face the very high costs and uncertainty associated with emerging technology, including development and commercialization. Conversely, the firms using mature patent portfolios face more limited opportunity to exploit these commercially. However, the firms with patents closer to the median age (in the relevant patent classes) have reduced much of the very high cost and uncertainty associated with the newer patents, but retain significant scope for further development and commercialization. Therefore one lesson may be for firms to more carefully manage the age profile of their patents, and to focus exploitation on a specific time window. This is not simply about being a fast follower, which implies some degree of imitation, but another argument for closer integration between technological and market strategies.

Source: Heeley, M.B. and R. Jacobson (2008) The recency of technological inputs and financial performance. *Strategic Management Journal*, **29**, 723–44.

licensing or sales. There may also be a different kind of spillover internal to the firm. Some products fail, but their R&D is still useful. For example, the large sums spent by IBM on the (failed) Stretch computer in the1960s (only a few were sold) led to the successful 360 series. Spillover from innovations between closely related sectors is not as great as previous research has suggested with regard to R&D spending.[11] Rather, there is spillover between producers and users.[22] This is presumably because the innovation itself is too firm-specific to show much spillover effect, whereas the information shared with R&D spillover is less firm-specific. Although firms are increasingly drawing upon external sources of innovation, few have yet to systematically scan outside their own sector.[23] A particular form of spillover occurs when the economy, as a whole, benefits more from an innovation than is appropriated as profits. A difference, then, occurs between the private rate of return and the social rate of return, and in general the social benefits of innovation far exceed the private returns to individual firms.[24]

The limitations of R&D and patents, as surrogates for innovation, have led to more recent studies turning to less robust but market-based measures, such as new product announcements and innovation counts. One study related the number of new chemical entities discovered in the US pharmaceutical industry to constant price R&D and other variables.[18] A nonlinear (convex) relationship with R&D was discovered and there was some indication that when R&D was interacted with sales in a large firm, it was more effective. Another study examined the strength of the relationship from patents to innovations in order to judge whether patents can be used as an innovation indicator. The results are striking in that at the four-digit industry level, there is a strong relationship. This disappears when the firm-level data is analysed. Indeed, the best predictor of a firm innovation is the patent intensity of the industry it is in.[25] Subsequent studies have analysed innovations announced in all major US publications, others have restricted the scope to leading financial publications such as the *Wall Street Journal*.[26] These studies indicate that innovation tends to be concentrated in larger firms, in less concentrated industries and is strongly affected by joint investment in advertising and R&D.[27] At the industry level, patent intensity and new product announcements are strongly related, with 60% of the variance in the new product sample being explained by patent intensity. However, at the level of the firm, the relationship is very weak, and only 2% of the variance of individual firm-level new product activity appears to be explained by patenting activity.[25]

The ratio of R&D/value added has been used as a proxy for innovation output in research. This is because identical R&D expenditures in different industries do not necessarily indicate identical innovation activity, and also R&D thresholds will be different for different industries, some being far more capital-intensive than others.[28] Similarly, an 'innovation ratio' has been developed, based on the ratio of cash outlay to cash return, as well as the ratio of development time to market life of specific development projects. The idea is that when, or if, a company with a portfolio of different products reaches steady state, the innovation ratio will be equivalent to the ratio of innovation spending to value added. On this basis, it is possible to calculate an innovation ratio for specific sectors and companies. For example, the ratio for the UK mechanical engineering sectors is around 14%. As the value added for that sector is some 50% of turnover, this suggests that at least 7% of revenue should be devoted to innovation in order to sustain intangible assets.[29] Conceptually, this ratio is similar to the depreciation charge for tangible assets.

Analysis of the SPRU database of innovations and company accounts shows that the profit margin of innovators is higher than non-innovators, controlling for other influences, although the effect is rather small.[12] The relationship between profitability and lagged indicators of capital input, marketing expenses and R&D reveals that the rate of return to R&D is about 33%, with an average lag of about five years. Process innovation has four times the rate of return as product innovation, but is more risky with more variable returns.[30]

The impact of R&D on the stock market is more difficult to judge as one needs a prior position on the efficiency or, otherwise, of financial markets before setting up a testable hypothesis. Some key studies find a significant (though noisy) effect.[31] These also raise an important worry about whether stock-market valuations of innovation are consistent, as the valuation of R&D capital collapsed from a value of unity to a quarter over the 1980s. However, the relationship between patents and the market value of the firm are not significant, with the exception of the pharmaceutical industry.[32] In contrast, product announcements have a positive effect on the share price of the originating firm.[27] The impact of the announcement on share price depends on two factors: first, an assessment of the probability of success of the new product; second, an evaluation of the level of future earnings from the product. The study found that firms introducing new products accrue around 0.75% excess market return over three days, beginning one day before the formal announcement. The average value of each new product announcement was found to be $26 million (in 1972 dollars). Of course, the precise return and value of each product announcement depends on the industry sectors: the highest returns were found to be in food, printing, chemicals and pharmaceuticals, computers, photographic equipment and durable goods. Excess returns due to new product announcements suggest that past and current accounting data have little predictive value.

The P/E (price/earnings) ratio may be a better indicator of (future) innovation performance. The average P/E ratio of the firms making new product announcements is almost twice that of the firms which make no new product announcements. This implies that the stock market is valuing the long-term stream of future earnings generated by the innovative firms at a much higher rate than the non-innovators. However, profitability declines as the market evolves over time for a number of reasons. First, product and service differentiation tend to be reduced. Second, competition tends to shift to price and rates of return fall. Third, at least in the manufacturing and production sectors, capital intensity tends to increase, driving returns down even further. More specifically, the real rate of market growth is associated with profitability. At the extremes, a real annual rate of growth of 10% or more has a ROI four points higher than markets declining at rates of 5% or more. High rates of market growth are associated with:[33]

- high gross margins
- high marketing costs
- rising productivity
- rising value added per employee
- rising investment
- low or negative cash flow.

Market differentiation measures the degree to which all competitors differ from one another across a market. Therefore, market differentiation is related to market segmentation and is a measure of market attractiveness. Customers in different market segments will value different product attributes. The joint effect of relative quality and market differentiation is significant. Markets in which there is little differentiation and no significant difference in the relative quality of competitors are characterized by low returns. High relative quality is a strong predictor of high profitability in any market conditions. Nevertheless, a niche business may achieve high returns in a market with high differentiation without high relative quality. A combination of both high market differentiation and high perceived relative quality yields very high ROI, typically in excess of 30%. The importance of market share varies with industry. Intuition would suggest that share would be most important in capital-intensive manufacturing and production industries, where economies of scale are required. However, PIMS suggests that market share has a much stronger impact on profitability in innovative sectors, that is, those industries characterized by high R&D and/or marketing expenditure. For the R&D and marketing-intensive businesses, the ROI of the market leader is on average 26 points higher than the average small share business. In the manufacturing-intensive businesses, the corresponding difference is only 12 points. This suggests that scale effects are more important in R&D and marketing than in manufacturing.

Our own study of the relationship between innovation and performance examined 40 companies, representing five different sectors.[9] We chose companies to provide a range of R&D intensity in each of the five sectors. Analysis of the data confirms that expenditure on R&D, as a proportion of sales, has a significant positive effect on value added, but also the number of new product announcements made. This suggests that R&D contributes both to increasing the number of new products introduced as well as their value. The introduction of a term to represent the interaction of research and development with sales indicates diminishing efficiency of innovation with firm size, in other words, larger firms introduce more new products, but not in proportion to their size. The results suggest that the financial markets do value expenditure on research and development. The coefficient of about 0.3 on R&D/sales may be used to estimate an elasticity of the dependent variable at the mean value of R&D/sales of 2.9. A 1% rise in R&D/sales would translate into a 0.08% rise in the market-to-book value. This suggests that the financial markets may somewhat undervalue R&D expenditure. If we use ratio of new products introduced/absolute R&D as a proxy for research efficiency, we find that the efficiency of research also has a significant positive effect on the market-to-book value. This suggests that the market values the past efficiency of R&D (that is, track record), as well as the expenditure on R&D.

11.3 Exploiting knowledge and intellectual property

In this section we discuss how individuals and organizations identify 'what they know' and how best to exploit this. We examine the related fields of knowledge management, organizational learning and intellectual property. Key issues include the nature of knowledge, for example explicit versus tacit knowledge; the locus of knowledge, for example individual versus organizational; and the distribution of knowledge across an organization. More narrowly, knowledge management is concerned with identifying, translating, sharing and exploiting the knowledge within an organization. One of the key issues is the relationship between individual and organizational learning, and how the former is translated into the latter, and ultimately into new processes, products and businesses. Finally, we review different types of formal intellectual property, and how these can be used in the development and commercialization of innovations.

In essence managing knowledge involves five critical tasks:

1. Generating and acquiring new knowledge.
2. Identifying and codifying existing knowledge.
3. Storing and retrieving knowledge.
4. Sharing and distributing knowledge across the organization.
5. Exploiting and embedding knowledge in processes, products and services.

Generating and acquiring knowledge

Organizations can acquire knowledge by experience, experimentation or acquisition. Of these, learning from experience appears to be the least effective. In practice, organizations do not easily translate experience into knowledge. Moreover, learning may be unintentional or it may not result in improved effectiveness. Organizations can incorrectly learn, and they can learn that which is incorrect or harmful, such as learning faulty or irrelevant skills or self-destructive habits. This can lead an organization to accumulate experience of an inferior technique, and may prevent it from gaining sufficient experience of a superior procedure to make it rewarding to use, sometimes called the 'competency trap'.

Experimentation is a more systematic approach to learning. It is a central feature of formal R&D activities, market research and some organizational alliances and networks. When undertaken with intent, a strategy of learning through incremental trial and error acknowledges the complexities of existing technologies and markets, as well as the uncertainties associated with technology and market change and in forecasting the future. The use of alliances for learning is less common and requires an intent to use them as an opportunity for learning, a receptivity to external know-how and partners of sufficient transparency. Whether the acquisition of know-how results in organizational learning depends on the rationale for the acquisition and the process of acquisition and transfer. For example, the cumulative effect of outsourcing various technologies on the basis of comparative transaction costs may limit future technological options and reduce competitiveness in the long term.

A more active approach to the acquisition of knowledge involves scanning the internal and external environments. As we discussed in Chapter 5, scanning consists of searching, filtering and evaluating potential opportunities from outside the organization, including related and emerging technologies, new market and services, which can be exploited by applying or combining with existing competencies. Opportunity recognition, which is a precursor to entrepreneurial behaviour, is often associated with a flash of genius, but in reality is probably more often the end result of a laborious process of environmental scanning. External scanning can be conducted at various levels. It can be an operational initiative with market- or technology-focused managers becoming more conscious of new developments within their own environments, or a top-driven initiative where venture managers or professional capital firms are used to monitor and invest in potential opportunities.

Identifying and codifying knowledge

It is useful to begin with a clearer idea of what we mean by 'knowledge' (Box 11.1). It has become all things to all people, ranging from corporate IT systems to the skills and experience of individuals. There is no universally accepted typology, but the following hierarchy is helpful:

BOX 11.1   Identifying different types of knowledge

The concept of disembodied knowledge can become a very abstract idea, but it can be assessed in practice. Here are some aspects and types of knowledge identified in a study of the biotechnology and telecommunications industries:

- variety of knowledge
- depth of knowledge
- source of knowledge, internal and external
- evaluation of knowledge and awareness of competencies
- knowledge management practices, the capability to identify, share and acquire knowledge
- use of IT systems to store, share and reuse knowledge
- identification and assimilation of external knowledge
- commercial knowledge of markets and customers
- competitor knowledge, current and potential
- knowledge of supplier networks and value chain
- regulatory knowledge
- financial and funding stakeholder knowledge
- knowledge of intellectual property (IPR), own and others
- knowledge practices, including documentation, intranets, work organization and multidisciplinary teams and projects.

The study concluded that each of these contributed to the intellectual assets and innovative performance of companies, but in different ways. In general, the less tangible and more tacit knowledge of individuals, groups and practices is necessary to exploit the more explicit and tangible types of knowledge, such as R&D and IPR, and these in turn can lead to better use and access to external sources of knowledge, due to a strengthening of position, reputation and trust.

Source: Derived from Marques, D.P., F.J.G. Simon and C.D. Caranana (2006) The effect of innovation on intellectual capital: an empirical evaluation in the biotechnology and telecommunications industries. *International Journal of Innovation Management*, **10** (1), 89–112.

- *Data* are a set of discrete raw observations, numbers, words, records and so on. Typically easy to structure, record, store and manipulate electronically.
- *Information* is data that has been organized, grouped or categorized into some pattern. The organization may consist of categorization, calculation or synthesis. This organization of data endows information with relevance and purpose, and in most cases adds value to data.
- *Knowledge* is information that has been contextualized, given meaning and therefore made relevant and easier to operationalize. The transformation of information into knowledge involves making comparisons and contrasts, identifying relationships and inferring consequences. Therefore knowledge is deeper and richer than information, and includes framed expertise, experience, values and insights.

There are essentially two different types of knowledge, each with different characteristics:

- *Explicit knowledge*, which can be codified, that is expressed in numerical, textual or graphical terms, and therefore is more easily communicated, for example, the design of a product.
- *Tacit or implicit knowledge*, which is personal, experiential, context-specific and hard to formalize and communicate, for example, how to ride a bicycle.

Note that the distinction between explicit and tacit is not necessarily the result of the difficulty or complexity of the knowledge, but rather how easy it is to express that knowledge. Blackler develops a finer typology of knowledge, which identifies five types:[34]

- *Embrained* knowledge, which depends on conceptual skills and cognitive abilities, and emphasizes the value of abstract knowledge.
- *Embodied* knowledge, which is action oriented but likely to be only partly explicit, for example problem-solving ability and learning by doing, and is highly context-specific.
- *Encultured* knowledge, which is the process of achieving shared understanding and meaning. It is socially constructed and open to negotiation, and involves socialization and acculturation.
- *Embedded* knowledge, which resides in systematic routines and processes. It includes resources and relationships between roles, procedures and technologies and is related to the notion of organizational capabilities or competencies.
- *Encoded* knowledge, which is represented by symbols and signs, and includes designs, blueprints, manuals and electronic media.

None of these types of knowledge is inherently superior, and the most relevant type will be contingent upon the organizational and environmental needs. It is also possible to add a sixth type of knowledge, *commodified* knowledge, which is embodied in the outputs of an organization, for example products and services. This is a critical point, because much of the writing and practice of knowledge management treats the creation and sharing of knowledge as an end in itself. However, in most organizations, perhaps with the exceptions of (some) schools and universities, this is not the case. Knowledge is simply an input or means to achieve some organizational goal.

It is useful to distinguish between learning 'how' and learning 'why'. Learning 'how' involves improving or transferring existing skills, whereas learning 'why' aims to understand the underlying logic or causal factors with a view to applying the knowledge in new contexts. For example, consider the case of Xerox and its range of knowledge management programmes.

Neither form of learning is inherently superior, and each will be important in different circumstances. For example, learning 'how' is more relevant where speed or quality is critical, but learning 'why' will be necessary to apply skills and know-how in new situations.

Much of the research on innovation management and organizational change has failed to address the issue of organizational learning. Instead, it has focused on learning by individuals within organizations: '. . .it is important to recognize that organizations do not learn, but rather the people in them do';[35] 'an organization learns in only two ways: (i) by the learning of its members; or (ii) by ingesting new members. . .' .[36]

Clearly, individuals do learn within the context of organizations. This context affects their learning which, in turn, may affect the performance of the organization. However, individuals and organizations are very different entities, and there is no reason why organizational learning should be conceptually or empirically the same as learning by individuals or individuals learning within organizations. Existing theory and research on organizational learning has been dominated by a weak metaphor of human learning and cognitive development, but such simplistic and inappropriate anthropomorphizing of organizational characteristics has contributed to confused research and misleading conclusions.

Using the dimensions of individual versus collective knowledge, and routine versus novel tasks, it is possible to identify four organizational configurations (Figure 11.1). This framework is useful because rather than advocate a simplistic universal trend towards 'knowledge workers', it allows different types of knowledge to be mapped on to different organizational and task requirements.

For example, this framework suggests that under conditions of environmental uncertainty embrained and encultured knowledge are more relevant than embedded or embodied knowledge. The choice between the two approaches will depend on the organizational culture and context. We might expect a small, entrepreneurial firm to rely more on embrained knowledge, and a large established firm on encultured knowledge.

As we have seen, knowledge can be embodied in people, organizational culture, routines and tools, technologies, processes and systems. Organizations consist of a variety of individuals, groups and functions with different cultures, goals and frames of reference. Knowledge management consists of identifying and sharing knowledge across these disparate entities. There is a range of integrating mechanisms which can help to do this. Mobilizing and managing knowledge should become a primary task and many of the recipes offered for achieving this depend upon mobilizing a much higher level of participation in innovative problem solving and on building such routines into the fabric of organizational life.

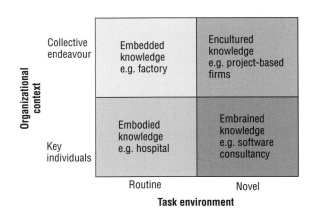

FIGURE 11.1: Task, organizational context and knowledge types
Source: Derived from Blackler, F. (1995) Knowledge, knowledge work and organizations: an overview and interpretation. *Organization Studies*, **16** (60), 1021–46

Nonaka and Takeuchi argue that the conversion of tacit to explicit knowledge is a critical mechanism underlying the link between individual and organizational knowledge. They argue that all new knowledge originates with an individual, but that through a process of dialogue, discussion, experience sharing and observation such knowledge is amplified at the group and organizational levels. This creates an expanding community of interaction, or *knowledge network*, which crosses intra- and inter-organizational levels and boundaries. Such knowledge networks are a means to accumulate knowledge from outside the organization, share it widely within the organization, and store it for future use. This transformation of individual knowledge into organizational knowledge involves four cycles:[37]

- *Socialization* – tacit to tacit knowledge, in which the knowledge of an individual or group is shared with others. Culture, socialization and communities of practice are critical for this.
- *Externalization* – tacit to explicit knowledge, through which the knowledge is made explicit and codified in some persistent form. This is the most novel aspect of Nonaka's model. He argues that tacit knowledge can be transformed into explicit knowledge through a process of conceptualization and crystallization. Boundary objects are critical here.
- *Combination* – explicit to explicit knowledge, where different sources of explicit knowledge are pooled and exchanged. The roles of organizational processes and technological systems are central to this.
- *Internalization* – explicit to tacit knowledge, whereby other individuals or groups learn through practice. This is the traditional domain of organizational learning.

Max Boisot has developed the similar concept of C-space (culture space) to analyse the flow of knowledge within and between organizations. It consists of two dimensions: codification, the extent to which information can be easily expressed; and diffusion, the extent to which information is shared by a given population. Using this framework he proposes a social learning cycle which involves four stages: scanning, problem solving, diffusion and adsorption (Figure 11.2).[38]

C-space (culture space) is a useful conceptual framework for this analysis. It focuses on the structuring and flow of knowledge within and between organizations. It consists of two dimensions: *codification* and *diffusion*. Codifying knowledge involves taking information that human agents carry in their heads and find hard to articulate, and structuring it in such a way that its complexity is reduced (Box 11.2). This enables it to be incorporated into physical objects or described on paper. Once this has occurred, it will develop a life of its own and can diffuse quite rapidly and extensively. Knowledge moves around the C-space in a cyclical fashion as shown in Figure 11.2.

This methodology can be used to map knowledge in an organization or industry. This framework can help define what an organization needs to do over time to maintain and renew resources and competencies. Effective management is about knowing where to locate knowledge resources and the organizational linkages that integrate them together to create competencies. The objectives of the framework are:

1. To enable an organization to map its resources and the key linkages between them on to the C-space.
2. To act as an elicitation device to facilitate a discussion about the meaning and action required – in terms of core competencies and knowledge resources.

Storing and retrieving knowledge

Storing knowledge is not a trivial problem, even now that the electronic storage and distribution of data is so cheap and easy. The biggest hurdle is the codification of tacit knowledge. The other common problem is

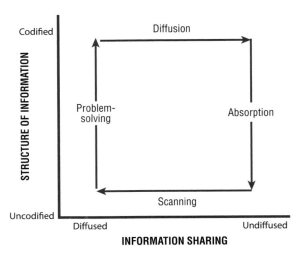

FIGURE 11.2: A model of knowledge structuring and sharing
Source: Adapted from Boisot, M. and D. Griffiths (2006) Are there any competencies out there? Identifying and using technical competencies. In J. Tidd, ed., *From Knowledge Management to Strategic Competence,* second edition (pp. 249–307), Imperial College Press, London

to provide incentives to contribute, retrieve and reuse relevant knowledge. Many organizations have developed excellent knowledge intranet systems, but these are often underutilized in practice (Case study 11.3).

In practice, there are two common but distinct approaches to knowledge management. The first is based on investments in IT, usually based on groupware and intranet technologies. This is the

---

BOX 11.2   **An example of codification and diffusion scales**

Codified → Uncodified
- can be totally automated
- can be partially automated
- can be systematically described
- can be described and put down on paper
- can be shown and described verbally
- can be shown
- inside someone's head

Diffused → Undiffused
- known by all firms in all industries
- known by many firms in all industries
- known by many firms in many industries
- known by many firms in a few industries
- known by a handful of firms in a few industries
- known by only a handful of firms in one industry
- known only by one firm in one industry

CASE STUDY 11.3

Knowledge management at Arup

Arup is an international engineering consultancy firm which provides planning, designing, engineering and project management services. The business demands the simultaneous achievement of innovative solutions and significant time compression imposed by client and regulatory requirements.

Since 1999 the organization has established a wide range of knowledge management initiatives to encourage sharing of know-how and experience across projects. These initiatives range from organizational processes and mechanisms, such as cross-functional communications meetings and skills networks, to technology-based approaches such as the Ovebase database and intranet.

To date, the former has been more successful than the latter. For example, a survey of engineers in the firm indicated that in design and problem solving, discussions with colleagues were rated as being twice as valuable as knowledge databases, and consequently engineers were four times as likely to rely on colleagues. Two primary reasons were cited for this. First, the difficulty of codifying tacit knowledge. Engineering consultancy involves a great deal of tacit knowledge and project experience which is difficult to store and retrieve electronically. Second, the complex engineering and unique environmental context of each project limits the reuse of standardized knowledge and experience.

favoured approach of many management consultants. But introducing knowledge management into an organization consists of much more than technology and training. It can require fundamental changes to organizational structure, processes and culture. The second approach is more people and process based, and attempts to encourage staff to identify, store, share and use information throughout the organization. Research suggests that, as in previous cases of process innovation, the benefits of the technology are not fully realized unless the organizational aspects are first dealt with.[39]

Therefore the storage, retrieval and reuse of knowledge demands much more than good IT systems. It also requires incentives to contribute to and use knowledge from such systems, whereas many organizations instead encourage and promote the generation and use of new knowledge.

Organizational memory is the process by which knowledge is stored for future use. Such information is stored either in the memories of members of an organization or in its operating procedures and routines. The former suffers from all of the shortcomings of human memory, with the additional organizational problem of personnel loss or turnover. However, over time, these behavioural routines create and are reinforced by artefacts such as organizational structures, procedures and policies. In these terms, competencies become highly firm-specific combinations of behavioural routines and artefacts. This specificity questions the validity of the current fashion for benchmarking 'best-practice' processes and structures: what works for one firm may not work for another. Conversely, the difficulty in anticipating future needs.

Richard Hall goes some way towards identifying the components of organizational memory. His main purpose is to articulate intangible resources and he distinguishes between intangible assets and intangible competencies. Assets include intellectual property rights and reputation. Competencies include the skills and know-how of employees, suppliers and distributors, as well as the collective attributes which constitute organizational culture. His empirical work, based on a survey and case studies, indicates that managers believe that the most significant of these intangible resources are the company's reputation and employees' know-how, both of which may be a function of organizational culture. These include:[40]

- *Intangible*, off balance sheet, assets, such as patents, licences, trademarks, contracts and protectable data.
- *Positional*, which are the result of previous endeavour, that is, with a high path dependency, such as processes and operating systems, and individual and corporate reputation and networks.
- *Functional*, which are either individual skills and know-how or team skills and know-how, within the company, at the suppliers or distributors.
- *Cultural*, including traditions of quality, customer service, human resources or innovation.

The key questions in each case are:

1. Are we making the best use of this resource?
2. How else could it be used?
3. Is the scope for synergy identified and exploited?
4. Are we aware of the key linkages which exist between the resources?

Sharing and distributing knowledge

In practice, large organizations often do not know what they know. Many organizations now have databases and groupware to help store, retrieve and share data and information, but such systems are often confined to 'hard' data and information, rather than more tacit knowledge. As a result functional groups or business units with potentially synergistic information may not be aware of where such information could be applied.

Knowledge sharing and distribution is the process by which information from different sources is shared and, therefore, leads to new knowledge or understanding. Greater organizational learning occurs when more of an organization's components obtain new knowledge and recognize it as being of potential use. Tacit knowledge is not easily imitated by competitors because it is not fully encoded, but for the same reasons it may not be fully visible to all members of an organization. As a result, organizational units with potentially synergistic information may not be aware of where such information could be applied. The speed and extent to which knowledge is shared between members of an organization is likely to be a function of how codified the knowledge is.

There are a large number of permutations of the processes required for converting and connecting knowledge from different parts of an organization:[41]

- *Converting data and information to knowledge* – for example identifying patterns and associations in databases.
- *Converting text to knowledge* – through synthesis, comparison and analysis.
- *Converting individual to group knowledge* – sharing knowledge requires a supportive culture, appropriate incentives and technologies.
- *Connecting people to knowledge* – for example through seminars, workshops or software agents.

- *Connecting knowledge to people* – pushing relevant information and knowledge through intranets, agent systems.
- *Connecting people to people* – creating expert and interest directories and networks, mapping who knows what and who knows who.
- *Connecting knowledge to knowledge* – identifying and encouraging the interaction of different knowledge domains, for example through common projects.

This process of conversion and connection is underpinned by *communities of practice*. A community of practice is a group of people related by a shared task, process or the need to solve a problem, rather than by formal structural or functional relationships.[42] Through practice, a group within which knowledge is shared becomes a community of practice through a common understanding of what it does, of how to do it, and how it relates to other communities of practice.

Within communities of practice, people share tacit knowledge and learn through experimentation. Therefore the formation and maintenance of such communities represents an important link between individual and organizational learning. These communities naturally emerge around local work practice and so tend to reinforce functional or professional silos, but also can extend to wider, dispersed networks of similar practitioners.

The existence of communities of practice facilitates the sharing of knowledge within a community, due to both the sense of collective identity, and the existence of a significant common knowledge base. However, the sharing of knowledge between communities is much more problematic, due to the lack of both these elements. Thus the dynamics of knowledge sharing within and between communities of practice are likely to be very different, with the sharing of knowledge between communities typically much more complex, difficult and problematic.

Taking the issue of identity first, differences between different communities of practice will complicate the process of knowledge sharing because of perceived or real differences of interest between communities, resulting in potential conflict. We discussed the benefits and drawbacks of conflict in Chapter 3. If conflict is too high, you may see information hoarding, open aggression, or people lying or exaggerating about their real needs. These conditions could be caused by power struggles of both a personal and professional nature. However, if conflict is too low, individuals and groups may lack motivation or interest in their tasks, and meetings are about one-way communication or reporting, rather than discussion and debate.

The other factor which can prevent the sharing of knowledge between communities of practice is the distinctiveness of different knowledge bases, and the lack of common knowledge, goals, assumptions and interpretative frameworks. These differences significantly increase the difficulty not just of sharing knowledge between communities, but appreciating the knowledge of another community.

There are a few proven mechanisms to help knowledge transfer between different communities of practice:[43]

1. An organizational *translator*, who is an individual able to express the interests of one community in terms of another community's perspective. Therefore the translator must be sufficiently conversant with both knowledge domains and trusted by both communities. Examples of translators include the 'heavyweight product manager' in new product development, who bridges different technical groups and the technical and marketing groups.
2. A knowledge *broker*, who differs from a translator in that they participate in different communities rather than simply mediate between them. They represent overlaps between communities, and are

typically people loosely linked to several communities through weak ties who are able to facilitate knowledge flows between them.[44] An example might be a quality manager responsible for the quality of a process that crosses several different functional groups.

3. A *boundary object or practice*, which is something of interest to two or more communities of practice. Different communities of practice will have a stake in it, but from different perspectives. A boundary object might be a shared document, for example a quality manual; an artefact, for example a prototype; a technology, for example a database; or a practice, for example a product design. A boundary object provides an opportunity for discussion, debate (and conflict) and therefore can encourage communication between different communities of practice.

For example, formally appointed 'knowledge brokers' can be used to systematically scavenge the organization for old or unused ideas, to pass these around the organization and imagine their application in different contexts. For example, Hewlett-Packard created a SpaM group to help identify and share good practice among its 150 business divisions. Before the new group was formed, divisions were unlikely to share information because they often competed for resources and were measured against each other. Similarly, Skandia, a Swedish insurance company active in overseas markets, attempts to identify, encourage and measure its intellectual capital, and has appointed a 'knowledge manager' who is responsible for this. The company has developed a set of indicators that it uses both to manage knowledge internally, and for external financial reporting. For example, the company Joint Solutions acts as an intermediary between medical professionals with ideas for innovations and companies which develop and manufacture medical devices. The business model is not simply based on IPR, but also provides analysis, advice and brokering between the relevant parties.

More generally, cross-functional team-working can help to promote this inter-communal exchange. Functional diversity tends to extend the range of knowledge available and increase the number of options considered, but also can have a negative effect on group cohesiveness and the cost of projects and efficiency of decision making. However, a major benefit of cross-functional team working is the access it provides to the bodies of knowledge that are external to the team. In general a high frequency of knowledge sharing outside of a group is associated with improved technical and project performance, as gatekeeper individuals pick up and import vital signals and knowledge. In particular, cross-functional composition in teams is argued to permit access to disciplinary knowledge outside. Therefore cross-functional team working is a critical way of promoting the exchange of knowledge and practice across disciplines and communities.

A wide range of strategies for introducing knowledge management are available, and no single approach will be appropriate in all circumstances. The most appropriate strategy will depend on the existing organizational culture, structure, processes and culture, the nature of knowledge, and the availability of resources and urgency of action.

It follows from this that developing a climate conducive to knowledge sharing is not a simple matter since it consists of a complex web of behaviours and artefacts. And changing this culture is not likely to happen quickly or as a result of single initiatives, such as restructuring or mass training in a new technique. Given this, it is clear that management cannot directly change culture but it can intervene at the level of artefacts – by changing structures or processes – and by providing models and reinforcing preferred styles of behaviour. Instead, building a culture supportive of knowledge management involves systematic development of organizational structures, communication policies and procedures, reward and recognition systems, training policy, accounting and measurement systems and deployment of strategy. Knowledge management tools can help, but are only effective in a supportive climate.

TABLE 11.1	Knowledge management implementation strategies		
Strategy	**Characteristics**	**Requirements**	**Risks**
Ripple	Bottom-up, continuous improvement, e.g. quality management	Process tools, sustained motivation	Isolation from technical excellence
Integration	Integration of functional knowledge within processes, e.g. product development	Improved interfaces, early involvement, overlapping phases	Conformity, coordination burden
Embedding	Coupling of systems, products and services, e.g. enterprise resource planning (ERP)	Common information systems and technology, motivation and rewards	Loss of autonomy, system complexity
Bridge	New knowledge by novel combination of existing competencies, e.g. architectural innovations	Common language and objectives	High control needs, technical feasibility, market failure
Transfer	Exploiting existing knowledge in a new context, e.g. related diversification	New market knowledge	Inappropriate technology, customer support and service

Source: Adapted from Friso den Hertog, J. and E. Huizenga (2000) *The Knowledge Enterprise*, Imperial College Press, London.

One useful way of understanding the advantages and disadvantages of different ways of implementing knowledge management is to identify five different strategies for introducing knowledge management to an organization (Table 11.1):[45]

- ripple
- flow
- embedding
- bridge
- transfer.

The *ripple* approach is the most basic, and consists of a knowledge centre or core of one specific discipline, technology or skill, which is developed incrementally over time. An example might be quality management, or the experience curve in mass production, or robust designs. The impact over time can be great, but the danger is that the knowledge will become detached from market needs and technological opportunities.

The *flow* approach involves projects being handed from one knowledge centre to another, often sequentially. This is similar to the traditional new product or service development process, and one of the biggest problems is managing the interfaces and integration between the knowledge centres, for example, the design, production and marketing functions.

The *embedding* approach brings different knowledge centres into a broader framework, without any major changes to the centres. An example would be the electronic data interchange (EDI) between a supplier and retailer to reduce stocks and improve responsiveness. Potential problems include asymmetric cost and benefits between the centres, and fear of control or leakage of information.

The *bridge* approach merges two or more different knowledge centres to create a whole new knowledge domain. This may be a merger of disciplines, for example mechanical and electrical engineering to form mechatronics, which is sometimes referred to as *technology fusion*, or may involve the combination of two organizations in a joint venture or merger. This is a very risky strategy, as such bridges typically have significant technological, organizational and commercial uncertainties, but when successful can result in radically new knowledge and high rewards.

The *transfer* approach is more selective, and consists of taking a useful element of one knowledge domain and adapting it for use in another. The knowledge transferred might be technology, market knowledge or organizational know-how or processes. Process benchmarking is an example of a knowledge transfer strategy.

This framework is useful because it helps us to understand better the needs and limits of different approaches to knowledge management, beyond the usual, but often unsuccessful 'technology and training' approach.

Converting knowledge into innovation

Knowledge management has all the characteristics of a management fad or fashion. However, successful management practice is never fully reproducible. In a complex world, neither the most scrupulous practising manager nor the most rigorous management scholar can be sure of identifying – let alone evaluating – all the necessary ingredients in real examples of successful management practice. In addition, the conditions of any (inevitably imperfect) reproduction of successful management practice will differ from the original, whether in terms of firm, country, sector, physical conditions, state of technical knowledge, or organizational skills and cultural norms. Therefore in real life there are no easily applicable recipes for successful management practice. This is one of the reasons why there are continuous swings in management fashion.

However, innovation and entrepreneurship are about knowledge – creating new possibilities through combining different knowledge sets. These can be in the form of knowledge about what is technically possible or what particular configuration of this would meet an articulated or latent need. Such knowledge may already exist in our experience, based on something we have seen or done before. Or it could result from a process of search – research into technologies, markets, competitor actions, etc. And it could be in explicit form, codified in such a way that others can access it, discuss it, transfer it, etc. – or it can be in tacit form, known about but not actually put into words or formulae.

The process of weaving these different knowledge sets together into a successful new process, product or business is one which takes place under highly uncertain conditions. We often don't know about what the final configuration will look like, or precisely how we will get there. In such cases managing knowledge is about committing resources to reduce the uncertainty.

A key contribution to our understanding here of the kinds of knowledge involved in different kinds of innovation is that innovation rarely involves dealing with a single technology or market but rather a bundle of knowledge which is brought together into a configuration. Successful innovation management requires that we can get hold of and use knowledge about *components* but also about how those can be put together – what they termed the *architecture* of an innovation (see Chapter 1). For example, change at the component level in building a flying machine might involve switching to newer metallurgy or composite materials for the wing construction or the use of fly-by-wire controls instead of control lines or hydraulics. But the underlying knowledge about how to link aerofoil shapes, control systems, propulsion systems, etc. at the *system* level is unchanged – and being successful at both requires a different and higher order set of competencies.

One of the difficulties with this is that innovation knowledge flows – and the structures which evolve to support them – tend to reflect the nature of the innovation. So if it is at component level then the relevant people with skills and knowledge around these components will talk to each other – and when change takes place they can integrate new knowledge. But when change takes place at the higher system level – 'architectural innovation' – then the existing channels and flows may not be appropriate or sufficient to support the innovation and the firm needs to develop new ones. This is another reason why existing incumbents often fare badly when major system-level change takes place – because they have the twin difficulties of learning and configuring a new knowledge system and 'unlearning' an old and established one.

A variation on this theme comes in the field of 'technology fusion', where different technological streams converge, such that products which used to have a discrete identity begin to merge into new architectures. An example here is the home automation industry, where the fusion of technologies like computing, telecommunications, industrial control and elementary robotics is enabling a new generation of housing systems with integrated entertainment, environmental control (heating, air conditioning, lighting, etc.) and communication possibilities.

Similarly, in services a new addition to the range of financial services may represent a component product innovation, but its impacts are likely to be less far-reaching (and the attendant risks of its introduction lower) than a complete shift in the nature of the service package – for example, the shift to direct-line systems instead of offering financial services through intermediaries.

David Tranfield and his colleagues map the different phases of the innovation process to identify the knowledge routines in each of three innovation phases – discovery, realization and nurture (Figure 11.3, Table 11.2):[46]

- *Discovery* – scanning and searching the internal and external environments, to pick up and process signals about potential innovation. These could be needs of various kinds, opportunities arising from research activities, regulative pressures, or the behaviour of competitors.
- *Realization* – how the organization can successfully implement the innovation, growing it from an idea through various stages of development to final launch as a new product or service in the external market place or a new process or method within the organization. Realization requires selecting from this set of potential triggers for innovation those activities to which the organization will commit resources.
- *Nurturing* the chosen option by providing resources, developing (either by creating through R&D or acquiring through technology transfer) the means for exploration. It involves not only codified

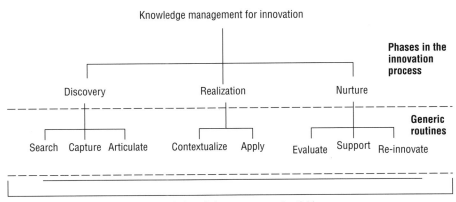

FIGURE 11.3: Process model of knowledge management for innovation

Source: Tranfield, D., M. Young, D. Partington, J. Bessant and J. Sapsed (2006) Knowledge management routines for innovation projects: developing a hierarchical process model. In J. Tidd, ed., *From Knowledge Management to Strategic Competence,* second edition (pp. 126–49), Imperial College Press, London. Reproduced with permission

TABLE 11.2	**Process model linking innovation phase to knowledge management activities**		
Phase in the innovation process	**Generic routines**	**Description**	**Examples of detailed knowledge management activities**
Discovery	Search	The passive and active means by which potential knowledge sources are scanned for items of interest	Active environmental scanning (technological, market, social, political, etc.)
			Active future scanning Experiment – R&D, etc.
	Capture	The means by which knowledge search outcomes are internalized within the organization	Picking up relevant signals and communicating them within and across the organization to relevant players
	Articulate	The means by which captured knowledge is given clear expression	Concept definition – what might we do?
			Strategic and operational planning cycles – from outline feasibility to detailed operational plan

(continued)

| TABLE 11.2 | (Continued) | | |

Phase in the innovation process	Generic routines	Description	Examples of detailed knowledge management activities
Realization	Contextualize	The means by which articulated knowledge is placed in particular organizational contexts	Resource planning and procurement – inside and outside the organization Prototyping and other concept refining activities Early mobilization across functions – design for manufacture, assembly, quality, etc.
	Apply	The means by which contextualized knowledge is applied to organizational challenges	Project team mobilization Project planning cycles Project implementation and modification – 'cycles of mutual adaptation' in technological, market, organizational domains Launch preparation and execution
Nurture	Evaluate	The means by which the efficacy of knowledge applications is assessed	Post-project review Market/user feedback Learning by using/making/etc.
	Support	The means by which knowledge applications are sustained over time	Feedback collection Incremental problem solving and debugging
	Re-innovate	The means by which knowledge and experience are reapplied elsewhere within the organization	Pick up relevant signals to repeat the cycle Mobilize momentum for new cycle

Source: Tranfield, D., M. Young, D. Partington, J. Bessant and J. Sapsed (2006) Knowledge management routines for innovation projects: developing a hierarchical process model. In J. Tidd, ed., *From Knowledge Management to Strategic Competence*, second edition (pp. 126–49), Imperial College Press, London. Reproduced with permission.

knowledge formally embodied in technology, but also tacit knowledge in the surrounding social linkage which is needed to make the innovation work. The nurture phase involves maintaining and supporting the innovation through various improvements and also reflecting upon previous phases and reviewing experiences of success and failure in order to learn about how to manage the process better, and capture relevant knowledge from the experience. This learning creates the conditions for beginning the cycle again, or 're-innovation'.

Exploiting intellectual property

In some cases knowledge, in particular in its more explicit or codified forms, can be commercialized by licensing or selling the intellectual property rights (IPR), rather than the more difficult and uncertain route of developing new processes, products or businesses.

For example, in one year IBM reported license income of US$1 billion, and in the USA the total royalty income of industry from licensing is around US$100 billion. Much of this is from payments for licenses to use software, music or films. For example, in 2005 the global sales of legal music downloads exceeded US$1 billion (although illegal downloads are estimated to be worth three to four times this figure), still only around 5% of all music company revenue, with music downloaded to mobile phones accounting for almost a quarter of this. Patterns of use vary by country, for example, in Japan 99.8% of all music downloads are to mobile phones, rather than to dedicated MP3 players. However, despite the growth of legal sites for downloading music and an aggressive programme of pursuing users of illegal file-sharing sites, the level of illegal downloads has not declined.

This clearly demonstrates two of the many problems associated with intellectual property: these may provide some legal rights, but such rights are useless unless they can be effectively enforced; and once in the public domain, imitation or illegal use is very likely. For these reasons, secrecy is often a more effective alternative to seeking IPR. However, IPR can be highly effective in some circumstances, and as we will argue later, can be used in less obvious ways to help to identify innovations and assess competitors. A range of IPR exists, but those most applicable to technology and innovation are patents, copyright and design rights and registration.

Patents

All developed countries have some form of patent legislation, the aim of which is to encourage innovation by allowing a limited monopoly, usually for 20 years, and more recently many developing and emerging economies have been encouraged to sign up to the TRIPS (Trade Related Intellectual Property System). Legal regimes differ in the detail, but in most countries the issue of a patent requires certain legal tests to be satisfied:

- *Novelty* – no part of 'prior art', including publications, written, oral or anticipation. In most countries the first to file the patent is granted the rights, but in the USA it is the first to invent. The US approach may have the moral advantage, but results in many legal challenges to patents, and requires detailed documentation during R&D.
- *Inventive step* – 'not obvious to a person skilled in the art'. This is a relative test, as the assumed level of skill is higher in some fields than others. For example, Genentech was granted a patent for the plasminogen activator t-PA which helps to reduce blood clots, but despite its novelty, a Court of Appeal revoked the patent on the grounds that it did not represent an inventive step because its development was deemed to be obvious to researchers in the field.

- *Industrial application* – utility test requires the invention to be capable of being applied to a machine, product or process. In practice a patent must specify an application for the technology, and additional patents sought for any additional application. For example, Unilever developed Ceramides and patented their use in a wide range of applications. However, it did not apply for a patent for application of the technology to shampoos, which was subsequently granted to a competitor.
- *Patentable subject* – for example, discoveries and formulae cannot be patented, and in Europe neither can software (the subject of copyright) or new organisms, although both these are patentable in the USA. For example, contrast the mapping of the human genome in the USA and Europe: in the USA the research is being conducted by a commercial laboratory which is patenting the outcomes, and in Europe by a group of public laboratories which is publishing the outcomes on the Internet.
- *Clear and complete disclosure* – note that a patent provides only certain legal property rights, and in the case of infringement the patent holder needs to take the appropriate legal action. In some cases secrecy may be a preferable strategy. Conversely, national patent databases represent a large and detailed reservoir of technological innovations which can be interrogated for ideas.

Apart from the more obvious use of patents as IPR, they can be used to search for potential innovations, and to help identify potential partners or to assess competitors. For example, the TRIZ system developed by Genrich Altshuller identifies standard solutions to common technical problems distilled from an analysis of 1.5 million patents, and applies these in different contexts. Many leading companies use the system, including 3M, Rolls-Royce and Motorola.

Patents can also be used to identify and assess innovation, at the firm, sector or national level. However, great care needs to be taken when making such assessments, because patents are only a partial indicator of innovation.

The main advantages of patent data are that they reflect the corporate capacity to generate innovation, are available at a detailed level of technology over long periods of time, are comprehensive in the sense that they cover small as well as large firms, and are used by practitioners themselves. However, patenting tends to occur early in the development process, and therefore can be a poor measure of the output of development activities, and tells us nothing about the economic or commercial potential of the innovation.

Crude counts of the number of patents filed by a firm, sector or country reveal little, but the quality of patents can be assessed by a count of how often a given patent is cited in later patents. This provides a good indicator of its technical quality, albeit after the event, although not necessarily commercial potential. Highly cited patents are generally of much greater importance than patents which are never cited, or are cited only a few times. The reason for this is that a patent which contains an important new invention – or major advance – can set off a stream of follow-on inventions, all of which may cite the original, important invention upon which they are building.

Using such patent citations, the quality distribution of patents tends to be very skewed: there are large numbers of patents that are cited only a few times, and only a small number of patents cited more than 10 times. For example, half of patents are cited two or fewer times, 75% are cited five or fewer times, and only 1% of the patents are cited 24 or more times. Overall, after 10 or more years, the average cites per patent is around six.[47]

The most useful indicators of innovation based on patents are (Table 11.3):

1. *Number of patents*. Indicates the level of technology activity, but crude patent counts reflect little more than the propensity to patent of a firm, sector or country.
2. *Cites per patent*. Indicates the impact of a company's patents.

TABLE 11.3 **Patent indicators for different sectors**			
	Current impact index (expected value 1.0)	**Technology life cycle (years)**	**Science linkage (science references/patents)**
Oil and gas	0.84	11.9	0.8
Chemicals	0.79	9.0	2.7
Pharmaceuticals	0.79	8.1	7.3
Biotechnology	0.68	7.7	14.4
Medical equipment	2.38	8.3	1.1
Computers	1.88	5.8	1.0
Telecommunications	1.65	5.7	0.8
Semiconductors	1.35	6.0	1.3
Aerospace	0.68	13.2	0.3

Source: Narin, F. (2006) Assessing technological competencies. In J. Tidd, ed., *From Knowledge Management to Strategic Competence*, second edition (pp. 179–219), Imperial College Press, London.

3. *Current impact index (CII)*. This is a fundamental indicator of patent portfolio quality, it is the number of times the company's previous five years of patents, in a technology area, were cited from the current year, divided by the average citations received.
4. *Technology strength (TS)*. Indicates the strength of the patent portfolio, and is the number of patents multiplied by the current impact index, that is, patent portfolio size inflated or deflated by patent quality.
5. *Technology cycle time (TCT)*. Indicates the speed of invention, and is the median age, in years, of the patent references cited on the front page of the patent.
6. *Science Linkage (SL)*. Indicates how leading edge the technology is, and is the average number of science papers referenced on the front page of the patent.
7. *Science Strength (SS)*. Indicates how much the patent applies basic science, and is the number of patents multiplied by science linkage, that is, patent portfolio size inflated or deflated by the extent of science linkage.

Companies whose patents have above-average current impact indices (CII) and science linkage indicators (SL) tend to have significantly higher market-to-book ratios and stock-market returns. However, having a strong intellectual property portfolio does not, of course, guarantee a company's success. Many additional factors influence the ability of a company to move from quality patents to innovation and financial and market performance. The decade of troubles at IBM, for example, is certainly illustrative of this, since IBM has always had very high quality and highly cited research in its laboratories.

At the firm level, rather than at the industry level, there is a lot of variability in the productivity of technological inputs, that is, how effectively these are translated into technological outputs. Research suggests at least three reasons for the differences in the ability of firms to translate inputs into outputs: scale; technological opportunity; and organization and management. However, few managers are interested in improved measures of technological inputs, and instead need ways to assess the *efficiency* and *effectiveness* of the innovation process: efficiency in the sense of how well companies translate technological and commercial inputs into new products, processes and businesses; effectiveness in the sense of how successful such innovations are in the market and their contribution to financial performance. Our own research, using various combinations of inputs, outputs and indicators of performance suggests that some ratio of outputs (e.g. new product announcements) to input (e.g. patents or R&D) provides a good proxy for innovation efficiency, and is associated with a range of financial and market measures of performance, such as value added and market-to-book value.

Therefore care needs to be taken when using patent data as an indicator of innovation. The main advantages of patents are:

1. Patents represent the output of the inventive process, specifically those inventions which are expected to have an economic benefit.
2. Obtaining patent protection is time consuming and expensive. Hence applications are only likely to be made for those developments which are expected to provide benefits in excess of these costs.
3. Patents can be broken down by technical fields, thus providing information on both the rate and direction of innovation.
4. Patent statistics are available in large numbers and over very long time series.

The main disadvantages of patents as indicators of innovation are:

1. Not all inventions are patented. Firms may choose to protect their discoveries by other means, such as through secrecy. It has been estimated that firms apply for patents for 66–87% of patentable inventions.
2. Not all innovations are technically patentable – for example, software development (outside the USA), and some organisms.
3. The propensity to patent varies considerably across different sectors and firms. For example, there is a high propensity to patent in the pharmaceutical industry, but a low propensity in fast-moving consumer goods.
4. Firms have a different propensity to patent in each national market, according to the attractiveness of markets.
5. A large proportion of patents are never exploited, or are applied for simply to block other developments. It has been estimated that between 40–60% of all patents issued are used.

There are major inter-sectoral differences in the relative importance of patenting in achieving its prime objective, namely, to act as a barrier to imitation. For example, patenting is relatively unimportant in automobiles, but critical in pharmaceuticals. Moreover, patents do not yet fully measure technological activities in software since copyright laws are often used as the main means of protection against imitation, outside the USA.

There are also major differences among countries in the procedures and criteria for granting patents. For this reason, comparisons are most reliable when using international patenting or patenting in one country. The US patenting statistics are a particularly rich source of information, given the rigour and fairness of criteria and procedures for granting patents, and the strong incentives for firms to get IPR in the world's largest market. More recently, data from the European Patent Office are also becoming more readily available.

Copyright

Copyright is concerned with the expression of ideas, and not the ideas themselves. Therefore the copyright exists only if the idea is made concrete, for example, in a book or recording. There is no requirement for registration, and the test of originality is low compared to patent law, requiring only that 'the author of the work must have used his own skill and effort to create the work'. Like patents, copyright provides limited legal rights for certain types of material for a specific term. For literary, dramatic, musical and artistic works copyright is normally for 70 years after the death of the author, 50 in the USA, and for recordings, film, broadcast and cable programmes 50 years from their

---

RESEARCH NOTE   Using patents strategically

Each year, some 400 000 patents are filed around the world. However, only a small proportion of these are ever exploited by the owners, and many are not renewed. Based on a review of the research and case studies of 14 firms from different sectors, the study identified a range of different patent strategies:

- *Offensive* – multiple patents in related fields to limit or prevent competition.
- *Defensive* – specific patents for key technologies which are intended to be developed and commercialized, to minimize imitation.
- *Financial* – primary role of patents is to optimize income through sale or license.
- *Bargaining* – patents designed to promote strategic alliances, adoption of standards or cross-licensing.
- *Reputation* – to improve the image or position of a company, for example, to attract partners, talent or funding, or to build brands or enhance market position.

In practice firms may combine different strategies, or more likely have no explicit strategy for patenting (which is our experience outside the pharmaceutical and biotechnology sectors). The European Patent Office (EPO) suggest only two alternatives: patenting as a cost centre, i.e. to provide the necessary legal support; or as a profit centre, to generate income. However, this ignores the more strategic positioning possibilities patents can provide if they are viewed as more than just a legal or income issue.

Source: Gilardoni, E. (2007) Basic approaches to patent strategy. *International Journal of Innovation Management*, **11** (3), 417–40.

---

creation. Typographical works have 25 years copyright. The type of materials covered by copyright include:

- 'Original' literary, dramatic, musical and artistic works, including software and in some cases databases.
- Recordings, films, broadcasts and cable programmes.
- Typographical arrangement or layout of a published edition.

Design rights

Design rights are similar to copyright protection, but mainly apply to three-dimensional articles, covering any aspect of the 'shape' or 'configuration', internal or external, whole or part, but specifically excludes integral and functional features, such as spare parts. Design rights exist for 15 years and 10 years if commercially exploited. Design registration is a cross between patent and copyright protection, is cheaper and easier than patent protection, but more limited in scope. It provides protection for up to 25 years, but covers only visual appearance – shape, configuration, pattern and ornament. It is used for designs that have aesthetic appeal, for example, consumer electronics and toys. For example the knobs on top of LEGO bricks are functional, and would therefore not qualify for design registration, but were also considered to have 'eye appeal', and therefore granted design rights.

Licensing IPR

Once you have acquired some form of formal legal IPR, you can allow others to use it in some way in return for some payment (a licence), or sell the IPR outright (or assign it). Licensing IPR can have a number of benefits:

- reduce or eliminate production and distribution costs and risks
- reach a larger market
- exploit in other applications
- establish standards
- gain access to complementary technology
- block competing developments
- convert competitor into defender.

Considerations when drafting a licensing agreement include degree of exclusivity, territory and type of end use, period of licence and type and level of payments – royalty, lump sum or cross-licence. Pricing a licence is as much an art as a science, and depends on a number of factors such as the balance of power and negotiating skills. Common methods of pricing licences are:

- Going market rate – based on industry norms, e.g. 6% of sales in electronics and mechanical engineering.
- 25% rule – based on licensee's gross profit earned through use of the technology.
- Return on investment – based on licensor's costs.
- Profit sharing – based on relative investment and risk. First, estimate total life-cycle profit. Next, calculate relative investment and weight according to share of risk. Finally, compare results to alternatives, e.g. return to licensee, imitation, litigation.

There is no 'best' licensing strategy, as it depends on the strategy of the organization and the nature of the technology and markets (see Case studies 11.4 and 11.5). For example, Celltech licensed its asthma treatment to Merck for a single payment of $50 million, based on sales projections. This isolated Celltech from the risk of clinical trials and commercialization, and provided a much-needed cash injection. Toshiba, Sony and Matsushita license DVD technology for royalties of only 1.5% to encourage its adoption as the industry standard. Until the recent legal proceedings, Microsoft applied a 'per processor' royalty to its OEM (original equipment manufacturer) customers for Windows to discourage them from using competing operating systems.

CASE STUDY 11.4

Open Source Software

Proprietary software usually restricts imitation by retaining the source code and by enforcing intellectual property rights such as patents (mainly the USA) or copyright (elsewhere). However, Open Source Software (OSS) has many characteristics of a public good, including non-excludability and non-rivalry, and developers and users of OSS have a joint interest in making OSS free and publicly available. The open software movement has grown since the 1980s when the programmer Richard Stallman founded the Free Software Foundation, and the General Public License (GPL) is now widely used to promote the use and adaptation of OSS. The GPL forms the legal basis of three-quarters of all OSS, including Linux.

Therefore firms active in the field of OSS have to create value and appropriate private benefits in different ways. The ineffectiveness of traditional intellectual property rights in such cases means that firms are more likely to rely on alternative ways of appropriating the benefits of innovation, such as being first to the market or by using externalities to create value. More generic strategies include product and service approaches:

- *Products* – adding a proprietary part to the open code and licensing this, or black-boxing by combining several pieces of OSS into a solution package.
- *Services* – consultancy, training or support for OSS.

Linux is a good example of a successful OSS that firms have developed products and services around. It has been largely developed by a network of voluntary programmers, often referred to as the 'Linux community'. Linus Torvalds first suggested the development of a free operating system to compete with the DOS/Windows monopoly in 1991, and quickly attracted the support of a group of volunteer programmers: *'having those 100 part-time users was really great for all the feedback I got. They found bugs that I hadn't because I hadn't been using it the way they were...after a while they started sending me fixes or improvements... this wasn't planned, it just happened.'* Thus Linux grew from 10 000 lines of code in 1991 to 1.5 million lines by 1998. Its development coincided with and fully exploited the growth of the Internet and later web forms of collaborative working. The provision of the source code to all potential developers promotes continuous incremental innovation, and the close and sometimes indistinguishable developer and user groups promote concurrent development and debugging. The weaknesses are potential lack of support for users and new hardware, availability of compatible software and forking in development.

By 1998 there were estimated to be more than 7.5 million users and almost 300 user groups across 40 countries. Linux has achieved a 25% share of the market for server operating systems, although its share of the PC operating system market was much lower, and Apache, a Linux application web server program, accounted for half the market. Although Linux is available free of charge, a number of businesses have been spawned by its development. These range from branding and distribution of Linux, development of complementary software and user support and consultancy services. For example, although Linux can be downloaded free of charge, RedHat Software provides an easier installation program and better documentation for around US$50, and in 1998 achieved annual revenues of more than US$10 million. Red Hat was floated in 1999. In China, the lack of legacy systems, low costs and government support have made Linux-based systems popular on servers and desktop applications. In 2004 Linux began to enter consumer markets, when Hewlett-Packard launched its first Linux-based notebook computer, which helped to reduce the unit cost by US$60.

Source: L. Dahlander (2005) Appropriation and appropriability in Open Source Software. *International Journal of Innovation Management,* **9** (3), 259–86.

CASE STUDY 11.5

ARM Holdings

ARM Holdings designs and licenses high-performance, low-energy-consumption 16- and 32-bit RISC (reduced instruction set computing) chips, which are used extensively in mobile devices such as mobile phones, cameras, electronic organizers and smart cards. ARM was established in 1990 as a joint venture between Acorn Computers in the UK and Apple Computer. Acorn did not pioneer the RISC architecture, but it was the first to market a commercial RISC processor in the mid-1980s. Perhaps ironically, the first application of ARM technology was in the relatively unsuccessful Apple Newton PDA (personal digital assistant). One of the most recent successful applications has been in the Apple iPod. ARM designs but does not manufacture chips, and receives royalties of between 5 cents and US$2.50 for every chip produced under licence. Licensees include Apple, Ericsson, Fujitsu, HP, NEC, Nintendo, Sega, Sharp, Sony, Toshiba and 3Com. In 1999 it announced joint ventures with leading chip manufacturers such as Intel and Texas Instruments to design and build chips for the next generation of hand-held devices. It is estimated that ARM-designed processors were used in 10 million devices in 1996, 50 million in 1998, 120 million devices sold in 1999, and a billion sold in 2004, and more than 2 billion in 2006, representing around 80% of all mobile devices. In 1998 the company was floated in London and on the NASDAQ in New York, and it achieved a market capitalization of £3 billion in December 1999, with an annual revenue growth of 40% to £15.7 million. The company employs around 400 staff, 250 of which are based in Cambridge in the UK, with an average age of 27. It spends almost 30% of revenues on R&D. The company has created 30 millionaires amongst its staff.

Since the mid-1980s universities have increasingly used IPR in an effort to commercialize technology and increase income. Changes in funding and law have clearly encouraged many more universities to establish licensing and technology transfer departments, but whilst the level of patenting has increased significantly as a result, the income and impact has been relatively small. The number of patents granted to US universities doubled between 1984 and 1989, and doubled again between 1989 and 1997. In 1979 the number of patents granted to US universities was only 264, compared to 2436 in 1997.

There are a number of explanations for this significant increase in patent activity. Changes in government funding and intellectual property law played a role, but detailed analysis indicates that the most significant reason was technological opportunity. For example, there is strong evidence that the scientific and commercial quality of patents has fallen since the mid-1980s as a result of these policy changes, and that the distribution of activity has a very long tail.

Measured in terms of the number of patents held or exploited, or by income from patent and software licenses, commercialization of technology is highly concentrated in a small number of elite universities which were highly active prior to changes to funding policy and law: the top 20 US universities account for 70% of the patent activity. Moreover, at each of these elite universities a very small number of key patents account for most of the licensing income, the five most successful patents typically account for 70–90% of total income. The average income from a university licence is only around $60 000, whereas the average return from a university spin-out firm was more than 10 times this. This suggests that a (rare) combination of research excellence and critical mass is required to succeed in the commercialization of technology. Nonetheless, technological opportunity has reduced some of the barriers to commercialization. Specifically, the growing importance of developments in the biosciences and software present new opportunities for universities to benefit from the commercialization of technology.

The successful exploitation of IPR also incurs costs and risks:

- Cost of search, registration and renewal.
- Need to register in various national markets.
- Full and public disclosure of your idea.
- Need to be able to enforce.

In most countries the basic registration fee for a patent is relatively modest, but in addition applying for a patent includes the cost of professional agents, such as patent agents, translation for foreign patents, official registration fees in all relevant countries and renewal fees. As a result the lifetime cost for a single non-pharmaceutical patent in the main European markets would be around £80 000, and the addition of the USA and Japan some £40 000 more. Patents in the other Asian markets are cheaper, at up to £5000 per country, but the cumulative cost becomes prohibitive, particularly for lone inventors or small firms. Pharmaceutical patents are much more expensive, up to five times more, due to the complexity and length of the documentation. In addition to these costs, firms must consider the competitive risk of public disclosure, and the potential cost of legal action should the patent be infringed (Figure 11.4). Costs vary by country, because of the size and attractiveness of different national markets, and also because of differences in government policy. For example, in many Asian countries the policy is to encourage patenting by domestic firms, so the process is cheaper.

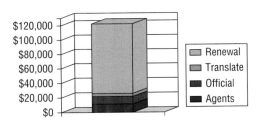

FIGURE 11.4: Typical lifetime cost of a single patent from the European Patent Office

11.4 Broader economic and social benefits

So far we have focused mainly on how firms can better capture the benefits of innovation, but arguably innovation has an even more profound influence on fundamental economic and social development. In this section we review briefly some of the relationships between innovation and economic and social development, and argue that there is much potential for innovation to make a more significant, positive contribution to emerging economies, social service and sustainability.

Innovation and economic development

In his best-selling book, *The World is Flat: The Globalized World in the 21st Century* (Penguin, 2006), Thomas Friedman argues that developments in technology and trade, in particular information and communications technologies (ICTs), are spreading the benefits of globalization to the emerging economies, promoting their development and growth. This optimistic thesis is appealing, but the evidence suggests the picture is rather more complex.

Firstly, because technology and innovation are not evenly distributed globally, and are not easily packaged and transferred across regions or firms. For example, only about a quarter of the innovative activities of the world's largest 500, technologically active firms are located outside their home countries.[48] Secondly, different national contexts influence significantly the ability of firms to absorb and exploit such technology and innovation. For example, state ownership and availability of venture capital both influence entrepreneurship.[49] Thirdly, the position of firms in international value chains can constrain profoundly their ability to capture the benefits of their innovation and entrepreneurship. Many firms in emerging economies have become trapped in dependent relationships as low-cost providers of low-technology, low-value manufactured goods or services, and have failed to develop their own design or new products.[50]

Therefore development of firms from emerging economies is much more than simply 'catching up' with those in the more advanced economies, and is not (only) the challenge of moving from 'followers' to 'leaders'. Global standards and position in international value chains can constrain the ability of firms based in emerging economies to upgrade their capabilities and appropriate greater value, but they also present ways in which these firms can innovate to overcome these hurdles, for example, by using international standards as a catalyst for change, or by repositioning themselves in local clusters or global networks. By position, we refer to the current endowment of technology and intellectual property of a firm, as well as its relations with customers and suppliers.

Innovation and enterprise are central to the development and growth of emerging economies, and yet their contribution is usually considered in terms of the most appropriate national policy and institutions, or the regulation of international trade. Macroeconomic issues are important, and national systems of innovation, including formal policy, institutions and governance, can have a profound influence on the degree and direction of innovation and enterprise in a country or region, but it is also critical to consider a more micro perspective, in particular innovation by firms and the entrepreneurship of individuals. Firms in emerging economies may pursue different routes to upgrading through innovation:[51]

- *Process upgrading* – incremental process improvements to adapt to local inputs, reduce costs or to improve quality.
- *Product upgrading* – through adaptation, differentiation, design and product development.
- *Capability upgrading* – improving the range of functions undertaken, or changing the mix of functions, for example, production versus development or marketing.
- *Inter-sectoral upgrading* – moving to different sectors, for example, to those with higher value added.

To some extent firms in emerging economies face a 'reverse product–process innovation life cycle'. We saw in Chapter 2 that the most common pattern of evolution of technological innovation in the industrialized world has been from product to process innovation on the one hand, and from radical to incremental innovation, on the other. Initially a series of different radical product innovations emerge and compete in the market, but as the innovations and markets evolve together a 'dominant design' begins to emerge, and the locus of innovation shifts from product to process, and from radical to more incremental improvements in cost and quality. However, in emerging economies, the path of evolution is often reversed, and begins with incremental process innovations, to produce an existing product at a lower cost or at a lower quality for different market needs (See Case study 11.6, which illustrates the rapid progress of China along this path). As firms improve their capabilities they may then begin to

CASE STUDY 11.6

Manufacturing innovation in China

Since economic reform began in 1978, the Chinese economy has grown by about 9–10% each year, compared to 2–3% for the industrialized countries. As a result its GDP overtook Italy in 2004, France and the UK in 2005 and is expected to overtake Germany in 2008. China has a population of around 1.3 billion, and an economy valued at $2.3 trillion in 2006 (for comparison, the UK was $2.1 trillion, the USA $11.7 trillion, and Japan $4.9 trillion). China now has the world's second-largest economy after the USA on a purchasing power (PPP) basis.

The Chinese government has followed a twin-track policy of exporting relatively low-technology products, while using various measures to protect its domestic economy, and providing subsidies to support selected state-owned firms, to build technological capability. This activist technology policy has been more tightly constrained since the completion of entry to the World Trade Organization in 2005, and implementation of TRIPS (Trade Related Intellectual Property System) in 2006. These will require stricter intellectual property laws and their enforcement, and limit subsidies and interference with trade.

After more than two decades of providing the world economy with inexpensive labour, China is now starting to become a platform for innovation, research and development. The actual formal R&D expenditure is still comparatively small, about 1.3% of GDP (compared to an average of 2.3% of GDP in the advanced economies of the OECD, although Japan exceeds 3%), but the Chinese government aims to make China an 'innovation nation' by 2010, and a scientific power by 2050, and in 2006 increased government funding in R&D by 25% to $425 million. It plans to increase R&D expenditure to 2.5% of GDP by 2020, in line with expenditure in developed economies. China's science and technology output is already increasing, and was ranked fifth globally in terms of science papers produced between 2002 and 2005, which is impressive given the language disadvantage.

China's policy has followed the East Asian model in which success has depended on techno-logical and commercial investment by and collaboration with foreign firms. Typically companies in the East Asian tiger economies such as South Korea and Taiwan developed technological capa-bilities on a foundation of manufacturing competence based on low-tech production, and devel-oped higher levels of capability such as design and new product development, for example, through OEM (Own Equipment Manufacture) production for international firms. However, the flow of technology and development of capabilities are not automatic. Economists refer to 'spillovers' of know-how from foreign investment and collaboration, but this demands a signifi-cant effort by domestic firms.

Most significantly, China has encouraged foreign multinationals to invest in China, and these are now also beginning to conduct some R&D in China. Motorola opened the first foreign R&D lab in 1992, and estimates indicate there were more than 700 R&D centres in China in 2005, although care needs to be taken in the definitions used. The transfer of technology to China, especially in the manufacturing sector, is considered to be a major contributor to its recent economic growth. Around 80% of China's inward foreign direct investment (FDI) is 'technology' (hardware and soft-ware), and FDI inflows have continued to grow, to US$72 billion in 2005 (for comparison, this is around 10 times that attracted by India, whereas some advanced economies continue to attract sig-nificant FDI, for example, $165 billion was invested in the UK in 2005). However, we must distin-guish between technology transferred by foreign companies into their wholly or majority-owned subsidiaries in China, versus the technology acquired by indigenous enterprises. It is only through the successful acquisition of technological capability by indigenous enterprises, many of which still remain state-owned, that China can become a really innovative and competitive economic power.

The import of foreign technology can have a positive impact on innovation, and for large enter-prises, the more foreign technology is imported, the more conducive to its own patenting. However, for the small- and medium-sized enterprises this is not the case. This probably implies that larger en-terprises possess certain absorptive capacity to take advantage of foreign technology, which in turn leads to an enhancement of innovation capacity, whereas the small- and medium-size enterprises are more likely to rely on foreign technology due to the lack of appropriate absorptive capacity and the possibly huge gap between imported and its own technology. Buying bundles of technology has been encouraged. These included embodied and codified technology: hardware and licences. If innova-tion expenditure is broken down by class of innovative activity, the costs of acquisition for embodied technology, such as machines and production equipment, account for about 58% of the total innovation expenditures, compared with 17% internal R&D, 5% external R&D, 3% marketing of new product, 2% training cost and 15% engineering and manufacturing start-up.

It is clear that the large foreign MNCs are the most active in patenting in China. Foreign patenting began around 1995, and since 2000 patent applications have increased annually by around 50%. MNCs' patenting activities are highly correlated with total revenue, or the overall Chinese market size. This strongly supports the standpoint that foreign patents in China are largely driven by demand factors. China's specialization in patenting does not correspond to its export specialization. Automobiles, household durables, software, communication equipments, computer peripherals, semiconductors, telecommunication services are the primary areas. The semiconductor industry in 2005, for example, was granted as many as fourfold inventions of the previous year. Patents by foreign MNCs account for almost 90% of all patents in China, the most active being firms from Japan, the USA and South Korea. Thirty MNCs have been granted more than 1000 patents, and eight of these each have more than 5000: Samsung, Matsushita, Sony, LG, Mitsubishi, Hitachi, Toshiba and Siemens. Almost half of these patents are for the application of an existing technology, a fifth for inventions, and the rest for industrial designs. Among the 18 000 patents for inventions with no prior overseas rights, only 924 originate from Chinese subsidiaries of these MNCs, accounting for only 0.75% of the total. The average lag between patenting in the home country and in China is more than three years, which is an indicator of the technology lag between China and MNCs.

One reason for this pattern is the very low level of industry-funded R&D, as opposed to public-funded, but there has also been a failure of corporate governance in the large state-owned enterprises selected for support. When the economic reform programme began in 1978 it inherited the advantages and disadvantages of Maoist autarchy. China had enterprises producing across a very wide range of products, having spent heavily from the late 1950s to give itself a high degree of technological independence. The main disadvantage was that its technologies were out of date. The government promoted FDI through joint ventures, 51% owned by a 'national team' of about 120 large domestic state-owned enterprises. Pressures on and incentives for management in state-owned firms have encouraged them to rely on external sources of technology, rather than to develop their own internal capabilities. At the same time private Chinese firms have been constrained by a shortage of finance. However, in 2000 the government reviewed its policy and began to restructure the state-owned firms and to support the most successful private firms. There is a clear link between such restructuring and the development of capabilities.

Examples of companies which have gone through significant changes in governance or financial structure include Xiali, which was transformed into a joint venture with Toyota, TPCO, where debt funding was changed into equity and shareholding, which allowed higher investment in production capacity and technology development; and Tianjin Metal Forming, restructured to remove debt and in a stronger position to invest and be a more attractive candidate for a foreign investment. Private firms like Lenovo, TCL, (Ningbo) Bird and Huawei have since prospered and with belated government help, are successful overseas: Huawei, in 2004 gained 40% of its over $5 billion revenues outside China; Haier has overseas revenues of over $1 billion from its home appliances; Lenovo bought IBM's PC division in 2005; TCL made itself the largest TV maker in the world by buying Thomson of France's TV division in 2004; Wanxiang, a motor components manufacturer started by a farmer's son as a bicycle repair shop, had by 2004 $2 billion annual sales.

However, there are significant differences of innovation and entrepreneurial activity in different areas of China. The eastern coastal region is higher than the other regions, especially in Shanghai, Beijing and Tianjin, whose entrepreneurial activity level is higher and continues to grow. Beijing and the Tianjin Region, Yangtze River Delta Region (Shanghai, Jiangsu, Zhejiang), Zhu Jiang Delta Region (Guangdong) are the most active regions. Shanghai ranks first in most surveys, followed by Beijing, but the disparity of the two areas has been expanding. For example, the local city government in Shanghai provides $12 million each year to fund 'little giants', small high-technology firms which it hopes will contribute annual sales of $12 billion by 2010. In the middle region and the northeast region, entrepreneurial activity level is lower than the eastern coastal region, but is increasing. The western and north-west region is the lowest and least improving area for entrepreneurial activity level, and shows little change. Econometric models indicate that the main determinants for entrepreneurial activity are explained by regional market demand, industrial structure, availability of financing, entrepreneurial culture and human capital. Technology innovation and rate of consumption growth have not had significant effects on entrepreneurship in China.

Although some 200 million Chinese still live on $1 a day, China is also the largest market in the world for luxury goods. China is estimated to have 300 000 dollar millionaires, 400 entrepreneurs valued at $60 million each, and seven billionaires. Such disparities in income can create huge social and political tensions, and may result in a reaction against further growth unless governance and distribution are improved further.

Sources: East meets West: 15th International Conference on Management of Technology, Beijing, May 2006; *R&D Management* (2004) Special issue on innovation in China, **34** (4).

make product adaptations and changes in design, and eventually move towards more radical product innovation. This has important implications for the type of capabilities firms need to develop. For example, at first, the emphasis should be on incremental process improvement and development, which suggests innovation in production and organization, rather than technological development or formal R&D (see Case study 11.7 for examples of service innovation in India). This suggests a hierarchy of capabilities or learning, each adding greater value.

Innovation and social change

There are many definitions of social innovation and entrepreneurship, but most include two critical elements:

1. The aim is to create social change and value, rather than commercial innovation and financial value. Conventional commercial entrepreneurship often results in new products and services and growth in the economy and employment, but social benefits are not the explicit goal.
2. It involves business-, public- and third-sector organizations to achieve this aim. Conventional commercial entrepreneurship tends to focus on the individual entrepreneur and new venture, which occupy the business sector, although organizations in the public or third sectors may be stakeholders or customers.

CASE STUDY 11.7

Service innovation in India

India has a population of around 1.1 billion, a large proportion of which is English-speaking, a relatively stable political and legal regime, and a good national system of education, especially in science and engineering. It has some 250 universities and listed 1500 R&D centres (although care needs to be taken in the definitions used in both cases), and this has translated into international strengths in the fields of biotechnology, pharmaceuticals and software. As a result Indian firms have benefited greatly by the increasing international division of labour in some services and the support and development of software and services. India is now a global centre for outsourcing and off-shoring. Until the mid-1980s the software industry was dominated by government and public research organizations, but the introduction of export processing zones provided tax breaks and allowed the import of foreign computer technology for the first time. The market liberalization of 1991 accelerated development and inward investment, and in 2005 India attracted inward investment of $6 billion (significant, but still only around a tenth of that attracted by China). Since then the software and services industry in India has grown by around 50% each year to reach US$8.3 billion by 2000, and employing 400 000, second only to the USA. The industry is forecast to grow to $50 billion by 2008. Unusually for India, which has historically pursued a policy of national self-reliance, the industry is very export-oriented, with around 70% of output being traded internationally.

There are three broad types of software firms in India. First, those that specialize in a specific sector or domain, for example accounting, gaming or film production, and these develop capabilities and relationships specific to those users. Second, those that develop methods and tools to provide low-cost and timely software support and solutions. The majority of the industry is in this lower value-added part of the supply chain, and is involved in low-level coding, maintenance and design, and relies on a large pool of English-speaking talent which costs around 10% of those in the USA or EU. However a third segment of firms is emerging, which is more involved with new product and service development.

India's version of Silicon Valley is around the southern city of Bangalore. This is home to a large number of firms from the USA, as well as indigenous Indian firms. Large employers include Infosys, and call and service centres here employ 250 000 operatives, including support services for firms such as Cisco, Microsoft and Dell. IBM, Intel, Motorola, Oracle, Sun Microsystems, Texas Instruments and GE all now have technology centres here. Texas Instruments was one of the few major foreign firms to start up a development unit in 1985, prior to the opening up of the India economy in 1991. GE Medical Systems followed in the late 1980s, and established a development centre in Bangalore in 1990, which later resulted in a joint venture with the India firm Wipro Technologies. GE now employs 20 000 people in India, who generate sales of $500 million. IBM was one of the first investors in India, but later withdrew because of the onerous government policy and restrictions in the 1980s. It returned after the government liberalized the economy, and its Indian operations contributed $510 million in sales in 2005, employing 43 000 in India following the acquisition of the Indian outsourcing company Daksh in 2004. In 2006 announced that it would triple its investment from $2 billion to $6 billion by 2009, including further service delivery centres to support computer networks worldwide and a new telecommunications research centre. Similarly, Adobe is to invest $50 million in India over the next five years, and to recruit

300 software developers. Each year Adobe India contributes 10 of the 60 patents which Adobe files each year.

One of the challenges of the software and services industry in India is to increase value added through product and service development. To date the impressive growth has been based on winning more outsourcing business from overseas and employing more staff, rather than by increasing the value added by new services and products. For example, the Indian software and service firm Tata plans to increase the proportion of its revenue from new products from around 5% to 40%, to make it less reliant on low-cost human capital, which is likely to become more expensive, and more mobile. Ramco Systems developed an ERP system in the 1990s, which cost a billion rupees to develop and involved 400 developers. By 2000 the company was profitable, with 150 customers, half overseas. It has established sales and support offices in the USA, Europe and Singapore. In 2006 the Indian outsourcing company Genpact (40% owned by GE of the USA) launched a joint venture with New Delhi Television (NDTV) to digital video editing, post-production and archiving services to media firms. The industry is worth $1 trillion, and 70% of all media work is now digital.

Based on patent citations, Indian firms rely much more on linkages with the science base and technology from the developed countries, whereas China has a broader reliance which includes its Asian neighbours in other emerging economies, and specializes on more applied fields of technology. Indian firms rely on technologies from USA firms most – about 60% of all patent citations, followed by (in order of importance), Japan, Germany, France and the UK. In many cases these linkages have been reinforced by inward investment by MNCs, but in other cases they are the result of Indians trained or employed overseas who have returned to India to create new ventures.

Infosys was one of the first and now one of the largest software and IT services firms in India. It was created by entrepreneur N.R. Narayana Murthy with six colleagues in 1981 with only US$250, but by 2006 it was worth $13.7 billion, with annual profits of $345 million. Murthy believes that '*entrepreneurship is the only instrument for countries like India to solve the problem of its poverty . . . it is our responsibility to ensure that those who have not made that kind of money have an opportunity to do so*'.

Sources: Forbes, N. and D. Wield (2002) *From Followers to Leaders: Managing Technology and Innovation*, Routledge, London; IEEE (2006) International Conference on Management of Innovation and Technology, Singapore; T.L. Friedman (2006) *The World is Flat: The Globalized World in the Twenty-First Century*, Penguin, London.

Examples of applications of social innovation and entrepreneurship include:

- poverty relief
- community development
- health and welfare
- environment and sustainability
- arts and culture
- education and employment.

However, social innovation is not simply innovation in a different context. Traditional public- and third-sector organizations have often failed to deliver improvement or change because of the constraints of organization, culture, funding or regulation. For example, in many public- and third-sector organizations the needs of the funders or employees may become more important to satisfy than the needs of their target community.

Therefore social entrepreneurs share most of the characteristics of entrepreneurs (see Chapter 10), but are different in some important respects:

- *Motives and aims* – less concerned with independence and wealth, and more on social means and ends.
- *Timeframe* – less emphasis on short-term growth and longer term harvesting of the venture, and more concern on long-term change and enduring heritage.
- *Resources* – less reliance on the firm and management team to execute the venture, and greater reliance on a network of stakeholders and resources to develop and deliver change.

Key characteristics which appear to distinguish social entrepreneurs from their commercial counterparts include a high level of empathy and need for social justice. The concept of empathy is complex, but includes the ability to recognize and emotionally share the feelings and needs of others, and is associated with a desire to help. However, whilst empathy and a need for social justice may be necessary attributes of a social entrepreneur, they are not sufficient. These may make a social venture desirable, but not necessarily feasible.[52] The feasibility will be influenced by the more conventional personal characteristics of an entrepreneur, such as background and personality, but also some contextual factors more common in public- and third-sector organizations (see Case study 11.8 for an example). Potential barriers to social entrepreneurship:

- Access to and support of local networks of social and community-based organizations, e.g. relationships and trust in informal networks.
- Access to and support of government and political infrastructure, e.g. nationality or ethnic restrictions.

Of course it is not simply a matter of individuals and start-up ventures. As we've seen throughout the book entrepreneurial behaviour can be found in any organization and is central to the ability to develop and reinvent. In the field of social entrepreneurship a growing number of businesses are recognizing the possibilities of pursuing parallel and complementary trajectories, targeting both conventional profits and also social value creation. The case of Carmel McConnell illustrates the potential for social innovation, moving beyond the conventional charity model. With her background in big business and MBA training, she has created a franchise model for social businesses.

Social innovation is also an increasingly important component of 'big business', as large organizations realize that they can only secure a licence to operate if they can demonstrate some concern for the wider communities in which they are located. (The recent backlash against the pharmaceutical firms as a result of their perceived policies in relation to drug provision in Africa is an example of what can happen if firms don't pay attention to this agenda.) 'Corporate social responsibility' (CSR) is becoming a major function in many businesses and many make use of formal measures – such as the 'triple bottom line' – to monitor and communicate their focus on more than simple profit making.

By engaging stakeholders directly, companies are also better able to avoid conflicts, or to resolve them when they arise. In some cases, this involves directly engaging activists who are leading campaigns or protests against a company. For example, Starbucks responded to customers' concerns and activist

Marc Koska and Star Syringe

Marc Koska founded Star Syringe in 1996 to design and develop disposable, single-use or so-called 'auto-disable syringes' (ADS) to help prevent the transmission of diseases like HIV/AIDS. For example, over 23 million infections of HIV and hepatitis are given to otherwise healthy patients through syringe reuse every year.

Marc had no formal training in engineering, but had relevant design experience from previous jobs in modelling and plastics design. He designed the ADS according to the following basic principles:

- Cheap: the same price as a standard disposable plastic syringe.
- Easy: manufactured on existing machinery, to cut set-up costs.
- Simple: used as closely as possible in the same way as a standard disposable plastic syringe.
- Scalable: licensed to local manufacturers, leveraging resources in a sustainable way.

The ADS is not manufactured in house, but by Star licensees based all over the world. The technology is now licensed to international aid agencies and is recognized by the UNICEF and the World Health Organization (WHO). Star Alliance is the network which connects the numerous manufacturing licensees to the global marketplace. The Alliance includes 19 international manufacturing partners, and serves markets in over 20 countries. The combined capacity of the alliance licensees is close to 1 billion annual units.

Koska's dedication and persistent drive over the last 20 years have earned him respect from leaders in state health services as well as industry: in February 2005 for example the Federal Minister for Health in Pakistan presented Marc with an award for Outstanding Contribution to Public Health for his work on safer syringes, and in 2006 the company won the UK Queen's Award for Enterprise and International Trade.

Sources: www.starsyringe.com; web.mac.com/marckoska/.

protests about the impact of coffee growing on songbirds by partnering with leading activist groups to improve organic, bird-friendly coffee production methods, setting up a pilot sourcing programme, and further increasing public awareness. The conflict was resolved, and Starbucks established itself as a leader on this issue.

Ahold, the largest retailer in the Netherlands, has also used stakeholder engagement to enable it to expand its operations into under-served urban areas. The company realized that on its own it would not be able to operate successfully and would need to work with government and other companies to create a 'sound investment climate' locally. With the local government and nine other retailers it developed a comprehensive development plan for the Dutch town of Enschede.

Sometimes there is scope for social entrepreneurship to spin out of mainstream innovative activity. Procter & Gamble's PUR water purification system offers radical improvements to point-of-use drinking water delivery. Estimates are that it has reduced intestinal infections by 30–50%. The product grew out of research in the mainstream detergents business but the initial conclusion was that the market potential of the product was not high enough to justify investment; by reframing it as a development aid the company has improved its image but also opened up a radical new area for working.

In some cases the process begins with an individual but gradually a trend is established which other players see as relevant to follow, in the process bringing their resources and experience to the game. Examples here might include 'Fair Trade' products, which were originally a minority idea but have now become a mainstream item in any supermarket, or the wind-up radio which provided a model that highlighted the needs – but also the opportunities – for communications in developing countries.

There is also increasing pressure on established businesses to work to a more socially responsible agenda – with many operating a key function around corporate social responsibility. The concept is simple – firms need to secure a 'licence to operate' from the stakeholders in the various constituencies in which they work. Unless they take notice of the concerns and values of those communities they risk passive, and increasingly active, resistance and their operations can be severely affected. CSR goes beyond public relations in many cases with genuine efforts to ensure social value is created alongside economic value, and that stakeholders benefit as widely as possible and not simply as consumers. CSR thinking has led to the development of formal measures and frameworks like the 'triple bottom line' which many firms use as a way of expanding the traditional company reporting framework to take into account not just financial outcomes but also environmental and social performance.

It is easy to become cynical about CSR activity, seeing it as a cosmetic overlay on what are basically the same old business practices. But there is a growing recognition that pursuing social entrepreneurship-linked goals may not be incompatible with developing a viable and commercially successful business. For example, in 2004 a survey by the consultants Arthur D. Little of around 40 technology firms in Europe, Japan and the USA suggested that a focus on the sustainability question was beginning to be recognized as a key way of creating new market space, products and processes. In particular 95% felt that it had potential to bring business value and almost a quarter felt it definitely would deliver such value. This value is in both intangible domains like brand and reputation but increasingly in bottom-line benefits like market share and product/service innovation. Significantly there has been considerable acceleration in these trends compared to the last time the survey was conducted, in 1999. When asked where they saw the benefits coming in five years' time 90% believed they would come through new products and services and 75% in new markets and new business models.

The A.D. Little survey suggests that an increasing number of firms are looking to develop new opportunities via social innovation. They use the metaphor of a journey which begins with simple compliance innovation – the 'licence to operate' argument. Many companies have now moved into the 'foothills' of the 'beyond compliance' area where they are realizing that they have to deal with key stakeholders and that in the process some interesting innovation opportunities can emerge (see Case study 11.9). But the real challenge is to move on to the innovation high ground of full-scale stakeholder innovation, *'creating new products and services, processes and markets which will respond to the needs of future as well as current customers'*.[53]

The process happens through seeking out opportunities – often new or different combinations which no one else has seen, and working them up into viable concepts which could be taken forward. It's then a matter of persuading various people – venture capitalists, senior management, etc. – to choose to put resources behind the idea rather than backing off or backing something else. If we get past this hurdle the next step is beginning to transform the idea into reality, weaving together a variety of different knowledge and resource streams before finally launching the new thing – product, process or service – on to a market. Whether they choose to adopt and use it, and spread the word to others so the

CASE STUDY 11.9

Public and private healthcare services

The Danish pharmaceutical firm Novo Nordisk is deploying stakeholder innovation through expansion and reframing of the role of its corporate stakeholder relations (CSR) activities. It has been consistently highly rated on this, not least because it is a board-level strategic responsibility (specified in the company's articles of association) with significant resources committed to projects to sustain and enhance good practice. It was one of the first companies to introduce the concept of the triple bottom line performance measurement, recognizing the need to take into account wider social and societal concerns and to be clear about its values.

But there is now growing recognition that this investment is also a powerful innovation resource. It offers a way of complementing the compound pipeline R&D. As we've seen, the questions here are:

- How does the organization pick up on emergent phenomena?
- How do they get in the game early?
- And if they do manage that, how might they position themselves to shape the emergent new game?

Investing in stakeholder relations represents a powerful way of doing this by involving the company closely in learning from a wide range of actors. Two examples will help highlight this process.

(i) The DAWN (Diabetes Attitudes, Wishes and Needs) programme

The objective of DAWN, initiated in 2001, was to explore attitudes, wishes and needs of both diabetes sufferers and healthcare professionals to identify critical gaps in the overall care offering. Its findings showed in quantitative fashion how people with diabetes suffered from different types of emotional distress and poor psychological well-being, and that such factors were a major contributing factor to impaired health outcomes. Insights from the programme opened up new areas for innovation across the system. For example, a key focus was on the ways in which healthcare professionals presented therapeutic options involving a combination of insulin treatment and lifestyle elements – and on developing new approaches to this.

A DAWN Summit in 2003 brought together representatives from 31 countries and key agencies such as the World Health Organization; it was widely publicized in specialist and non-specialist journals and via the International Diabetes Federation (IDF). The result has been to establish a common framework within which an understanding of the issues is combined with relationships with key players who could become involved in the design and delivery of relevant innovations. DAWN's value is as an independent, evidence-based platform on which extended discussion and exploration can take place around the future of diabetes management as a holistic system – not simply the treatment via insulin or other specific therapies. It has helped mobilize a global community of practice across which there is significant sharing of learning and interactive changing of perspectives.

Søren Skovlund, senior adviser, Corporate Health Partnerships, sees the key element as '. . . *the use of the DAWN study as a vehicle to get all the different people round the same table . . . to bring patients, health professionals, politicians, payers, the media together to find new ways to work more effectively together on the same task . . . You can't avoid getting some innovation because you're bringing together different baskets of knowledge in the room!'*

Why do it? One reason is a growing sense that the rules of the game around chronic disease management are shifting. For example the WHO estimate that diabetes is a bigger killer than AIDS with around 3.2 million deaths attributable to the disease – and its complications – every year. In developing countries the figures are particularly alarming where 1 in 10 deaths of adults aged 35 to 64 are due to diabetes (in some countries the figure is as high as 1 in 5). Chronic diseases like diabetes represent a time bomb around which major activity is likely to happen in the near future. Healthcare systems are increasingly focusing their efforts on reducing the socioeconomic burden of disease through reorganization of the care process and structure. These major shifts pose the risk that the product-focused pharmaceutical industry is falling behind.

DAWN is a learning investment for Novo Nordisk about the whole system of diabetes care, not just the drug side. It opens up possibilities around emergent models – for example, in integrated service solutions provision around chronic healthcare management.

(ii) National Diabetes Programmes

DAWN provides an input to a set of activities operated by Novo Nordisk under the banner of National Diabetes Programmes (NDPs). These programmes bring the company into close and continuing proximity with key and diverse players in that field. Beyond the PR value of showing the company's commitment to improving diabetes care it creates presence/positioning for emergence.

This initiative began in 2001 when the company set about building a network of relationships in key geographical areas helping devise and configure relevant holistic care programmes. Rather than a product focus, NDPs offer a range of inputs, for example, supporting education of healthcare professionals or establishing clinics for care of diabetic ulcers. CEO, Lars Rebien Sørensen argues that *'only by offering and advocating the right solutions for diabetes care will we be seen as a responsible company. If we just say "drugs, drugs, drugs", they will say "give us a break!"'* This is clearly good CSR practice – but the potential learning about new approaches to care, especially under resource-constrained conditions, also represents an important 'hidden R&D' investment.

Typically the NDP process involves identifying needs with key partners and developing a National Diabetes Healthcare Plan – with Novo Nordisk providing resources to help with implementation. The NDPs are closely linked to another initiative, the World Diabetes Foundation, established in 2003 with an initial pledge of $100 million over a 10-year period. It operates in over 40 countries trying to raise awareness and improve care especially in areas – such as India and China – where diabetes is seriously under-diagnosed.

The core underlying principle is one of developing and testing generic prototype plans which can then be 'customized' for a variety of other countries. For example, Tanzania was an early pilot. It was initially difficult to convince authorities to take chronic diseases like diabetes into account since they had no budget for them and were already fighting hard with infectious diseases. With little likelihood of new investment Novo Nordisk began working with local diabetes associations to establish demonstration projects. It set up clinics in hospitals and villages, trained staff and provided relevant equipment and materials. This gave visibility to the possibilities in a chronic disease management approach – for example before the programme someone with diabetes might have had to travel 200 km to the major hospital in Dar-es-Salaam whereas now they can be dealt with locally. The value to the national health system is significant in terms of savings on the costs of treating complications such as blindness and amputations, which are tragic and expensive results of poor and delayed treatment. As a result the Ministry of Health is able to deal with diabetes management without the need for new investment in hospital capacity or recruitment of new doctors and nurses. Novo Nordisk is essentially a facilitator here – but in the process is very much centrally involved in an emerging and shifting healthcare system.

NDPs represent an experience-sharing network across over 40 countries. Much of the learning is about the context of different national healthcare systems and how to work within them to bring about significant change – essentially positioning the company for co-evolution. One of the big lessons has been the recognition of the problem of under-diagnosis. Typically around 80% of diabetes sufferers in developing countries remain undiagnosed, and as a result most attention (of the healthcare system and the pharmaceutical companies working with them) goes on the 20% who are identified. The move is now towards finding the undiagnosed and developing ways to manage their diabetes in such a way that they don't get complications which is where the major costs arise. This has implications not only for expanding the potential market for insulin treatment but also moving the company into much broader areas of healthcare management and delivery.

innovation diffuses depends a lot on how we manage using other knowledge and resource streams to understand, shape and develop the market. We also know that the whole process is influenced and shaped by having clear strategic direction and support, an underlying innovative and enthusiastic organization willing to commit its creativity and energy, and extensive and rich links to other players who can help with the knowledge and resource flows we need. Fuelling the whole is the underlying creativity, drive, foresight and intuition to make it happen – entrepreneurship – to undertake and take the risks.

So how does this play out in the case of social innovation and entrepreneurship? Table 11.4 gives some examples of the challenges and potential responses.

Table 11.4	**Challenges of social innovation**
What has to be managed . . .	**Challenges in social entrepreneurship**
Search for opportunities	Many potential social entrepreneurs (SEs) have the passion to change something in the world – and there are plenty of targets to choose from, like poverty, access to education, healthcare and so on. But passion isn't enough – they also need the classic entrepreneur's skill of spotting an opportunity, a connection, a possibility which could develop. It's about searching for new ideas which might bring a different solution to an existing problem – for example, the micro-finance alternative to conventional banking or street-level moneylending.

(continued)

Table 11.4	**(Continued)**
What has to be managed . . .	**Challenges in social entrepreneurship**
	As we've seen elsewhere in the book the skill is often not so much discovery – finding something completely new – as connection – making links between disparate things. In the SE field the gaps may be very wide – for example, connecting rural farmers to high-tech international stock markets requires considerably more vision to bridge the gap than spotting the need for a new variant of futures trading software. So SEs need both passion and vision, plus considerable broking and connecting skills.
Strategic selection	Spotting an opportunity is one thing – but getting others to believe in it and, more importantly, back it is something else. Whether it's an inventor approaching a venture capitalist or an internal team pitching a new product idea to the strategic management in a large organization, the story of successful entrepreneurship is about convincing other people. In the case of SE the problem is compounded by the fact that the targets for such a pitch may not be immediately apparent. Even if you can make a strong business case and have thought through the likely concerns and questions, who do you approach to try and get backing? There are some foundations and non-profit organizations but in many cases one of the important skill sets of a SE is networking, the ability to chase down potential funders and backers and engage them in their project. Even within an established organization the presence of a structure may not be sufficient. For many SE projects the challenge is that they take the firm in very different directions, some of which fundamentally challenge its core business. For example, a proposal to make drugs cheaply available in the developing world might sound a wonderful idea from an SE perspective – but it poses huge challenges to the structure and operations of a large pharmaceutical firm with complex economics around R&D funding, distribution and so on.
	It's important to build coalitions of support – securing support for social innovation is very often a distributed process – but power and resources are often not concentrated in the hands of a single decision-maker. There may also not be a 'board' or venture capitalist to pitch the ideas to – instead it is a case of building momentum and groundswell.
	It's very important to provide practical demonstrations of what otherwise might be seen as idealistic 'pipedreams'. Role

(continued)

Table 11.4	(Continued)
What has to be managed . . .	**Challenges in social entrepreneurship**
	of pilots which then get taken up and gather support is well-proven – for example, the Fair Trade model or micro-finance.
Implementation	Social innovation requires extensive creativity in getting hold of the diverse resources to make things happen – especially since the funding base may be limited. Networking skills become critical here – engaging different players and aligning them with the core vision.
Innovation strategy	Here the overall vision is critical – the passionate commitment to a clear vision can engage others – but social entrepreneurs can also be accused of idealism and head in the clouds. Consequently there is a need for a clear plan to translate the vision step by step into reality.
Innovative organization	Social innovation depends on loose and organic structures where the main linkages are through a sense of shared purpose. At the same time there is a need to ensure some degree of structure to allow for effective implementation.
Rich linkages	The history of many successful social innovations is essentially one of networking, mobilizing support and accessing diverse resources through rich networks. This places a premium on networking and broking skills.

Innovation and sustainability

Social and political concerns about the environment and sustainability present a critical, but often subtle, influence on the *rate*, and more importantly *direction*, of innovation. Science and technology do have their own internal logics, but development paths and applications are influenced and shaped by broader political, social and commercial imperatives. In most cases there are numerous potential technological trajectories, most of which will not be pursued, or will fail to become established. For example, nuclear power as a technological innovation has evolved in very different ways in countries like the USA, the UK, France and Japan. Similarly, innovation in genetically modified crops and foods has taken radically different paths in the USA and Europe, mainly due to public concerns and pressure. Case study 11.10 discusses some of the more general issues related to managing sustainable innovation.

The most conventional approach to innovation and sustainability focuses on how to influence the development and application of innovations through regulation and control. In this approach, formal policies are used in an attempt to direct innovation by using systems of regulation, targets, incentives,

Managing innovation for sustainability

In their review of the field, Frans Berkhout and Ken Green argue that 'technological and organizational innovation stands at the heart of the most popular and policy discourses about sustainability. Innovation is regarded as both a cause and solution...yet, very little attempt has been made in the business and environment, environmental management and environmental policy literatures to systematically draw on the concepts, theories and empirical evidence developed over the past three decades of innovation studies.' They identify a number of limitations in the innovation literature, and suggest potential ways to link innovation and sustainability research, policy and management:

1. A focus on managers, the firm, or the supply chain is too narrow. Innovation is a distributed process across many actors, firms and other organizations, and is influenced by regulation, policy and social pressure.
2. A focus on a specific technology or product is inappropriate. Instead the unit of analysis must be on technological systems or regimes, and their evolution rather than management.
3. The assumption that innovation is the consequence of coupling technological opportunity and market demand is too limited. It needs to include the less obvious social concerns, expectations and pressures. These may appear to contradict stronger but misleading market signals.

They present empirical studies of industrial production, air transportation and energy to illustrate their arguments, and conclude that 'greater awareness and interaction between research and management of innovation, environmental management, corporate social responsibility and innovation and the environment will prove fruitful.'

Source: From Berkhout, F. and K. Green (eds) (2002) Special issue on managing innovation for sustainability. *International Journal of Innovation Management*, **6** (3), 227–232.

and usually punishments for non-compliance. This can be effective, but is a rather blunt instrument to encourage change, and can be slow and incremental.

A more balanced and effective approach tries to understand how technology, markets and society co-evolve through a process of negotiation, consultation and experimentation with new ways of doing things. This perspective demands a better appreciation of how firms and innovation work, and highlights the need to better understand all the organizations involved – the policy-makers, consumers, firms, institutions and other stakeholders that can influence the rate and direction of innovation.[54] By focusing on policy and regulation the innovation–environment debate and research has not really fully understood or engaged with the motivations and actions of individual entrepreneurs or innovative organizations.

Innovation is often presented as a major contribution to the degradation of the environment, through its association with increased economic growth and consumption.[55] However, innovation must also be a large part of any potential solution to a range of environmental issues, including:

- *Cleaner products* – with a lower environmental impact over their life cycle.
- *More efficient processes* – to minimize or treat waste, to reuse or recycle.

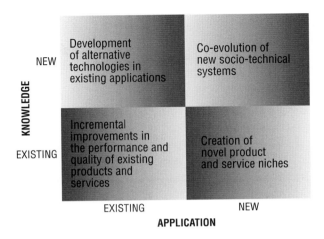

FIGURE 11.5: A typology of sustainable innovations

- *Alternative technologies* – to reduce emissions, provide renewable energy.
- *New services* – to replace or reduce consumption of products.
- *Systems innovation* – to measure and monitor environmental impact, new socio-technical systems.

Figure 11.5 presents a typology of the different ways in which innovation can contribute to sustainability.[56] One dimension is the novelty of the knowledge, and the other dimension is the novelty of the application of that knowledge. In the bottom left quadrant the innovation focuses on the improvement of existing technologies, products and services. This is not necessarily incremental, and may at times involve radical innovation, but the goals and performance criteria remain the same, for example, increasing the fuel efficiency of a power station or car engine. This is the most common type of innovation, and we have discussed this throughout this book. The top left-hand quadrant represents the development of new knowledge, but its application to existing problems. This includes alternative materials, processes or technologies used in existing products. For example, in energy production and packaging of goods there are often many alternative competing technologies, with very different properties and benefits. In food packaging, glass, different plastics, aluminium and steel are all viable alternatives, but each has different energy requirement over their life cycle in their production and reuse or recycling.

Moving to the right-hand column, the bottom quadrant represents the application of existing knowledge to create new market niches. These are sometime called architectural innovations, because they reuse different components and subsystems in new configurations. These are very important for sustainable innovation, as typically such innovations emerge and are developed in niches which initially coexist with the existing mass market, but these niches can mature and grow to influence demand and development in the dominant market (Case study 11.11). For example, in the car industry safety was not a significant feature until the early 1980s. Up until that point the assumption was that 'safety did not sell', and manufacturers were reluctant to develop such features. Corning was initially unable to convince any USA manufacturer to adopt laminated windscreens (windshields). However, local demand for improved safety in Scandinavia, especially Sweden, encouraged local manufacturers such as Volvo and Saab to develop and incorporate new safety technologies. These slowly became popular in overseas markets, and competing manufacturers had to respond with similar features. As a result today almost all

CASE STUDY 11.11

The evolution of electric and hybrid cars

The car industry is an excellent example of a large complex socio-technical system which has evolved over many years, such that the current system of firms, products, consumers and infrastructure interact to restrict the degree and direction of innovation. Since the 1930s the dominant design has been based around a gasoline (petrol)- or diesel-fuelled reciprocating combustion engine/Otto cycle, mass-produced in a wide variety of relatively minimally differentiated designs. This is no industrial conspiracy, but rather the almost inevitable industrial trajectory, given the historical and economic context. This has resulted in car companies spending more on marketing than on research and development. However, growing social and political concerns over vehicle emissions and their regulation have forced the industry to reconsider this dominant design, and in some cases to develop new capabilities to help to develop new products and systems. For example, zero and low emissions targets and legislation have encouraged experimentation with alternatives to the combustion engine, whilst retaining the core concept of personal, rather than collective or mass travel.

For example, the zero-emission law passed in California in 1990 required manufacturers selling more than 35 000 vehicles a year in the state to have 2% of all vehicle sales zero-emission by 1998, 5% by 2001 and 10% by 2003. This most affected GM, Ford, Chrysler, Toyota, Honda and Nissan, and potentially BMW and VW, if their sales increased sufficiently over that period. However, the US automobile industry subsequently appealed, and had the quota reduced to a maximum of 4%. As fuel cells were still very much a longer term solution, the main focus was on developing electric vehicles. At first sight this would appear to represent a rather 'autonomous' innovation, that is the simple substitution of one technology (combustion engine) for another (electric). However, the shift has implications for related systems such as power storage, drive-train, controls, weight of materials used and the infrastructure for refuelling/recharging and servicing. Therefore it is much more of a 'systemic' innovation than it first seems. Moreover, it challenges the core capabilities and technologies of many of the existing car manufacturers. The US manufacturers struggled to adapt, and early vehicles from GM and Ford were not successful. However, the Japanese were rather more successful in developing the new capabilities and technologies, and new products from Toyota and Honda have been particularly successful.

However, zero-emissions legislation was not adopted elsewhere, and more modest emission reduction targets were set. Since then, hybrid petrol-electric cars have been developed to help to reduce emissions. These are clearly not long-term solutions to the problem, but do represent valuable technical and social prototypes for future systems such as fuel cells. In 1993, Eiji Toyoda, Toyota's chairman and his team embarked on the project code named G21. G stands for global and 21, the twenty-first century. The purpose of the project was to develop a small hybrid car that could be sold at a competitive price in order to respond to the growing needs and eco awareness of many consumers worldwide. A year later a concept vehicle was developed called the 'Prius', taken from the Latin for 'before'. The goal was to reduce fuel consumption by 50%, and emissions by more than that. To find the right hybrid system for the G21, Toyota considered 80 alternatives before narrowing the list to four. Development of the Prius required the integration of different technical capabilities, including, for example, a joint venture with Matsushita Battery.

The prototype was revealed at the Tokyo Motor Show in October 1995. It is estimated that the project cost Toyota US$1 billion in R&D. The first commercial version was launched in Japan in December 1997, and after further improvements such as battery performance and power source management, introduced to the US market in August 2000. For urban driving the economy is 60 MPG, and 50 for motorways – the opposite consumption profile of a conventional vehicle, but roughly twice as fuel efficient as an equivalent Corolla. From the materials used in production, through driving, maintenance, and finally its disposal, the Prius reduced CO_2 emissions by more than a third, and has a recyclability potential of approximately 90%. The Prius was launched in the USA at a price of $19995, and sales in 2001 were 15556 in the USA, and 20119 in 2002. However, industry experts estimate that Toyota was losing some $16000 for every Prius it sold because it costs between $35000 and $40000 to produce. Toyota did make a profit on its second-generation Prius launched in 2003, and other hybrid cars such as the Lexus range in 2005, because of improved technologies and lower production costs.

The Hollywood celebrities soon discovered the Prius: Leonardo DiCaprio bought one of the first in 2001, followed by Cameron Diaz, Harrison Ford and Calista Flockhart. British politicians took rather longer to jump on the hybrid bandwagon, with the leader of the opposition David Cameron driving a hybrid Lexus in 2006. In 2005, 107897 cars were sold in the USA, about 60% of global Prius sales, and four times more than the sales in 2000, and twice as many in 2004. Toyota plans to sell a million hybrids by 2010.

In addition to the direct income and indirect prestige the Prius and other hybrid cars have created for Toyota, the company has also licensed some of its 650 patents on hybrid technology to Nissan and Ford, which are expected to develop hybrid vehicles for 2009, and Ford plans to sell 250000 hybrids by 2010. Mercedes-Benz showed a diesel-electric S-class at the Frankfurt auto show in autumn 2005, and Honda has developed its own technology and range of hybrid cars, and is also probably the world leader in fuel cell technology for vehicles.

Sources: A. Pilkington and Dyerson, R. (2004) Incumbency and the disruptive regulator: the case of the electric vehicles in California. *International Journal of Innovation Management*, **8** (4), 339–54; *The Economist* (2004) Why the future is hybrid, 4 December; *Financial Times* (2005) Too soon to write off the dinosaurs. 18 November; *Fortune* (2006) Toyota: the birth of the Prius, 21 February.

cars have a range of active and passive safety technologies, such as airbags, side-impact protection, crumple zones, anti-lock brakes and electronic stability systems.

The top-right quadrant is probably the most fundamental contribution of innovation to sustainability. It is here that new socio-technical systems co-evolve. Developers and users of innovation interact more closely, and many more actors are involved in the process of innovation. In this case firms are not the only, or even the most important, actor, and the successful development and adoption of such systems innovation demand a range of 'externalities', such as supporting infrastructure, complementary products and services, finance and new training and skills. For example, the micro-generation of energy requires much more than technological innovation and product development. It requires changes in energy pricing and regulation, an infrastructure to allow the sale of energy back to the grid, and new skills and services in the installation and service of generators. Such innovations typically evolve by a combination of top-down policy change and coordination, and bottom-up social change and firm behaviour.

Innovation networks can exist at any level: global, national, regional, sector, organizational or individual. Whatever the level of analysis, the most interesting attribute of an innovation network is the degree and type of interaction between actors, which results in a dynamic but inherently unstable set of relationships. Innovation networks are an organizational response to the complexity or uncertainty of technology and markets, and as such innovations are not the result of any linear process. This makes it very difficult, if not impossible, to predict the path or nature of innovation resulting from network interactions. The generation, application and regulation of an innovation within a network is unlike the trial-and-error process within a single firm or venture, or variation and selection within a market. Instead, actors in an innovation network attempt to reduce the uncertainty associated with complexity through a process of recursive learning and testing. For example, analysis of the entire value and supply chain of a business can reveal opportunities for innovation for sustainability, which are less obvious but much more effective than simply regulating outcomes.

Summary and further reading

In this chapter we have attempted to develop a broad view of innovation and its more fundamental financial, economic and social benefits. Most accounts of innovation and performance adopt a rather narrow perspective, typically focusing on how firms appropriate the benefits from innovation, usually by means of intellectual property rights, standards or first-mover advantages. An exception is the excellent collection of papers in *Research Policy*, volume 35, 2006, in honour of the seminal paper by David Teece on the subject (see below).

The generation, acquisition, sharing and exploitation of knowledge are central to successful innovation. However, there is a wide range of different types of knowledge, and each plays a different role. Tacit knowledge is critical, but is difficult to capture, and draws upon individual expertise and experience. Therefore, where possible, tacit knowledge needs to be made more explicit and codified to allow it to be more readily shared and applied to different contexts. One of the key challenges is to identify and exchange knowledge across different groups and organizations, and a number of mechanisms can help, mostly social in nature, but supported by technology. In limited cases, codified knowledge can form the basis of legal IPR, and these can form a basis for the commercialization of knowledge. However, care needs to be taken when using IPR, as these can divert scarce management and financial resources, and can expose organization to imitation and illegal use of IPR.

Knowledge management and intellectual property are both very large and complex subjects. For knowledge management, we would recommend the books by Friso den Hertog, *The Knowledge Enterprise* (Imperial College Press, 2000), for applications and examples, and for theory Nonaka's *The Knowledge Creating Company* (Oxford University Press, 1995). We provide a good combination of theory, research and practice of knowledge management in *From Knowledge Management to Strategic Competence*, edited by Joe Tidd (Imperial College Press, 2006, second edition), which examines the links between knowledge, innovation and performance. More critical accounts of the concept and practice of knowledge management can be found in the editorial by Jackie Swan and Harry Scarbrough, (2001) 'Knowledge management: concepts and controversies', *Journal of Management Studies*, **38** (7), 913–21; J. Storey and E. Barnett (2000) 'Knowledge management initiatives: learning from failure', *Journal of Knowledge Management*, **4** (2), 145–56; and C. Pritchard, R. Hull, M. Chumer and H. Willmott, *Managing Knowledge: Critical Investigations of Work and Learning* (Macmillan, 2000). Harry Scarbrough also edits *The Evolution of Business Knowledge* (Oxford University Press, 2008), which reports the findings of the UK national research programme on the relationships between business and knowledge (including one of our research projects).

For a recent technical review of intellectual property, see W. van Caenegem's *Intellectual Property Law and Innovation* (Cambridge University Press, 2007). For understanding the role and limitations of intellectual property, we like the theoretical approach adopted by David Teece, for example, in his book *The Transfer and Licensing of Know-how and Intellectual Property* (World Scientific, 2006), or for a more applied treatment of the topic see *Licensing Best Practices: Strategic, Territorial and Technology Issues*, edited by Robert Goldscheider and Alan Gordon (John Wiley & Sons, Ltd, 2006), which includes practical case studies of licensing from many different countries and sectors. The Open Source movement is covered widely, but often in a partisan way, and a good balanced discussion which links this to innovation can be found in *Open Source: A Multidisciplinary Approach*, by Moreno Muffatto (Imperial College Press, 2006), which focuses on the management issues, and *Innovation without Patents*, edited by Uma Suthersansen, Graham Dutfield and Kit Boey Chow (Edward Elgar, 2007), which examines the policy aspects, especially for developing economies.

Web links

Here are the full details of the resources available on the website flagged throughout the text:

Case studies:
    Joint Solutions
    Green Supply Chain

Interactive exercises:
    Identifying innovative capabilities
    Knowledge mapping

Tools:
    Value analysis
    Value stream analysis
    TRIZ

Video podcast:
    Xerox

Audio podcast:
    Carmel McConnell on social businesses

References

1. **Arrow, K.** (1962) Economic welfare and the allocation of resources for invention. In R. Nelson, ed., *The Rate and Direction of Inventive Activity*, Princeton University Press, Princeton, NJ.
2. **Teece, D.** (1986) Profiting from technological innovation: implications for integration, collaboration, licensing and public policy. *Research Policy*, **13**, 343–73.

3. **Dosi, G., L. Marengo and C. Pasquali** (2006) How much should society fuel the greed of innovators? On the relations between appropriability, opportunities and rates of innovation. *Research Policy*, **35**, 1110–21.

4. **Pianta, M. and A. Vaona** (2007) Innovation and productivity in European industries. *Economics of Innovation and New Technology*, **16** (7), 485–99.

5. **Mansury, M.A. and J.H. Love** (2008) Innovation, productivity and growth in US business services: a firm-level analysis. *Technovation*, **28**, 52–62.

6. **Hurmelinna, P., K. Kylaheiko and T. Jauhianen** (2007) The Janus face of the appropriability regime in the protection of innovations: theoretical re-appraisal and empirical analysis. *Technovation*, **27**, 133–44.

7. **Jacobides, M.G., T. Knudsen and M. Augier** (2006) Benefiting from innovation: value creation, value appropriation and the role of industry architectures. *Research Policy*, **35**, 1200–21.

8. **Hurmelinna-Laukkanen, P. and K. Puumalainen** (2007) Nature and dynamics of appropriability: strategies for appropriating returns on innovation. *R&D Management*, **37** (2), 95–110.

9. **Tidd, J.** (2006) *From Knowledge Management to Strategic Competence*, second edition, Imperial College Press, London; **Tidd, J., C. Driver and P. Saunders** (1996) Linking technological, market and financial indicators of innovation. *Economics of Innovation and New Technology*, **4**, 155–72.

10. **Griliches, Z. and A. Pakes** (1984) *Patents R&D and Productivity*, University of Chicago Press, Chicago; **Stoneman, P.** (1983) *The Economic Analysis of Technological Change*, Oxford University Press, Oxford.

11. **Geroski, P.** (1991) Innovation and the sectoral sources of UK productivity growth. *Economic Journal*, **101**, 1438–51.

12. **Geroski, P.** (1994) *Market Structure, Corporate Performance and Innovative Activity*, Oxford University Press, Oxford.

13. **Pavitt, K.** (1988) Uses and abuses of patent statistics, In A.F.J. Van Raen, ed., *Handbook of Quantitative Studies of Science and Technology*, North Holland, Amsterdam; **Silberston, A.** (1989) *Technology and Economic Progress*, Macmillan, London.

14. **Levin, R., W. Cohen and D. Mowery** (1985), R&D, appropriability, opportunity, and market structure: new evidence on the Schumpeterian hypothesis. *American Economic Review*, **75**, 20–4; **Levin, R.C., A. Klevorick, R. Nelson and S. Winter** (1987) Appropriating the returns from industrial research and development. *Brookings Papers on Economic Activity*, **3**, 783–831.

15. **Pavitt, K. and P. Patel** (1988) The international distribution and determinants of technological activities. *Oxford Review of Economic Policy*, **4** (4).

16. **Hay, D.A. and D.J. Morris** (1991) *Industrial Economics and Organisation*, Oxford University Press, Oxford.

17. **Cohen, W. and R. Levin** (1989) Empirical studies of innovation and market structure. In R. Schmalensee and R. Willig, eds, *The Handbook of Industrial Organisation*, volume 1, North Holland, Amsterdam.

18. **Jensen, E.** (1987) Research expenditures and the discovery of new drugs. *Journal of Industrial Economics*, **XXXVI** (1), 83–96.

19. **Blundell, R., R. Griffith and S. Van Reenen** (1993) Knowledge stocks, persistent innovation and market dominance. Paper given to SPES Discussion Group, Brussels, September.

20. **Freeman, C.** (1982) *The Economics of Industrial Innovation*, Pinter, London.

21. **Mansfield, E.** (1986) Patents and innovation: an empirical study. *Management Science*, **32**, 173–81; **Griliches, Z., B.H. Hall and A. Pakes** (1991) R&D, patents and market value revisited. *Economics of Innovation and New Technology Journal*, **1** (3), 183–202.

22. **Jaffe, A.B.** (1986) Technological opportunity and spillovers of R&D: evidence from firms' patents, profits and market values. *American Economic Review*, **76**, 948–99.

23. **Tidd, J. and M. Trewhella** (1997) Organisational and technological antecedents for knowledge acquisition and learning. *R&D Management*, **27** (4), 359–75.

24. **Mansfield, E.** (1990) *Managerial Economics: Theory, Application and Cases*, sixth edition, W.W. Norton.

25. **Devinney, T.M.** (1993) How well do patents measure new product activity? *Economics Letters*, **41**, 447–50.

26. **Acs, Z. and D.B. Audretsch** (1990) *Innovation and Small Firms*, MIT Press, Cambridge, MA; (1988) Innovation in large and small firms: an empirical analysis. *American Economic Review*, **78**, 678–90.

27. **Chaney, R., T. Devinney and R. Winer** (1992) The impact of new product introductions on the market value of firms. *Journal of Business*, **64** (4), 573–610.

28. **Walker, W.B.** (1979) *Industrial Innovation and International Trading Performance*, JAI Press, New York.

29. **Budworth, D.W.** (1993) Intangible assets and their renewal. *Foundation for Performance Measurement*, UK National Meeting, London, October.

30. **Scherer, F.** (1965) Firm size, market structure, opportunity and the output of patented inventions. *American Economic Review*, **55**, 1097–125; (1983) The propensity to patent. *International Journal of Industrial Organisation*, **50** (1), 107–28.

31. **Pakes, A.** (1985) On patents, R&D and the stock market rate of return. *Journal of Political Economy*, **93**, 390–409; **Hall, R.** (1993) A framework linking intangible resources and capabilities to sustainable competitive advantage. *Strategic Management Journal*, **14**, 607–18.

32. **Griliches, Z., B.H. Hall and A. Pakes** (1991) R&D, patents and market value revisited. *Economics of Innovation and New Technology Journal*, **1** (3), 183–202.

33. **Buzell, R.D. and B. Gale** (1987) *The PIMS Principle*, Free Press, New York.

34. **Blackler, F.** (1995) Knowledge, knowledge work and organizations: an overview and interpretation. *Organization Studies*, **16** (60), 1021–46.

35. **Bessant, J.** (2003) *High-Involvement Innovation*, John Wiley & Sons, Ltd, Chichester.

36. **Simon, H.A.** (1996) Bounded rationality and organizational learning. In M.D. Cohen and L.S. Sproull, eds, *Organizational Learning* (pp. 175–87), Sage, London.

37. **Nonaka, I. and H. Takeuchi** (1995) *The Knowledge Creating Company*, Oxford University Press, Oxford.

38. **Boisot, M. and D. Griffiths** (2006) Are there any competencies out there? Identifying and using technical competencies. In J. Tidd, ed., *From Knowledge Management to Strategic Competence*, second edition (pp. 249–307), Imperial College Press, London.

39. **Crespi, G., C. Criscuolo and J. Haskel** (2006) Information technology, organisational change and productivity growth: evidence from UK firms. *The Future of Science, Technology and Innovation Policy: Linking Research and Practice*, SPRU 40th Anniversary Conference, Brighton, UK, September.

40. **Hall, R.** (2006) What are strategic competencies? In J. Tidd, ed., *From Knowledge Management to Strategic Competence*, second edition (pp. 26–49), Imperial College Press, London.

41. **O'Leary, D.** (1998) Knowledge management systems: converting and connecting. *IEEE Intelligent Systems*, **13** (3), 30–3; **Becker, M.** (2001) Managing dispersed knowledge: organizational problems, managerial strategies and their effectiveness. *Journal of Management Studies*, **38** (7), 1037–51.

42. **Brown, J.S. and P. Duguid** (2001) Knowledge and organization: a social practice perspective. *Organization Science*, **12** (2), 198–213; (1991) Organizational learning and communities of practice: towards a unified view of working, learning and organization. *Organizational Science*, **2** (1), 40–57;

Hildreth, P., C. Kimble and P. Wright (2000) Communities of practice in the distributed international environment. *Journal of Knowledge Management*, **4** (1), 27–38.

43. Star, S.L. and J.R. Griesemer (1989) Institutional ecology, translations and boundary objects. *Social Studies of Science*, **19**, 387–420; Carlile, P.R. (2002) A pragmatic view of knowledge and boundaries: boundary objects in new product development. *Organization Science*, **13** (4), 442–55.

44. Granovetter, M. (1976) The strength of weak ties. *American Journal of Sociology*, 1360–80; Cummings, J.N. (2004) Work groups, structural diversity, and knowledge sharing in a global organization. *Management Science*, **50** (3), 352–64.

45. den Hertog, J.F. and E. Huizenga (2000) *The Knowledge Enterprise*, Imperial College Press, London.

46. Tranfield, D., M. Young, D. Partington, J. Bessant and J. Sapsed (2006) Knowledge management routines for innovation projects: developing a hierarchical process model. In J. Tidd, ed., *From Knowledge Management to Strategic Competence*, second edition (pp. 126–49), Imperial College Press, London; Coombs, R. and R. Hull (1998) Knowledge management practices and path-dependency in innovation. *Research Policy*, 237–53.

47. Narin, F. (2006) Assessing technological competencies. In J. Tidd, ed., *From Knowledge Management to Strategic Competence*, second edition (pp. 179–219), Imperial College Press, London.

48. Cantwell, J. and J. Molero (2003) *Multinational Enterprises, Innovative Systems and Systems of Innovation*, Edward Elgar, Cheltenham; Granstrand, O., L. Hêakanson and S. Sjèolander (1992) *Technology Management and International Business: Internationalization of R&D and Technology*, John Wiley & Sons, Ltd, Chichester, especially the chapter by P. Patel and K. Pavitt, Large firms in the production of the world's technology: an important case of non-globalization.

49. Kim, L. and R.R. Nelson (2000) *Technology, Learning and Innovation: Experiences of Newly Industrializing Economies*, Cambridge University Press, Cambridge; Viotti, E.B. (2002) National Learning Systems: a new approach on technological change in late industrializing economies and evidences from the cases of Brazil and South Korea. *Technological Forecasting and Social Change*, **69**, 653–80; Bell, M. and K. Pavitt (1993) Technological accumulation and industrial growth: contrasts between developed and developing countries. *Industrial and Corporate Change*, **2** (2), 157–210.

50. Schimtz, H. (2004) *Local Enterprises in the Global Economy*, Edward Elgar, Cheltenham; Sahay, A. and D. Riley (2003) The role of resource access, market conditions, and the nature of innovation in the pursuit of standards in the new product development process. *Journal of Product Innovation Management*, **20**, 338–55.

51. Forbes, N. and D. Wield (2002) *From Followers to Leaders: Managing Technology and Innovation*, Routledge, London.

52. Mair, J., J. Robinson and K. Hockets (2006) *Social Entrepreneurship*, Palgrave Macmillan, Basingstoke. An edited book which discusses the definitions, boundaries and some of the problems of research and practice in the emerging field of social entrepreneurship.

53. Arthur D. Little Ltd (2003) *The Business Case for Corporate Responsibility*, December, Cambridge.

54. Geels, F.W. (2002) Technological transitions as evolutionary reconfiguration processes: a multi-level perspective and a case study. *Research Policy*, **31** (8–9), 1257–74.

55. Porter, M.E. and C. van der Linde (1995) Green and competitive: ending the stalemate. *Harvard Business Review*, **73** (5), 120–34.

56. Smith, A. (2004) Alternative technology niches and sustainable development. *Innovation: Management, Policy and Practice*, **6** (2), 220–35; Smith, A., A. Stirling, and F. Berkhout (2005) The governance of sustainable socio-technical transitions. *Research Policy*, **34** (10), 1491–510.

CHAPTER 12

Capturing learning from innovation

One of the common metaphors used to describe innovation is that of a journey – a complex, fitful travel through uncertain territory involving false starts, wrong directions, blind alleys and unexpected problems. Successful innovation implies the completion of this risky adventure and – through widespread adoption and diffusion of the new idea as a product, service or process – a happy ending with valuable returns on the original investment. But it also provides an opportunity to reflect on the journey and to take stock of the knowledge acquired through an often difficult experience. It's worth doing this because the knowledge gained through such reflection can provide a powerful resource to help with the next innovation journey.

Not all innovation is, of course, successful – but the opportunities for learning from failure are also considerable. Understanding what doesn't work on a technological level, or recognizing the difficulties in a particular marketplace that led to non-adoption is useful information to take stock of and use when planning the next expedition. Experience is an excellent teacher – but its lessons will only be of value if there is a systematic and committed attempt to learn them.

This chapter reviews the ways in which learning can be captured from the innovation experience.

12.1 What have we learned about managing innovation?

It will be useful to briefly take stock of the key themes we have been covering in the book. We can summarize these as follows:

- Learning and adaptation are essential in an inherently uncertain future – thus innovation is an imperative.
- Innovation is about interaction of technology, market and organization.
- Innovation can be linked to a generic process which all enterprises have to find their way through.
- Different firms use different routines with greater or lesser degrees of success. There are general recipes from which general suggestions for effective routines can be derived – but these must be customized to particular organizations and related to particular technologies and products.
- Routines are learned patterns of behaviour which become embodied in structures and procedures over time. As such they are hard to copy and highly firm-specific.
- Innovation management is the search for effective routines – in other words, it is about managing the learning process towards more effective routines to deal with the challenges of the innovation process.

We have also argued that innovation management is not a matter of doing one or two things well, but about good all-round performance. There are no, single, simple magic bullets but a set of learned behaviours. In particular we have identified four clusters of behaviour which we feel represent particularly important routines:

- Successful innovation is strategy-based.
- Successful innovation depends on effective internal and external linkages.
- Successful innovation requires enabling mechanisms for making change happen.
- Successful innovation only happens within a supporting organizational context.

In the *strategy* domain there are no simple recipes for success but a capacity to learn from experience and analysis is essential. Research and experience point to three essential ingredients in innovation strategy:

1. The *position* of the firm, in terms of its products, processes, technologies and the national innovation system in which it is embedded. Although a firm's technology strategy may be influenced by a particular national system of innovation it is not determined by it.
2. The technological *paths* open to the firm given its accumulated competencies. Firms follow technological trajectories, each of which has distinct sources and directions of technological change and which define key tasks for strategy.
3. The organizational *processes* followed by the firm in order to integrate strategic learning across functional and divisional boundaries.

Within the area of *linkages*, developing close and rich interaction with markets, with suppliers of technology and other organizational players, is of critical importance. Linkages offer opportunities for learning – from tough customers and lead users, from competitors, from strategic alliances and from alternative perspectives. The theme of 'open innovation' is increasingly becoming recognized as relevant to an era in which networking and inter-organizational behaviour is the dominant mode of operation.

In order to succeed organizations also need *effective implementation mechanisms* to move innovations from idea or opportunity through to reality. This process involves systematic problem solving and works best within a clear decision-making framework, which should help the organization to stop as well as to progress development if things are going wrong. It also requires skills in project management and control under uncertainty and parallel development of both the market and the technology streams. And it needs to pay attention to managing the change process itself, including anticipating and addressing the concerns of those who might be affected by the change.

Finally, innovation depends on having a *supporting organizational context* in which creative ideas can emerge and be effectively deployed. Building and maintaining such organizational conditions are a critical part of innovation management, and involve working with structures, work organization arrangements, training and development, reward and recognition systems and communication arrangements. Above all, the requirement is to create the conditions within which a learning organization can begin to operate, with shared problem identification and solving and with the ability to capture and accumulate learning about technology and about management of the innovation process.

Throughout the book we have tried to consider the implications of managing innovation as a generic process but also to look at the ways in which approaches need to take into account two key challenges in the twenty-first century – those of managing 'beyond the steady state' and 'beyond boundaries'. The same basic recipe still applies but there is a need to configure established approaches and to learn to develop new approaches to deal with these challenges.

12.2 How can we continue to learn to manage innovation?

To answer this question we need to focus on two dimensions of learning. First there is the acquisition of new knowledge to add to the stock of knowledge resources which the organization possesses. These can be technological or market knowledge, understanding of regulatory and competitive contexts, etc. As we've seen throughout the book, innovation represents a key strategy for developing and sustaining competitiveness in what are increasingly 'knowledge economies' – but being able to deploy this strategy depends on continuing accumulation, assimilation and deployment of new knowledge. Firms that exhibit competitive advantage – the ability to win and to do so continuously – demonstrate *'timely responsiveness and rapid product innovation, coupled with the management capability to effectively coordinate and redeploy internal and external competencies'*.[1]

And second there is knowledge about the innovation process itself – the ways in which it can be organized and managed, the bundle of routines which enable us to plan and execute the innovation journey. We have been speaking throughout the book about the idea of '*innovation capability*' – the ability to organize and manage the process. Figure 12.1 reminds us of the model we have been using as an explanatory framework and '*innovation capability*' refers to our ability to create and operate such a framework in our organizations.

But in a constantly changing environment that capability may not be enough – faced with moving targets along several dimensions (markets, technologies, sources of competition, regulatory rules of the game) we have to be able to adapt and change our framework. This process of constant modification and development of our innovation capability – adding new elements, reinforcing existing ones and sometimes letting go of older and no longer appropriate ones – is the essence of what is called '*dynamic capability*'.[1]

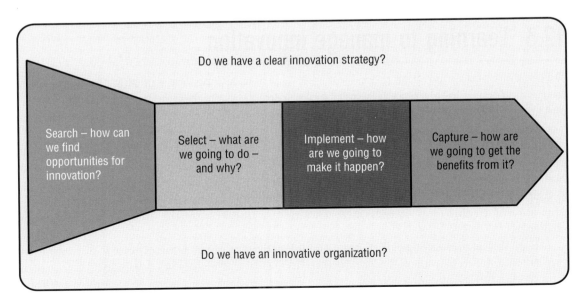

FIGURE 12.1: Simplified model of the innovation process

The lack of such capability can explain many failures, even amongst large and well-established organizations. For example:

- The failure to recognize or capitalize on new ideas that conflict with an established knowledge set – the 'not-invented-here' problem.[2]
- The problem of being too close to existing customers and meeting their needs too well – and not being able to move into new technological fields early enough.[3]
- The problem of adopting new technology – following technological fashions – without an underlying strategic rationale.[4]
- The problem of lack of codification of tacit knowledge.[5]

The costs of not managing learning – of lacking the dynamic capability – can be high. At the least it implies a blunting of competitive edge, a slipping against previously strong performance. For example, 3M was for many years in the top three of *Business Week*'s list of innovative companies, but following a change in CEO and a shift in emphasis away from breakthrough innovation and towards incremental improvement linked to a 'six sigma' programme, their position fell to seventh in 2006 and 22nd in 2007. This prompted significant debate both within the company and in its wider stakeholder community and a refocusing of efforts around developing their core innovation capabilities further. In some cases the fall accelerates and eventually leads to terminal decline – as the fate of companies like Digital, Polaroid or Swissair, once feted for their innovative prowess, indicates.

So we need to look hard at the ways in which organizations can learn – and learn to learn in conscious and strategic fashion. This is why routines play such an important role in managing innovation – they represent the firm-specific patterns of behaviours that enable a firm to solve particular problems.[6] In other words, they embody what an organization (and the individuals within it) has learned about how to learn.

12.3 Learning to manage innovation

We can think of the innovation process in Figure 12.1 as a learning loop – picking up signals which trigger a response. In that sense it is an 'adaptive' learning system, helping the organization survive and grow within its environment. But making sure that this adaptive system works well also requires a second learning loop, one which can 're-program' the system to tune it better to a changing environment and as a result of lessons learned about how well it works. (It's a little like a central heating or air conditioning system – there is an adaptive loop which responds when the temperature gets hotter or colder in the room by modifying the output of the heater or air conditioning unit. But we also need someone to think about – and reset – the thermostat to suit the changing conditions.) This kind of 'double loop' or generative learning is at the heart of the innovation management challenge.[7–9]

All of this argues strongly that firms should undertake some form of review of innovation projects in order to help them develop both technological and managerial capability.[10] One way of representing the learning process that can take place in organizations is to use a simple model of a learning cycle (Figure 12.2). Here learning is seen as requiring:[11]

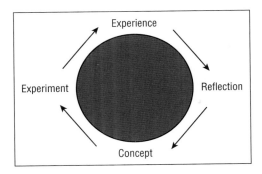

FIGURE 12.2: Kolb's cycle of experiential learning

- Structured and challenging reflection on the process – what happened, what worked well, what went wrong, etc.
- Conceptualizing – capturing and codifying the lessons learned into frameworks and eventually procedures to build on lessons learned.
- Experimentation – the willingness to try and manage things differently next time, to see if the lessons learned are valid.
- Honest capture of experience (even if this has been a costly failure) – so we have raw material on which to reflect.

Effective learning from and about innovation management depends on establishing a learning cycle around these themes.

We should also recognize the problem of *unlearning*. Not only is learning to learn a matter of acquiring and reinforcing new patterns of behaviour – it is often about forgetting old ones.[12] Letting go in this way is by no means easy, and there is a strong tendency to return to the status quo or equilibrium position – which helps account for the otherwise surprising number of existing players in an industry who find themselves upstaged by new entrants taking advantage of new technologies and emerging markets of new business models. Managing discontinuous innovation requires the capacity to cannibalize and look for ways in which other players will try and bring about 'creative destruction' of the rules of the game. Jack Welch, former CEO of General Electric is famous for having sent out a memo to his senior managers asking them to tell him how they were planning to destroy their businesses! The intention was not, of course, to execute these plans but rather to use the challenge as a way of focusing on the need to be prepared to let go and rethink – to unlearn.[13]

12.4 Tools to help capture learning

If we are to extract useful learning from successful – or unsuccessful – innovation activities, then we need to look at the range of tools which might help us with the task. In the following section we'll look briefly at some of the possible approaches to this task.

Post-project reviews (PPRs) are structured attempts to capture learning at the end of an innovation project – for example, in a project debrief. This is an optional stage and many organizations fail to carry out any kind of review, simply moving on to the next project and running the risk of repeating mistakes made in previous projects. Others do operate some form of structured review or post-project audit, however, this does not of itself guarantee learning since emphasis may be more on avoiding blame and trying to cover up mistakes.

On the positive side, they work well when there is a structured framework against which to examine the project, exploring the degree to which objectives were met, the things which went well and those which could be improved, the specific learning points raised and the ways in which they can be captured and codified into procedures which will move the organization forward in terms of managing technology in the future.[14]

Such reviews depend on establishing a climate in which people can honestly and objectively explore issues that the project raises. For example, if things have gone badly the natural tendency is to cover up mistakes or try and pass the blame around. Meetings can often degenerate into critical sessions with little being captured or codified for use in future projects.

The other weakness of PPRs is that they are best suited to distinct projects, for example, developing a new product or service or implementing a new process.[15] They are not so useful for the smaller scale, regular incremental innovation which is often the core of day-to-day improvement activity. Instead we need some form of *systematic capture*. Variations on the standard operating procedures approach can be powerful ways of capturing learning – particularly in translating it from tacit and experiential domains to more codified forms for use by others.[16] They can be simple, for example, in many Japanese plants working on 'total productive maintenance' programmes, operators are encouraged to document the operating sequence for their machinery. This is usually a step-by-step guide, often illustrated with photographs and containing information about 'know-why' as well as 'know-how'. This information is usually contained on a single sheet of paper and displayed next to the machine. It is constantly being revised as a result of continuous improvement activities, but it represents the formalization of all the little tricks and ideas which the operators have come up with to make that particular step in the process more effective.[17]

On a larger scale, capturing knowledge into procedures also provides a structured framework within which to operate more effectively. Increasingly organizations are being required by outside agencies and customers to document their processes and how they are managed, controlled and improved, for example, in the quality area under ISO9000, in the environmental area under ISO14000 and in an increasing number of customer/supplier initiatives such as Ford's QS9000.

Once again there are strengths and weaknesses in using procedures as a way of capturing learning. On the plus side there is much value in systematically trying to reflect on and capture knowledge derived from experience – it is the essence of the learning cycle. But it only works if there is commitment to learning and a belief in the value of the procedures and their subsequent use. Otherwise the organization simply creates procedures which people know about but do not always observe or use. There is also the risk that, having established procedures, the organization then becomes resistant to changing them – in other words, it blocks out further learning opportunities.

Benchmarking is the general name given to a range of techniques which involve comparisons – for example, between two variants of the same process or two similar products – so as to provide opportunities for learning.[18–20] Benchmarking can, for example, be used to compare how different companies manage the product development processes: where one is faster than the other there are learning opportunities in trying to understand how they achieve this.

Benchmarking works in two ways to facilitate learning. First, it provides a powerful motivator since comparison often highlights gaps which – if they are not closed – might well lead to problems in competitiveness later. In this sense it offers a structured methodology for learning and is widely used by external agencies who see it as a lever with which to motivate particularly smaller enterprises to learn and change.[21] It provides a powerful focus for the operation of 'learning networks' (described in Chapter 6) since it offers a framework around which shared learning can be targeted and monitored and across which experiences can be exchanged.[22]

Benchmarking also provides a structured way of looking at new concepts and ideas. It can take several forms:

- Between similar activities within the same organization.
- Between similar activities in different divisions of a large organization.
- Between similar activities in different firms within a sector.
- Between similar activities in different firms and sectors.

The last group is often the most challenging since it brings completely new perspectives. By looking at, for example, how a supermarket manages its supply chain a manufacturer can gain new insights into logistics. Looking at how an engineering shop can rapidly set up and changeover between different products can help a hospital use its expensive operating theatres more effectively.

For example, Southwest Airlines achieved an enviable record for its turnaround speed at airport terminals. It drew inspiration from watching how industry carried out rapid changeover of complex machinery between tasks – and, in turn, those industries learned from watching activities like pit-stop procedures in the Grand Prix motor racing world. In similar fashion Kaplinsky reports on dramatic productivity and quality improvements in the healthcare sector, drawing on lessons originating in inventory management systems in manufacturing and retailing.[23]

Building on the success of benchmarking as an organizational development tool there has been increasing use of what can be termed 'capability maturity models'.[24] The auditing and reviewing process central to benchmarking does not necessarily have to be done in comparison with another organization but can usefully be done against ideal-type or normative models of good practice. This found particular expression during the 'quality revolution' of the 1990s where benchmarking frameworks such as the Malcolm Baldrige Award in the USA, the Deming Prize in Japan and the European Quality Award all used sophisticated benchmarking frameworks.[25] The approach has been extended to a number of other domains – for example, software development processes, project management, IT implementation and new product development.[24] In the UK a framework for benchmarking and auditing manufacturing performance was developed and offered as a national service, with special emphasis on assisting smaller firms improve their performance.[26,27]

12.5 Innovation auditing

Such capability/maturity auditing offers another structured way of reflecting on the process of innovation and how it is managed. The analogy can be drawn with financial auditing where the health of the company and its various operations can be seen through auditing its books. The principle is simple: using what we know about successful and unsuccessful innovation and the conditions which bring it about, we can construct a checklist of questions to ask of the organization. We can then score its performance against some model of 'best practice' and identify where things could be improved.

This auditing approach has considerable potential relevance for the practice of innovation management and a number of frameworks have been developed to support it. Back in the 1980s the UK

National Economic Development Office developed an 'innovation management tool kit' which has been updated and adapted for use as part of a European programme aimed at developing better innovation management amongst small- and medium-sized enterprises (SMEs). Another framework, originally developed at London Business School, was promoted by the UK Department of Trade and Industry and led to the development of a series of frameworks including the 'living innovation' model which was jointly promoted with the Design Council.[26,28] Francis offers an overview of a number of these.[29] Other frameworks have been developed which cover particular aspects of innovation management, such as creative climate, continuous improvement and product development.[30–32] With the increasing use of the Internet have come a number of sites which offer interactive frameworks for assessing innovation management performance as a first step towards organization development.

In each case the purpose of such auditing is not to score points or win prizes but to enable the operation of an effective learning cycle through adding the dimension of structured reflection. It is the process of regular review and discussion that is important rather than detailed information or exactness of scores. The point is not simply to collect data but to use these measures to drive improvement of the innovation process and the ways in which it is managed. As the quality guru, W. Edwards Deming, pointed out, '*If you don't measure it you can't improve it!*' (see Box 12.1).

In reviewing innovative performance we can look at a number of possible measures and indicators – as Box 12.1 indicates. We can look at measures of specific outputs of various kinds, for example, patents and scientific papers as indicators of knowledge produced, or number of new products introduced (and percentage of sales and/or profits derived from them) as indicators of product innovation success.[33] And we can use measures of operational or process elements, such as customer satisfaction surveys to measure and track improvements in quality or flexibility.[34] We can also try to assess the strategic impact where the overall business performance is improved in some way and where at least some of the benefit can be attributed directly or indirectly to innovation, for example, growth in revenue or market share, improved profitability, higher value added.[35]

We could also consider a number of more specific measures of the internal workings of the innovation process or particular elements within it. For example, we could monitor the number of new ideas (product/service/process) generated at the start of the innovation system, failure rates – in the development process, in the marketplace or the number or percentage of overruns on development time and cost budgets. In process innovation we might look at the average lead time for introduction or use measures of continuous improvement, for example, suggestions per employee, number of problem-solving teams, savings accruing per worker, cumulative savings.

BOX 12.1   **Measuring innovation**

The problem with audit frameworks and benchmarks of this kind is that they often provide an indication of how a system and its components are performing but they fail to take into account the final piece of the puzzle – why are they successful? For example, in the total quality field in the USA much interest was shown in the self-assessment framework surrounding the Malcolm Baldrige Award, and many firms used this benchmarking and assessment framework to improve their quality performance. However, one of the winners, Florida Power and Light, whilst undoubtedly doing many of the right things, was none the less forced into receivership; this prompted the addition of a performance category to the assessment framework.

There is also scope for measuring some of the influential conditions supporting or inhibiting the process, for example, the 'creative climate' of the organization or the extent to which strategy is clearly deployed and communicated. And there is value in considering *inputs* to the process, for example, percentage of sales committed to R&D, investments in training and recruitment of skilled staff.

VIEWS FROM THE FRONT LINE

Key lessons learned about managing innovation

- Innovation capability is difficult to create and easy to destroy. It is not a 'fix and forget' thing. It needs constant nourishment and protection when operating in a business environment that is focused on exploitation and where compliance with rules is seen as paramount. It also needs constant attention to keep the momentum going – as if it were an aeroplane, always needing to keep moving forward in order to remain in the air. Managing innovation requires an innovative approach.

Do:
- Be very visible and very active in promoting innovation.
- Encourage senior management to take an active role in promoting innovation.
- Encourage people to challenge and question.
- Allow experimentation.
- Allow individuality to take over at times.
- Protect from the corporate bureaucracy.
- Remember that it takes time to develop an innovation capability.
- Continuously monitor innovation performance.
- Make sure that the team has a clear objective, an end point rather than a tightly specified outcome.
- Allow the people involved latitude to try things out for themselves.
- Promote innovation across the whole business.

Don't:
- Lose focus on the objective – what is the innovation for?
- Use your innovation capability and resource as a quick fix in cost reduction situations.
- Be prescriptive in how results have to be achieved.
- Force conformity on the innovation team.
- Allow excess resource or time, as this will dilute the pressure to come up with a solution.
- Try to manage innovation with a rule book.

Patrick McLaughlin, Managing Director, Cerulean

Do:
- Build a project-based organization.
- Build a good portfolio management structure.
- Build a funnel or stage-gate system, with gates where projects pass through.
- Ensure a large enough human resource base allocated to innovation-related activities.

Don't:
- Put people in functional positions only.
- Lose track of whether projects are rightly being continued in the innovation funnel.

Wouter Zeeman, CRH Insulation Europe

- Don't over-manage people, people generally want to do a good job.
- Get the best team that you can around you in particular people that are better than you.
- Learn from your team, don't be afraid for them to learn from you.
- Look for the simple, not the complex. Things often don't need to be so difficult.
- Don't try and measure everything: the key is customer first all else is secondary.

John Tregaskes, Technical Specialist Manager, Serco

- Focus on a clearly articulated 'outcome', i.e. the result you are trying to achieve, and channel the scarce resources and creative talent you have toward finding innovative ways of delivering on this outcome.

Do:
- Leverage and institutionalize the use of tools.
- Make it fun.
- Engage diverse groups of people.
- Get off-site if you can.
- Value and encourage contributions, keep it simple to begin with.
- Focus on innovation driven from large programmes as well as bottom-up engagement of the line.
- Deliver some early successes and publicize the hell out of them to gain management attention and traction.
- Have a creative process in mind and a means of narrowing to get to solution.

Don't:
- Just put a mechanism in place and expect miracles.
- Let our interpretation of regulatory constraints get in the way (be compliant, but explore the interpretation we have made of the underlying regulations).
- Sit in your office – get out there.
- Underestimate the impact of peer pressure.
- Personal risk taking/willingness to think outside the box.

John Gilbert, Head of Process Excellence, UBS

- Front end of innovation process must be detached from standard development process, e.g. stage-gate model.
- Dedicated people for dedicated tasks to reduce the risk of 'fluffiness'.
- Difficult to maintain full attention from senior management on innovation projects over several years and acceptance from senior management that radical innovation projects will have a higher risk than incremental projects.

John Thesmer, Managing Director, Ictal Care, Denmark

- **Do** talk frequently with end users of your technology and understand the other constraints that might make your innovation less than practical for them.

- My biggest lesson with regards to managing innovation – at least in the oil and gas industry – is that the human issues and change management dimensions of technology deployment are much bigger than most people think. This tends to be the 'Achilles' heel' that dooms many innovations to failure in this sector. One has to remember that most of the people working in an average *Fortune 500* company are focused on making money for their company by using today's technologies and methods. When an innovator shows up with a new gizmo, the deployment process is typically perceived by many as an intrusion to their day-to-day workflows and procedures. Innovators seem to be born with an instinct that new technologies are inherently better than whatever they are replacing, but this is not a perspective that one's co-workers will always share. Accordingly, getting a new technology deployed into the energy industry takes a surprising amount of salesmanship, convincing other people, and tenacity. The 'big lesson,' therefore, is that most of your non-R&D colleagues won't necessarily look at new technologies through the same lens as you do.
- **Don't** assume that people will naturally want to use your innovation. It may take years before they feel this way.
- **Do** everything in your power to make a technology successful, but **don't** feel like a failure if it doesn't take root. If you're never failing, you're not pushing the envelope.

Rob Perrons, Shell Exploration, USA

12.6 Developing innovation management capability

A great deal of research effort has been devoted to the questions of what and how to measure in innovation. The risk is that we become so concerned with these questions that we lose sight of the practical objective, which is to reflect upon and improve the management of the process. Having established in this book some of the factors which appear to influence success and failure in innovation in the experience of others, we can begin to develop a tool for assessing and developing innovative performance in organizations. We might begin with a simple checklist of factors and assign a score to each of them so as to develop a profile of innovation performance.

So, for example, an organization with no clear innovation strategy, with limited technological resources and no plans for acquiring more, with weak project management, with poor external links and with a rigid and unsupportive organization would be unlikely to succeed in innovation. By contrast, one that was focused on clear strategic goals, had developed long-term links to support technological development, had a clear project management process, which was well supported by senior management and which operated in an innovative organizational climate, would have a better chance of success.

An example of such a scale might be:

1 = Innovation not even thought about, rarely happens.
2 = Some awareness but random and occasional responses, informal systems.
3 = Awareness and formal systems in place – but could still be improved.
4 = Highly developed and effective systems including provision for improvement and development.

Of course no organization starts with a perfectly developed capability to organize and manage innovation. It goes through the process of trial and error learning, slowly finding out which behaviours work and which do not and gradually repeating and reinforcing them into a pattern of 'routines'. Developing innovation capability involves establishing and reinforcing those routines, and reviewing and checking that they are still appropriate or whether they need replacing or modifying. Key questions are:

- What do we need to do more of, strengthen?
- What do we need to do less of, or stop?
- What new routines do we need to develop?

We saw in Chapter 2 that innovation capability could be represented as a series of stages in development and we can make use of this model as a 'road-map' for considering the question of *how* organizations can learn to manage innovation better (Figure 12.3).

Box 12.2 gives an example of an outline 'innovation audit' which could be used to focus attention on some of the issues flagged in this book and help begin the process of auditing innovation management capability. The responses to these questions describe 'the way we do things around here' – the pattern of behaviour which describes how the organization handles the question of innovation. These represent the tip of an iceberg but can help focus attention on areas where there is room for further development and where more detailed questions need to be asked.

Based on the model we have used throughout the book, any organization needs to ask itself questions in five key areas:

- Do we have effective enabling mechanisms for the innovation process – to search, to select, to implement and to capture?

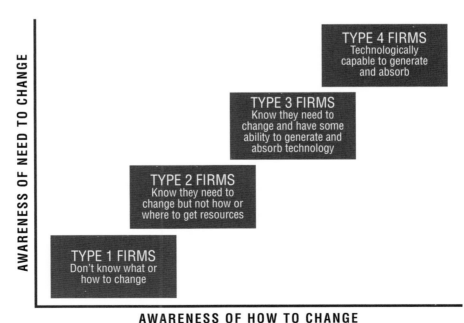

FIGURE 12.3: Groups of firms according to innovation capability

- Do we have a clear innovation strategy and is it communicated and deployed effectively?
- Do we have an innovative organization, one which provides a supportive climate for innovation?
- Do we build and manage rich external linkages to enable 'open innovation'?
- Do we capture learning to help us develop improved innovation management capability?

Kao, Electroco, 3M and Corning provide examples of companies mapped on to this framework.

In carrying out auditing of this kind there is clearly no such thing as an absolute score. None the less it is possible to develop a number of indicators which give some underpinning to what will otherwise be rather subjective judgements about the innovation management capability of a company. For example, a firm that spends 10% of its turnover on R&D is likely to be better on resourcing innovation than one that does no R&D at all.

Box 12.2   **How well do we manage innovation?**

This simple self-assessment tool focuses attention on some of the important areas of innovation management. Below you will find statements which describe 'the way we do things around here' – the pattern of behaviour that describes how the organization handles the question of innovation. For each statement simply put a score from 1 (= not true at all) to 7 (= very true).
Around here . . .

Statement	Score (1 = not true at all to 7 = very true)
1. People have a clear idea of how innovation can help us compete	
2. We have processes in place to help us manage new product development effectively from idea to launch	
3. Our organization structure does not stifle innovation but helps it to happen	
4. There is a strong commitment to training and development of people	
5. We have good 'win–win' relationships with our suppliers	
6. Our innovation strategy is clearly communicated so everyone knows the targets for improvement	
7. Our innovation projects are usually completed on time and within budget	
8. People work well together across departmental boundaries	
9. We take time to review our projects to improve our performance next time	

Statement	Score (1 = not true at all to 7 = very true)
10. We are good at understanding the needs of our customers/end users	
11. People know what our distinctive competence is – what gives us a competitive edge	
12. We have effective mechanisms to make sure everyone (not just marketing) understands customer needs	
13. People are involved in suggesting ideas for improvements to products or processes	
14. We work well with universities and other research centres to help us develop our knowledge	
15. We learn from our mistakes	
16. We look ahead in a structured way (using forecasting tools and techniques) to try and imagine future threats and opportunities	
17. We have effective mechanisms for managing process change from idea through to successful implementation	
18. Our structure helps us to take decisions rapidly	
19. We work closely with our customers in exploring and developing new concepts	
20. We systematically compare our products and processes with other firms	
21. Our top team have a shared vision of how the company will develop through innovation	
22. We systematically search for new product ideas	
23. Communication is effective and works top-down, bottom-up and across the organization	
24. We collaborate with other firms to develop new products or processes	
25. We meet and share experiences with other firms to help us learn	
26. There is top management commitment and support for innovation	
27. We have mechanisms in place to ensure early involvement of all departments in developing new products/processes	
28. Our reward and recognition system supports innovation	
29. We try to develop external networks of people who can help us, e.g., with specialist knowledge	

Statement	Score (1 = not true at all to 7 = very true)
30. We are good at capturing what we have learned so that others in the organization can make use of it	
31. We have processes in place to review new technological or market developments and what they mean for our firm's strategy	
32. We have a clear system for choosing innovation projects	
33. We have a supportive climate for new ideas – people don't have to leave the organization to make them happen	
34. We work closely with the local and national education system to communicate our needs for skills	
35. We are good at learning from other organizations	
36. There is a clear link between the innovation projects we carry out and the overall strategy of the business	
37. There is sufficient flexibility in our system for product development to allow small 'fast-track' projects to happen	
38. We work well in teams	
39. We work closely with 'lead users' to develop innovative new products and services	
40. We use measurement to help identify where and when we can improve our innovation management	

When you have finished, add the totals for the questions in the following way:

Quest. no.	Scores	Quest. no.	Scores	Quest. no.	Scores	Quest. no.	Scores	Quest. no.	Scores
1		2		3		5		4	
6		7		8		10		9	
11		12		13		14		15	
16		17		18		19		20	
21		22		23		24		25	

Quest. no.	Scores	Quest. no.	Scores	Quest. no.	Scores	Quest. no.	Scores	Quest. no.	Scores
26		27		28		29		30	
31		32		33		34		35	
36		37		38		39		40	
Sum									
Total									
Divide by 8									
Total score for	Strategy		Processes		Organization		Linkages		Learning

Now plot a profile for the five dimensions.

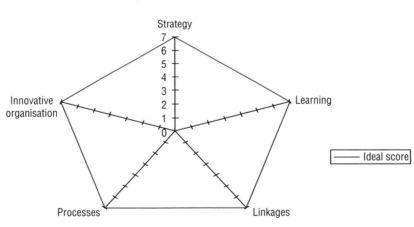

Innovation audit

12.7 Using the framework

This simple audit provides a framework and a brief checklist of questions which might enable an assessment of innovation management to be undertaken. It is not exhaustive, but it does indicate the balance of facts and subjective judgements which would need to be considered to make a realistic response to the question, 'How well does this organization manage innovation?'. The website contains several cases which illustrate how firms have used innovation auditing to develop their capability, specifically Coloplast and Cerulean.

The format of the particular tool is not important – what is needed is the ability to use it to make a wide-ranging review of the factors affecting innovation success and failure, and how management of the process might be improved. Innovation audits of this kind offer:

- An audit framework to see what you did right and wrong in the case of particular innovations or as a way of understanding why things happened the way they did.
- A checklist to see if you are doing the right things.
- A benchmark to see if you are doing them as well as others.
- A guide to continuous improvement of innovation management.
- A learning resource to help acquire knowledge and provide inspiration for new things to try.
- A way of focusing on subsystems with particular problems and then working with the owners of those processes and their customers and suppliers to see if the discussion cannot improve on things.

(The website contains further detail to help you interpret your particular scores and think about what you might do next in terms of organizational development for innovation.)

12.8 Variations on a theme

Throughout the book we have stressed that whilst the challenge in innovation management is generic there are specific issues around which specific responses need to be configured. We might, for example, look at the case of service innovation and focus our audit questions around themes that might be particularly relevant in thinking about managing such innovation. See Box 12.3.

Box 12.3   Measuring service innovation

The organization and management of new service development and delivery can be assessed by five components: strategy, process, organization, tools/technology and system (SPOTS). This framework has been developed and tested by analysing more than 100 firms in the USA and the UK, and validated during the course of conducting a total of 27 cases studied from 18 companies.

Each of the five factors plays a different role in the performance of service innovation. *Strategy* provides focus; *process* provides control; *organization* provides coordination of people; *tools and technologies* provide transformation/transaction capabilities, and *system* provides integration. Performance is analysed as a total index and as three subscales: (i) innovation and quality; (ii) time compression in development and cost reduction in development/delivery; and (iii) service delivery. The first two factors roughly correspond to generic strategic alternatives, differentiation versus cost. The third factor is conceptually important because it distinguishes the service delivery process from product features. Delivery processes often comprise a significant proportion of value added by services, especially if interpersonal exchanges are involved.

The scores and comparisons with those of other companies in the database allow a company to identify its strengths and weaknesses. For example:

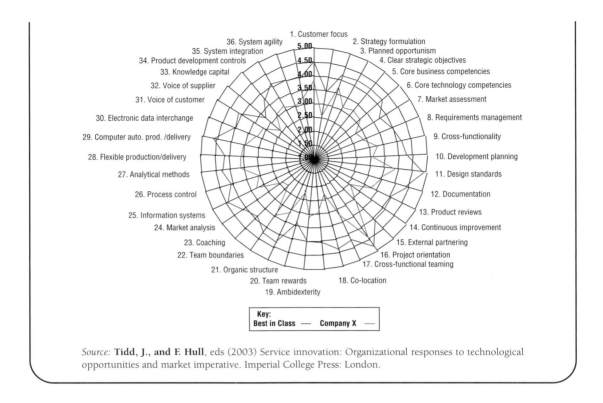

Source: **Tidd, J., and F. Hull**, eds (2003) *Service innovation: Organizational responses to technological opportunities and market imperative.* Imperial College Press: London.

Similarly we have been arguing that there are conditions – beyond the steady state – where we need to take a different approach to managing innovation and to introduce new or at least complementary routines to those helpful in dealing with 'steady-state' innovation. Again we can develop specific audit questions to help facilitate this kind of reflection. Box 12.4 gives an example, using the same approach as we saw earlier.

| Box 12.4 | **How well do we manage discontinuous innovation?** |

As with the 'steady-state' audit, simply put a score from 1 (= not true at all) to 7 (= very true) against each statement. The same score sheet and profile can be used.
Around here . . .

Statement	Score (1 = not true at all to 7 = very true)
1. We deploy 'probe and learn' approaches to explore new directions in technologies and markets	
2. We actively explore the future, making use of tools and techniques like scenarios and foresight	

Statement	Score (1 = not true at all to 7 = very true)

3. Our organization allows some space and time for people to explore 'wild' ideas

4. We make connections across industry to provide us with different perspectives

5. We make regular use of formal tools and techniques to help us think 'out of the box'

6. We have alternative and parallel mechanisms for implementing and developing radical innovation projects which sit outside the 'normal' rules and procedures

7. We have capacity in our strategic thinking process to challenge our current position – we think about 'how to destroy the business'!

8. We have mechanisms to bring in fresh perspectives, e.g., recruiting from outside the industry

9. We make use of formal techniques for looking and learning from outside our sector

10. We focus on 'next practices' as well as 'best practices'

11. We have mechanisms for managing ideas that don't fit our current business, e.g., we license them out or spin them off

12. We use some form of technology scanning/intelligence gathering – we have well-developed technology antennae

13. We have mechanisms to identify and encourage 'intrapreneurship' – if people have a good idea they don't have to leave the company to make it happen

14. We have extensive links with a wide range of outside sources of knowledge – universities, research centres, specialized agencies – and we actually set them up even if not for specific projects

15. We make use of simulation, etc. to explore different options and delay commitment to one particular course

16. We work with 'fringe' users and very early adopters to develop our new products and services

17. We allocate a specific resource for exploring options at the edge of what we currently do – we don't load everyone up 100%

18. We have reward systems to encourage people to offer their ideas

Statement	Score (1 = not true at all to 7 = very true)

19. We have well-developed peripheral vision in our business

20. We use technology to help us become more agile and quick to pick up on and respond to emerging threats and opportunities on the periphery

21. We have 'alert' systems to feed early warning about new trends into the strategic decision-making process

22. We have strategic decision-making and project selection mechanisms which can deal with more radical proposals outside of the mainstream

23. We value people who are prepared to break the rules

24. We practice 'open innovation' – rich and widespread networks of contacts from whom we get a constant flow of challenging ideas

25. We learn from our periphery – we look beyond our organizational and geographical boundaries

26. We are organized to deal with 'off-purpose' signals (not directly relevant to our current business) and don't simply ignore them

27. We deploy 'targeted hunting' around our periphery to open up new strategic opportunities

28. We have high involvement from everyone in the innovation process

29. We have an approach to supplier management which is open to strategic alliances

30. We are good at capturing what we have learned so that others in the organization can make use of it

31. We have processes in place to review new technological or market developments and what they mean for our firm's strategy

32. Management create 'stretch goals' that provide the direction but not the route for innovation

33. Peer pressure creates a positive tension and creates an atmosphere to be creative

34. We have active links into a long-term research and technology community – we can list a wide range of contacts

Statement	Score (1 = not true at all to 7 = very true)
35. We create an atmosphere where people can share ideas through cross-fertilization	
36. There is sufficient flexibility in our system for product development to allow small 'fast-track' projects to happen	
37. We are not afraid to 'cannibalize' things we already do to make space for new options	
38. Experimentation is encouraged	
39. We recognize users as a source of new ideas and try and 'co-evolve' new products and services with them	
40. We regularly challenge ourselves to identify where and when we can improve our innovation management	

We can also develop audits for particular aspects of the innovation process, for example, is there a 'creative climate' within which ideas can flourish and be built upon? Or are there structures and processes in place to enable high involvement of employees in the innovation process? Are there conditions – beyond the steady state – where we need to take a different approach to managing innovation and to introduce new or at least complementary routines to those helpful in dealing with 'steady state' innovation? A variety of frameworks exist and more details and downloadable versions can be found on the website.

12.9 Final thoughts

We have repeatedly said that innovation is complex, uncertain and almost (but not quite) impossible to manage. That being so, we can be sure that there is no such thing as the perfect organization for innovation management; there will always be opportunities for experimentation and continuous improvement. In the end innovation management is not an exact or predictable science but a craft, a reflective practice in which the key skill lies in reviewing and configuring to develop dynamic capability.

Throughout the book we have tried to consider the implications of managing innovation as a generic process but also to look at the ways in which approaches need to take into account two key challenges in the twenty-first century – those of managing 'beyond the steady state' and 'beyond boundaries'. The same basic recipe still applies but there is a need to configure established approaches and to learn to develop new approaches to deal with these challenges.

Summary and further reading

In this chapter we have looked at the ways in which organizations can capture learning and build capability in innovation management. The major requirement is for a commitment to undertake such learning but it can also be enabled by the use of tools and reflection aids. In particular the chapter looks at various approaches to innovation auditing and offers some templates for reviewing and developing capability across the process as a whole and in particular key areas.

A wide range of books and online reviews of innovation now offer some form of audit framework including the Pentathlon model from Cranfield University (K. Goffin and R. Mitchell, *Innovation Management*, Pearson, 2005). For other examples see M. Dodgson *et al.*, *The Management of Technological Innovation*, Oxford University Press, 2008; P. Trott, *Innovation Management and New Product Development*, Fourth Edition Prentice-Hall, 2006; B. Von Stamm, *The Innovation Wave*, 2003 and *Managing Innovation, Design and Creativity*, 2008, John Wiley & Sons, Ltd. Websites include www.innovationdoctor.htm, www.thinksmart. htm, www.jpb.com/services/audit.php, www.innovation-triz.com/innovation/, www .cambridgestrategy.com/page_c5_summary.htm, and www.innovationwave.com/

Web links

Here are the full details of the resources available on the website flagged throughout the text:

 Case studies:
>   Kao
>   Electroco
>   3M
>   Corning
>   Coloplast
>   Cerulean

 Tools:
>   Post-project reviews
>   Benchmarking
>   Business Excellence models
>   Capability maturity models
>   Innovation audits (various)

References

1. **Teece, D., G. Pisano and A. Shuen** (1997) Dynamic capabilities and strategic management. *Strategic Management Journal*, **18** (7), 509–33.
2. **Utterback, J.** (1994) *Mastering the Dynamics of Innovation*, Harvard Business School Press, Boston, MA.